TRENDS IN MATHEMATICS

Trends in Mathematics is a series devoted to the publication of volumes arising from conferences and lecture series focusing on a particular topic from any area of mathematics. Its aim is to make current developments available to the community as rapidly as possible without compromise to quality and to archive these for reference.

Proposals for volumes can be sent to the Mathematics Editor at either

Birkhäuser Verlag
P.O. Box 133
CH-4010 Basel
Switzerland

or

Birkhäuser Boston Inc.
675 Massachusetts Avenue
Cambridge, MA 02139
USA

Material submitted for publication must be screened and prepared as follows:

All contributions should undergo a reviewing process similar to that carried out by journals and be checked for correct use of language which, as a rule, is English. Articles without proofs, or which do not contain any significantly new results, should be rejected. High quality survey papers, however, are welcome.

We expect the organizers to deliver manuscripts in a form that is essentially ready for direct reproduction. Any version of TeX is acceptable, but the entire collection of files must be in one particular dialect of TeX and unified according to simple instructions available from Birkhäuser.

Furthermore, in order to guarantee the timely appearance of the proceedings it is essential that the final version of the entire material be submitted no later than one year after the conference. The total number of pages should not exceed 350. The first-mentioned author of each article will receive 25 free offprints. To the participants of the congress the book will be offered at a special rate.

Mathematical Finance

Workshop of the Mathematical Finance Research Project, Konstanz, Germany, October 5–7, 2000

Michael Kohlmann
Shanjian Tang
Editors

Springer Basel AG

Editors' addresses:

Michael Kohlmann
Fakultät für Mathematik und Informatik
Universität Konstanz
Postfach 5560
D-78434 Konstanz

michael.kohlmann@uni-konstanz.de

Shanjian Tang
Fudan University
Department of Mathematics
Shanghai 200433
China

currently research fellow at the University of Konstanz
e-mail: tang@fmi.uni-konstanz.de

2000 Mathematical Subject Classification 91B28; 91B02

A CIP catalogue record for this book is available from
the Library of Congress, Washington D.C., USA

Deutsche Bibliothek Cataloging-in-Publication Data

Mathematical finance / Workshop of the Mathematical Finance Research Project,
Konstanz, Germany, October 5–7, 2000. Michael Kohlmann ; Shanjian Tang ed. -
Basel ; Boston ; Berlin : Birkhäuser, 2001
(Trends in mathematics)
ISBN 978-3-0348-9506-4 ISBN 978-3-0348-8291-0 (ebook)
DOI 10.1007/978-3-0348-8291-0

Originally published by Birkhäuser Verlag, Basel, Switzerland in 2001
Softcover reprint of the hardcover 1st edition 2001
Printed on acid-free paper produced from chlorine-free pulp. TCF ∞

ISBN 978-3-0348-9506-4

9 8 7 6 5 4 3 2 1

www.birkhasuer-science.com

Preface

In the centenary year of the publication of Bachelier's thesis, what today is considered as the foundation of modern finance, we had the opportunity to invite experts in this relatively new field in mathematics to participate in a meeting at the University of Konstanz, Germany. This could be the place to consider the historical development, but as Professor Girlich presented a remarkable lecture on the past of what now is known as mathematical finance, we refer the reader to the article in this volume.
Instead we take the opportunity to express our thanks to those colleagues who made this workshop possible: to Professor Sondermann, University of Bonn, Germany, who in a sense initiated the idea, to our friends Mark Davis, Robert Elliott, and Xun Yu Zhou, whose advices were extremely helpful in establishing the program of the conference, and to the invited lecturers Nicole El Karoui, Eckhard Platen, and Stan Pliska for carefully preparing their lectures.
Also we gratefully appreciate our students' help during the conference and Mrs Weisser's successful efforts in providing our guests with visa and hotel reservations. Last we should mention the students of the business department who organized a great conference dinner on Mainau island and who documented the meeting in a lot of photos.
Finally we thank our wives Evi and Jie for giving us leave from home (-work) to organize the conference.

Konstanz, in October 2000, Michael Kohlmann and Shanjian Tang

Table of Contents

Note: in the titles of co-authored papers the lecturer's name is in bold face)

Participants workshop on mathematical finance, Konstanz, October 5-7, 2000

Volbert Alexander
University of Giessen, Giessen, Germany

J.S.F Barbachan
Universidade Catolica de Braslia, Brasilia, Brazil
pepe@impa.br

Juergen W. Bauer
Germany
Dr.JuergenW.Bauer@trinkaus.de

Christian Bender
University of Konstanz, Konstanz, Germany
christian.bender@uni-konstanz.de

Fred Espen Benth
University of Oslo, Oslo, Norway
fredb@math.uio.no

Francesca Biagini
University of Bologna, Bologna, Italy
biagini@dm.unibo.it

Frederik Boetius
University of Konstanz, Konstanz, Germany
frederik.boetius@uni-konstanz.de

Anton Buchmeier
Siemens AG, München, Germany
anton.buchmeier@mchp.siemens.de

Volker Bürkel
University of Konstanz, Konstanz, Germany
volker.buerkel@uni-konstanz.de

Mark Davis
Imperial College, London, UK
m.davis@ic.ac.uk

Karim Djaidja
University of Konstanz, Konstanz, Germany

Thomas Domenig
Zürcher Kantonalbank, Zurich, Switzerland

G. Dozeman
SEE, Amstelveen, The Netherlands

Dana Düvelmeyer
TU Chemnitz, Chemnitz, Germany
dana.duevelmeyer@hrz.tu-chemnitz.de

Nicole El Karoui
Polytechnique Paris,Paris,France
elkaroui@cmapx.polytechnique.fr

Robert Elliott
UAlta, Edmonton, Edmonton, Canada
r.elliott@ualberta.ca

Jean-Pierre Fouque
North Carolina State University, Raleigh, USA
fouque@math.ncsu.edu

Rüdiger Frey
University of Zuerich, Zuerich, Switzerland
freyr@isb.unizh.ch

Vladimir Gazda
University of Bratislava, Bratislava, Slovakia
gazda@euke.sk

Hans-Joachim Girlich
University of Leipzig, Leipzig, Germany
girlich@mathematik.uni-leipzig.de

Rudolf Gorenflo
Free University of Berlin, Berlin, Germany
gorenflo@math.fu-berlin.de

Peter Grandits
TU Wien, Wien, Austria
pgrand@fam.tuwien.ac.at

Emmanuel Haven
Concordia University, Cote St. Luc, Canada
hanene@alcor.concordia.ca

Vicky Henderson
University of Warwick, Warwick,UK
vh@math.ethz.ch

Juri Hinz
University of Tuebingen, Tuebingen, Germany
juri.hinz@uni-tuebingen.de

Markus Chr. Holder
HSBC Trinkaus und Burkhardt KGaA
markus.holder@epost.de

Ralf Jäger
University of Marburg, Marburg, Germany
jagger@mathematik.uni-marburg.de

Jan Kallsen
Universtity of Freiburg, Freiburg i. Br., Germany
kallsen@stochastik.uni-freiburg.de

Anne Kandler
TU Chemnitz, Chemnitz, Germany
anne.kandler@mathematik.tu-chemnitz.de

Ioannis Karatzas
Columbia University, New York, USA
ik@math.columbia.edu

Moritz Kassmann
University of Bonn, Bonn, Germany

Claudia Klüppelberg
TU Muenchen, Muenchen, Germany
cklu@mathematik.tu-muenchen.de

Marek Kocinski
Warsaw Agricultural University, Warsaw, Poland
kocinski@delta.sggw.waw.pl

Michael Kohlmann
University of Konstanz, Konstanz, Germany
Michael.Kohlmann@uni-konstanz.de

Christoph Krischanitz
UNIQA Versicherungen AG, Wien, Austria
christoph.krischanitz@uniqa.at

K. Suresh Kumar
University of Twente, Enschede, Netherlands

Yuriy Krvavych
Kyiv University, Kyiv, Ukraine
krvavych@yahoo.com

LanYun Lao
Fudan University, Shanghai, China
lao@wiener.iam.uni-bonn.de

Johannes Leitner
University of Konstanz, Konstanz, Germany
johannes.leitner@uni-konstanz.de

Serguei Levendorski
Rostov State University, Rostov-on-Don, Russia
leven@ns.rnd.runnet.ru

Andrew Lim
Columbia University, New York,USA
andrew@se.cuhk.edu.hk

Karsten Linowsky

Arne Lokka
University of Oslo, Oslo, Norway
alokka@math.uio.no

Enrico Melchioni
FMR Consulting, Voghera, Italy
info@fmrcons.com

A.V. Melnikov
Steklov Institute, Moscow, Russia
melnikov@mi.ras.ru

Leonid Mytnik
Technion, Israel Institute of Technology, Technion City, Haifa, Israel
leonid@ie.technion.ac.il

Sergey Nagornii, SEE
sergey.nagornii@southernenergy-europe.nl

Bernt Oeksendal
University of Oslo, Oslo, Norway
oksendal@math.uio.no

Andrea Pallavicini
FMR Consulting, Voghera, Italy
info@fmrcons.com

Bernhard Peisl
University of Konstanz, Konstanz, Germany
bernhard.peisl@uni-konstanz.de

Eckhard Platen
University of Technology, Sydney, Australia
eckhard.platen@uts.edu.au

Stanley Pliska
University of Illinois Chicago, Chicago, USA
srpliska@uic.edu

Miklos Rasonyi
University of Budapest, Budapest, Hungary
rasonyi@sztaki.hu

Paul Ressel
University of Eichstaett, Eichstaett, Germany
paul.ressel@ku-eichstaett.de

Wolfgang Runggaldier
University of Padova, Padova, Italy
runggal@math.unipd.it

Joern Sass
University of Kiel, Kiel, Germany

M. Schäl
University of Bonn, Bonn, Germany
schael@wiener.iam.uni-bonn.de

Julia Schmelz
TU Muenchen, Muenchen, Germany
schmelz@mathematik.tu-muenchen.de

Thorsten Schmidt
University of Gießen, Gießen, Germany
Thorsten.Schmidt@math.uni-giessen.de

Stefan Scholz

Michael Schroeder
University of Mannheim, Mannheim, Germany
schroeder@math.uni-mannheim.de

Martin Schweizer
TU Berlin, Berlin, Germany
mschweiz@math.tu-berlin.de

Mark Shackleton
Lancaster University, Lancaster, UK
m.shackleton@lancaster.ac.uk

Mikhail Soloveitchik
Dresdner Bank, Germany
mikhail.soloveitchik@dresdner-bank.com
soloveit@de.ibm.com

Thomas Sparla
University of Dortmund, Dortmund, Germany
sparla@statistik.uni-dortmund.de

Srdjan Stojanovic
University of Cincinnati, Cincinnati, OH, USA
srdjan@math.uc.edu

Wolfgang Stummer
University of Ulm, Ulm, Germany
ws@mathematik.uni-ulm.de

W. Stute
University of Siegen, Siegen, Germany
winfried.stute@math.uni-giessen.de

Barbara Swart
University of Pretoria, Pretoria, South Africa
bswart@scientia.up.ac.za

Shanjian Tang
Fudan University, Shanghai, China
tang@fmi.uni-konstanz.de

Dirk Tasche
Munich University of Technology, Muenchen, Germany
tasche@ma.tum.de

Luisa Tibiletti
University of Torino, Torino, Italy
tibiletti@econ.unito.it

Tiziano Vargiolu
University of Padova, Padova, Italy
vargiolu@galileo.math.unipd.it

O. Weick
University of Freiburg, Freiburg, Germany
weicko@uni-freiburg.de

Ralf Wilke
University of Dortmund, Dortmund, Germany
wilke@statistik.uni-dortmund.de

Rafal Wojakowski
Lancaster University, Lancaster, UK
r.wojakowski@lancaster.ac.uk

Uwe Wystup
Commerzbank, Frankfurt, Germany
uwe.wystup@commerzbank.com

Aleksandar Zatezalo
Siemens AG, Mnchen, Germany
aleksandar.zatezalo.external@mchp.siemens.de

Xuelei Zhao
Fudan University, Shanghai, China
zhao@wiener.iam.uni-bonn.de

X.Y. Zhou
CUHK, Hong Kong, China

Helmut Zink
University of Berne, Berne, Switzerland

Trends in Mathematics, © 2001 Birkhäuser Verlag Basel/Switzerland

On-line portfolio strategy with prediction

Sergio Albeverio[1)], LanJun Lao[2)] and XueLei Zhao[3)]

Abstract. We present an on-line portfolio selection strategy with prediction taking transaction costs into account. The new prediction method is based on the new idea of "cross rate" for the sequence of price relative vectors. It is proved that the new portfolio strategy is profitable almost surely under certain mild assumption. The performance of our algorithm is tested on real data from the London Stock Exchange.

1. Introduction

On-line portfolios in the absence of transaction costs have been discussed in the literature, see, e.g., [2], [3], [4], [7], [13]. We present in this paper a new on-line portfolio strategy using techniques of prediction for stock prices and taking transaction costs into account.

1.1. On-line portfolio strategies

Consider a market consisting of m different securities, say stocks, the performance of which are modelled via their price relatives. Let the vector of price relatives of the m stocks at the k-th trading day be denoted by $\mathbf{x}_k = (x_{k,1},\ x_{k,2},\ \ldots,\ x_{k,m})^T$ with

$$x_{k,i} = \frac{\text{the closing price of the } i\text{th stock on trading day } k}{\text{the opening price of the } i\text{th stock on trading day } k} \geq 0,$$

$i = 1,\ \ldots,\ m, k \in \mathbb{N}$. We measure time in days in this paper, but the time scale can of course also be adjusted to any trading period. We assume that the stocks are perfectly divisible and that the closing stock price of day k is the same as the opening one of day $k+1$.

A **portfolio vector** $\mathbf{b}_k = (b_{k,1},\ b_{k,2},\ \ldots,\ b_{k,m}) \in \mathcal{B}$, where

$$\mathcal{B} = \left\{ \mathbf{b} = (b_1, \ldots, b_m) \in R_+^m : \sum_{i=1}^{m} b_i = 1, b_i \geq 0, i = 1,\ \ldots,\ m \right\}, \qquad (1)$$

denotes the proportion of the current wealth invested in each stock at the beginning of the kth trading day. An **on-line portfolio strategy** for N trading days is given by (an algorithm for determining) a sequence of portfolio vectors $\{\mathbf{b}_k\}_{1 \leq k \leq N}$, where $\mathbf{b}_k$ is determined in terms of the previous information $\mathbf{b}_1, \mathbf{b}_2, \cdots, \mathbf{b}_{k-1}$;

$\mathbf{x}_1, \mathbf{x}_2, \cdots, \mathbf{x}_{k-1}$. Given $\mathbf{x}_k$, an investor using $\mathbf{b}_k$ increases (or decreases) his wealth during trading day k by the factor

$$\mathbf{b}_k\mathbf{x}_k = \sum_{i=1}^{m} b_{k,i}x_{k,i}. \tag{2}$$

$\mathbf{b}_k\mathbf{x}_k$ is thus the return of the portfolio on the kth day.

In this paper, we restrict our attention to the class of on-line portfolios in discrete setting. Also other kinds of dynamic portfolios have been under discussions in recent years, see, e.g., [14], [17] and the references therein.

Denote by T_k the total amount of transactions costs paid at the beginning of the kth trading day, by W_k' the invested wealth after making the transactions and by W_k the wealth reached at the end of day k with the initial wealth $W_0 = 1$. A portfolio strategy is **self-financing** if, according to it,

$$W_k' = W_{k-1} - T_k, \tag{3}$$

$$W_k = W_k'\mathbf{b}_k\mathbf{x}_k, \ 1 \le k \le N. \tag{4}$$

1.2. Transaction costs

Assuming that transaction costs are proportional, with equal proportion $0 < c < 1$, being same for both sale and purchase, to the ammounts of transactions, we give the following proposition to reveal the relation between the transaction costs and a distance.

Proposition 1.1. *For a self-financing strategy, assuming* $\mathbf{b}_k\mathbf{x}_k > 0$, *we have*

$$\frac{c}{1+c}W_k\mathbf{d}(\mathbf{b}_{k+1}, \tilde{\mathbf{b}}_k) \le T_{k+1} \le \frac{c}{1-c}W_k\mathbf{d}(\mathbf{b}_{k+1}, \tilde{\mathbf{b}}_k), \tag{5}$$

where

$$\mathbf{d}(\mathbf{b}_{k+1}, \tilde{\mathbf{b}}_k) = \sum_{i=1}^{m} |b_{k+1,i} - \tilde{b}_{k,i}|. \tag{6}$$

$$\tilde{\mathbf{b}}_k = (\tilde{b}_{k,1}, \ \tilde{b}_{k,2}, \ \ldots, \ \tilde{b}_{k,m}), \ \tilde{b}_{k,i} = \frac{b_{k,i}x_{k,i}}{\mathbf{b}_k\mathbf{x}_k}, \quad i = 1, \ 2, \ \ldots, \ m. \tag{7}$$

Proof. At the beginning of the $(k+1)$th trading day, the transaction amount related to the stock i

$$Q_{k+1,i} = |W_{k+1}'b_{k+1,i} - W_k'b_{k,i}x_{k,i}|. \tag{8}$$

Then the total amount of trading wealth

$$Q_{k+1} = \sum_{i=1}^{m} Q_{k+1,i} = \sum_{i=1}^{m} |(W_k - T_{k+1})b_{k+1,i} - W_k'b_{k,i}x_{k,i}|. \tag{9}$$

Therefore,

$$T_{k+1} = cQ_{k+1} = c\sum_{i=1}^{m} |W_kb_{k+1,i} - W_k'b_{k,i}x_{k,i} - T_{k+1}b_{k+1,i}|. \tag{10}$$

From this, the triangle inequality and the fact that $\mathbf{b}_{k+1} \in \mathcal{B}$ and $T_{k+1} \geq 0$ we obtain

$$T_{k+1} \geq c \sum_{i=1}^{m} |W_k b_{k+1,i} - W_k' b_{k,i} x_{k,i}| - cT_{k+1} \tag{11}$$

and

$$T_{k+1} \leq c \sum_{i=1}^{m} |W_k b_{k+1,i} - W_k' b_{k,i} x_{k,i}| + cT_{k+1}. \tag{12}$$

(5) follows from (11), (12) and (4). □

Remark 1.2. $\tilde{\mathbf{b}}_k$ is actually the portfolio of the m stocks in the market reached at the end of day k. In fact, the ith entry $\tilde{b}_{k,i}$ of the vector $\tilde{\mathbf{b}}_k$ is the proportion of the wealth of stock i to the whole wealth, which is automatically reached at the end of the trading day k due to changes of the stock prices.

Remark 1.3. Proposition 1.1 shows the relation, in terms of lower and upper bounds, between T_{k+1} and $\mathbf{d}(\mathbf{b}_{k+1}, \tilde{\mathbf{b}}_k)$, the "distance" of $\mathbf{b}_{k+1}$ and $\tilde{\mathbf{b}}_k$. When this distance is equal to zero, there are no transaction costs. The bigger the distance, the bigger the sum of transaction costs. Reducing transaction costs is hence equivalent to shortening this distance.

In (14), the relative entropy is used to shorten this distance, thereby to reduce the transaction costs, and to derive the portfolio algorithm (13). Using the relative entropy to shorten the distance is motivated by [7] and [12].

We thank professor M. Schäl for his useful discussions. The financial supports of SFB 256, SFB 237, BiBoS, CERFIM, Acc. Arch. (USI), AvH Foundation and NSF of China are gratefully acknowledged.

2. The new on-line portfolio strategy

We present a portfolio strategy with prediction being taken into the update rule.

2.1. The update rule

The portfolio update rule forms as follows

$$b_{k+1,i} = \frac{\tilde{b}_{k,i} \exp(\lambda \hat{x}_{k+1,i})}{\sum_{j=1}^{m} \tilde{b}_{k,j} \exp(\lambda \hat{x}_{k+1,j})}, \tag{13}$$

where $\hat{x}_{k+1,i}$ is the prediction of $x_{k+1,i}$ $i = 1, 2, \ldots, m$, $k \in \mathbb{N}$. This is obtained by maximizing

$$F(\mathbf{b}_{k+1}) \equiv \lambda \mathbf{b}_{k+1} \hat{\mathbf{x}}_{k+1} - \mathbf{d}_{re}(\mathbf{b}_{k+1}, \tilde{\mathbf{b}}_k) \tag{14}$$

where

$$\mathbf{d}_{re}(\mathbf{b}_{k+1}, \tilde{\mathbf{b}}_k) = \sum_{i=1}^{m} (b_{k+1,i} \log \frac{b_{k+1,i}}{\tilde{b}_{k,i}}), \tag{15}$$

which is relative entropy being widely used in information theory, economics modeling, biology etc., see, e.g., [8] and [9]. The first term of $F(\mathbf{b}_{k+1})$ is the Factor of Wealth Growth on the $(k+1)$th trading day under the prediction of $\hat{\mathbf{x}}_{k+1}$, see (2). The second term measures the Transaction Costs. (We refer to (13) as the FWGTC algorithm.) It is easy to know that $F(\mathbf{b}_{k+1})$ is concave in $\mathbf{b}_{k+1}$. Lagrange's method was used to maximizing $F(\mathbf{b}_{k+1})$ under the constraint $\mathbf{b}_{k+1} \in \mathcal{B}$.

2.2. The prediction of price relatives

Many methods can be found, e.g., in [5], [6], [10], [11] and [12], to predict the values of price relatives of the next trading day, usually by minimizing a certain kind of loss function or error function. We present a new prediction method, which is aimed at predicting the order of price relatives instead of their values.

When an investor wants to change his portfolio, he would like to know which stocks will behave well in the future. This, however, can only be predicted from the past data. The investor likes to put more wealth on the expected better stocks than on the expected worse ones. It seems thus appropriate to pay more attention to the "order" of the stocks than to the values of the stock prices themselves.

2.2.1. Cross Rate To forecast the order, we introduce the notion of 'cross rate'. For simplicity, we only consider the case of two assets.

Consider a sequence of price relatives $X(N) = \{\mathbf{x}_k = (x_{k,1}, x_{k,2})', k = 1, \ldots, N\}$. The **order** of price relative vector $\mathbf{x}_k$ is defined by

$$O(\mathbf{x}_k) \equiv \begin{cases} 1, & \text{if } x_{k,1} > x_{k,2} \\ 2, & \text{if } x_{k,1} < x_{k,2} \\ 0, & \text{if } x_{k,1} = x_{k,2} \end{cases} \tag{16}$$

We call the sequence **strictly unequal** if

$$O(\mathbf{x}_k) \neq 0, \tag{17}$$

holds for all $k \in \{1, \ldots, N\}$.

We will discuss the case of a strictly unequal sequence. The case of a sequence being not strictly unequal can be handled in a similar way, see [1]. We say a **cross** occurs in the sequence $X(N)$ if there exists some $k \in \{1, \ldots, N\}$ such that $O(\mathbf{x}_k) \neq O(\mathbf{x}_{k-1})$. We refer to k as a **cross position**.

For any segment $X(J,K) \equiv \{\mathbf{x}_k, k = J, J+1, \ldots, K\}$, where J, K are integers and $N \geq K > J > 1$, we define the **cross number** of $X(J,K)$ by

$$V_{J,K} \equiv \#\{k : O(\mathbf{x}_{k-1}) \neq O(\mathbf{x}_k),\ J \leq k \leq K\}, \tag{18}$$

which is the number of crosses happening during the trading period from day J to day K. Further the **cross rate** of $X(J,K)$ is defined by,

$$R_{J,K} \equiv \frac{V_{J,K}}{K-J+1}. \tag{19}$$

We split the sequence of price relative vectors of two stocks $\{\mathbf{x}_k, k \in \mathbb{N}\}$ into segments of the identical length $L > 0$, $\{X_n(L) = X(((n-1)L+1), nL), n \in \mathbb{N}\}$, and define its corresponding cross rate sequence $\{R_n(L), n \in \mathbb{N}\}$ with reference to the length L as follows

$$R_n(L) \equiv R_{(n-1)L+1,nL}, \ n \in \mathbb{N} \tag{20}$$

2.2.2. CROSS-RATE METHOD We now present a method for predicting price relative vectors of two stocks in terms of their cross rate (CR). Let

$$P_{AA}(n) \equiv P(R_{n+1}(L) \in A, R_n(L) \in A), \tag{21}$$
$$P_{AB}(n) \equiv P(R_{n+1}(L) \in B, R_n(L) \in A), \tag{22}$$
$$P_{BA}(n) \equiv P(R_{n+1}(L) \in A, R_n(L) \in B), \tag{23}$$
$$P_{BB}(n) \equiv P(R_{n+1}(L) \in B, R_n(L) \in B), \tag{24}$$

where $A = [0, \frac{1}{2})$, $B = [\frac{1}{2}, 1]$. Suppose that $P_{AA}(n), P_{AB}(n), P_{BA}(n)$, and $P_{BB}(n)$ do not depend on n, being denoted by $P_{AA}, P_{AB}, P_{BA}, P_{BB}$ respectively. Using above notation we have

$$P_{AA} + P_{AB} + P_{BA} + P_{BB} = 1. \tag{25}$$

Recall that $\{X_n(L) = X(((n-1)L+1), nL), n \in \mathbb{N}\}$, and the corresponding cross rate sequence is denoted by $\{R_n(L), \ n \in \mathbb{N}\}$. We predict a $\mathbf{x}_{k+1} \in X_{n+1}(L)$ in three steps: to estimate $R_{n+1}(L)$, to predict the order of $\mathbf{x}_{k+1}$ and to estimate $\mathbf{x}_{k+1}$.

Step 1: To estimate $R_{n+1}(L)$ using $R_1(L), \ R_2(L), \ \ldots, \ R_n(L)$. We suggest the following methods for predicting the cross rate (MPCR), **MPCR1:**

$$\hat{R}_{n+1}(L) = R_n(L), \ n \in \mathbb{N} \tag{26}$$

for a cross rate sequence with $P_{AA} + P_{BB} \geq \frac{1}{2}$ and **MPCR2:**

$$\hat{R}_{n+1}(L) = \begin{cases} c_A, & \text{if } R_n(L) \in B \\ c_B, & \text{if } R_n(L) \in A \end{cases} \quad n \in \mathbb{N} \tag{27}$$

for a cross rate sequence with $P_{AB} + P_{BA} \geq \frac{1}{2}$, where constants $c_A \in A$, $c_B \in B$.

Step 2: To predict $O(\mathbf{x}_{k+1})$, see(16), according to $\hat{R}_{n+1}(L)$. We present two methods for predicting the order (MPO), **MPO1:**

$$\hat{O}(\mathbf{x}_{k+1}) = \begin{cases} O((x_{k,2}, \ x_{k,1})'), & \text{if } \hat{R}_{n+1}(L) \in [\frac{1}{2}, 1] \\ O(\mathbf{x}_k), & \text{if } \hat{R}_{n+1}(L) \in [0, \frac{1}{2}) \end{cases} \tag{28}$$

and **MPO2:**

$$\hat{O}(\mathbf{x}_{k+1}) = \begin{cases} O(\mathbf{x}_{k-1}), & \text{if } \hat{R}_{n+1}(L) \in [\frac{1}{2}, 1] \\ O(\mathbf{x}_k), & \text{if } \hat{R}_{n+1}(L) \in [0, \frac{1}{2}) \end{cases} \tag{29}$$

where $\hat{O}(\mathbf{x}_{k+1})$ stands for the predicted value of $O(\mathbf{x}_{k+1})$. In Lemma 3.1 we show that both MPO1 and MPO2 can forecast correctly at least half of the orders of the price relative vectors in segment $X_{n+1}(L)$ when $\hat{R}_{n+1}(L)$ belongs to the same part, $[0, \frac{1}{2})$ or $[\frac{1}{2}, 1]$ in interval $[0, 1]$, as $R_{n+1}(L)$.

Step 3: To assign a $\hat{\mathbf{x}}_{k+1}$ so that $\hat{\mathbf{x}}_{k+1}$ has the right order, i.e. $O(\hat{\mathbf{x}}_{k+1}) = \hat{O}(\mathbf{x}_{k+1})$. Recalling our idea that order rather than the concrete values should matter, we have some choices for evaluating $\hat{\mathbf{x}}_{k+1}$. We can, e.g., simply take the following $\hat{\mathbf{x}}_{k+1}$

$$\hat{\mathbf{x}}_{k+1} = \begin{cases} (x_{k,2},\ x_{k,1})', & \text{if } \hat{R}_{n+1}(L) \in [\frac{1}{2}, 1] \\ \mathbf{x}_k, & \text{if } \hat{R}_{n+1}(L) \in [0, \frac{1}{2}) \end{cases} \tag{30}$$

for MPO1, and

$$\hat{\mathbf{x}}_{k+1} = \begin{cases} \mathbf{x}_{k-1}, & \text{if } \hat{R}_{n+1}(L) \in [\frac{1}{2}, 1] \\ \mathbf{x}_k, & \text{if } \hat{R}_{n+1}(L) \in [0, \frac{1}{2}) \end{cases} \tag{31}$$

for MPO2.

We denote a CR method by CR(MPCR, MPO, $\hat{\mathbf{x}}$) to mean we get $\hat{\mathbf{x}}_{k+1}$ for $\mathbf{x}_{k+1} \in X_{n+1}(L)$ with the pair of MPCR and MPO.

3. Profitability of the FWGTC with CR(MPCR, MPO,$\hat{\mathbf{x}}$)

For the segment $X_n(L)$, we define the **success rate** α_n of the CR(MPCR, MPO, $\hat{\mathbf{x}}$) as following

$$\alpha_n(\text{MPCR, MPO, } \hat{\mathbf{x}}) \equiv \frac{\#\{\hat{O}(\mathbf{x}_k) :\ \hat{O}(\mathbf{x}_k) = O(\mathbf{x}_k), \mathbf{x}_k \in X_n(L)\}}{L}. \tag{32}$$

A CR(MPCR, MPO, $\hat{\mathbf{x}}$) is said to be **effective** for the segment $X_n(L)$ if

$$\alpha_n(\text{MPCR, MPO, } \hat{\mathbf{x}}) \geq \frac{1}{2}. \tag{33}$$

Using the FWGTC update rule (13) with an effective CR(MPCR, MPO,$\hat{\mathbf{x}}$) for the segment $X_n(L)$, investors can move their wealth in the correct direction, i.e. from the worse stock to the better stock, for more than half of the total times L in $X_n(L)$.

We call the FWGTC with CR(MPCR, MPO,$\hat{\mathbf{x}}$) a **profitable strategy** for the whole sequence $X(N)$, if the CR(MPCR, MPO,$\hat{\mathbf{x}}$) is effective for more than half of the segments of $X(N)$, i.e.

$$\begin{aligned} &\beta(X(N), \text{MPCR, MPO, } \hat{\mathbf{x}}_{k+1}) \\ &\equiv \frac{\#\{X_n(L) : \text{CR(MPCR, MPO, } \hat{\mathbf{x}}) \text{ is effective for } X_n(L)\}}{\#\{X_n(L)\}} \\ &\geq \tfrac{1}{2}. \end{aligned} \tag{34}$$

Lemma 3.1. *CR(MPCRi, MPOj, $\hat{\mathbf{x}}$), i, j =1, 2 are all effective for the segment $X_{n+1}(L)$, if*

$$\hat{R}_{n+1}(L),\ R_{n+1}(L) \in [0, \frac{1}{2}) \tag{35}$$

or

$$\hat{R}_{n+1}(L),\ R_{n+1}(L) \in [\frac{1}{2}, 1] \tag{36}$$

holds.

Proof. We prove CR(MPCR1, MPO1, $\hat{\mathbf{x}}$) is effective for the segment $X_{n+1}(L)$ when (35) holds, the proof for the other parts of this lemma is similar.

(35) means that both $\hat{R}_{n+1}(L)$ and $R_{n+1}(L)$ are smaller than $\frac{1}{2}$. Therefore more than half points of the set $\{nL+k,\ k=1,\ 2,\ \ldots,\ L\}$ are not cross positions. We take $\hat{O}(\mathbf{x}_{k+1}) = O(\mathbf{x}_k)$ for $\mathbf{x}_{k+1} \in X_{n+1}$ according to MPO1, which is correct if $\mathbf{x}_{k+1}$ is not a cross position. Consequently we forecast correctly at least half of the orders of the segment $X_{n+1}(L)$, i.e.

$$\alpha_{n+1}(\text{MPCR1, MPO1, } \hat{\mathbf{x}}) \geq \frac{1}{2}. \tag{37}$$

□

Remark 3.2. The proof of Lemma 3.1 shows that CR(MPCR1, MPO1, $\hat{\mathbf{x}}$) is effective for the segment $X_{n+1}(L)$ if the prediction $\hat{R}_{n+1}(L)$ is correct in the sense that the prediction of the cross rate $\hat{R}_{n+1}(L)$ is in the same part as the real cross rate $R_{n+1}(L)$ in $[0,1]$, the left part $[0,\frac{1}{2})$ or the right part $[\frac{1}{2},1]$. It is not necessary for $\hat{R}_{n+1}(L)$ to take exactly the same value as $R_{n+1}(L)$.

A real number sequence $\{\xi_n\}_{n\geq 1}$ is called **strictly stationary** (s.s.) if for any $n_1,\cdots,n_k,h \in \mathbb{N}$, $\xi_{n_1}\cdots\xi_{n_k}$ and $\xi_{n_1+h}\cdots\xi_{n_k+h}$ have the same finite-dimensional distribution. The sequence $\{\xi_n\}_{n\geq 1}$ is called **asymptotic independent** (a.i.) if $\lim_{r\to\infty} P(\xi_n > a, \xi_{n+r} > b) = \lim_{r\to\infty} P(\xi_n > a) P(\xi_{n+r} > b), a, b \in \mathbb{R}$, for any $n \in \mathbb{N}$.

Lemma 3.3. *Suppose $\{\xi_n\}_{n\geq 1}$ is a s.s. and a.i. sequence and $E\xi_1 \geq \beta$ for some constant β, then*

$$\lim_{N\to\infty} \frac{1}{N}\sum_{n=1}^{N} \xi_n \geq \beta, \qquad a.s. \tag{38}$$

Proof. The sequence $\{\xi_n\}_{n\geq 1}$ is ergodic since the mixing condition holds, see,e.g., Chapter IIV in [19]. Therefore,

$$\lim_{N\to\infty} \frac{1}{N}\sum_{n=1}^{N} \xi_n = E\xi_1 \geq \beta, \qquad a.s.$$

□

The profitability of the FWGTC strategy is handled in the following theorem.

Theorem 3.4. *Suppose the sequence $\{\mathbf{x}_k\}$ is s.s. and a.i.*

(1) The FWGTC with CR(MPCR1, MPO1, $\hat{\mathbf{x}}$) or CR(MPCR1, MPO2, $\hat{\mathbf{x}}$) is a profitable strategy almost surely when N tends to ∞, if

$$P_{AA} + P_{BB} \geq \frac{1}{2}. \tag{39}$$

(2) The FWGTC with CR(MPCR2, MPO1, $\hat{\mathbf{x}}$) or CR(MPCR2, MPO2, $\hat{\mathbf{x}}$) is a profitable strategy almost surely when N tends to ∞, if

$$P_{AB} + P_{BA} \geq \frac{1}{2}. \tag{40}$$

Proof. We only prove that the FWGTC with CR(MPCR1, MPO1, $\hat{\mathbf{x}}$) is a profitable strategy almost surely in the case (1). The proof for the other parts of this theorem is analogous. Let

$$\xi_n = \begin{cases} 1, & \text{if } \alpha_n(\text{MPCR1, MPO1, } \hat{\mathbf{x}}) \geq \frac{1}{2}, \\ 0, & \text{if } \alpha_n(\text{MPCR1, MPO1, } \hat{\mathbf{x}}) < \frac{1}{2}, \end{cases} \tag{41}$$

$n = 1, 2, \ldots$ · It can be shown that $\{\xi_n\}$ is also a s.s. and a.i. sequence. With CR(MPCR1, MPO1,$\hat{\mathbf{x}}$), when condition (39) holds, we have

$$\beta_n \equiv P(\xi_n = 1) \geq P_{AA} + P_{BB} \geq \frac{1}{2}, \tag{42}$$

because it is sure that CR(MPCR1, MPO1,$\hat{\mathbf{x}}$) is effective for the segment $X_n(L)$ when (35) or (36) holds, see Lemma 3.1, and one has

$$P(\hat{R}_{n+1}(L),\ R_{n+1}(L) \in A) + P(\hat{R}_{n+1}(L),\ R_{n+1}(L) \in B) = P_{AA} + P_{BB}. \tag{43}$$

¿From Lemma 3.3, we have

$$\lim_{N\to\infty} \beta(X(N), \text{MPCR, MPO, } \hat{\mathbf{x}}) = \lim_{n\to\infty} \frac{1}{n} \sum_{k=1}^{n} \xi_k \geq \frac{1}{2}, \text{ a.s.} \tag{44}$$

□

For any sequence $\{\mathbf{x}_k\}$, at least one of (39) and (40) holds. We can of course also discuss the MPCR for the other two possible situations: $P_{AA} + P_{BA} \geq \frac{1}{2}$ and $P_{AB} + P_{BB} \geq \frac{1}{2}$.

4. An example and conclusion

We did experiments to test the FWGTC on five-year (1305-trading-day) long real data from the London Stock Exchange. We give graphically an example to show the effectiveness of the FWGTC, see Fig.1. FTO (the Fortune Oil) and TCC (the Trace Comps.) are two stocks with the final returns (i.e. W_{1305} with $W_0 = 1$) 0.619 and 3.119 respectively. The final return of the FWGTC portfolio is 9.959.

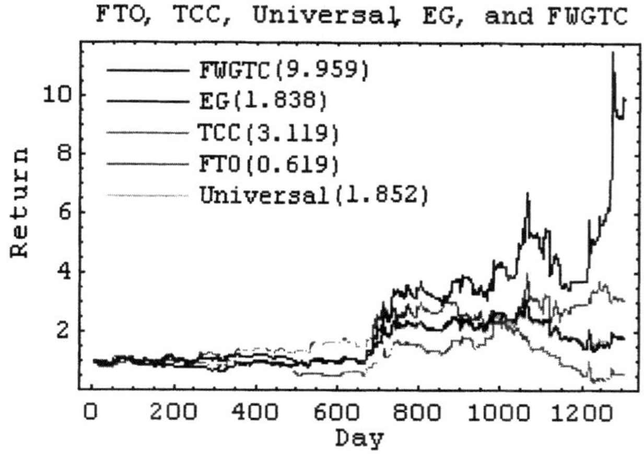

Fig. 1 The performance of FWGTC with FTO and TCC.

We presented a strategy with prediction for on-line portfolio management taking transaction costs into account. The new prediction method proves to be suitable and useful for portfolio selection. Our experiments show that our algorithm performs quite well in the comparison with Cover's ([3]) universal portfolio, the EG(η) update algorithm of Helmbold et al. ([7]). Even our algorithms have chances to outperform the best constant-rebalanced portfolio, which is impossibly reached by other mentioned algorithms.

References

[1] S. Albeverio, L.J. Lao and X.L. Zhao, On-line portfolio selection strategy with prediction in the presence of transaction costs, Preprint of Institute of Applied Mathematics, University of Bonn, 2000.

[2] A. Blum and A. Kalai, Universal portfolios with and without transaction costs, *in* "Proceedings of the Tenth Annual Conference on Computational Learning Theory," pp. 309-313, ACM Press, 1997.

[3] T. M. Cover, Universal Portfolios, *Mathematical Finance* **1**(1) (1991), 1-29.

[4] T. M. Cover and E. Ordentlich, Universal Portfolios with Side Information. *IEEE Trans. Info. Theory* **42**(2) (1996), 348-363.

[5] P. Embrechts, C. Klüppelberg and T. Mikosch, "Modelling Extremal Events: for Insurance and Finance," Springer, Berlin Heidelberg, 1997.

[6] C. Gouriéroux, "ARCH Models and Financial Applications," Springer, New York, 1997.

[7] D. P. Helmbold, R. E. Schapire, Y. Singer and M. K. Warmuth, On-line portfolio selection using multiplicative updates, *Mathematical Finance* **8**(4) (1998), 325–347.

[8] G. Jumarie, "Relative Information," Springer, New York, 1990.

[9] J. N. Kapur, "Maximum Entropy Models in Science and Engineering," John Wiley & Sons, New Delhi, India, 1993.

[10] T. Kimoto, A. Kazuo, M. Yoda and Takeoka, Stock market prediction system with modular neural networks, *in "Proceedings of International Joint Conference on Neural Networks," Vol. 1, pp. 1-6, 1989.*

[11] J. Kingdon, "Intelligent Systems and Financial Forecasting," Springer, London, 1997.

[12] J. Kivinen and M. K. Warmuth, Exponentiated gradient versus gradient descent for linear predictors, *Info. Computation* **132**(1) (1997), 1-63.

[13] R. Korn, "Optimal Portfolios," World Scientific, Singapore, 1997.

[14] R. Korn and M. Schäl, On value preserving and growth optimal portfolios, *Math. Meth. Oper. Res.* **50** (1999), 189-218.

[15] L. J. Lao, An additive update rule for on-line portfolio selection, in preparation, 2000.

[16] R. C. Merton, Optimum consumption and portfolio rules in a continuous time model, *J. Econ. Theory* **3** (1971), 373-413.

[17] A. J. Morton and S. R. Pliska, Optimal portfolio management with fixed transaction costs, *Mathematical Finance* **5.4**(1995), 337-356.

[18] I. Pikovsky and I. Karatzas, Anticipative portfolio optimization. *Adv. Appl. Prob.* **28** (1996), 1095-1122.

[19] M. Rosenblatt, *Random Processes*, Springer, New York Heidelberg Berlin, 1974.

[1)2)]Institut für Angewandte Mathematik
Universität Bonn, Wegelerstrasse 6
D-53115 Bonn, Germany
E-mail address: [1)]`albeverio@uni-bonn.de`
E-mail address: [2)]`lao@wiener.iam.uni-bonn.de`

[3)]Institute of Mathematics,
Fudan University, Handan Road 220
200433 Shanghai, P.R.China
E-mail address: `xlzhao@fudan.edu.cn`

Trends in Mathematics, © 2001 Birkhäuser Verlag Basel/Switzerland

Continuous time financial market, transaction cost and transaction intensity

Sergio Albeverio[1)], LanJun Lao[2)] and XueLei Zhao[3)]

Abstract. An investor who makes transaction according to a Poisson point processes is considered. Some transaction strategies in terms of two different standards are presented. We introduce a utility function for the opportunity brought by transactions and prove that there exists an optimal transaction strategy for the corresponding optimization problem in the framework of functional analysis. The optimal transaction intensity is characterized in terms of numerical equations. We also give an asymptotic property for such a problem.

1. Introduction

A portfolio usually consists of several kinds of risky or riskless financial instruments. In this paper, we consider the model consistent of m risky stocks whose prices follow geometric Brownian motions, along with a single riskless bond growing at rate r.

With Z_t^k denoting the price at time t of stock k, $(1 \leq k \leq m)$, and Z_t^0 denoting the price at t of the riskless bond, we have the following Black-Scholes equations

$$dZ_t^0 = r(t) Z_t^0 dt$$

and

$$dZ_t^k = Z_t^k \left(\mu_k(t)dt + \sum_{i=1}^{m} \lambda_{ki}(t) dW_t^i \right),$$

where $\mu(\cdot)$ is an m-vector variable, $\lambda_{ij}(\cdot)$ is the (i,j) entry in the m by m matrix $\Lambda(\cdot)$, and $W^1, \ldots, W^m$ are independent Brownian motion processes. We assume that the variance-covariance matrix $M := \Lambda(\cdot)\Lambda^T(\cdot)$ is of full rank and assume all coefficients $r(\cdot)$, $\mu(\cdot)$, $\Lambda(\cdot)$ are adapted to a complete filtration $\mathcal{F}_t$, $t \geq 0$, which contains the filtration generated by the Brownian motions and with respect to which the Brownian motion is still a martingale. This kind of filtrations indeed exist. For example, $\mathcal{F}_t$, the augmentation of the Brownian motion and an independent random process, is such a filtration.

We consider the influences of transactions on an investor's wealth process in the presence of transaction costs. We will focus on the interplay between the transaction costs and the increase of total wealth in the case where transactions are made according to the jumping times of a Poisson point process with intensity measure ν. We refer to this intensity measure as **transaction intensity**, indicating how much and when the investor makes transactions. The connection between transaction intensity and the financial market will be discussed in Section 2, on the basis of observation extracted from the data published by NYSE and Nasdaq. The transaction strategies are analyzed in Section 3. We propose two kinds of transaction strategies according to the two standards. Some utility functions are introduced to analyze the opportunity brought by transactions in Section 4.

2. On the transaction intensity

In financial markets, transactions and price evolutions influence each other. Usually the larger volatility a stock has the bigger transaction volume the investors make for this stock. Moreover, the moments when the stock price changes are closely related to those when transactions are made. In fact this can be observed by charts from real markets, e.g., NasdaqNM: COKE and NasdaqNM: SUNW on July 31, 2000 in the end of this note.

Even though the transaction time does not exactly coincide with the moment of stock price change and the transaction volume has an ambiguous connection with the range of price fluctuations no one doubt the close relationship between the transactions and the price fluctuations. In our model we assume the transaction measure ν depends on the coefficients in the financial market. In other words, ν, $r(\cdot)$, $\mu(\cdot)$ and $\Lambda(\cdot)$ can be formally modeled by

$$\nu((0,t]) = Eg(s, r(s), \mu(s), \Lambda(s), 0 \leq s \leq t), \tag{1}$$

where g is a non-negative functional. Of course, to investigate the explicit form of g will be a challenging problem.

3. Transaction strategy

3.1. General setting

We now describe the set of admissible trading strategies. Let the portfolio value be V_0 at the initial time $t_0 = 0$. The investor allocates V_0 in the proportions $b = (b^1, \ldots, b^m)^T$ for some $m \in N$, where $b^i \geq 0$ for all i and $\sum_i b^i \leq 1$ (which we write as $1^T b \leq 1$, with 1 denoting a column of ones). More precisely, the investor puts an amount $b_0^i V_0$ in stock i ($1 \leq i \leq m$), and an amount $(1 - \sum_i b_0^i)V_0$ into the riskless bond. He then holds this portfolio until some time t (the fraction of wealth held in each set will fluctuate as prices change). Let V_t denote the portfolio value at t. The portfolio value is given by

$$\begin{aligned} V_t &= V_0\Bigg((1-1^Tb)\exp\left(\int_0^t r(s)ds\right) \\ &+b^1\exp\left(\int_0^t\left(\mu_1(u)-\frac{1}{2}\sum_{j=1}^m\lambda_{1j}^2(u)\right)du+\sum_{j=1}^m\int_0^t\lambda_{1j}(u)dW_u^j\right) \\ &+\cdots \qquad (2) \\ &+b^m\exp\left(\int_s^t\left(\mu_m(u)-\frac{1}{2}\sum_{j=1}^m\lambda_{mj}^2(u)\right)du+\sum_{j=1}^m\int_s^t\lambda_{mj}(u)dW_u^j\right)\Bigg). \end{aligned}$$

Now we consider an investor who makes transactions in a self-financing way according to a sequence of stopping times $\{\tau_n\}_{n\geq 1}$ which are the jumping times of a Poisson point process, $N(t)$, over the positive line $(0,\infty)$ with a (σ-finite) intensity measure ν. Throughout we assume that, whenever a trade is made, the investor pays a transaction cost equal to a fraction of his entire portfolio value. This fraction is usually a random variable in $[0,1]$, denoted by c, which is termed "portfolio management fee" (see Duffie and Sun 1990, Merton and Pliska 1995). If, as before, the investor has an initial portfolio value of V_0 at time $t_0=0$, he allocates V_0 in the proportions b_0, and holds this portfolio until the stopping time τ_1. At τ_1 he pays the transaction fee in the rate c_1, whose expectation is equal to Ec, so that $V_{\tau_1}=(1-c_1)V_{\tau_1-}$ (we observe the convention that V_{τ_1} is the value just after the fee has been paid). He then allocates V_{τ_1} in the portfolio proportions b_1, observes price changes until τ_2, pays transaction fee in c_2, allocates his wealth in the portfolio proportions b_2, and so forth. In this way a sequence of random variables $\{c_i\}_{i\geq 1}$ is obtained. Each of them has the *same expectation as random variable* c. Assuming $c_0=0$, $\tau_0=0$ and noticing that the investor makes transactions for $N(t)$ times up to t, we have

$$\begin{aligned} V_t &= V_0\prod_{i=0}^{N(t)}(1-c_i)\left((1-1^Tb_i)\exp\left(\int_{\tau_i\wedge t}^{\tau_{i+1}\wedge t}r(s)ds\right)\right. \\ &\left.+b_i^1\frac{Z^1_{\tau_{i+1}\wedge t}}{Z^1_{\tau_i\wedge t}}+\cdots+b_i^m\frac{Z^m_{\tau_{i+1}\wedge t}}{Z^m_{\tau_i\wedge t}}\right). \qquad (3) \end{aligned}$$

Thus for t between stopping times, for example $\tau_n\leq t<\tau_{n+1}$ (i.e., $N(t)=n$), we have

$$\begin{aligned} \ln V_t &= \ln V_0+\sum_{i=0}^n\ln(1-c_i)+\sum_{i=0}^{n-1}\ln\left((1-1^Tb_i)\exp\left(\int_{\tau_i}^{\tau_{i+1}}r(s)ds\right)\right. \\ &\left.+b_i^1\frac{Z^1_{\tau_{i+1}}}{Z^1_{\tau_i}}+\cdots+b_i^m\frac{Z^m_{\tau_{i+1}}}{Z^m_{\tau_i}}\right) \qquad (4) \\ &+\ln\left((1-1^Tb_n)\exp\left(\int_{\tau_n}^t r(s)ds\right)+b_n^1\frac{Z_t^1}{Z^1_{\tau_n}}+\cdots+b_n^m\frac{Z_t^m}{Z^m_{\tau_n}}\right). \end{aligned}$$

In these expressions, the per-dollar return on stock i between s and t, $t > s$, is

$$\frac{Z_t^i}{Z_s^i} = \exp\left(\int_s^t \left(\mu_i(u) - \frac{1}{2}\sum_{j=1}^m \lambda_{ij}^2(u)\right) du + \sum_{j=1}^m \int_s^t \lambda_{ij}(u) dW_u^j\right).$$

3.2. Constant coefficients

When the coefficients r, μ, Λ are constants the above formula can be rewritten as

$$\begin{aligned} \ln V_t &= \ln V_0 + \sum_{i=0}^n \ln(1-c_i) + \sum_{i=0}^{n-1} \ln\Bigg((1-1^T b_i)\exp(r(\tau_{i+1}-\tau_i)) \\ &+b_i^1 \frac{Z_{\tau_{i+1}}^1}{Z_{\tau_i}^1} + \cdots + b_i^m \frac{Z_{\tau_{i+1}}^m}{Z_{\tau_i}^m}\Bigg) \\ &+ \ln\left((1-1^T b_n)\exp(r(t-\tau_n)) + b_n^1 \frac{Z_t^1}{Z_{\tau_n}^1} + \cdots + b_n^m \frac{Z_t^m}{Z_{\tau_n}^m}\right), \end{aligned} \tag{5}$$

and the per-dollar return on stock i between s and t, $t > s$, is

$$\frac{Z_t^i}{Z_s^i} = \exp\left(\left(\mu_i - \frac{1}{2}\sum_{j=1}^m \lambda_{ij}^2\right)(t-s) + \sum_{j=1}^m \lambda_{ij}(W_t^j - W_s^j)\right),$$

which has the same distribution as

$$\exp\left(\left(\mu_i - \frac{1}{2}\sum_j \lambda_{ij}^2\right)(t-s) + \sum_j \lambda_{ij} W_{t-s}^j\right).$$

This model has been investigated by many authors such as Merton and Pliska (1995) for maximizing the $\ln V_T$ for some finite horizon T. The best portfolio in this case is the proportions $\tilde{b}$ which solve the quadratic programming problem

$$\begin{cases} \max\limits_b (1-1^T b)r + b^T\mu - \frac{1}{2} b^T \Lambda\Lambda^T b \\ \text{s.t. } 1^T b \le 1,\ b^k \ge 0,\ k = 1, \ldots, m. \end{cases}$$

Clearly the solution is independent of the time because the parameters are constant in this model. The investors usually prefer to reballance a portfolio at each transaction moment. Thus we assume the investors to choose a common portfolio,

$\tilde{b}$. By the concavity of the logarithm function formula (5) therefore becomes

$$\begin{aligned}
\ln V_t &\geq \ln V_0 + \sum_{i=0}^{N(t)} \ln(1-c_i) + (1 - 1^T\tilde{b})rt \\
&\quad +\tilde{b}^1 \left(\left(\mu_1 - \frac{1}{2}\sum_j \lambda_{1j}^2 \right) t + \sum_j \lambda_{1j} W_t^j \right) + \cdots \\
&\quad +\tilde{b}^m \left(\left(\mu_m - \frac{1}{2}\sum_j \lambda_{mj}^2 \right) t + \sum_j \lambda_{mj} W_t^j \right) \\
&=: \ln V_0 + G(t, N(t), \{c_i\}). \qquad (6)
\end{aligned}$$

In order to enlarge $E\ln V_t$ we can enlarge the function $EG(t, N(t), \{c_i\})$. For this purpose we assume that $\{c_i\}$ is an sequence of independent random variables with common expectation and this sequence is independent of $N(t)$, then we have

$$\begin{aligned}
EG(t, N(t), \{c_i\}) &= E\ln(1-c)\nu((0,t]) + (1 - 1^T\tilde{b})rt \\
&\quad +\tilde{b}^1 \left(\mu_1 - \frac{1}{2}\sum_j \lambda_{1j}^2 \right) t + \cdots \\
&\quad +\tilde{b}^m \left(\mu_m - \frac{1}{2}\sum_j \lambda_{mj}^2 \right) t.
\end{aligned}$$

¿From the standard $EG(t, N(t), \{c_i\}) \geq 0$ we can conclude that

Proposition 3.1. (Strategy I) *Suppose the investor invests in the market in the above manner and all coefficients in Black-Scholes equations are constant, then* $E\ln V_t \geq \ln V_0$ *if the transaction intensity* ν *satisfies*

$$\begin{aligned}
\nu((0,t]) &\leq -\frac{1}{E\ln(1-c)} \left((1 - 1^T\tilde{b})rt + \tilde{b}^1 \left(\mu_1 - \frac{1}{2}\sum_j \lambda_{1j}^2 \right) t + \cdots \right. \\
&\quad \left. +\tilde{b}^m \left(\mu_m - \frac{1}{2}\sum_j \lambda_{mj}^2 \right) t \right). \qquad (7)
\end{aligned}$$

However, the above criteria is obtained from a loose standard. We now change an idea and assume the investor like to optimize EV_t instead of $E\ln V_t$. Since $N(\cdot)$, $\{\tau_i\}$ and the coefficients of the markets $r(\cdot)$, $\mu(\cdot)$, $\Lambda(\cdot)$ are generally dependent (see Section 2), in the general setting it is impossible to directly obtain good results on EV_t. To illustrate the "interplay" between transactions and the evolution of stocks we now consider the simple case of one stock

$$dZ_t = Z_t(\mu dt + \lambda dW_t),$$

and its solution is given by

$$Z_t = Z_0 \exp\left(\left(\mu - \frac{1}{2}\lambda^2 \right) t + \lambda W_t \right). \qquad (8)$$

In this case V_t described in (3) is

$$V_t = V_0 \exp\left(\left(\mu - \frac{1}{2}\lambda^2\right) t + \lambda W_t\right) \prod_{i=0}^{N(t)} (1 - c_i). \tag{9}$$

We have that

Proposition 3.2. (Strategy II) *In the setting of (9), assume as before that $\{c_i\}$ is a sequence of independent random variables with common expectation Ec, and assume that $N(t)$ and W_t are mutually independent. Then $EV_t \geq V_0$ if the transaction intensity ν satisfies*

$$\nu((0,t]) \leq \mu t / Ec. \tag{10}$$

Proof. ¿From the martingale property of $\exp\left(-\frac{1}{2}\lambda^2 t + \lambda W_t\right)$ and some calculations we have that

$$\begin{aligned} EV_t &= V_0 \exp(\mu t) \sum_{n=0}^{\infty} P(N(t) = n) \prod_{i=0}^{n} (1 - Ec_i) \\ &= V_0 \exp(\mu t - \nu((0,t]) Ec). \end{aligned}$$

Therefore,

$$\mu t - \nu((0,t]) Ec \geq 0$$

is a sufficient and necessary condition for $EV_t \geq V_0$. □

We now turn to consider the risk fluctuation after a transaction under the assumption that $c_i \equiv c$, a constant. We have

Proposition 3.3. *Under the above assumptions, let the risk fluctuation with transaction costs be $DV_t := E(V_t - EV_t)^2$ and the risk fluctuation without transaction costs be $DU_t := E(U_t - EU_t)^2$ (where $U_t := \frac{V_0}{Z_0} Z_t$). Suppose that $\nu((0,\infty)) > 0$, then $DV_t \geq DU_t$ if $t \leq t_0$ and $DV_t < DU_t$ if $t > t_0$, where t_0 is the solution of equation*

$$\exp\{-2c\nu((0,t])\}(\exp\{\lambda^2 t + c^2 \nu((0,t])\} - 1) = \exp\{\lambda^2 t\} - 1. \tag{11}$$

Proof. It is easy to obtain that

$$DV_t = V_0^2 \exp\{2\mu t - 2c\nu((0,t])\}\Big\{ \exp\{\lambda^2 t + c^2 \nu((0,t])\} - 1 \Big\}$$

and

$$DU_t = V_0^2 \exp\{2\mu t\}\Big\{ \exp\{\lambda^2 t\} - 1 \Big\}$$

Let us consider the following elementary function

$$g(t) := e^{-2c\nu((0,t])}(e^{c^2\nu((0,t]) + \lambda^2 t} - 1) - (e^{\lambda^2 t} - 1)$$

It is easy to see that $g(0) = 0$ and under the assumption that $\nu((0,\infty)) > 0$ there exists $t_0 > 0$ such that $g(t) \geq 0$ when $t \leq t_0$ and $g(t) < 0$ when $t > t_0$. In fact this t_0 is given by the equation (11). This implies the required assertion. □

Remark 3.4. *We now remark that* **the transaction reducing the deviation of the wealth is usually not true.** *In fact, Proposition 3.3 give us a criteria to judge when the deviation becomes smaller.*

4. A utility function for the opportunity brought by transaction

We know from previous section that a transaction with intensity measure ν reduces the growth of wealth to

$$\exp\{-c\nu((0,t])\}$$

times of the total wealth during the period of $[0,t]$. It is also clear that transactions can provide some chance or opportunity for investors. ¿From this point of view we prefer to introduce a mathematical formula to describe the benefit of transactions. A reasonable choice in economics is the one given by a utility function (c.f., e.g., Duffie 1992, Korn 1997). A continuous concave function U on $[0,\infty)$ is called a **utility function** if U is continuously differentiable with derivative U' and $\lim_{u\to 0+} U'(u)$ exists and satisfies $0 < \lim_{u\to 0+} U'(u) \leq +\infty$, $\lim_{u\to\infty} U'(u) = 0$. In this section we only consider the family of the following form utility functions:

$$\mathcal{U} = \{u(x),\ x > 0;\ u(x) = \alpha(ax+b)^{\gamma} \text{ or } \alpha\ln(ax+b),\ a,\ b,\ \alpha \in R_+,\ \gamma < 1\},$$

which has been extensively considered (see Korn 1997).

Consider

$$F(t,x,y):\ R_+ \times R_+ \times R_+ \mapsto R_+$$

satisfying F *is a utility function with respect to each single variable the other two variables being held fixed.* We now restrain our attention in this class of F:

$$\frac{F(t,x,y)}{u_1(t)u_2(x)u_3(y)} \leq C, \text{ for some } u_i \in \mathcal{U}, i = 1,2,3, \tag{12}$$

for some constant $C > 0$ and introduce

$$\mathcal{C}(t,c,\nu,F) := \exp\{-c\nu((0,t])\}F(t,\nu((0,t]),V_t),\ \nu \in M, \tag{13}$$

where M is the set of all measures on $(0,\infty)$. We shall discuss the following two questions:

- Does there exist a **local optimal transaction mean strategy**, $\nu \in M$, in the sense that the expectation of (13) is maximal for fixed $T > 0$, $c > 0$, and F, i.e.,
$$\nu_0 = \arg\max_{\nu\in M} E\mathcal{C}(T,c,\nu,F)? \tag{14}$$
- What is the asymptotic behavior of $\mathcal{C}(T,c,\nu',F)$ as $T \to \infty$?

To the first question, we have that

Proposition 4.1. *A local optimal transaction mean strategy exists for problem (14) when F satisfies (12).*

Proof. We first remark that $C(T,c,\nu,F)$ is finite almost surely for all $T \geq 0,\ c \geq 0,\ \nu \in M$, and $EC(T,c,\nu,F)$ is finite from condition (12). In fact, since V_t depends on $N(t)$ and W_t and the latter two processes are independent, we have

$$\begin{aligned} EC(T,c,\nu,F) &= \sum_{j=0}^{\infty} \frac{\nu((0,T])^j}{j!} e^{-c\nu((0,T])} EF\Bigg(T, \nu((0,T]), \\ &\qquad V_0 \exp\left(\left(\mu - \frac{1}{2}\lambda^2\right)T + \lambda W_T\right) \prod_{i=0}^{j} (1-c)\Bigg). \end{aligned}$$

¿From the continuity of F and the mapping $\nu \mapsto \nu((0,T])$ it is easy to show the continuity of $EC(T,c,\nu,F)$ in ν by noticing that the series in the right hand converges uniformly in $\nu :\ \nu([0,T]) \leq K,\ (K > 0)$ under the assumption on F. On the other hand, it is sufficient to prove the desired assertion for $\nu \in M :\ \nu([0,T]) \leq K$ for a large enough constant $K > 0$, because $F(T,\ \cdot,\ V)$ is a utility function, which increases much slower than the exponential function $\exp\{c\cdot\}$. However, since the restriction $M|_{[0,T]}^K$ of $\{\mu \in M :\ \nu([0,T]) \leq K\}$ to the interval $[0,T]$ is a weakly compact set (see Yosida 1968 or Zhao 2000). Combining the above argument, we know that there must exist a $\nu_0 \in M|_{[0,T]}$ such that $EC(T,c,\nu_0,F) = \max_{\mu \in M|_{[0,T]}} EC(T,c,\mu,F)$. □

Usually we can give an explicit expression for the optimal value of ν when the form of F is given.

Example 4.2. *Let $F(t,x,y) = \tilde{F}(t,x)$ and let*

$$S := \{(t,\nu);\ \tilde{F}_x(t,\nu((0,t])) = cF(t,\nu((0,t]))\},$$

where F_x means the partial derivative of $F(t,x)$ in x. If S is not empty, then there exists a local optimal transaction mean strategy ν_0 of problem (13) such that $(t_0,\nu_0) \in S$ for some t_0. If S is empty, then the zero measure is the only local optimal transaction mean strategy ν.

In fact, the local optimal transaction mean strategy is usually not unique. For instance, let $F(t,x,y) = x$, the maximum point of problem (13) becomes $\nu((0,T]) = \frac{1}{c}$, which constitutes a continuum set in M.

We now consider the asymptotic behavior of $C(T,c,\nu,F)$ as $T \to \infty$ for the particular case where

$$F(t,x,y) = \ln(1+t)\ln(1+x)y^\eta, \quad \text{for } \eta \in (0,1]. \tag{15}$$

We can do similarly for other F described in (12).

When the investor permits to make transaction arbitrarily, we have that

Proposition 4.3. *Under the above assumption one has*

$$\limsup_{T\to\infty} \frac{EC(T,c,\nu,F)}{V_0^\eta \ln(1+T)\exp\{\eta(\mu - \frac{1}{2}(1-\eta)\lambda^2)T\}} \leq \max_{x \in R_+} I(x) < \infty, \tag{16}$$

where $I(x) = \exp\{-((1+c)-(1-c)^\eta)x\}\ln(1+x)$.

The best transaction intensity $\nu_0 \in M$ is characterized by $x := \nu_0((0,\infty))$ which satisfies

$$(1+x)\ln(1+x) = \frac{1}{(1+c)-(1-c)^\eta}. \tag{17}$$

Proof. We can rewrite (13) as

$$\begin{aligned} \mathcal{C}(T,c,\nu,F) &= \exp\Big\{-c\nu((0,T])\Big\}\ln(1+T)\ln\Big(1+\nu((0,T])\Big) \\ &\quad \Big(V_0(1-c)^{N(T)}\exp\Big\{(\mu-\frac{1}{2}\lambda^2)t+\lambda W_T\Big\}\Big)^\eta. \end{aligned}$$

It is easy to show that

$$\begin{aligned} E\mathcal{C}(T,c,\nu,F) &= V_0^\eta \ln(1+T)\exp\{\eta(\mu-\frac{1}{2}(1-\eta)\lambda^2)T\} \\ &\quad \exp\{-((1+c)-(1-c)^\eta)\nu((0,T])\}\ln(1+\nu((0,T])). \end{aligned}$$

To solve the optimization problem

$$\max_{\nu\in M}\exp\{-((1+c)-(1-c)^\eta)\nu((0,T])\}\ln(1+\nu((0,T])) \tag{18}$$

for fixed $T>0$, the solution ν_0 such that $x := \nu_0((0,T])$ satisfies (17). □

The solution of (17) means that the investor is required to make only finite transactions in average during infinite time interval $(0,\infty)$, according to the local optimal transaction mean strategy. However, it is not the case in reality. It is reasonable to consider the case where $\lim_{t\to\infty}\nu((0,t])/h(t)$ exists for some increasing positive function $h(t)$ with $\lim_{t\to\infty}h(t)=\infty$. In this direction there are a lot of interesting questions. We have to stop here for the limited pages.

Acknowledgments We thank Prof. Manfred Schäl for many helpful and stimulating discussions. The financial support of SFB 237 and NSF of China is gratefully acknowledged.

References

[1] S. Albeverio, L. J. Lao and X. L. Zhao, *On-Line Portfolio Selection Strategy with Predictions in the Presence of Transaction Costs,* (1999), in preparation.

[2] S. Albeverio and V. Steblovskaya, *A Model of Financial Market with Several Interacting Assets (Extending the Multidimensional Black-Scholes Model)*, Preprint of Institute of Applied Mathematics, University of Bonn, No. 609, 1999.

[3] A. Cadenillas and S. R. Pliska, *Optimal trading of a security when there are taxes and transaction costs*, Finance and Stochastics, **2** (1999), 137–165.

[4] D. Duffie, *Dynamic Asset Pricing Theory*, Princeton University Press, New Jersey **1992**.

[5] D. Duffie and T. Sun, *Transaction costs and portfolio choice in a discrete-continuous time setting*, J. Econ. Dyn. Cont., **14** (1990), 35–51.

[6] I. Karatzas, *Lectures on the Mathematics of Finane*, American Mathematical Society, Rhode Island, **1997**.

[7] R. Korn, *Optimal Portfolios*, World Scientific, Singapore, **1997**.

[8] A. J. Merton and S. R. Pliska, *Optimal portfolio management with fixed transaction costs*, Mathematical Finance, **5.4** (1995), 337-356.

[9] K. Yosida, *Functional Analysis*, Springer-Verlag, Berlin, **1968**.

[10] X. L. Zhao, *Introduction to Measure-Valued Branching Processes* (Chinese), Science Press, Beijing, **2000**.

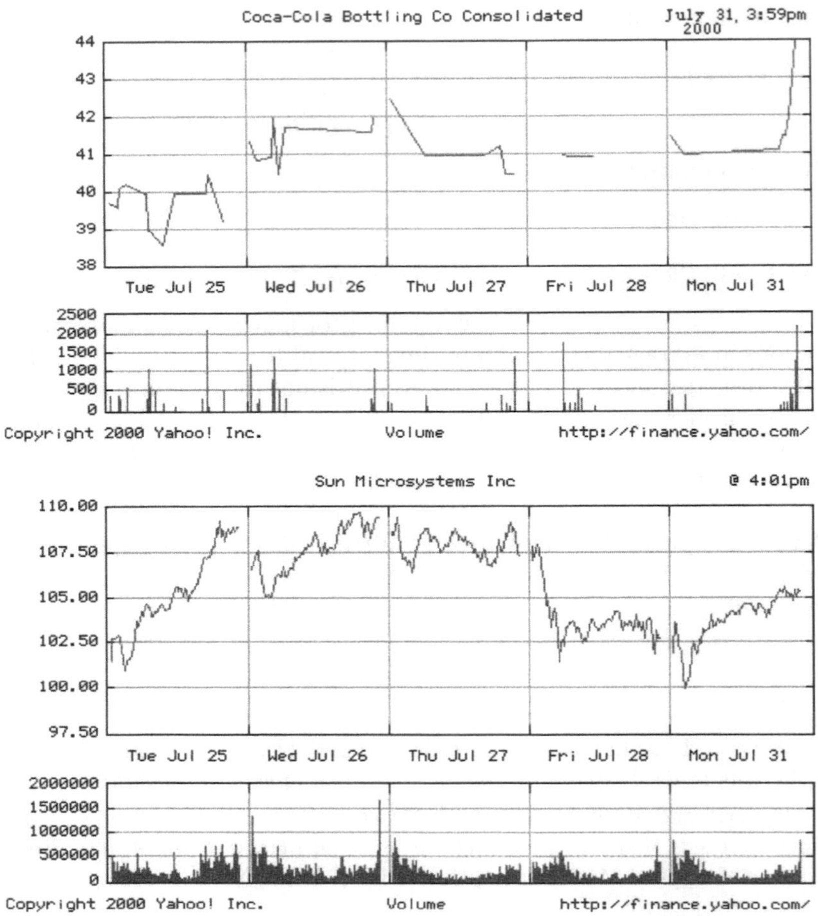

FIGURE 1. *Price value process and transaction volume for COKE and SUNW from July 27 to July 31, 2000*

[1)2)]Institut für Angewandte Mathematik
Universität Bonn, Wegelerstrasse 6
D-53115 Bonn, Germany
E-mail address: [1)]`albeverio@uni-bonn.de`
E-mail address: [2)]`lao@wiener.iam.uni-bonn.de`

[3)]Institute of Mathematics,
Fudan University, Handan Road 220
200433 Shanghai, P.R.China
E-mail address: `xlzhao@fudan.edu.cn`

Trends in Mathematics, © 2001 Birkhäuser Verlag Basel/Switzerland

Demand Heterogeneity and Price Volatility

David R. Alexander[1] and Emmanuel E. Haven[2]

Abstract. In this note, we propose a notion of financial market heterogeneity based on the rates at which agents adjust their demand for a risky asset. These rates can be interpreted as the speeds at which agents process market information and act on it. We indicate through several simple examples how this notion of heterogeneity is connected to the volatility of aggregate excess demand for the risky asset. We also indicate how heterogeneity and volatility in excess demand can be transmitted to that of the risky asset price, possibly providing another framework for modeling asset price volatility and its statistical estimation.

1. The Economic Assumptions

Consider a financial market populated by a large but finite set $\mathcal{A}$ of agents who are buying and selling a risky financial asset, such as a stock. Assume that every agent $a \in \mathcal{A}$ has wealth $W^a \equiv \{W_t^a\}_{t\in[0,T]}$ and risky asset demand $D^a \equiv \{D_t^a\}_{t\in[0,T]}$ which are both continuous time stochastic processes on a complete probability space $(\Omega, \mathcal{F}, \mathbf{P})$ and which both have continuous sample paths. Also, we assume that for all $a \in \mathcal{A}$, $t \in [0,T]$ and $\omega \in \Omega$ we have $W^a(\omega)_t \in [0,\infty)$, a no bankruptcy condition, and $D_t^a(\omega) \in (-\infty, \infty)$, a finite demand condition. If $D_t^a(\omega) > 0$, agent a wishes to invest the wealth $D_t^a(\omega) > 0$ in the risky asset. If $D_t^a(\omega) < 0$, agent a wishes to sell off $-D_t^a(\omega)$ of the risky asset. The aggregate wealth in the economy is defined by $\mathcal{W}_t(\omega) = \sum_{a\in\mathcal{A}} W_t^a(\omega)$ and we assume that $\mathcal{W}_t(\omega) \in (0,\infty)$. We also assume agents are so numerous that $-\mathcal{W}_t(\omega) \ll D_s^a(\omega) \ll \mathcal{W}_t(\omega)$ for all $s,t \in [0,T]$, that is, $D_s^a(\omega)/\mathcal{W}_t(\omega)$ is very small. Note that we allow individual agents to short sell the stock; all we are imposing with this last condition is that a single agent's demand is insignificant compared to the aggregate wealth in the economy and so agents act, approximately, as price takers.

---

1. The financial help from NSERC (Natural Sciences and Engineering Research Council of Canada) and McGill University in the form of a Max Bell Fellowship is gratefully acknowledged.
2. The financial help from SSHRC (Social Sciences and Humanities Research Council of Canada), FCAR (Fonds pour les Chercheurs et l'Aide à la Recherche, Quebec) and the IFM2 (Institut de Finance Mathématiques, Quebec) is gratefully acknowledged.

2. A Sinusoidal Approximation of Individual Demand

For any $N \in \mathbb{N}$, subdivide the time interval $[0,T]$ into N subintervals of equal width $\frac{T}{N}$ as $[0,T] = \bigcup_{k=1}^{N}[t_{k-1},t_k]$ where $t_k = k\left(\frac{T}{N}\right)$ for $k = 0,\ldots,N$ and consider the $N+1$ points $\{(t_k, D^a_{t_k}(\omega))\}_{k=0}^{N}$ on agent a's demand path. Instead of using a polygonal approximation of D^a, obtained by connecting $\{(t_k, D^a_{t_k}(\omega))\}_{k=0}^{N}$ by N lines, we use an approximation which connects these points by N sinusoidal arcs of minimal curvature. We connect the kth pair of points $(t_{k-1}, D^a_{t_{k-1}}(\omega))$ and $(t_k, D^a_{t_k}(\omega))$ by a sinusoidal arc of the form "$A\sin(\lambda t+\phi)$", $t \in [t_{k-1},t_k]$, which we call a λ-sinusoid, and select appropriate values for the amplitude, frequency, and phase. Since $\mathcal{W}_{t_{k-1}}(\omega) > 0$ and $-\mathcal{W}_{t_{k-1}}(\omega) \ll D^a_t(\omega) \ll \mathcal{W}_{t_{k-1}}(\omega)$ for $t \in [t_{k-1},t_k]$ we can select $\mathcal{W}_{t_{k-1}}(\omega)$ as the amplitude. The frequency and the phase of the kth approximating arc are then characterized as the unique solution $\{\lambda^a_{t_{k-1}}(\omega), \phi^a_{t_{k-1}}(\omega)\}$, in terms of $\{D^a_{t_{k-1}}(\omega), \mathcal{W}_{t_{k-1}}(\omega), D^a_{t_k}(\omega)\}$, to the system

$$\begin{aligned} D^a_{t_{k-1}}(\omega) &= \mathcal{W}_{t_{k-1}}(\omega)\sin(\lambda^a_{t_{k-1}}(\omega)t_{k-1} + \phi^a_{t_{k-1}}(\omega)) \\ D^a_{t_k}(\omega) &= \mathcal{W}_{t_{k-1}}(\omega)\sin(\lambda^a_{t_{k-1}}(\omega)t_k + \phi^a_{t_{k-1}}(\omega)) \end{aligned}$$

such that $\lambda^a_{t_{k-1}}(\omega)$ is the smallest non-negative frequency and $\phi^a_{t_{k-1}}(\omega) \in (-\pi,\pi]$. The kth approximating arc is $\mathcal{W}_{t_{k-1}}(\omega)\sin(\lambda^a_{t_{k-1}}(\omega)t + \phi^a_{t_{k-1}}(\omega))$ for $t \in [t_{k-1},t_k]$. Temporarily suppressing references to the sample point ω, the process defined by $D^{a,N}_t = \mathcal{W}_0 \sin\phi^a_0 1_{\{0\}}(t) + \sum_{k=1}^{N} \mathcal{W}_{t_{k-1}}\sin(\lambda^a_{t_{k-1}}t + \phi^a_{t_{k-1}})1_{(t_{k-1},t_k]}(t)$ is what we shall call a sinusoidal approximation of $D^a_t(\omega)$.

3. A Financial Interpretation of Frequency and Phase

For $t \in [t_{k-1},t_k]$ we have $D^a_t(\omega) \approx \mathcal{W}_{t_{k-1}}(\omega)\sin(\lambda^a_{t_{k-1}}(\omega)t + \phi^a_{t_{k-1}}(\omega))$. Since we assume $\mathcal{W}_{t_{k-1}}(\omega) > 0$ and $-\mathcal{W}_{t_{k-1}}(\omega) \ll D^a_t(\omega) \ll \mathcal{W}_{t_{k-1}}(\omega)$ we know that

$$\frac{D^a_t(\omega)}{\mathcal{W}_{t_{k-1}}(\omega)} \approx \sin(\lambda^a_{t_{k-1}}(\omega)t + \phi^a_{t_{k-1}}(\omega)) \approx 0$$

and hence $\lambda^a_{t_{k-1}}(\omega)t + \phi^a_{t_{k-1}}(\omega) \approx \pi n$ for some integer n. Now, for large enough N, the time interval $\Delta t = t_k - t_{k-1}$ is small and, if the paths of D^a are sufficiently smooth (possibly differentiable), we then have that

$$\begin{aligned} \frac{\Delta D^a_t(\omega)}{\Delta t} &\approx \frac{d}{dt}\left[\mathcal{W}_{t_{k-1}}(\omega)\sin(\lambda^a_{t_{k-1}}(\omega)t + \phi^a_{t_{k-1}}(\omega))\right] \\ &= \mathcal{W}_{t_{k-1}}(\omega)\lambda^a_{t_{k-1}}(\omega)\cos(\lambda^a_{t_{k-1}}(\omega)t + \phi^a_{t_{k-1}}(\omega)) \\ &\approx \mathcal{W}_{t_{k-1}}(\omega)\lambda^a_{t_{k-1}}(\omega)\cos(\pi n) \\ &= \pm\mathcal{W}_{t_{k-1}}(\omega)\lambda^a_{t_{k-1}}(\omega) \end{aligned}$$

and so we see that the rate of change of D^a with respect to time is proportional to the frequency, the proportionality factor being aggregate wealth, a rather stable quantity. Thus, all agents with demand approximated by a λ-sinusoid will be adjusting their demands at one of the two rates $\pm\lambda\mathcal{W}_{t_{k-1}}(\omega)$, approximately. The frequency λ can therefore be interpreted as the magnitude of the rate of change of individual demand relative to aggregate wealth. Also, $\lambda^a_{t_{k-1}}(\omega)t + \phi^a_{t_{k-1}}(\omega) \approx \pi n$ implies that all agents with demand following a λ-sinusoid will have phase angles $\phi^a_{t_{k-1}}(\omega) \approx \pi n - \lambda$ which all differ by a multiple of π, approximately, and hence, the approximating λ-sinusoids are either approximately in phase or approximately out of phase. Since we have restricted our phase angles to $(-\pi, \pi]$, all phases differ by either 0 or π, i.e., the phase angles are the same or are different by π so there are only two angles, approximately. Any two agents having their demands on λ-sinusoids with phases differing by 0 are in phase and will have a common rate of demand change, both being $\lambda \mathcal{W}_{t_{k-1}}(\omega)$ or both being $-\lambda \mathcal{W}_{t_{k-1}}(\omega)$. Any two agents having λ-sinusoids with phases differing by π are out of phase and will have rates of demand change which are equal in magnitude but opposite in sign; one will have a rate of change of $\lambda \mathcal{W}_{t_{k-1}}(\omega)$ and the other $-\lambda \mathcal{W}_{t_{k-1}}(\omega)$. The sign of the rate of change is determined by the phase angle and so the financial interpretation of phase is that it indicates whether an agent is choosing to increase or decrease current demand at one of the two rates $\pm\lambda \mathcal{W}_{t_{k-1}}(\omega)$.

4. A Spectral Representation of Aggregate Excess Demand

For all a, ω, and t, the approximating process $D^{a,N}_t(\omega)$ converges to $D^a_t(\omega)$ as $N \to \infty$, yielding the representation

$$D^a_t(\omega) = \mathcal{W}_t(\omega) \sin(\lambda^a_t(\omega)t + \phi^a_t(\omega))$$

for limiting processes $\lambda^a_t(\omega)$ and $\phi^a_t(\omega)$. Define $\mathcal{A}_{\lambda,t}(\omega) = \{a \in \mathcal{A} : \lambda^a_t(\omega) = \lambda\}$, the group of agents whose time t, state ω demand moves infinitesimally along a λ-sinusoid, and consider $\mathcal{S}_\lambda = \{D^a_t(\omega) = \mathcal{W}_t(\omega)\sin(\lambda t + \phi^a_t(\omega)) : a \in \mathcal{A}_{\lambda,t}(\omega)\}$, the collection of all their demand sinusoids. As discussed in the previous section, the rates of demand change for agents in $\mathcal{A}_{\lambda,t}(\omega)$ are approximately $\pm\mathcal{W}_t(\omega)\lambda$, the sign being determined by the phase. Aggregating the demands of this group of agents yields

$$\begin{aligned}
D_{\lambda,t}(\omega) &= \sum_{a\in\mathcal{A}_{\lambda,t}(\omega)} D^a_t(\omega) \\
&= \mathcal{W}_t(\omega) \sum_{a\in\mathcal{A}_{\lambda,t}(\omega)} \sin(\lambda t + \phi^a_t(\omega)) \\
&= \mathcal{W}_t(\omega) N_{\lambda,t}(\omega) \sin(\lambda t + \phi_{\lambda,t}(\omega))
\end{aligned}$$

where, in the case that $\mathcal{A}_{\lambda,t}(\omega) = \emptyset$, we take empty sums to be zero. Since the sinusoids in $\mathcal{S}_\lambda$ all have the same amplitude and are either in phase or out of phase, approximately, many cancel in the sum through destructive interference. $N_{\lambda,t}(\omega)$ is therefore the approximate number of sinusoids which do not cancel and $\phi_{\lambda,t}(\omega)$ is their necessarily common phase. In financial terms, $N_{\lambda,t}(\omega)$ is the number of agents in $\mathcal{A}_{\lambda,t}(\omega)$ whose demand and rate of change of demand is not met equally and oppositely by some other agent in the same group, i.e., it counts the discrepancy between the number of potential buyers and potential sellers in the group. Hence, $D_{\lambda,t}(\omega)$ is the excess demand of the agents in $\mathcal{A}_{\lambda,t}(\omega)$; it is the demand that is not met within the group and can only be met by agents from other groups, if at all.

Summing up all group excess demands we obtain a representation for aggregate excess demand $\mathcal{D} = \{\mathcal{D}_t\}_{t\in[0,T]}$ as

$$\mathcal{D}_t(\omega) = \mathcal{W}_t(\omega) \sum_{\lambda:\mathcal{A}_{\lambda,t}(\omega)\neq\emptyset} A_{\lambda,t}(\omega)\sin(\lambda t + \phi_{\lambda,t}(\omega))$$

which, using the measure defined for each t and ω by $A_{\lambda,t}(\omega) = \sum_{0\leq\mu\leq\lambda} N_{\mu,t}(\omega)$, together with standard identities and a change of variables, can easily be put into the form

$$\mathcal{D}_t(\omega) \quad = \quad \mathcal{W}_t(\omega)\int_{-\infty}^{\infty} e^{i\lambda t}\left[\frac{1}{2i}\text{sign}(\lambda)e^{i\text{sign}(\lambda)\phi_{|\lambda|,t}(\omega)}\right]dA_{|\lambda|,t}(\omega) \qquad (1)$$

In summary, we view the evolution of an individual agent's demand as following one sinusoid for an infinitesimal time period and then switching to another. For each fixed λ we consider the group of agents $\mathcal{A}_{\lambda,t}(\omega)$ whose time t, state ω demand follows a λ-sinusoid and then aggregate their contributions by superposition to obtain the excess demand $D_{\lambda,t}(\omega)$ for the group. However, since an individual agent is jumping between sinusoids, say from a λ_1-sinusoid at time t_1 to a λ_2-sinusoid at time t_2, the agent is also switching groups from $\mathcal{A}_{\lambda_1,t_1}(\omega)$ to $\mathcal{A}_{\lambda_2,t_2}(\omega)$. Thus, for a given λ, the agents comprising $\mathcal{A}_{\lambda,t}(\omega)$ change over time; some agents transfer onto a λ-sinusoid and some transfer off so the group excess demand component $D_{\lambda,t}(\omega)$ associated with frequency λ is due to different agents at different times and states. Finally, we emphasize that in arriving at the spectral representation of aggregate excess demand (1), extremely mild economic/financial assumptions have been made: the risky asset is infinitely divisible, individual wealth and demand follow continuous trajectories, individual demand is finite, aggregate excess demand is strictly positive and finite, and, an individual agent's demand is much smaller than aggregate wealth.

5. Stationary Excess Demand & Constant Volatility

For purely illustrative purposes, we consider a very restrictive special case: aggregate excess demand is a zero mean, covariance stationary process. Such a process has the form

$$\mathcal{D}_t(\omega) = \int_{-\infty}^{\infty} e^{i\lambda t}dZ_\lambda(\omega) \qquad (2)$$

where Z is a zero mean process with uncorrelated increments and spectral measure $dF(\lambda) = \mathbf{E}[|dZ_\lambda|^2] = \mathbf{Var}[dZ_\lambda]$. We assume for simplicity that $dF(\lambda) = f(\lambda)d\lambda$ for some spectral density function $f : \mathbb{R} \to \mathbb{R}_+$, which of course requires a continuum of frequencies and agents. The autocovariance function $\gamma(h)$ of $\mathcal{D}$ has the spectral representation

$$\gamma(h) = \int_{-\infty}^{\infty} e^{i\lambda h} f(\lambda) d\lambda$$

and the variance-frequency decomposition for the variance of $\mathcal{D}$ is

$$\mathbf{Var}[\mathcal{D}_t] = \gamma(0) = \int_{-\infty}^{\infty} f(\lambda) d\lambda = \int_{-\infty}^{\infty} \mathbf{Var}[dZ_\lambda] \tag{3}$$

indicating that the variance of $\mathcal{D}$ is constant. By Fourier theory, we can obtain the spectral density from the autocovariance

$$f(\lambda) = \frac{1}{2\pi} \int_{-\infty}^{\infty} e^{-ih\lambda} \gamma(h) dh$$

we have the associated Parseval relation

$$\int_{-\infty}^{\infty} \gamma^2(h) dh = \int_{-\infty}^{\infty} f^2(\lambda) d\lambda = 4K$$

for some constant $K > 0$ and we have the uncertainty principle

$$\left(\int_{-\infty}^{\infty} h^2 \gamma^2(h) dh\right) \left(\int_{-\infty}^{\infty} \lambda^2 f^2(\lambda) d\lambda\right) \geq K$$

We define measures heterogeneity H and memory M associated with aggregate excess demand for the risky asset by the quantities

$$H = \int_{-\infty}^{\infty} \lambda^2 f^2(\lambda) d\lambda \qquad M = \int_{-\infty}^{\infty} h^2 \gamma^2(h) dh$$

and so the uncertainty relation becomes $MH \geq K$. The quantity M is a measure of dispersion of the correlogram $\gamma(h)$ for $\mathcal{D}$ and H is a measure of dispersion of the spectrum $f(\lambda)$ of $\mathcal{D}$.

To interpret the relation $MH \geq K$ we return to the financial interpretation of λ as the rate of demand adjustment relative to aggregate wealth. If all agents are adjusting their individual demands at similar rates, we refer to this market condition as one of excess demand homogeneity, which is indicated by small values of H. If, on the other hand, the rates of demand adjustment vary widely among agents, we refer to this market condition as one of excess demand heterogeneity which is indicated by large values of H.

As is well known, since $\gamma(h)$ and $f(\lambda)$ are related by Fourier transforms, there is an inverse relationship between M and H and the uncertainty relation $HM \geq K$ puts a lower bound on how small both can be simultaneously. If the frequencies in the spectral representation of the process are highly dispersed, H is large, the autocovariance function is highly concentrated, indicating that the aggregate excess demand process has a limited memory, characterized by a low M value. On the other hand, if H is small so that the spectrum is highly concentrated then $\mathcal{D}$ has a highly dispersed correlogram, with a large M value, and therefore has relatively long range dependence. Thus, the uncertainty relation rules out the possibility of limited memory under conditions of high homogeneity which has an interesting interpretation when one considers issues of market efficiency. Also, it is intuitively appealing to view the volatility of $\mathcal{D}$ as being the result of agents adjusting their demands at different rates; a large (small) number of demand adjustment rates together with many (few) agents making such demand adjustments leads to large and frequent (small and infrequent) corrections of aggregate demand and, hence, high (low) levels of excess demand volatility.

6. Semi-Stationary Excess Demand & Deterministic Volatility

In the prior illustrative example, the spectral density $f(\lambda)$ is constant in time and hence $\mathbf{Var}[\mathcal{D}_t]$ is constant in time. Here, we consider a larger class of excess demand processes than stationary processes, as in (2), to what Priestley defines as semi-stationary processes ([8],[9],[10],[11],[12]). The form of such processes is

$$\mathcal{D}_t(\omega) = \int_{-\infty}^{\infty} e^{i\lambda t} C(\lambda, t) dZ_\lambda(\omega) \tag{4}$$

Where $C(\lambda, t)$ is a function such that the family $\{C(\lambda, t)e^{i\lambda t}\}$ is of oscillatory type. Using the measure defined by $d\mu(\lambda) = \mathbf{E}[|dZ_\lambda|^2]$, the autocovariance function is

$$\gamma(s,t) = \int_{-\infty}^{\infty} e^{i(t-s)} C(\lambda, t)\overline{C}(\lambda, s) d\mu(\lambda)$$

and the variance-frequency decomposition is

$$\mathbf{Var}[\mathcal{D}_t] = \gamma(t,t) = \int_{-\infty}^{\infty} |C(\lambda,t)|^2 d\mu(\lambda) = \int_{-\infty}^{\infty} \mathbf{Var}[C(\lambda,t)dZ_\lambda] \tag{5}$$

indicating that the variance of $\mathcal{D}$ is a deterministic function of time. The time dependent spectral density for $\mathcal{D}$ is therefore

$$f(\lambda,t)d\lambda = |C(\lambda,t)|^2 d\mu(\lambda) = \mathbf{Var}[C(\lambda,t)dZ_\lambda]$$

which is composed of a stationary component dZ_λ modified by a deterministically time dependent component $C(\lambda, t)$.

7. A More General Excess Demand & Stochastic Volatility

We can generalize the structure of the excess demand process in (4) even further by introducing a sample point dependence. Consider, for each fixed λ, a process $C_{\lambda,t}(\omega)$ which is adapted to some filtration $\mathbb{F} = \{\mathcal{F}_t\}_{t\in[0,T]}$ on $(\Omega, \mathcal{F}, \mathbf{P})$. Also assume that the process Z is independent of $\mathbb{F}$. Generalizing (4) this way yields

$$\mathcal{D}_t(\omega) = \int_{-\infty}^{\infty} e^{i\lambda t} C_{\lambda,t}(\omega) dZ_\lambda(\omega) \tag{6}$$

Denote the mean and variance conditional on $\mathbb{F}$ at time t by $\mathbf{E}_t[.]$ and $\mathbf{Var}_t[.]$. Taking $\mathcal{F}_0$ to consist only of sets of probability 0 or 1, conditioning on $t = 0$

yields the corresponding unconditional quantities. Using the measure defined by $d\mu(\lambda) = \mathbf{E}[|dZ_\lambda|^2]$, the conditional autocovariance function of $\mathcal{D}$ is (for $a, b \geq 0$)

$$\gamma(t+a, t+b|t) = \int_{-\infty}^{\infty} e^{i(a-b)} \mathbf{E}_t[C_{\lambda,t+a}\overline{C}_{\lambda,t+b}] d\mu(\lambda)$$

and the conditional variance-frequency decomposition is

$$\begin{aligned} \mathbf{Var}_t[\mathcal{D}_{t+a}] = \gamma(t+a, t+a|t) &= \int_{-\infty}^{\infty} \mathbf{E}_t\left[|C_{\lambda,t+a}|^2\right] d\mu(\lambda) \qquad (7) \\ &= \int_{-\infty}^{\infty} \mathbf{Var}_t[C_{\lambda,t+a} dZ_\lambda] \end{aligned}$$

Conditioning on $t = 0$ yields the unconditional variance:

$$\begin{aligned} \mathbf{Var}[\mathcal{D}_t] = \gamma(t, t|0) &= \int_{-\infty}^{\infty} \mathbf{E}\left[|C_{\lambda,t}|\right]^2 d\mu(\lambda) \\ &= \int_{-\infty}^{\infty} \mathbf{Var}[C_{\lambda,t} dZ_\lambda] \end{aligned}$$

Thus, $\mathcal{D}$ is unconditionally heteroskedastic, but deterministically so, whereas it is conditionally heteroskedastic and stochastic. The spectral density for $\mathcal{D}$ is

$$f(\lambda, t)d\lambda = \mathbf{E}[|C_{\lambda,t}|]^2 d\mu(\lambda) = \mathbf{Var}[C_{\lambda,t} dZ_\lambda]$$

and the conditional spectral density is

$$f(\omega, \lambda, t+a|t)d\lambda = \mathbf{E}_t[|C_{\lambda,t+a}|^2]d\mu(\lambda) = \mathbf{Var}_t[C_{\lambda,t} dZ_\lambda]$$

8. A Simple Tâtonnement Example

Consider the use of the unconditionally and conditionally heteroskedastic form of excess demand given in the last section. Suppose the dynamics of the risky asset price $\{S_t(\omega)\}_{t\in[0,T]}$ were determined by

$$dS_t(\omega) = \alpha \mathcal{D}_t(\omega)$$

for some $\alpha > 0$, which, for simplicity, we take to be 1. Thus,

$$\begin{aligned} \Delta S_t(\omega) = S_t(\omega) - S_0(\omega) &= \int_0^t \mathcal{D}_s(\omega) ds = \int_0^t \left[\int_{-\infty}^{\infty} e^{i\lambda s} C_{\lambda,s}(\omega) dZ_\lambda(\omega)\right] ds \\ &= \int_{-\infty}^{\infty} \left[\int_0^t e^{i\lambda s} C_{\lambda,s}(\omega) ds\right] dZ_\lambda(\omega) \end{aligned}$$

Now, the unconditional variance of the asset price is

$$\mathbf{Var}[\Delta S_t] = \int_{-\infty}^{\infty} \mathbf{Var}\left[\left(\int_0^t e^{i\lambda s} C_{\lambda,s}(\omega) ds\right) dZ_\lambda\right]$$

and the conditional variance is

$$\mathbf{Var}_t[\Delta S_{t+a}] = \int_{-\infty}^{\infty} \mathbf{Var}_t\left[\left(\int_0^{t+a} e^{i\lambda s} C_{\lambda,s}(\omega) ds\right) dZ_\lambda\right]$$

From this we can obtain the spectral density and the conditional spectral density of the asset price:

$$f_{\Delta S}(\lambda, t)d\lambda = \mathbf{Var}\left[\left(\int_0^t e^{i\lambda s} C_{\lambda,s}(\omega)ds\right) dZ_\lambda\right]$$

$$f_{\Delta S}(\lambda, t+a|t)d\lambda = \mathbf{Var}_t\left[\left(\int_0^{t+a} e^{i\lambda s} C_{\lambda,s}(\omega)ds\right) dZ_\lambda\right]$$

With this representation, we have expressed both unconditional and conditional asset price heteroskedasticity in terms of a frequency decomposition where the frequency can be given a microeconomic interpretation in terms of individual demand adjustment rates.

9. Further Work

As this note is just a report on preliminary results, there are many things left to do. First, it is essential to find appropriate economic and financial assumptions needed for the general form of demand (1) to reduce to the one of the forms (4) or (6). In the case that (4) results, statistical techniques exist for the estimation of the time dependent spectral density. It would then be of interest to test the usefulness of this simple model. In the case that (6) results, testing procedures need to be developed. Other asset price models to consider could involve a noise term which is only correlated to demand rather than an integral of demand. For instance, one could consider models of the form

$$dS_t(\omega) = S_t(\omega)[\mu_t(\omega) + N_t(\omega)]dt$$

where N is one component of a bivariate process $\mathbf{X}_t(\omega) = [\mathcal{D}_t(\omega), N_t(\omega)]^T$. This approach allows for the specification of the cross spectral parameters (coherence and phase lag) on a frequency by frequency basis. Since we have an interpretation of frequency, it may be possible to base these parameters on microeconomic principles. Other tasks include applications to the pricing of derivatives, and, directly modeling the supply and demand for derivatives and determining the resulting price. Other possible applications we are considering are modeling the interaction between noise traders and fundamental traders in an attempt to explain market herding. Finally, the possibility exists of treating asset price processes based on stable distributions since this spectral approach involves Fourier transforms of distributions.

References

[1] Bouchaud, J.P. and Cont, R., Herd Behavior and Aggregate Fluctuations in Financial Markets, working paper, Science & Finance Research Group, Levallois, France, 1998.

[2] Chacko, G. and Visceira, L., Spectral GMM Estimation of Continuous Time Processes, Working paper, Graduate School of Business, Harvard University, 1999.

[3] Fama, E., Mandelbrot and the Stable Paretian Hypothesis, Journal of Business, 36(4):420-429, 1963.

[4] Follmer, H. and Schweizer, M., A Microeconomic Approach to Diffusion Models for Stock Prices, Mathematical Finance, 3(1):1-23, 1993.

[5] Granger, C. and Morgenstern, O., Spectral Analysis of New York Stock Market Prices, Kyklos, 16:1-27, 1963.

[6] Lux, T., Time Variation of Second Moments in a Noise Trader/Infection Model, Journal of Economic Dynamics and Control, 22:1-38, 1997.

[7] Mandelbrot, B., The Variation of Certain Speculative Prices, Journal of Business, 36(4):394-419, 1963.

[8] Priestley, M.B., Evolutionary Spectra and Non-stationary Processes, Journal of the Royal Statistical Society Series B, 27:204-237, 1965.

[9] Priestley, M.B., Power Spectral Analysis of Non-stationary Random Processes, Journal of Sound and Vibration, 6:86-97, 1967.

[10] Priestley, M.B., Time-dependent Spectral Analysis and its Applications in Prediction and Control, Journal of Sound and Vibration, 17:139-156, 1971.

[11] Priestley, M.B., *Spectral Analysis and Time Series*, Academic Press, 1981.

[12] Priestley, M.B., *Non-linear and Non-stationary Time Series Analysis*, Academic Press, 1988.

[13] Samorodnitsky, G. and Taqqu, M.S., *Stable Non-Gaussian Random Processes*, Chapman and Hall, 1994.

David R. Alexander
Department of Mathematics & Statistics, McGill University and
Department of Finance, Concordia University,
Montreal, Quebec, Canada.
dralex@math.mcgill.ca
dalexander@mercato.concordia.ca

Emmanuel E. Haven
Department of Finance, University of Essex,
Colchester, Essex, UK and
Department of Finance, Concordia University,
Montreal, Quebec, Canada
ehaven@essex.ac.uk

Trends in Mathematics, © 2001 Birkhäuser Verlag Basel/Switzerland

Optimal default boundary in a discrete time setting

Agata Altieri and Tiziano Vargiolu

Abstract. In this paper we solve the problem of determining the default time of a firm in such a way as to maximize its total value, which includes bankruptcy costs and tax benefits, with the condition that the value of equity must be nonnegative. By applying dynamic programming in discrete time, we find results which extends those of Leland (1994) and Leland-Toft (1996).

1. Introduction

The aim of this work is to find the optimal time of default of a firm which has issued a coupon bond with a given maturity. The default time is optimal in the sense that it maximises the total value of the firm, given by its net value plus tax benefits of the coupons minus bankruptcy costs, with the constraint that equity, given by the difference between the total value of the firm minus the debt, must have a positive value at all times prior to default.

In the field of credit risk, valuation models can be divided into two categories which differ from one another in modelling default time (for a concise description of the two approaches, see [7]). In the first category (*intensity based* models), the default time τ is typically modeled as a first jump time of a Poisson process; this captures the idea that the time of a default takes the bondholders by surprise. In the second category (models based on the *value of the firm*), the default time is determined by an underlying process V describing the value of the firm; default occurs when this process hits a certain boundary (like in a barrier option), and typically represents the fact that the firm cannot repay a debt with a given maturity, so the default time is of the form $\tau = \inf\{t \mid V_t \leq f(t)\}$. In this last category, the prices of credit derivatives as corporate bonds and swaps depend on the shape of the default boundary $f(t)$. One can see different choices of f in the works [2], [5], [8], [9], [10], [11]. The choice of this default boundary in the different models seems as arbitrary as the choice of a term structure of risk-free interest rates is, and is usually exogenously suggested by personal taste. The first example of boundary built in an endogenous way is in the works [8] and [9], where a constant boundary K is determined such that the total value (including tax benefits on the debt) of the firm is maximised, with the constraint that equity must have a positive value in all the times prior to default.

Our model generalises Leland-Toft's approach by searching for a default boundary f (not of the specific kind $f(t) \equiv K$) such that the total value of the firm is maximised, with the same constraint as Leland-Toft. In order to obtain explicit results, we choose to work in discrete time. From an economic point of view, this could reflect the fact that default can be decided by the stockholders only at particular times (that could be the maturities of the coupons of the debt); from a mathematical point of view, this can be the first step for a continuous time formulation: in fact the solution of a continuous time problem of this kind can be approximated by the solutions of discrete time problems (see [6]).

We formulate this problem using the tools of stochastic control, so we write it as an optimal stopping problem with a constraint. While unconstrained optimal stopping can be treated by methods that are by now classical (see for example [14]), we found that there are no specific references for our problem, so we developed an ad-hoc method of solution that makes use of the dynamic programming principle. We find that the optimal default time is the first time at which equity falls to zero, where equity turns out to be a deterministic function of the value of the firm. In the case of a debt with no coupons, we find nice parallels between our model and Merton's: namely, under assumptions more general than the ones in [11], we find out that it is optimal for the firm to wait until the final maturity of the debt.

The paper is organised as follows. In Section 2 we describe our model and present the default-dependent quantities relevant for our work. In Section 3 we formulate and solve the optimal stopping problem. In Section 4 we apply our results to the binomial model.

The authors wish to thank Stefan Jaschke, Nicole El Karoui, Monique Jeanblanc, Wolfgang Runggaldier and Jerzy Zabczyk, that gave useful contributions at various stages of the work.

2. The model

We consider a market in which the primary assets are a riskless asset B, called the *money market account*, and a risky asset V which represents the total value of the firm. We represent the value of the firm as a stochastic process $V = (V_t)_{t\in[0,N]}$ on the probability space $(\Omega, \mathcal{F}, \mathbb{P})$, where $N \in \mathbb{N}$ is a given maturity.

We assume that the market is arbitrage free and complete. This is equivalent to assume that there exists a unique martingale measure $\mathbb{Q}$ and every contingent claim can be priced via the expected value of its discounted final payoff under $\mathbb{Q}$. This situation occurs for example if V is a binomial process (i.e. the sample paths of V are pathwise constant and they can change assuming only two values) or if V is a diffusion driven by a 1-dimensional Brownian motion. We will be interested in particular in some activities related to the firm (corporate debt, equity, tax benefits, bankruptcy costs, etc.), that will be viewed as a derivative asset of the value of the firm.

We also assume that default can take place only at discrete times $n = 0, 1, \ldots, N$ where $N > 0$ is a given maturity. Since we are interested only in finding which is the optimal default time among the dates $n = 0, 1, \ldots, N$, we represent the dynamics of B and V as

$$B_{n+1} = B_n(1+r) ,$$

where $r > 0$ is the deterministic risk-free interest rate, and as

$$V_{n+1} = V_n \omega_n , \quad n = 0, \ldots, N ,$$

where $\{\omega_n\}_{n \in \{0,1,\ldots,N\}}$ is a sequence of positive i.i.d. random variables on $(\Omega, \mathcal{F}, \mathcal{F}_n, \mathbb{P})$ and $(\mathcal{F}_n)_n$ is the filtration generated by V. Under these assumptions, V is a Markov chain with transition operator T defined by

$$Tf(v) = \mathbb{E}_{\mathbb{Q}}[f(V_{n+1}) \mid V_n = v]$$

(since the $(\omega_n)_n$ are i.i.d., the right hand side does not depend on n) for every $f : \mathbb{R}^+ \to \mathbb{R}$ measurable and bounded.

Let τ denote the time at which bankruptcy occurs. As said before, we are assuming that τ takes values in the set $\{0, \ldots, N\}$. There is also the possibility that default does not take place before time N; when this happens, we will give τ the value $N+1$. This is not a real date (because the terminal one is N) but it indicates that the firm has arrived at N without declaring bankruptcy.

Following [8] and [9], we consider three particular claims depending on the value of the firm and on default time. Since there exists a unique martingale measure $\mathbb{Q}$, the value of these contingent claims will be given by the discounted expectation under $\mathbb{Q}$ of the payments of the claim.

1. **Bankruptcy costs.** We assume that the costs connected with default are equal to αV_τ, where $\alpha \in (0,1)$ is a fixed fraction. These costs are equivalent to a claim that does not pay coupons and is worth αV_τ in the event of bankruptcy. Then the value of bankruptcy costs at time n is given by:

$$BC(V_n, \tau) = \mathbb{E}_{\mathbb{Q}} \left[\frac{\alpha V_\tau}{(1+r)^{\tau - n}} 1_{\{n \le \tau \le N\}} \middle| \mathcal{F}_n \right] .$$

2. **Debt.** We assume that debt issued by the firm pays coupon payments C_n at every date n prior to maturity or default and an amount P at the terminal date; in the event of default the bondholders receive the residual value of the firm minus bankruptcy costs. Thus the value of debt is:

$$\begin{aligned} D(V_n, \tau) &= \mathbb{E}_{\mathbb{Q}} \left[\sum_{i=n}^{\tau-1} \frac{C_n}{(1+r)^{i-n}} + \frac{P}{(1+r)^{\tau-n}} 1_{\{\tau = N+1\}} + \right. \\ &\quad \left. + \frac{(1-\alpha)V_\tau}{(1+r)^{\tau-n}} 1_{\{n \le \tau \le N\}} \middle| \mathcal{F}_n \right] . \end{aligned}$$

3. **Tax benefits.** The tax benefits associated with debt financing are proportional to the coupon C_n via a constant $\gamma \in (0,1)$. The value of tax benefits is:
$$TB(V_n,\tau) = \mathbb{E}_{\mathbb{Q}}\left[\sum_{i=n}^{\tau-1} \frac{\gamma C_n}{(1+r)^{i-n}} \middle| \mathcal{F}_n\right] .$$

The total value of the firm is given by:
$$v(V_n,\tau) = V_n + TB(V_n,\tau) - BC(V_n,\tau) .$$

By the Modigliani-Miller theorem, the value of equity is the total value of the firm minus the value of debt:
$$E(V_n,\tau) = v(V_n,\tau) - D(V_n,\tau) .$$

3. The optimal stopping problem

We now suppose that the firm can choose the time τ at which it declares bankruptcy and we assume that the firm does it in such a way as to maximize its total value with the condition that equity must be nonnegative. Hence the problem we want to solve is
$$W_0(V_0) = \max_{\tau \in \mathcal{T}_0} v(V_0,\tau) , \tag{1}$$
where $\mathcal{T}_n$ is the set of the $\{n,\dots,N+1\}$-valued stopping times such that $\{\tau > k\} \subseteq \{E(V_k,\tau) > 0\}$ for all $k = n,\dots,N$. This problem is not a classical optimal stopping problem (see for instance [14]) because of the presence of a constraint for the stopping time. Therefore we present an ad-hoc method of solution that makes use of dynamic programming.

Theorem 3.1. *Given the problem (1) and defining successively the functions $h_N, h_{N-1},\dots,h_0$, $\overline{W}_N, \overline{W}_{N-1},\dots,\overline{W}_0$ as:*
$$h_N(V) = (V - P - (1-\gamma)C_N)^+ ,$$
$$\overline{W}_N(V) = \begin{cases} V(1-\alpha) & \text{if } h_N(V) = 0 , \\ V + \gamma C_N & \text{if } h_N(V) > 0 , \end{cases}$$
$$h_k(V) = \left(\frac{1}{1+r} T h_{k+1}(V) - (1-\gamma)C_k\right)^+ , \qquad k = N-1,\dots,0 ,$$
$$\overline{W}_k(V) = \begin{cases} \gamma C_k + \dfrac{1}{1+r} T W_{k+1}(V) & \text{if } h_k(V) > 0 , \\ V(1-\alpha) & \text{if } h_k(V) = 0 , \end{cases}$$
the optimal stopping time is
$$\hat{\tau} = \begin{cases} \inf\{j \mid V_j \le v_j^*\} , \\ N+1 \quad \text{if the set above is empty} , \end{cases} \tag{2}$$

where v_j^* is the maximal solution of the equation

$$Th_{j+1}(v_j^*) = (1+r)(1-\gamma)C_j \tag{3}$$

and we have:

$$\begin{aligned} \overline{W}_0(V_0) &= W_0(V_0) = v(V_0, \hat{\tau}) \, , \\ E(V_0, \hat{\tau}) &= h_0(V_0) \, . \end{aligned}$$

The proof proceeds in this way. For each time k, we assume that the optimal stopping time for the time period $\{k+1, \ldots, N\}$ is given by Equation (2). The function h_k is related to the constraint $\tau \in \mathcal{T}_k$ in this way: if $h_k(V_k) = 0$, then we are forced to stop and default; conversely, if $h_k(V_k) > 0$, then we can choose either to stop or to continue. We prove that in the second situation it is always optimal to continue, so we obtain the function $\bar{W}_k$. The interested reader can find the complete proof in [1].

Corollary 3.2. *If $C_i = 0$, $i = 1, \ldots, N$, and the random variables ω_n have support equal to $\mathbb{R}^+$, then the optimal stopping time is:*

$$\hat{\tau} = \begin{cases} N & \text{if } V_N \leq P \, , \\ N+1 & \text{if } V_N > P \, , \end{cases}$$

and we have:

$$\begin{aligned} W_n(V_n) &= V_n - \mathbb{E}_{\mathbb{Q}}\left[\frac{\alpha V_N}{(1+r)^{N-n}} 1_{\{V_N \leq P\}} \middle| \mathcal{F}_n \right] , \\ D_n(V_n) &= \mathbb{E}_{\mathbb{Q}}\left[\frac{(1-\alpha) V_N}{(1+r)^{N-n}} 1_{\{V_N \leq P\}} + \frac{P}{(1+r)^{N-n}} 1_{\{V_N > P\}} \middle| \mathcal{F}_n \right] , \\ BC_n(V_n) &= \mathbb{E}_{\mathbb{Q}}\left[\frac{\alpha V_N}{(1+r)^{N-n}} 1_{\{V_N \leq P\}} \middle| \mathcal{F}_n \right] , \\ TB_n(V_n) &= 0 \, , \\ E_n(V_n) &= V_n - \mathbb{E}_{\mathbb{Q}}\left[\frac{V_N}{(1+r)^{N-n}} 1_{\{V_N \leq P\}} + \frac{P}{(1+r)^{N-n}} 1_{\{V_N > P\}} \middle| \mathcal{F}_n \right] . \end{aligned}$$

Proof. By Theorem 3.1 we have

$$h_N(V) = (V-P)^+ \, ,$$

$$h_k(V) = \left(\frac{1}{1+r} Th_{k+1}(V) \right)^+ , \quad k = N-1, \ldots, 0 \, ,$$

and the optimal stopping time is

$$\hat{\tau} = \begin{cases} \inf\{j \mid h_j(V_j) = 0\} \, , \\ N+1 \text{ if the set above is empty} \, . \end{cases}$$

We prove by induction that the functions $Th_{k+1}(V)$ for $k = N-1, \dots, 0$, are all strictly positive. In fact, if we assume that $h_{k+1} \geq 0$ and is different from zero we have:

$$Th_{k+1}(V) = \int_0^{+\infty} h_{k+1}(y)\, p(V, dy) > 0 \quad \forall V \in \mathbb{R} ,$$

where p is the transition function of the Markov chain $\{V_n\}_{n\in\{0,\dots,N\}}$ under $\mathbb{Q}$. Since $h_N \geq 0$ and is different from the function zero, we deduce that

$$h_k(V) = \frac{1}{1+r} Th_{k+1}(V) > 0 \quad \forall k = N-1, \dots, 0 ,$$

so $h_j(V_j) > 0$ for all $j = N-1, \dots, 0$, and the only nontrivial condition to stop is $h_N(V_N) = 0$. Thus the optimal stopping time is

$$\hat{\tau} = \begin{cases} N & \text{if } \; V_N \leq P , \\ N+1 & \text{if } \; V_N > P . \end{cases}$$

□

This corollary draws a nice parallel between our model and the seminal one by Merton [11]. In fact it says that, if there are not intermediate coupons, then the optimal rule for the firm to default is to wait until the maturity and then to default if and only if its net value is worth less than the principal, as in Merton's model. Moreover, if we take $\omega_n = \exp((r - 1/2\sigma^2)T/N + \sigma W_N)$, where $W_n \sim N(0, T/N)$, then we obtain exactly the same quantities for D and E as Merton.

Remark 3.3. If the random variables ω_n have a support smaller than $\mathbb{R}^+$ (such as in the case when the possible values are a finite number as in the binomial model which we will discuss in the next section), then the results of Corollary 3.2 are still valid if we take the admissible default times in the set $\tilde{\mathcal{T}}_n$ of the $\{n, \dots, N+1\}$-valued stopping times such that $\{\tau > k\} \subseteq \{E(V_k, \tau) \geq 0\}$ for all $k = n, \dots, N$. This corresponds to allowing the equity to be exactly zero in some periods of time. This could be quite unpleasant to the intuition, but it could be justified by the fact that in our model the primary asset is not the equity but the value of the firm.

4. The binomial model

In this section we apply our results, collected in Theorem 3.1, to the case of the binomial model. The purpose is to approximate a lognormal model or a pure jump model, in order to obtain results similar to the ones in [8] and [9].

Let us suppose that the ω_n are random variables which take only the two values $1+u$ and $1+d$ with probability p and $1-p$ $(0 < p < 1)$, respectively, under the equivalent martingale measure $\mathbb{Q}$. We want the discounted value of V to be a $\mathbb{Q}$-martingale, so

$$p = \frac{r-d}{u-d} , \qquad 1-p = \frac{u-r}{u-d} ,$$

with $-1 < d < r < u$. The transition operator T this time is such that

$$T\varphi(x) = p\varphi(x(1+u)) + (1-p)\varphi(x(1+d)) .$$

We now apply Theorem 3.1 to the case $C_n = 0$. The functions h_n, $n = 0, \dots, N$ turn out to be pathwise affine increasing functions, and it is easy to determine the optimal default boundary v_n^*, $n = 0, \dots, N$.

Theorem 4.1. *If $C_n \equiv 0$, then the optimal default boundary is given by*

$$v_k^* = \frac{P}{(1+u)^{N-k}}, \quad k = 0, \dots, N ,$$

and the optimal stopping time is given by Equation (2).

Proof. We have $h_N(v) = (v - P)^+$. Since $C_n = 0$ and T sends nonnegative functions into nonnegative functions, then

$$h_{N-n}(v) = \frac{T^n h_N(v)}{(1+r)^n} .$$

Now we need to find the maximal solution v_j of the equation $Th_{j+1}(v) = 0$, that becomes $T^{N-j}h_N(v) = 0$. We claim that $v_j = P/(1+u)^{N-j}$ for all $j = 0, \dots, N$. This is true for $j = N$, because $h_N = 0$ on $(0, P]$ and $h_N > 0$ on $(P, +\infty)$. We now proceed by induction, assuming that it is true for j. This means that $h_j = 0$ on $(0, P/(1+u)^{N-j}]$ and $h_j > 0$ on $(P/(1+u)^{N-j}, +\infty)$. Then for $v \in (0, P/(1+u)^{N-j+1}]$:

$$h_{j-1}(v) = \frac{Th_j(v)}{1+r} = ph_j(v(1+u)) + (1-p)h_j(v(1+d)) = 0 ,$$

since $v(1+u), v(1+d) \in (0, P/(1+u)^{N-j}]$. For $v \in (P/(1+u)^{N-j+1}, +\infty)$,

$$h_{j-1}(v) = \frac{Th_j(v)}{1+r} = ph_j(v(1+u)) + (1-p)h_j(v(1+d)) > 0 ,$$

since $v(1+u) \in (P/(1+u)^{N-j}, +\infty)$. The result follows from Theorem 3.1. □

Now we apply Theorem 3.1 to the case $C_n \neq 0$. Also now the functions h_n, $n = 0, \dots, N$, turn out to be pathwise affine increasing functions, but now it is more difficult to find explicitly the optimal default boundary v_n^*, $n = 0, \dots, N$. However we find that if the condition

$$\sum_{i=0}^{N-k} \frac{(1-\gamma)C}{(1+r)^i} + \frac{P}{(1+r)^{N-k}} \le \frac{(1-\gamma)C(r-d)}{(1+d)(1+r)}, \quad k = 0, \dots, N , \tag{4}$$

holds, then it is possible to find explicitly v_n^*, $n = 0, \dots, N$.

Theorem 4.2. *If the condition in Equation (4) holds, then the optimal default boundary is given by*

$$v_k^* = \sum_{i=0}^{N-k} \frac{(1-\gamma)C}{(1+r)^i} + \frac{P}{(1+r)^{N-k}}, \quad k = 0, \dots, N , \tag{5}$$

and the optimal stopping time is given by Equation (2).

The proof proceeds in the same spirit as in Theorem 4.1, with some more algebra involved. Again, the interested reader can find the complete proof in [1].

The cases in the two previous theorems can be regarded as extreme cases: in fact, the optimal default boundary $(v_k^*)_k$ lies always between the two boundaries found in these cases.

Theorem 4.3. *The optimal default boundary $(v_k^*)_k$ satisfies the following bound:*

$$\frac{P}{(1+u)^{N-k}} \le v_k^* \le \sum_{i=0}^{N-k} \frac{(1-\gamma)C}{(1+r)^i} + \frac{P}{(1+r)^{N-k}}, \qquad k = 0, \dots, N,$$

and the optimal stopping time is given by Equation (2).

Proof. We have the following bound on h_N:

$$v - P - (1-\gamma)C \le h_N(v) \le (x-P)^+ .$$

Since the operator T is linear and monotone, this implies:

$$v - \frac{P + (1-\gamma)C}{1+r} - (1-\gamma)C \le \frac{Th_N(v)}{1+r} - (1-\gamma)C \le \frac{T(v-P)^+}{1+r} .$$

By taking the positive parts of the two last quantities, this implies that

$$v - \frac{P + (1-\gamma)C}{1+r} - (1-\gamma)C \le h_{N-1}(v) \le \frac{T(v-P)^+}{1+r} .$$

By induction we have that

$$v - \sum_{i=0}^{n} \frac{(1-\gamma)C}{(1+r)^i} - \frac{P}{(1+r)^n} \le h_{N-n}(v) \le \frac{T^n(v-P)^+}{(1+r)^n} .$$

By taking positive parts we have that

$$\left(v - \sum_{i=0}^{n} \frac{(1-\gamma)C}{(1+r)^i} - \frac{P}{(1+r)^n} \right)^+ \le h_{N-n}(v) \le \frac{T^n(v-P)^+}{(1+r)^n} .$$

The first function is equal to zero for

$$v \le \sum_{i=0}^{n} \frac{(1-\gamma)C}{(1+r)^i} + \frac{P}{(1+r)^n},$$

while by Theorem 4.1 the third is equal to zero for $v \le (1-\gamma)C/(1+u)^n$. This ends the proof. □

We notice that the default boundary v_k^* in general depends on time. In particular, it is bounded by two boundaries that evolve in an exponential way with respect to time.

Now we present two ideas of how we can use our results in order to find the optimal default boundary when the process V is a geometric Brownian motion or a geometric Poisson process.

4.1. Approximation of a geometric Brownian motion

We take a maturity T and we discretise the interval $[0, T]$ in N intervals of length T/N. If we let

$$C_n = \frac{C}{N}\,, \quad r_n = \frac{r}{N}\,, \quad u = \frac{r}{N} + \frac{\sigma}{\sqrt{N}}\,, \quad d = \frac{r}{N} - \frac{\sigma}{\sqrt{N}}\,,$$

then the process $(V_{[t/N]})_t$ (where $[u]$ denotes the greatest integer smaller or equal to u) approximates a geometric Brownian motion

$$\tilde{V}_t = V_0 \exp\left(\left(r - \frac{1}{2}\sigma^2\right) t + \sigma W_t\right)$$

(in particular for $N \to \infty$ we have convergence in law). The condition (4) can be rewritten as

$$(1-\gamma)\frac{C}{\sqrt{N}} > \frac{P\sigma}{1 + o(N^{-1/2})}\,,$$

so for $N \to \infty$ we have $P\sigma < 0$, that is never verified. This means that if we want to approximate a lognormal model with this discrete time model, for N sufficiently large the h_n become complex as n gets far from N.

4.2. Approximation of a geometric Poisson process

As before, we take a maturity T and we discretise the interval $[0, T]$ in N intervals of length T/N. If we let

$$C_n = \frac{C}{N}\,, \quad r_n = \frac{r}{N}\,, \quad u = b\,, \quad d = 0\,,$$

then the process $(V_{[t/N]})_t$ approximates a geometric Poisson process

$$\tilde{V}_t = V_0(1+b)^{N_t}$$

(in particular for $N \to \infty$ we have convergence in law) where N is a Poisson process having intensity r/b. This time the condition (4) can be rewritten as

$$(1-\gamma)C > Pr\,.$$

If this condition holds (we remind that in order not to have arbitrage we must have $C > Pr$), then the shape of the functions h_n remains stable, and we have an explicit solution. But if the condition above is not satisfied, for N sufficiently large the function h_n becomes complex as n gets far from N as in the approximation of a geometric Brownian motion.

References

[1] A. Altieri and T. Vargiolu, *Optimal default boundary in discrete time models*, preprint, Department of Pure and Applied Mathematics, Padova, (2000)

[2] F. Black and J. Cox, *Valuing corporate securities: Some effects of bond indenture provision*, The Journal of Finance, **31** (1976), 351–367.

[3] J. M. Harrison and S. R. Pliska, *Martingales and stochastic integrals in the theory of continuous trading*, Sthocastic Processes and their Applications, **11** (1981), 215–260.

[4] T. Ho and R. F. Singer, *The value of corporate debt with a sinking-fund provision*, The Journal of Business, **57** (1984), 315–336.

[5] H. Johnson and R. Stulz, *The pricing of options with default risk*, The Journal of Finance, **42** (1987), 267–279.

[6] H. J. Kushner, *Probability methods for approximations in stochastic control and for elliptic equations*, New York Academic Press, **1977**.

[7] D. Lando, *Modelling bonds and derivatives with credit risk*, in: M. Dempster, S. Pliska, Eds., Mathematics of Derivative Securities, (Cambridge University Press, Cambridge) (1997), 369–393.

[8] H. E. Leland, *Corporate debt value, bond covenants, and optimal capital structure*, The Journal of Finance, **49** (1994), 1213–1252.

[9] H. E. Leland and K. B. Toft, *Optimal capital structure, endogenous bankruptcy, and the term structure of credit spreads*, The Journal of Finance, **51** (1996), 987–1019.

[10] F. A. Longstaff and E. S. Schwartz, *A simple approach to valuing risky fixed and floating rate debt*, The Journal of Finance, **50** (1995), 789–819.

[11] R. C. Merton, *On the pricing of corporate debt: the risk structure of interest rates*, The Journal of Finance, **29** (1974), 449–470.

[12] F. Modigliani and M. Miller, *The cost of capital corporation finance, and the theory of investment*, American Economic Review, **48** (1958), 261–277.

[13] M. Musiela and M. Rutkowski, *Martingale methods in financial modelling*, Springer, **1997**.

[14] A. N. Ŝirjaev, *Statistical Sequential Analysis*, American Mathematical Society, **1973**.

Department of Pure and Applied Mathematics
University of Padova
via Belzoni 7
I-35131 Padova, Italy
E-mail address: altieri@cicrp.jussieu.fr
vargiolu@math.unipd.it

Trends in Mathematics, © 2001 Birkhäuser Verlag Basel/Switzerland

An Infinite Factor Model for the Interest Rate Derivatives

Arunabha Bagchi and K. Suresh Kumar

Abstract. In this paper we model the forward rate process as a stochastic partial differential equation in a Sobolev space. We establish the existence of a martingale measure. We also derive the price of a general contigent claim as the solution to a partial differential equation in an appropriate Hilbert space. Moreover we obtain an explicit formula for the price of the interest rate cap in the Gaussian framework.

1. Introduction

In finance, the importance of interest rate models is increasing rapidly. The reason behind this is the huge derivatives market based on interest rates. Note that swaption market is the biggest financial market. Because of this, finding 'good' models for interest rates is a very active research area. Unlike stocks, modeling interest rates is more complicated due to the fact that the model should describe the behaviour of the entire yield curve.

The first generation of interest rate models eg. Vasicek, Cox-Ingersoll-Ross, model short rate as a one factor stochastic differential equation and estimate the parameters involved to fit the initial term structure as closely as possible. These models are easily implementable for pricing derivatives but have the disadvantage of not fitting correctly into the initial term structure. This can lead to significant errors in the pricing process. The second generation models, eg. Ho-Lee, Heath-Jarrow-Morton, directly model forward rates as a finite factor stochastic differential equation using initial term structure as the input. These models avoid the problem of the first generation models but all the forward rates are driven by the same noises. This restricts the correlation between the forward rates and hence restricts the admissible shape of the yield curve. More precisely, in finite factor forward rate models, the correlation between the forward rates evolve in a very limited way directed by the initial term structure and the estimated volatilities. So we need to recalibrate the parameters frequently with new market term structure. One natural way to avoid this is to use different noise for each forward curve. This leads to the idea of modeling forward rates as a random field. One advantage of modeling forward rates as a random field is that we can avoid recalibration.

There have already seen some attempts to model the forward rates as random fields. Kennedy in 1994 modeled forward rate as a Gaussian random field. In 1997, Goldstein generalized this by taking the driving noise as an Ornstein-Uhlenbech sheet. In 1998, Santa-Clara and Sornette modeled forward rate with noise driven by a stochastic string of which Goldstein's model is a special case. In this work we propose a model for the forward rate as a nonlinear stochastic partial differntial equation(SPDE) which is a generalization of the string model. Besides being a much more general model, our model has the advantage of more mathematical flexibility. As an example, the existence of a martingale measure comes as a direct consequeness of Girsanov's theorem.

Our paper is organized as follows. In Section 2 we describe our model. In Section 3 we show the existence of martingale measure. In Section 4 we discuss the pricing of general contigent claim and obtain an equation which generalizes the PDE for the contigent claim in the HJM setup to our model. This section ends with an explicit formula for a forward rate cap in the Gaussian framework.

2. Description of the model

Fix $\hat{T} > 0$. For us $\hat{T}$ is the maturity of the longest bond in the market, e.g., 30 years.
Let U and H be Hilbert Spaces. We take U, H to be subspaces of $L^2(0, \hat{T})$ with suitable innerproduct structure depending on the particular model. Let $Q \in L(U)$ be nonnegative, symmetric and nuclear. There exists a complete orthonormal basis $\{e_n\}$ of U and $\lambda_n \geq 0$ such that

$$Q\, e_n = \lambda_n\, e_n,\ n \geq 1.$$

Set $U_0 = Q^{\frac{1}{2}}(U)$ with the innerproduct

$$< u, v >_0 = \sum_{k=1}^{\infty} \frac{1}{\lambda_k} < u, e_k >< v, e_k >$$

Then U_0 is a Hilbert space. Denote $L_2^0 = L_2(U_0, H)$, the set of all Hilbert-Schmit operators from U_0 to H. Then L_2^0 is Hilbert with the norm

$$\|\Phi\|^2_{L_2^0} = \sum_{i,j=1}^{\infty} \lambda_i | < \Phi\, e_i, \bar{e}_j > |^2 ,$$

where $\{\bar{e}_n\}$ is an orthonormal basis for H.
Let $\alpha : [0, \hat{T}] \times H \to H$, $\sigma : [0, \hat{T}] \times H \to L_2^0$ and $A : \mathcal{D}(A) \to H$ be the infinitesimal generator of a C_0-semigroup, $S(t)$ in H with $\mathcal{D}(A)$ dense in H.
Let $(\Omega, \mathcal{F}, \mathbf{P})$ denote the probability space underlying the market.
We model forward rate in the real world probability P as the SPDE

$$\begin{aligned} df(t,\cdot) &= (Af(t,\cdot) + \alpha(t, f(t,\cdot)))\, dt + \sigma(t, f(t,\cdot))\, dW(t), \\ f(0,\cdot) &= f_0 \in H, \end{aligned} \tag{1}$$

where $W(\cdot)$ is a U-valued Q-Wiener process. See [3] for the definition of Wiener process.

We assume that

(A1) The equation (2.1) has a unique strong solution.

Before going into the study of the financial issues with our model, we make a comparison of our model with the existing random field models of the forward rate.

Santa-Clara and Sornette modeled forward rate as

$$df(t,x) = \alpha(t,x)\,dt + \sigma(t,x)\,dZ(t,x)$$

where Z is a stochastic string given by the SPDE

$$\frac{\partial}{\partial x}\Big(\sqrt{j(x)}\,\frac{\partial Z(t,x)}{\partial t}\Big) = \sqrt{\frac{dj(x)}{dx}}\,\eta(t,x) \tag{2}$$

where η is the white noise in both time and space. Note that when $j(x) = e^{2kx}$, the equation (2.2) represents the Ornstein-Uhlenbech sheet.

The string $Z(t,\cdot)$ is a Gaussian process with covariance operator given by

$$Q(t)(h)(x) = t\int_0^{\hat{T}}\sqrt{\frac{j(x)\wedge j(y)}{j(x)\vee j(y)}}\,h(y)\,dy,\;\; h\in U, x\in(0,\hat{T})$$

Hence

$$Q(t) = t\,Q\,,$$

where

$$Q(h)(x) = \int_0^{\hat{T}}\sqrt{\frac{j(x)\wedge j(y)}{j(x)\vee j(y)}}\,h(y)\,dy,\;\; h\in U, x\in(0,\hat{T})$$

It is easy to verify that $Q \in L(U)$ is nonnegative, symmetric and nuclear. Since $Z(t,\cdot)$ is Gaussian, we can see that $Z(t,\cdot)$ is U-valued Q-Wiener process.

Hence by choosing the covariance operator suitably we can get a wide range of forward rate curves which include the models by Goldstein [1] and Santa-Clara etal. [4].

3. Existence of a martingale measure

Our market consists of a bank account $B(\cdot)$ and bonds of all maturities denoted by $\mathcal{P} = \{P(t,T)\,|\,0\le t\le T\le\hat{T}\}$, $P(t,T) = \exp\Big(-\int_0^{T-t} f(t,u)\,du\Big)$ and $B(\cdot)$ is given by

$$\begin{aligned} dB(t) &= r(t)\,B(t)\,dt, \\ B(0) &= B_0, \end{aligned} \tag{1}$$

where $r(\cdot)$ is the short rate. Recall that the probability measure $\mathbb{Q}\sim\mathbf{P}$ is a martingale measure if the discounted bond price $\overline{P}(t,T) = \dfrac{P(t,T)}{B(t)}$ is a $\mathbb{Q}$-martingale with respect to $\{\mathcal{F}_t(=\sigma(W(s) : s \leq t))\}$. First we have the following characterization of martingale measures.

Lemma 3.1. *Let $\mathbb{Q} \sim \mathbf{P}$ and under $\mathbb{Q}$, $f(t,\cdot)$ satisfies the SPDE*

$$\begin{aligned} df(t,\cdot) &= \beta(t, f(t,\cdot))\, dt + \sigma(t, f(t,\cdot))\, dW(t), \\ f(0,\cdot) &= f_0 \in H, \end{aligned} \tag{2}$$

where $\beta(t, f(t,\cdot)) = Af(t,\cdot) + \alpha(t, f(t,\cdot))$. Then $\mathbb{Q}$ is a martingale measure if and only if

$$\begin{aligned} &\beta(t, f(t,\cdot))(x) \\ &= \frac{\partial f(t,x)}{\partial x} + \sum_{k=1}^{\infty} \lambda_k \Big[\int_0^x \sigma(t, f(t,\cdot)) e_k(u) du \Big] \sigma(t, f(t,\cdot)) e_k(x) \end{aligned} \tag{3}$$

Proof. For simplicity in notation, we take σ and α to be independent of $f(\cdot,\cdot)$. The same arguments can be used for the general case.

Let $\mathbb{Q}$ be a martingale measure. Fix t. For $s \geq t$,

$$P(t,s) = \exp\Big(- \int_0^{s-t} f(t,u)\, du \Big).$$

Set $y(t,s) = - \int_0^{s-t} f(t,u)\, du$. Inview of (A1) and since

$$W(t) = \sum_{k=1}^{\infty} \sqrt{\lambda_k}\, e_k\, W_k(t),$$

where $W_k(\cdot)$ are $\mathbb{R}$-valued independent Wiener processes, we have

$$\begin{aligned} y(t,s) &= - \int_0^{s-t} f_0(u)\, du - \int_0^{s-t} \int_0^t [Af(\tau,\cdot)(u) + \alpha(\tau)(u)]\, d\tau\, du \\ &\quad - \sum_{k=1}^{\infty} \sqrt{\lambda_k} \int_0^{s-t} \int_0^t \sigma(\tau)\, e_k(u)\, dW_k(\tau)\, du. \end{aligned} \tag{4}$$

Add and substract $\int_0^t f(\tau, s-\tau)\,d\tau - \int_0^t f_0(s-\tau)\,d\tau$, we get

$$\begin{aligned}
y(t,s) &= -\int_0^{s-t} f_0(u)\,du - \int_0^t f_0(s-\tau)\,d\tau + \int_0^t f(\tau, s-\tau)\,d\tau \\
&\quad -\Big(\int_0^t \int_0^\tau [Af(u,\cdot)(s-\tau) + \alpha(u)(s-\tau)]\,du\,d\tau \\
&\qquad + \int_0^{s-t} \int_0^t [Af(\tau,\cdot)(u) + \alpha(\tau)(u)]\,d\tau\,du\Big) \qquad (5) \\
&\quad -\sum_{k=1}^{\infty} \sqrt{\lambda_k}\Big(\int_0^t \int_0^\tau \sigma(u)\,e_k(s-\tau)\,dW_k(u)\,d\tau \\
&\qquad + \int_0^{s-t} \int_0^t \sigma(\tau)\,e_k(u)\,dW_k(\tau)\,du\Big).
\end{aligned}$$

Stoshastic Fubini's theorem we can show that

$$\begin{aligned}
y(t,s) &= y(0,s) + \int_0^t f(\tau, s-\tau)\,d\tau - \int_0^t \int_0^{s-\tau} [Af(\tau,\cdot)(u) + \alpha(\tau)(u)]\,du\,d\tau \\
&\quad -\sum_{k=1}^{\infty} \int_0^t \int_0^{s-\tau} \sigma(\tau)\,e_k(u)\,du\,dW_k(\tau)
\end{aligned}$$

By an application of Ito's formula to the process $y(t,\cdot)$, we have

$$\begin{aligned}
dP(t,s) &= \Big[f(t, s-t) - \int_0^{s-t} [Af(t,\cdot)(u) + \alpha(t)(u)]\,du \\
&\quad + \frac{1}{2}\sum_{k=1}^{\infty} \lambda_k \Big(\int_0^{s-t} \sigma(t)\,e_k(u)\,du\Big)^2\Big]\,P(t,s)\,dt \qquad (6) \\
&\quad -\sum_{k=1}^{\infty} \sqrt{\lambda_k} \int_0^{s-t} \sigma(t)\,e_k(u)\,du\,P(t,s)\,dW_k(t).
\end{aligned}$$

Again Ito's formula implies that

$$\begin{aligned}
d\overline{P}(t,s) &= \Big[r(t) + f(t, s-t) - \int_0^{s-t} [Af(t,\cdot)(u) + \alpha(t)(u)]\,du \\
&\quad + \frac{1}{2}\sum_{k=1}^{\infty} \lambda_k \Big(\int_0^{s-t} \sigma(t)\,e_k(u)\,du\Big)^2\Big]\,P(t,s)\,B(t)\,dt \\
&\quad -\sum_{k=1}^{\infty} \sqrt{\lambda_k} \int_0^{s-t} \sigma(t)\,e_k(u)\,du\,P(t,s)\,dW_k(t).
\end{aligned}$$

Since $\overline{P}(t,s)$ is a martingale, we have

$$r(t) + f(t, s-t) - \int_0^{s-t} [Af(\tau,\cdot)(u) + \alpha(\tau)(u)]\, du + \frac{1}{2}\sum_{k=1}^{\infty} \lambda_k \left(\int_0^{s-t} \sigma(t) e_k(u) du\right)^2$$

$$= 0$$

Hence, by replacing $s-t$ by τ and taking the distributional derivative with respect to τ, we have

$$Af(t,\cdot)(\tau) + \alpha(t)(\tau)\beta(t,x) = \frac{\partial f(t,\tau)}{\partial \tau} + \sum_{k=1}^{\infty} \lambda_k \Big(\int_0^{\tau} \sigma(t)\, e_k(u)\, du\Big)\, \sigma(t)\, e_k(\tau)\,.$$

The converse is proved in the same fashion. □

We assume the following conditions to prove the non-arbitrageness of the model.

(A2) σ is independent of $f(t,\cdot)$ and $\mathrm{Im}\,\alpha \subseteq \mathrm{Im}\,\sigma$.

(A3) (Novikov's condition) For some $\delta > 0$,

$$\sup_{0 \le t \le \hat{T}} E \exp(\delta\, |\sigma(t)^{-1}\, \overline{\alpha}(t)(\cdot)|_0^2) < \infty$$

Set $\mathcal{U}_0 = L^2(0,T;U_0)$ and $\mathcal{H} = L^2(0,T;H)$. Define L, $\tilde{L}$ from $\mathcal{U}_0$ to $\mathcal{H}$ as follows:

$$Lu(t) = \int_0^t S(t-s)\sigma(s)u(s)\, ds\,, t \le \hat{T}\,,$$

$$\tilde{L}u(t) = \int_0^t \tilde{S}(t-s)\sigma(s)u(s)\, ds\,, t \le \hat{T}\,,$$

where $\tilde{S}$ is the semigroup of the operator $\frac{\partial}{\partial \tau}$.

(A4) (i) $\mathrm{Im}\, L = \mathrm{Im}\, \tilde{L}$

(ii) The operator $(\tilde{L}^{-1} L)\,(\tilde{L}^{-1} L)^* - I$ is Hilbert Schmidt.

It is important to note that, since $\sigma(t)$ in (A2) is a Hilbert-Schmit operator, its inverse does not exists in the usual sense. So here we use the pseudo-inverse. For the existence of the pseudo-inverse assumption (A2) is crucial. It is clear that (A2) restricts the forward rate to be $\mathrm{Im}\sigma$ valued. Hence, (A2) in the finite dimensional means that the number of maturities involved in the model of the forward rate is atmost the number of independent noises. This is almost a necessary condition for the existence of a martingale measure even for the stock price models. Intuitively, pseudo-inverse is obtained by reducing the dimension of the domain of the operator in a suitable way so as to make the operator an injection and take the inverse in the subspace. So essentially we are pretending that the operator is acting in a subspace where it is invertable. Note that the subspace may not be unique.

Theroem 3.2. *The market $(B(\cdot), \mathcal{P})$ has a martingale measure.*

Proof. Set

$$\beta_1(t)(\cdot) = \sum_{k=1}^{\infty} \lambda_k \Big[\int_0^{\cdot} \sigma(t)e_k(u)du\Big]\, \sigma(t)e_k(\cdot)\,.$$

In view of lemma 3.1, it is enough to show that under some probability measure $\mathbb{Q}$, the forward rate process given by (2.1) satisfies

$$df(t,\cdot) = [\frac{\partial f(t,\cdot)}{\partial \tau} + \beta_1(t)]\,dt + \sigma(t)\,d\hat{W}(t)\,, \tag{7}$$

where $\hat{W}(\cdot)$ is a Q-Wiener process.
Set $\psi_1(t) = -\sigma(t)^{-1}(\alpha(t))$. Then using the definition of the pseudo-inverse, $\psi(t)$ is a U_0-valued process. The $\mathcal{F}_t$-predictability follows from the predictability of $\alpha(t)$ and the measurability of the pseudo-inverse. Now, inview of (A3), by applying [[3], prop.10.17, p.295], we have

$$E\Big[\exp\Big(\int_0^{\hat{T}} <\psi_1(s), dW(s)>_0 - \frac{1}{2}\int_0^{\hat{T}} |\psi_1(s)|_0^2\,ds\Big)\Big] = 1\,.$$

Hence, using Girsanov's Theorem [[3], Theorem 10.14, p.290], the process given by

$$W_1(t) = W(t) - \int_0^t \sigma(s)^{-1}\alpha(s)(\cdot)\,ds$$

is a Q-Wiener process under the probability measure $\mathbb{Q}_1$ given by

$$d\mathbb{Q}_1 = \exp\Big(\int_0^{\hat{T}} <\sigma(s)^{-1}\alpha(s)(\cdot), dW(s)>_0 - \frac{1}{2}\int_0^{\hat{T}} |\sigma(s)^{-1}\alpha(s)(\cdot)|_0^2\,ds\Big)\,d\mathbb{P} \tag{8}$$

Under $\mathbb{Q}_1$, the forward rate have the dynamics

$$df(t,\cdot) = Af(t,\cdot)\,dt + \sigma(t)\,dW_1(t) \tag{9}$$

Now consider the SPDE,

$$d\tilde{f}(t,\cdot) = \frac{\partial \tilde{f}(t,\cdot)}{\partial \tau}\,dt + \sigma(t)\,dW_1(t)\,. \tag{10}$$

Then using (A4), we can see from [[3], Theorem 10.11, p.286] the law of (3.9) and the law of (3.10) are equivalent. Set $\mathbb{Q}_2$ given by

$$d\mathbb{Q}_2 = G\,d\mathbb{Q}_1\,,$$

where G the Radon-Nikodym derivative of the law of f with respect to the law of $\tilde{f}$. Then we can show that there exists a Q-Wiener process $W_2(\cdot)$ under $\mathbb{Q}_2$ such that f satisfies

$$df(t,\cdot) = \frac{\partial f(t,\cdot)}{\partial \tau}\,dt + \sigma(t)\,dW_2(t)$$

Set $\psi_2(t) = \sigma(t)^{-1}(\beta_1(t))$. Then as described above, the process given by

$$\hat{W}(t) = W_2(t) - \int_0^t \psi_2(s)\,ds$$

is a Q-Wiener process under the measure

$$d\mathbb{Q} = \exp\Big(\int_0^{\hat{T}} < \psi_2(s), dW_2(s) >_0 - \frac{1}{2}\int_0^{\hat{T}} |\psi_2(s)|_0^2 \, ds\Big) \, d\mathbb{Q}_2$$

Then under $\mathbb{Q}$, $f(t,\cdot)$ satisfies (3.7). □

4. Pricing of interest rate derivatives

In this section first we derive a partial differential equation which characterizes the price of general contigent claims. In the second part we derive an explicit formula for a forward rate cap in the Gaussian framework.

Consider a contigent claim of the form $C(t, P(t))$. i.e., $C : [0, \hat{T}] \times H \to \mathbb{R}$ We assume that C is twice continously differentiable in the second argument and continuously differentiable in the first.
Note that we define $P(t) \in H$ as

$$\begin{aligned} P(t,T) &= 1 \text{ if } T \leq t \\ &= \text{bond price at } t \text{ with maturity } T \text{ if } t \leq T. \end{aligned}$$

For a self-financing strategy (β, γ), the value

$$\pi(t) = \beta(t)\, B(t) + \int_t^{\hat{T}} P(t,T)\, \gamma(t)(dT)\,.$$

Set

$$V_1(t) = \frac{C(t)}{B(t)},\ V_2(t) = \frac{\pi(t)}{B(t)}\,.$$

By Ito's formula we have

$$\begin{aligned} dV_1(t) &= \frac{1}{B(t)}\Big[\frac{\partial C(t,P(t))}{\partial t} + \Big\langle_{H'} \frac{\partial C(t,P(t))}{\partial x}, r(t)\, P(t)\Big\rangle_H \\ &+ \frac{1}{2}\operatorname{Tr}\Big(\frac{\partial^2 C(t,P(t))}{\partial x^2}\Big(\overline{\sigma}(t,f(t,\cdot))Q^{\frac{1}{2}}\Big)\Big(\overline{\sigma}(t,f(t,\cdot))Q^{\frac{1}{2}}\Big)^* - r(t)\, C(t,P(t))\Big]\, dt \\ &+ \frac{1}{B(t)}\Big\langle_{H'} \frac{\partial C(t,P(t))}{\partial x}, \overline{\sigma}(t,f(t,\cdot))\, P(t)\, dW(t)\Big\rangle_H \end{aligned} \tag{1}$$

and

$$dV_2(t) = \frac{1}{B(t)}\int_t^{\hat{T}} \overline{\sigma}(t,f(t,\cdot))\, P(t,T)\, \gamma(t)(dT)\, dW(t)\,. \tag{2}$$

The portfolio of bonds replicates the option if

$$dV_1(t) = dV_2(t)\,.$$

Since $V_2(\cdot)$ is a martingale, it follows from (4.3) that, C satisfies the equation,

$$\frac{\partial C(t,P(t))}{\partial t} + \Big\langle_{H'} \frac{\partial C(t,P(t))}{\partial x}, r(t)\,P(t)\Big\rangle_H$$

$$+\frac{1}{2}\,\mathrm{Tr}\Big(\frac{\partial^2 C(t,P(t))}{\partial x^2}\Big(\overline{\sigma}(t,f(t,\cdot))Q^{\frac{1}{2}}\Big)\Big(\overline{\sigma}(t,f(t,\cdot))Q^{\frac{1}{2}}\Big)^*\Big) \tag{3}$$

$$= r(t)\,C(t,P(t))$$

Thus we have shown that the price of the contigent claim is a solution to the partial differential equation (4.3) in H. In [5], Sornette have obtained an integro-partial differential equation for the price of a contigent claim. Here we have an equation which takes specific forms depending on the space H. If we take $H = L^2(0,\hat{T})$, then we get a similar equation. Note that we have not considered the case when the contigent claim involves $r(\cdot)$.

We give an explicit formula for the price of a forward cap when the diffusion is deterministic.

Consider a forward cap with principal 1 settled at $T_i = T_0 + \Delta i$, $i = 1, 2, \cdots n$. K-cap rate and $L(T_{i-1},0)$(Δ-Libor rate)-floting rate at T_i.

Cash flow at T_i

$$\Big[\Delta L(T_{i-1},0) - \Delta K\Big]_+ = \Big[\exp\Big(\int_0^\Delta f(T_{i-1},u)\,du\Big) - 1 - \Delta K\Big]_+$$

Price of the cap at time 0

$$C = \sum_{i=1}^n E^{\mathbb{Q}}\Big\{\exp\Big(-\int_0^{T_i} r(s)\,ds\Big)\Big[\exp\Big(\int_0^\Delta f(T_{i-1},u)\,du\Big) - 1 - \Delta K\Big]_+\Big\}$$

Set

$$\mu_{t,x} = E^{\mathbb{Q}} f(t,x)$$

$$= f_0(t+x) + \sum_{k=1}^\infty \lambda_k \int_0^t \sigma(t)e_k(x+t-s)\int_0^{x+t-s}\sigma(s)e_k(u)\,du\,ds$$

$$\mathrm{cov}(f(t_1,x_1), f(t_2,x_2)) = \sum_{k=1}^\infty \lambda_k \int_0^{t_1\wedge t_2}\sigma(s)e_k(x_1+t_1-s)e_k(x_2+t_2-s)\,ds$$

Set

$$N_{1i} = \int_0^{T_i} r(s)\,ds,\;\; N_{2i} = \int_0^\Delta f(T_{i-1},u)\,du,\; i = 1, 2, \cdots n$$

Normally distributed with mean and variance

$$\mu_{1i} = \int_0^{T_i}\mu_{u,0}\,du,\;\; \sigma_{1i}^2 = \int_0^{T_i}\int_0^{T_i}\mathrm{cov}(r(x), r(y))\,dx\,dy$$

and

$$\mu_{2i} = \int_0^{\Delta} \mu_{T_i,u}\, du, \ \sigma_{2i}^2 = \int_0^{\Delta}\int_0^{\Delta} \mathrm{cov}(f(T_{i-1},x), f(T_{i-1},y))\, dx\, dy$$

Now

$$E^{\mathbb{Q}}\Big[\exp(-N_{1i})(\exp(-N_{2i}) - 1 - \Delta K)_+\Big]$$

$$= \exp(\mu_{2i} - \mu_{1i} + \tfrac{1}{2}\mathrm{var}(N_{2i} - N_{1i}))\,\Phi\Big(\tfrac{\mu_{2i} - \ln(1+\Delta K) + \sigma_{2i}^2 - \mathrm{cov}(N_{1i},N_{2i})}{\sigma_{2i}}\Big)$$

$$-(1+\Delta K)\exp(\tfrac{1}{2}\sigma_{1i}^2 - \mu_{1i})\Phi\Big(\tfrac{\mu_{2i} - \ln(1+\Delta K) + \sigma_{2i}^2 - \mathrm{cov}(N_{1i},N_{2i})}{\sigma_{2i}}\Big)$$

where Φ is the normal distribution function.
Now C is evaluated by suming up the above terms over i

5. Conclusions

We proposed a model for forward rates which includes the situation when all points in the forward rate curve is driven by noise. we proved that our model is arbitrage free under certain assumptions. There are a number of issues raised in the paper that need further investigation. Assumptions (A2)-(A4) need to be verified for specific models. The characterization of the price of a special class of contigent claims derived in the paper is in an abstract framework. It is important to see the concrete form in specific situations.

References

[1] R. Goldstein, The term structure of interest rates as a Random field, 1997, Preprint.

[2] D. P. Kennedy, The term structure of interest rates as a Gaussian Random field, Math. Finance, Vol4, No.3, 1994, 247-258.

[3] G. Da Prato and J. Zabczyk, Stochastic equations in infinite dimensions, Cambridge University Press, 1992.

[4] P. Santa-Clara and D. Sornette, The dynamics of the forward rate curve with Stochastic String Shocks, 1999, Preprint.

[5] D. Sornette, "String" formulation of the dynamics of the forward interest rate curve, 1998, Preprint.

Faculty of Mathematical Sciences,
University of Twente,
7500 AE Enschede,
The Netherlands
E-mail address: bagchi@math.utwente.nl

Trends in Mathematics, © 2001 Birkhäuser Verlag Basel/Switzerland

Arbitrage and Pricing with Collateral

José Fajardo Barbachan

Abstract. This work presents the implications of the absence of arbitrage in a two period incomplete markets economy where default is allowed, but it is required that all the assets be backed by a collateral bundle. This collateral can be exogenously given or can be determined by the sellers of assets, as in the Collateralized Mortgage Obligation (CMO) markets. Also we show the existence of equilibrium, but without bounded short-sales assumption in the endogenous case. Finally we present some efficency results.

1. Introduction

There is no doubt that default is an important fact in many markets, exmple of them are the Collateralized Mortgage Obligation (CMO) markets and the Emerging markets. Many model have been developed in order to capture the behaviour of these markets. In the context of general equilibrium the firt attempt are due to Dubey, Geanakoplos and Shubik (see [7]), they intriduce default in a two period economy where defaulters suffer utility penaltiesand anticipate that asset returns are discounted according to the average repayment rate. Araujo, Monteiro and Páscoa extended the model to an infinite-tree with a continuum of states at each node (see [2] and [1]). Dubey, Geanakoplos and Zame developed in [8] the collateral models, in which securities are backed by a collateral bundle and utility penalties are absent, but asset prices are affected by default. Since all the agents know the assets' payoffs, namely the minimum between the value of the claim and the value of the collateral bundle. With this charaterization of the true return we avoid adverse selection problems. In [8] it is proved that when collateral is fixed exogenously equilibrium exists. In [10] Fajardo studies the implications of absence of arbitrage and obtains a characterization of the arbitrage free asset prices. Recently, [4] and [3] have allowed the collateral to be fixed endogenously by borrowers, as happens in the CMO markets. In [4] Araujo, Orrillo and Páscoaprove that equilibrium exist assuming an specific form for asset prices and that the value of collateraL per unit of asset must excced the asset price by some arbitrary small amount exogenously fixed. In [3] Araujo, Fajardo and Páscoa avoid this ankward condition and prove existence of equilibrium using non arbitrage conditions. Moreover, they obtain an arbitrage pricing formula. In this work I present the implications of absence of arbitrage in collateral models. These results are contained in [3] and [10].

2. Exogenous Collateral Model

We consider an exchange economy over two periods. At the second period a finite number of states $s \in S = \{1, 2, ..., S\}$. can occur. There are H agents, J assets and L goods. In the first period, there is a market where physical commodities and assets are traded against each other. in the second period commodity trades take place again, and assets pay off. Denote by $\theta_j \geq 0$ the purchased units of asset j and by $\varphi_j \geq 0$ the sale units. This sale must be backed by a collection of goods (Collateral). Denote by $C_j \in I\!R^L_+$ the collateral coefficient that the lender is allowed to hold and borrower is obliged to hold for one unit of purchased and sale asset, respectively. In this way an asset j is defined by the promise it makes and the collateral backing it, (R_j, C_j). In this model, in order to mantain the anonimity of the market, is necessary to suppose that agents are not allowed to use their future endowment as collateral.

We let the goods be durable. Thus, one unit of commodity $l \in L$ becomes a vector of commodities next period by depending on how it used. More specifically we suppose that a x quantity of each commodity l is transformed into $Y^l_s x$ in each state $s \in S$ if it is used for consumption; in similar way for collateral. Also in order to preserve the anonimity of the market, is necessary to assume that durability is not affected by the identity of the user, or by intensity of its use.

The notation can be formalized as:

- $l \in L = \{1, \ldots, L\}$ set of commodities.
- $s \in S = \{1, 2, .., S\}$ set of states in the second period.
- $h \in H = \{1, \ldots, H\}$ set of agents.
- $e^h = (e^h_0, e^h_1, \ldots, e^h_S) \in I\!R^{(S+1)L}_+$, initial endowments of agent h
- $(p, q, \pi) \in I\!R^L_+ \times I\!R^{SL}_+ \times I\!R^J_+$ is the price system of the economy.
- $(R^j, C_j)_{j \in J}$ are real assets where $R^j : S \mapsto I\!R^L_+$ is the promise made by the asset j and C_j is the collateral that back it.
- $Y_j : S \mapsto I\!R^L$ are random variables that represent the durability of goods
- $D^h_j : S \mapsto I\!R^L_+$ is defined by: $D^h_j(s) := \min\{q(s)R^j(s), q(s)Y(s)C_j\}$ is the true return on asset j in state s.

Our economy is defined by

$$\mathcal{E} = ((U^h, e^h)_{h \in H}, (R^j, C_j)_{j \in J}, (Y^l)_{l \in L})$$

the budget constraints of each agent $h \in H$ are given by

$$p(x_o - e^h_o) + \pi(\theta - \varphi) + p \sum_{j \in J} C_j \varphi_j \leq 0, \tag{1}$$

$$q(s)(x(s) - e^h(s) - Y(s)x_o) - \sum_{j \in J} \varphi_j q(s)[Y(s)C_j] - \sum_{j \in J} (\theta_j - \varphi_j) D^h_j(s) \leq 0 \tag{2}$$

In this setting each agent $h \in H$ face the following problem:

$$\max_{(x,\theta,\varphi) \in B^h(p,q,\pi)} U^h(x_o + C\varphi, x_{-o}) \tag{3}$$

where $x_o + C\varphi$ is the total commodity consumption in the first period , $x_{-o} = (x_1, \ldots, x_S)$ is the consumption in the several states of nature in the second period and

$$B^h : I\!R^L_+ \times I\!R^{SL}_+ \times I\!R^J_+ \mapsto I\!R^{(S+1)L}_+ \times I\!R^J_+ \times I\!R^J_+$$

is defined by:

$$B^h(p, q, \pi) := \{(x, \theta, \varphi) \in I\!R^{(S+1)L}_+ \times I\!R^J_+ \times I\!R^J_+ : (1) \text{ and } (2) \text{ hold } \}$$

Or in a matrix form

$$P \cdot (x - \tilde{e}^h) \le A_{(C)}\Psi$$

where $P \cdot (x - \tilde{e}^h) = \big(p(x_o - e^h_o), q_1(x_1 - e^h_1 - Y_1 x_o), .., q_S(x_S - e^h_S - Y_S x_o)\big)$, $\Psi = (\theta, \varphi)$ and

$$A_{(C)} = \begin{bmatrix} -\pi & \pi - pC \\ D_1 & q_1 Y_1 C - D_1 \\ D_2 & q_2 Y_2 C - D_2 \\ \cdot & \cdot \\ \cdot & \cdot \\ D_S & q_S Y_S C - D_S \end{bmatrix}$$

Now we will define arbitrage in our context, assuming that agents preferences are monotonic.

2.1. Arbitrage

As in [10] we define arbitrage opportunities in a nontrivial context where $p >> 0, \forall s$ and $C_j \neq 0, \forall j$. Monotonicity of preferences determines already that the commmodity arbitrage opportunities derived from zero spot prices have to be ruled out.

Definition 2.1. *We say that there exists arbitrage opportunities if* $\exists\ (C, \Psi) \in I\!R^{3JL}_+$ *such that*

$$A_{(C)}\Psi > 0 \ \ a.s. \tag{4}$$

or if $\pi = 0$ *or* $pC - \pi = 0$. *We say that the economy is arbitrage free if there are not such opportunities.*

The case when $\pi^j = 0$ creates arbitrage opportunities since $C_j \neq 0$ and $q_s >> 0, \forall s$, impliying $D_s\theta_j > 0, \forall s$. The case when $pC^j - \pi^j = 0$ creates also arbitrage opportunities since even if $q_s Y_s C^j = D_s j$ for every s there would be unbounded utility gains from consumption of $C^j\varphi_j$ by choosing unbounded short sales of asset j. All the trading strategies that satisfy the budget constraints and (4) is called admissible and will be denoted by the set Θ. Now we will characterize the arbitrage free prices.

Theorem 2.2. *The economy is arbitrage free if and only if there exist*

$$\beta \in I\!R^{(S+2J+1)}_{++}$$

such that

$$\pi^j = \sum_{s=1}^{S} \beta_s D_s^j + \beta_{S+j} = (p - \sum_{s=1}^{S} \beta_s q_s Y_s) C^j + \sum_{s=1}^{S} \beta_s D_s^j - \beta_{S+J+j}$$

Proof. The necessity is an application of Stiemke's lemma, the sufficency comes from

$$pC_j - \pi^j = \sum_{s=1}^{S} \beta_s \left[q_s Y_s C_j - \min\{q_s R_s^j, q_s Y_s C_j\} \right] + \beta_{S+J+j} > 0$$

for more details see [10] . □

2.2. Equilibrium

The aim of this section is to give and alternative proof of existence of equilibrium to the one given by [8]. We use generalized games approach introduced by Debreu (see [6]). We enunciate the result as

Theorem 2.3. *For an economy $\mathcal{E}$ in which*

a. *For all agent $h \in H$, e^h belongs to $I\!R^L_{++}$;*
b. *The utility functions $U^h : X^h \to I\!R_+$ are continuous, strictly increasing, strictly quasi-concave;*
c. *$C^j \neq 0 \; \forall j \in J$,*

there exists an equilibrium.

Proof. Define a sequence of auxiliary economies namely

Truncated Economy

Denote by $(\mathcal{E}_n)_n$ the sequence of truncated economies such that the budget set of each agent h is

$$B_n^h(p, q, \pi, C) := \{(x_n^h, \theta_n^h, \varphi_n^h) \in [0, n]^{L(S+1)+2J} : (1), (2) \text{ hold } \}$$

For each $n \in \mathbb{N}$ we define the following generalized game played by consumers and $S+1$ auctioneers. Denote this game by $\mathcal{J}_n$ which is described by:

- Each consumer $h \in H$ maximizes U^h in the constrained strategy set $B_n^h(p, q, \pi, C)$.
- The auctioneer of the first period chooses $(p, \pi) \in \Delta^{L+J+S-1}$,in order to maximize

$$p \sum_h (x_o^h + \sum_j C_j \varphi_j^h - e_o^h) + \pi \sum_h (\theta^h - \varphi^h)$$

- The auctioneer of state s of the second period chooses $q_s \in \Delta^{L-1}$ in order to maximize $q_s \sum_h (x_s^h + \sum_j (Y_s C_j \varphi_j^h + Y_s x_o^h) - e_s^h)$.

As in [6] it is easy to prove that $\mathcal{J}_n$ has an equilibrium in pure strategies, since it is a convex problem and the objective functions of agents and auctioneers are continuous and quase-concave.
Now from these sequence of games we obtain an equilibrium for our auxiliary economies, as we state in the next lemma

Lemma 2.4. *An equilibrium in pure strategies of the generalized game $\mathcal{J}_n$ is an equilibrium for the truncated economy $\mathcal{E}_n$ for n large enough.*

Proof. :
From optimality conditions of the auctioneers' problems we obtain the market clearing conditions, for more details see [10]. □

In order to obtain an equilibrium in the original economy we have to study the case $n \to \infty$.

Asymptotics of truncated equilibria

Now let $\{((x_n^h, \theta_n^h, \varphi_n^h)_{\{h\in H\}}) \in ([0,n]^{S(L+1)+2J})^H, p_n, q_n, \pi_n\}$ be the sequence of equilibria corresponding to $\mathcal{E}_n$. Then by definition, for any n,

1. Markets clear:
$\sum_h (x_{on}^h + C^h \varphi_n^h) = \sum_h e_o^h$,
$\sum_h x_{sn}^h = \sum_h (e_s^h + Y_s C \varphi_n^h + Y_s x_{on}^h)$ $s \in S$ and
$\sum_h \theta_n^h = \sum_h \varphi_n^h$
2. $(x_n^h, \theta_n^h, \varphi_n^h)$ maximizes U^h on $B_n^h(p_n, q_n, \pi_n, C)$

Now it is easy to see that $\{\sum_h (x_n^h, \theta_n^h, \varphi_n^h\}$ is a bounded sequence, since by 1, $\sum_h x_{no}^h \le \sum_h e_o^h$ and $\sum_h C\varphi_n^h \le \sum_h e_o^h$.
So

$$\sum_h x_{ns}^h < \sum_h (e_s^h + 2Y_s e_o^h), \forall s \in S. \tag{5}$$

For each $l \in L$ the following holds

$$C_{lj} \sum_h \varphi_{nj}^h = C_{lj} \sum_h \theta_{nj}^h \le \sum_h (e_{lo}^h) \tag{6}$$

Therefore, $\sum_h \theta_{nj}^h = \sum_h \varphi_{nj}^h$ is bounded as long as $C \ne 0$.
From these facts we have that the sequence z_n^h has a cluster point. Therefore $\exists z$ such that

$$z^h \in \lim\{z_n^h\}$$

This imply that z^h is budget feasible at $(p, q, \pi) = \lim_{n\to\infty}(p^n, q^n, \pi_n)$(passing to a subsequence if necessary). Now we show that $(x^h, \theta^h, \varphi^h)$ maximizes U^h at the cluster point of (p^n, q^n, π^n). In fact, suppose that it is not optimal, i.e; $\exists \overline{z}^h \in B^h(p, q, \pi)$ such that $u^h(\overline{z}^h) > u^h(z^h)$. By applying the lower hemi-continuity of the budget set (see [10]), $\exists \overline{z}_n^h \in B^h(p^n, q^n, \pi^n)$ and $\overline{z}_n^h \to \overline{z}^h$. Now, for $n \ge n_o$ one has $\overline{z}_n^h \in B_n^h(p^n, q^n, \pi^n)$. Since u^h is continuous, the following holds

$$u^h(\overline{z}_n^h) > u^h(z_n^h), \forall n \ge n_1$$

Therefore for $n \geq \max\{n_o, n_1\}$, z_n^h is not optimal in the truncated economy $\mathcal{E}_n^c$, a contradiction. □

3. Endogenous Collateral Model

Now to endogenize the collateral we will let the borrowers (seller of assets) to choose the collateral coefficient for the different assets that he sells and we suppose that there exist anonymous collateral coefficients which will be taken as given by each buyer of assets. For each asset j denote by $M_j \in \mathbb{R}_+^L$ the choice of collateral coefficients. The anonymous collateral coefficients will be denoted by $C \in \mathbb{R}_+^{JL}$ and will be be taken as given. The collateral bundle choosen by borrower will be $M\varphi$ and his whole first period consumption bundle is $x_o + M\varphi$.
Let π_1 and π_2 be the purchase and sale prices of assets, respectively. Then, as in the exogenous case, we obtain a budget constraints for each agent:

$$p_o x_o + p_o M\varphi + \pi_1\theta \leq p_o e_o + \pi_2\varphi \tag{7}$$

$$q_s x_s + \sum_{j=1}^{J} D_{sj}\varphi_j \leq q_s e_s + \sum_{j=1}^{J} N_{sj}\theta_j + \sum_{j=1}^{J} q_s Y_s M_j \varphi_j + q_s Y_s x_o, \quad \forall s \in S \tag{8}$$

Here $D_{sj} = \min\{q_s R_s^j, q_s Y_s M_j\}$ and $N_{sj} = \min\{q_s R_s^j, q_s Y_s C_j\}$ are what he will paid and receive with the sale and purchase of one unit of asset j.

3.1. Arbitrage

As in the exogenous case we have to rule out the commmodity arbitrage opportunities derived from zero spot prices and we consider the case $M^j \neq 0$, other wise π_2^j would be equal to 0, because the true return of this promisse would be 0. Construct an analogous matrix $A(M)$ to the one constructed ($A(C)$) in section 2, then we can state the following

Definition 3.1. *We say that there exist arbitrage opportunities if* $\exists\ (M, \Psi) \in \mathbb{R}_+^{(2+L)J}$ *such that*

$$A_{(M)}\Psi > 0$$

or also when $\pi_1^j = 0$ *or* $pM_j - \pi_2^j = 0$ *for some* j.

The case when $pM_j - \pi_2^j = 0$ creates also arbitrage opportunities even if $q_s Y_s M_j = D_{sj}$ for every s since there would be unbounded utility gains from consumption of $M_j\varphi_j$ by choosing unbounded short sales of asset j. All trading strategies that satisfy (8),(7) and do not satisfy (3.1) we called admissible. Now we have the analogous of Theorem 2.1

Theorem 3.2. *There are no arbitrage opportunities if and only if there exists* $\beta \in \mathbb{R}_{++}^{S+2J}$ *such that for each* $j = 1, 2, .., J$

$$\pi_1^j = \sum_{s=1}^{S} \beta_s q_s R_s^j - \sum_{s=1}^{S} \beta_s (q_s R_s^j - q_s Y_s C_j)^+ + \beta_{S+j} \tag{9}$$

and

$$\pi_2^j = \sum_{s=1}^{S} \beta_s q_s R_s^j - \sum_{s=1}^{S} \beta_s (q_s R_s^j - q_s Y_s M_j)^+ + (pM_j - \sum_{s=1}^{S} \beta_s q_s Y_s M_j) - \beta_{S+J+j} \quad (10)$$

3.2. Equilibrium

As a result of endogenizing the collateral we have a non-convex problem. To circunvemt this problem [4] and [3] consider a continuum of agents $H = [0,1]$ modeled by the Lebesgue probability space $(H, \mathcal{B}, \lambda)$. This continuum of agents will let us to capture a important fact of CMO markets that is lenders trade indirectly with borrowers through a pool of mortgages.

The allocation of the commodities is an integrable map $x : H \to I\!R_+^{(S+1)L}$. The assets purchase and sale allocations are represented by two integral maps; $\theta : H \to I\!R_+^J$ and $\varphi : H \to I\!R_+^J$, respectively. The allocation of collateral coefficients chosen by borrowers is described by the function $M : H \to I\!R_+^J$. Each buyer of assets (lender) will take an anonymous collateral coefficient vector $C \in I\!R_+^{JL}$ as given. Let $x_{-o}^h = (x_1^h, \ldots, x_S^h)$ be the commodity consumption in the several states of the world in the second period.
Asset prices consist of a base price (common to the purchase and sale prices) and also a spread (varying across sellers in the case of the sale price). Let

$$\pi_1 = k - \sum_s \gamma_s (q_s R_s - q_s Y_s C)^+ \quad (11)$$

and

$$\pi_2 = k - \sum_s \gamma_s (q_s R_s - q_s Y_s M)^+ \quad (12)$$

Here k is understood as a base. The state prices γ_s are common to all agents and taken as given together with the base price k. Then the individual problem is

$$\max_{(x^h, \theta^h, \varphi^h, M^h) \in B^h} U^h(x_o^h + M^h \varphi^h, x_{-o}^h) \quad (13)$$

where B^h is the budget set of each agent $h \in H$ given by:

$$B^h(p, C, k, \gamma) = \left\{ (x, \theta, \varphi, M) \in I\!R^{L(S+1)+2J+JL} : (7) \text{ and } (8) \text{ hold} \right\}$$

Remark

- The form of asset prices considered above are derived from the non arbitrage valuation established,since we can take corrections factors in order to satisfy:

  $$(p_o - \sum_{s \in S} \beta_s q_s Y_s) M_j - \beta_{S+J+j} = \beta_{S+j}$$

  In general equilibrium prices given by (11) and (12) will have non arbitrage representations according to (9) and (10) for vectors β^h that vary across agents.

Definition 3.3. *An equilibrium is a vector* $((p,\pi_1,\pi_2,C),(x^h,\theta^h,\varphi^h,M^h)_{h\in H})$ *such that:*

- $$(x^h,\theta^h,\varphi^h,M^h)$$
  solves problem (13)

- $$\int_H \left(x_o^h + \sum_{j\in J} M_j^h\varphi_j^h\right) dh = \int_H e_o^h dh$$

$$\int_H x^h(s)dh = \int_H \left(e^h(s) + \sum_{j\in J}(Y_s M_j^h \varphi_j^h + Y_s x_o^h)\right) dh \tag{14}$$

- $$\int_H (\theta^h - \varphi^h)dh = 0 \tag{15}$$

- $$\int_H M_j^h\varphi_j^h dh = C_j \int_H \theta_j^h dh \ \ \forall\, j \in J \tag{16}$$

- $$\int_{h\in\mathcal{S}_s^j} (q_s R_s^j - q_s Y_s C_j)^+ \theta_j^h dh = \int_{h\in\mathcal{G}_s^j} (q_s R_s^j - q_s Y_s M_j^h)^+ \varphi_j^h dh \ \ \forall s \in S,\ \forall j \in J \tag{17}$$

  Where $\mathcal{S}_s^j = \{h \in H : q_s R_s^j > q_s Y_s C_j\}$ *is the set of agents that suffered default in state of nature* s *on asset* j *and* $\mathcal{G}_s^j = \{h \in H : q_s R_s^j > q_s Y_s M_j^h\}$ *is the set of agents that give default in state of nature* s *on asset* j. *Note that* $\mathcal{S}_s^j$ *is equal to* H *or* ϕ, *since* $q_s R_s^j$ *and* $q_s Y_s C_j$ *do not depend on* h.

The principal assumptions on preferences is.

Assumption : preferences are time and state separable, monotonic, representable by a smooth strictly concave utility function u^h satisfying:

i) Inada's condition
ii) $\frac{\partial u^h(z)}{\partial z_{0l}} z_{0l} \to \infty$ for any l, when $\min_l z_{0l} \to \infty$
iii) $\{u^h\}_{h\in H}$ and $\{Du^h\}_{h\in H}$ are equicontinuous.

Theorem 3.4. *If consumers's preferences satisfy above Assumption and the endowments allocation* e *belongs to* $L^\infty(H, I\!R_{++}^{(S+1)L})$, *then there exists an equilibrium where borrowers choose their respective collateral coefficients.*

The proof of this theorem is given in [3].

4. Efficency

In this section we show that an equilibrium allocation is is Pareto superior to the equilibrium obtained in [4], since we are not impossing any kind of bounded short sale condition. As in [12], we compare the equilibrium allocation with one feasible allocation whose portfolios do not necessarily result from trading competitively in asset markets. That is, in alternative allocations agents pay participation fees which may differ from the market portfolio cost. Equivalently, we allow for transfers across agents which are being added to the usual market portfolio cost.

Proposition 4.1. *Let* $((\overline{x},\overline{\theta},\overline{\varphi},\overline{M}),\overline{p},\overline{\pi}_1,\overline{\pi}_2,\overline{C})$ *be an equilibrium. The allocation* $(\overline{x},\overline{\theta},\overline{\varphi},\overline{M})$ *is efficient among all allocations* (x,θ,φ,M) *for which there are transfers* $T^h \in I\!R$ *across agents and a vector* $C \in I\!R^{JL}_{+}$, *such that*

(i) $\int_H (x^h_o + M^h\varphi^h)dh = \int_H e^h_o dh$, $\int_H x^h_s = \int_H (e^h_s + Y_s M^h \varphi^h + Y_s x^h_o)dh$, $\int_H \theta^h dh = \int_H \varphi^h dh$

(ii)

$$\overline{q}_s(x^h_s - e^h_s - Y_s x^h_o) + \sum_{j\in J} \min\{\overline{q}_s R^j_s, \overline{q}_s Y_s M^h_j\}\varphi^h_j$$

$$=\sum_{j\in J} \min\{\overline{q}_s R^j_s, \overline{q}_s Y_s C_j\}\theta^h_j + \sum_{j\in J} \overline{q}_s Y_s M^h_j \varphi^h_j, \quad \forall s, \quad a.e. \ h$$

(iii) $\overline{p}(x^h_o + M^h\varphi^h - e^h_o) + \overline{\pi}_1\theta^h - \overline{\pi}_2\varphi^h + T^h = 0$

(iv) $\int_H T^h dh = 0$

(v) $C_j \int_H \theta^h dh = \int_H M^h_j \varphi^h_j, \ \forall j$

where the equilibrium prices are given by

$$\overline{\pi}_1 = \overline{q} - \sum_s \overline{\gamma}_s \overline{g}_{1s}$$

and

$$\overline{\pi}_2 = \overline{q} \sum_s \overline{\gamma}_s \overline{g}_{2s}$$

Proof. See [3].□

References

[1] Araujo A., P. Monteiro and M. Páscoa, *Infinte Horizon Incomplete Markets, Continuum of states*, Mathematical Finance **6** (1998), 119–132.

[2] Araujo A., P. Monteiro and M. Páscoa, *Incomplete Markets, Continuum of states and Default*, Economic Theory **11** (1998), 205–213.

[3] Araujo A., J. Fajardo and M. Páscoa, *Arbitrage and Equilibrium in CMO Markets* (2000), Working Paper IMPA.

[4] Araujo A., J. Orrillo and M. Páscoa, *Equilibrium with Default and Endogenous Collateral*, Mathematical Finance **Vol 10** (2000), No 1. 1–21

[5] Cvitanić, J. and I. Karatzas, *Hedging Contingent Claims with Constrained Portfolios*, Annals of Applied Probability **3** (1993), 652–681.

[6] Debreu, G., *A Social Equilibrium Existence Theorem*, Proceedings of The National Academy of Sciences (1952), 38, 886-893.

[7] Dubey, P., J. Geanakoplos and M.Shubik, *Liquidity and bankruptcy with incomplete markets: Pure exchange* (1989), Cowels Foundation discussion paper 900.

[8] Dubey, P.,J. Geanakoplos and W. Zame, *Default, Collateral, and Derivatives* (1995), Yale University, mimeo.

[9] EL Karoui, N. and M. Quenez, *Dynamic Programming and Pricing of Contingent Claims in an Incomplete Market*, SIAM Journal of Control and Optimization **33** (1995), 29–66.

[10] Fajardo, J., *A Note on Arbitrage and Exogenous Collateral* (2000), Working Paper Catholic University of Brasília.

[11] Jouini,E. and H. Kallal, *Arbitrage in security markets with short-sales constraints*, Mathematical Finance **5(3)** (1995)

[12] Magill,M. and W. Shafer, *Incomplete Markets,* in W.Hindenbrand and H. Sonnenschein, Handbook of Mathematical economics, Vol. IV (1991). North Holland, Amsterdam.

[13] Zame, W., *Efficiency and the role of default when security markets are Incomplete,* Amer. Econ. Review **83** (1963), 1142–1164.

Department of Economics,
Catholic University of Brasília,
SGAN 916 Mod B. Asa Norte
70790-160 Brasília, D.F. Brazil
E-mail address: pepe@pos.ucb.br

Trends in Mathematics, © 2001 Birkhäuser Verlag Basel/Switzerland

On the existence of optimal controls for a singular stochastic control problem in finance

Fred E. Benth, Kenneth H. Karlsen, and Kristin Reikvam

Abstract. We prove existence of optimal investment-consumption strategies for an infinite horizon portfolio optimization problem in a Lévy market with intertemporal substitution and transaction costs. This paper complements our previous work [4], which established that the value function can be uniquely characterized as a constrained viscosity solution of the associated Hamilton-Jacobi-Bellman equation (but [4] left open the question of existence of optimal strategies). In this paper, we also give an alternative proof of the viscosity solution property of the value function. This proof exploits the existence of optimal strategies and is consequently simpler than the one proposed in [4].

1. Introduction

We prove existence of optimal controls for the singular stochastic control problem studied in Benth, Karlsen, and Reikvam [4] (see also [1, 2, 3] for related problems). An optimal consumption-investment problem over an infinite investment horizon in a market consisting of one risky asset (stock) and one risk-free asset (bank) is considered. The dynamics of the risky asset follows a geometric-type Lévy motion, generalizing the classical Black & Scholes model. Proportional transaction costs are incurred when selling or buying assets. In addition, following Hindy and Huang [9], the investor derives utility from an average of present and past consumption.

In Benth, Karlsen, and Reikvam [4], the value function of this optimization problem was characterized as the unique constrained viscosity solution [6] of the associated Hamilton-Jacobi-Bellman equation. Due to the singular controls and the Lévy dynamics of the risky asset, the Hamilton-Jacobi-Bellman equations takes the form of a second-order *integro-differential* variational inequality.

The existence of optimal controls was not addressed in [4]. To prove existence of an optimal control, we shall here use the convex analysis techniques described in Ekeland and Temam [10] together with the martingale methods of Cvitanić and Karatzas [7]. The main problem we are facing is the infinite investment horizon, which leads to unbounded controls. However, from Benth, Karlsen, and Reikvam

Benth is partially sponsored by MaPhySto - Centre for Mathematical Physics and Stochastics (University of Aarhus, Denmark), funded by a grant from the Danish National Research Foundation. Reikvam is supported by the Norwegian Research Council (NFR) under grant 118868/410.

[5], we have explicit bounds on the expected growth of the controls and these bounds enable us to introduce a time-weighted L^2 - space. This space provides the starting point for using the martingale methods in [7].

We end this paper by giving an alternative proof of the result in [4] stating that the value function is a constrained viscosity solution of the associated Hamilton-Jacobi-Bellman equation. Compared with the proof proposed in [4], the presènt proof is simpler since we exploit the existence of an optimal strategies.

2. Formulation of the control problem

Let $(\Omega, \mathcal{F}, \{\mathcal{F}_t\}_{t\geq 0}, P)$ be a filtered complete probability space satisfying the usual hypotheses. We consider a single investor who divides her wealth between one risk-free asset (bank account) paying a fixed interest rate $r > 0$ and a risky asset (stock). We denote by $B(t)$ the amount of money the investor has in the bank account and $S(t)$ the amount of money the investor has in the stock, at time $t \geq 0$. We assume that the holdings of the investor follow the dynamics

$$\begin{cases} B(t) = b_0 - C(t) + \int_0^t rB(s)\,ds - (1+\lambda)L(t) + (1-\mu)M(t), \\ S(t) = s + \int_0^t aS(s)\,ds + \int_0^t \sigma S(s)\,dW(s) \\ \qquad\qquad + \int_0^t \int_{\mathbb{R}\setminus\{0\}} \eta(z)S(s-)\,\tilde{N}(ds,dz) + L(t) - M(t), \end{cases} \tag{1}$$

where $a, \sigma > 0$ are constants, $C(t)$ is the cumulative consumption up to time t, $L(t)$ is the cumulative value of the shares *bought* up to time t, $M(t)$ is the cumulative value of the shares *sold* up to time t, and $\mu \in [0,1]$ and $\lambda \geq 0$ are the proportional transaction costs of respectively selling and buying shares from the stock. We assume $\mu + \lambda > 0$. In addition, $W(s)$ is a standard Brownian motion and $\tilde{N}$ is a compensated Poisson random measure independent of W with Lévy measure $\ell(dz)$. The function $\eta(z)$ is assumed to be Borel measurable on $\mathbb{R}\setminus\{0\}$ with the property $\eta(z) > -1$ to ensure that the stock holdings remains positive as long as we are not short of stocks. In addition, we require the following integrability conditions on the Lévy measure:

$$\int_{|z|<1} (\eta(z))^2\,\ell(dz) < \infty, \quad \int_{|z|\geq 1} |\eta(z)|\,\ell(dz) < \infty.$$

We assume throughout the paper that the expected rate of return a of the stock is greater than or equal to the risk-free interest rate r.

Introduce the process of average past consumption,

$$dY(t) = \beta\,dC(t) - \beta Y(t)\,dt, \qquad \beta > 0, \tag{2}$$

which has the explicit solution

$$Y(t) = y\mathrm{e}^{-\beta t} + \beta \mathrm{e}^{-\beta t} \int_{[0,t]} \mathrm{e}^{\beta s}\, dC(s).$$

The investor will derive utility from this average, rather than directly from present consumption [9].

The market considered here does not allow short-selling of stocks nor borrowing of money in the bank. In other words, the amount of money allocated in the bank account and the stocks must stay nonnegative. Hence the domain for the control problem is

$$\mathcal{D} = \Big\{ x = (b, s, y) \in \mathbb{R}^3 \mid x > 0 \Big\}.$$

We refer to $\Pi = (C, L, M)$ as a policy for investment and consumption if Π belongs to the set $\mathcal{A}_x$ of *admissible controls*. For $x \in \overline{\mathcal{D}}$, we say that $\Pi \in \mathcal{A}_x$ if the following conditions hold:

(C.1) $C(t), L(t), M(t)$ are adapted, nondecreasing, and right-continuous with left limits. Moreover, $C(0-) = M(0-) = L(0-) = 0$.

(C.2) The state process $X(t) = (B(t), S(t), Y(t))$ is a solution to the stochastic differential equations (1) and (2) and respects the state-space constraint $X(t) = X^{\Pi}(t) \in \overline{\mathcal{D}}$ for all $t \geq 0$.

Note that $0 \in \mathcal{A}_x$. The objective of the investor is to maximize her expected utility over an infinite investment horizon. The functional to be optimized is

$$\mathcal{J}(x; \Pi) = \mathbb{E}\Big[\int_0^{\infty} \mathrm{e}^{-\delta t} U(Y^{\Pi}(t))\, dt \Big], \quad x \in \overline{\mathcal{D}},$$

where U is the investor's utility function and $\delta > 0$ is the discount factor. We introduce the following assumptions on the utility function:

(U.1) $U(z)$ is a continuous, nondecreasing, and concave function on $[0, \infty)$ with $U(0) = 0$.

(U.2) There exist $\gamma \in (0, 1)$ and constant $K > 0$ such that $U(z) \leq K(1 + z)^{\gamma}$ for all $z \in [0, \infty)$.

Define the value function of the optimization problem to be

$$V(x) = \sup_{\Pi \in \mathcal{A}_x} \mathcal{J}(x; \Pi), \qquad x \in \overline{\mathcal{D}}. \tag{3}$$

Our singular stochastic control problem is to find an optimal control $\Pi^* \in \mathcal{A}_x$ such that

$$V(x) = \mathcal{J}(x; \Pi^*), \qquad x \in \overline{\mathcal{D}}. \tag{4}$$

Sections 3 and 4 are devoted to the existence of Π^*, while Section 5 is devoted to a brief discussion of the Hamilton-Jacobi-Bellman equation satisfied (in a suitable sense) by the value function (3).

In what follows (Sections 3 and 4), the point $x \in \overline{\mathcal{D}}$ will always be considered as fixed and not explicitly mentioned anymore.

3. Some estimates on the control and state processes

We recall some results from [5] which will be needed in Section 4. The first result states that the set of admissible controls is uniformly bounded in $L^2(P)$. Moreover, we can control the growth in time.

Proposition 3.1. *For every $t \geq 0$, the controls are uniformly bounded in $L^2(P)$:*

$$\sup_{\Pi \in \mathcal{A}_x} \mathbb{E}\big[C_t^2 + L_t^2 + M_t^2\big] \leq K e^{kt},$$

where K is a positive constant and

$$k = 2r + 2(a-r)(2K_1^2 + K_2) + 2K_2^2\Big(\sigma^2 + \int_{\mathbb{R}\setminus\{0\}} (\eta(\xi))^2 \ell(d\xi)\Big),$$

for $K_1 = (x_0 + (1+\lambda)x_1)/(\lambda + \mu)$ and $K_2 = (1+\lambda)/(\mu + \lambda)$.

Proof. This result is proven in [5], where the rate of the exponential growth is explicitly calculated. □

We have a similar uniform bound on the averaging process Y_t^Π:

Proposition 3.2. *For every $t \geq 0$,*

$$\sup_{\Pi \in \mathcal{A}_x} \mathbb{E}\big[(Y_t^\Pi)^2\big] \leq K(y + e^{kt}),$$

where K is a positive constant and k is as in Proposition 3.1.

Proof. This follows from the fact that $Y_t \leq y + C_t$ and the estimate on $\mathbb{E}[C_t^2]$. □

Via the growth of V proven in [4, Cor. 3.5], we have that the value function is well-defined, that is, the following result holds:

Proposition 3.3. *There exists a positive constant K such that*

$$0 \leq V(x) \leq K\big(1 + |x|\big)^\gamma.$$

4. Existence of optimal controls

In this section we prove existence of an optimal control. Let $\alpha > k$, for the k given in Prop. 3.1, and introduce the measure $m_\alpha(dt) = e^{-\alpha t}\, dt$ on $\mathbb{R}_+$. Notice that

$$\int_{\mathbb{R}_+} e^{ct}\, m_\alpha(dt) < \infty, \qquad \forall c \leq k. \tag{5}$$

Define the the (weighted Hilbert) space

$$\mathcal{H}_x^\alpha = \Big\{0 \leq H \in L^2(m_\alpha \otimes P) \;:\; \exists \Pi \in \mathcal{A}_x \text{ s. t. } H \leq Y^\Pi,\, m_\alpha \otimes P \text{ - a. e.}\Big\}.$$

The next two lemmas show that $\mathcal{H}_x^\alpha$ is a non-empty, bounded, convex, and closed subspace of $L^2(m_\alpha \otimes P)$:

Lemma 4.1. *$\mathcal{H}_x^\alpha$ is a non-empty, bounded, and convex subspace of $L^2(m_\alpha \otimes P)$.*

Proof. Since $Y_t^\Pi \geq 0$ for all $\Pi \in \mathcal{A}_x$, it follows that $0 \in \mathcal{H}_x^\alpha$. The uniform exponential bound of Y^Π in Prop. 3.2 proves that $\mathcal{H}_x^\alpha$ is a norm-bounded subspace of $L^2(m_\alpha \otimes P)$. Let $H^1, H^2 \in \mathcal{H}_x^\alpha$ be such that $H^1 \leq Y^{\Pi^1}$ and $H^2 \leq Y^{\Pi^2}$. Define the control $\Pi := \theta\Pi^1 + (1-\theta)\Pi^2$ for $\theta \in (0,1)$. From the proof of Prop. 5.1 in [4], it follows that $\Pi \in \mathcal{A}_x$. Moreover, by uniqueness of paths, $Y^\Pi = \theta Y^{\Pi^1} + (1-\theta)Y^{\Pi^2}$, and hence $H := \theta H^1 + (1-\theta)H^2 \leq Y^\Pi$. This proves the convexity of $\mathcal{H}_x^\alpha$. □

Lemma 4.2. *$\mathcal{H}_x^\alpha$ is closed in $L^2(m_\alpha \otimes P)$.*

Proof. Choose a sequence H^n in $\mathcal{H}_x^\alpha$ which converges to H in $L^2(m_\alpha \otimes P)$. We can associate a control Π^n to each H^n such that $H^n \leq Y^{\Pi^n}$. Moreover, by the uniform bound on the controls (see Section 3) and (5), we know that $\{\Pi^n\}_n$ is uniformly bounded in $L^1(m_\alpha \otimes P)$, and, hence by the Dunford-Pettis compactness criterion, there exists $\Pi \in L^1(m_\alpha \otimes P)$ and a subsequence, also denoted Π^n, such that $\Pi^n \to \Pi$ weakly in $L^1(m_\alpha \otimes P)$. The question is whether Π is an admissible control or not. Following the arguments of Karatzas and Shreve [11, Section 4], we can prove the existence of a version of Π, still denoted by Π, for which $\Pi \in \mathcal{A}_x$ and $\Pi^n \to \Pi$ weakly. In addition, as in Cvitanić and Karatzas [7, Appendix A], we can show that $B^{\Pi^n} \to B^\Pi$, $S^{\Pi^n} \to S^\Pi$, and $Y^{\Pi^n} \to Y^\Pi$ weakly. Since $H^n \leq Y^{\Pi^n}$ for every n, we conclude $H \leq Y^\Pi$, thereby proving the closedness of $\mathcal{H}_x^\alpha$. □

Define the functional $\mathcal{I} : \mathcal{H}_x^\alpha \to \mathbb{R}$ to be

$$\mathcal{I}(H) = -\mathbb{E}\Big[\int_0^\infty e^{-\delta t} U(H_s)\, ds\Big].$$

Lemma 4.3. *If $\delta > \alpha/2$, then $\mathcal{I}$ is proper, convex, and lower-semicontinuous with respect to the $L^2(m_\alpha \otimes P)$ - norm.*

Proof. From the growth property of V in Prop. 3.3 it follows that $\mathcal{I}$ is proper. Furthermore, since U is assumed to be concave, $\mathcal{I}$ is obviously convex.

To prove the lower semicontinuity of $\mathcal{I}$, we modify slightly the argument of Cvitanić and Karatzas. Let H^n be a sequence in $\mathcal{H}_x^\alpha$ that converges to H in $L^2(m_\alpha \otimes P)$. From the sublinear growth of U, there exist positive constants a and b such that $a + bH - U(H) \geq 0$ for all $H \in \mathcal{H}_x^\alpha$. Fatou's lemma gives us

$$\begin{aligned}\mathbb{E}\Big[\int_0^\infty e^{-\delta t}\Big(a + bH_t - U(H_t)\Big)\, dt\Big] &\leq \liminf_{n\to\infty} \mathbb{E}\Big[\int_0^\infty e^{-\delta t}\Big(a + bH_t^n - U(H_t^n)\Big)\, dt\Big] \\ &= \mathbb{E}\Big[\int_0^\infty e^{-\delta t}(a + bH_t)\, dt\Big] + \liminf_{n\to\infty} -\mathbb{E}\Big[\int_0^\infty e^{-\delta t} U(H_t^n)\, dt\Big].\end{aligned}$$

The last step holds true due to the assumption $\delta > \alpha/2$ since, from Hölder's inequality,

$$\begin{aligned}\Big|\mathbb{E}\Big[\int_0^\infty e^{-\delta t} H_t\, dt\Big] - \mathbb{E}\Big[\int_0^\infty e^{-\delta t} H_t^n\, dt\Big]\Big| &\leq \mathbb{E}\Big[\int_0^\infty e^{-\delta t}|H_t - H_t^n|\, dt\Big] \\ &\leq \Big(\int_0^\infty e^{-(2\delta-\alpha)t}\Big)^{1/2} \mathbb{E}\Big[\int_0^\infty e^{-\alpha t}|H_t - H_t^n|^2\, dt\Big]^{1/2} \to 0.\end{aligned}$$

In conclusion,

$$-\mathbb{E}\Big[\int_0^\infty e^{-\delta t}U(H_t)\,dt\Big] \le \liminf_{n\to\infty} -\mathbb{E}\Big[\int_0^\infty e^{-\delta t}U(H_t^n)\,dt\Big],$$

and hence the lower-semicontinuity of $\mathcal{I}$ follows. □

Remark 4.4. *We mention that if the utility function $U(\cdot)$ is Hölder continuous, then one can prove, under a slightly different assumption on δ, that the functional $\mathcal{I}$ is continuous (and not merely semicontinuous).*

Set

$$\mathcal{V}(x) = -\inf_{H\in\mathcal{H}_x^\alpha} \mathcal{I}(H).$$

From Ekeland and Temam [10] we can conclude that there exists a $H^* \in \mathcal{H}_x^\alpha$ such that $\mathcal{V}(x) = -\mathcal{I}(H^*)$. We claim that Π^* associated to H^* is an optimal control, that is,

$$V(x) = \mathbb{E}\Big[\int_0^\infty e^{-\delta t}U(Y_t^{\Pi^*})\,dt\Big].$$

First, observe that since $Y^\Pi \in \mathcal{H}_x^\alpha$ for every $\pi \in \mathcal{A}_x$, $\mathcal{A}_x$ can be naturally imbedded in $\mathcal{H}_x^\alpha$, which implies $V(x) \le \mathcal{V}(x)$. We next claim that $H^* = Y^{\Pi^*}$ $m_\alpha \otimes P$ - a.e. If not, $H^* < Y^{\Pi^*}$ on a set with positive measure. But this contradicts the optimality of H^* since $Y^{\Pi^*} \in \mathcal{H}_x^\alpha$ and U is nondecreasing. Hence,

$$V(x) \le \mathcal{V}(x) = \mathbb{E}\Big[\int_0^\infty e^{-\delta t}U(H_t^*)\,dt\Big] = \mathbb{E}\Big[\int_0^\infty e^{-\delta t}U(Y_t^{\Pi^*})\,dt\Big] \le V(x).$$

Summing up, we have proven the following main theorem:

Theorem 4.5. *Suppose $\delta > \alpha/2$. For each $x \in \overline{\mathcal{D}}$, there exists an optimal control $\Pi^* \in \mathcal{A}_x$ for the singular stochastic control problem* (3) *such that* (4) *holds.*

Remark 4.6. *Theorem 4.5 may easily be generalized to hold for n risky assets, as is the set-up in* [4].

5. The Hamilton-Jacobi-Bellman equation

Thm. 4.5 is only of theoretical interest since it says nothing about the structure of the optimal strategies or how we can compute them. A natural way to compute the optimal strategies is via the dynamic programming method, which is based on Bellman's principle of dynamic programming:

Proposition 5.1 (Dynamic programming principle). *For any stopping time τ and $t \ge 0$, the value function satisfies*

$$V(x) = \sup_{\Pi\in\mathcal{A}_x} \mathbb{E}\Big[\int_0^{t\wedge\tau} e^{-\delta s}U(Y^\Pi(s))\,ds + e^{-\delta(t\wedge\tau)}V(X^\Pi(t\wedge\tau))\Big]. \tag{6}$$

Thanks to the dynamic programming principle, the value function (3) can be associated with the so-called Hamilton-Jacobi-Bellman equation, which is the infinitesimal version of (6). For $x = (b, s, y) \in \overline{\mathcal{D}}$, define a second order degenerate elliptic integro-differential operator $\mathcal{A}$ by

$$\mathcal{A}v = -\beta y v_y + rbv_b + asv_s + \frac{1}{2}\sigma^2 s^2 v_{ss} + \int_{\mathbb{R}\setminus\{0\}} \Big(v(b, (1+\eta(z))s, y) - v(x) - \eta(z) s v_s(x) \Big) \, \ell(dz).$$

The Hamilton-Jacobi-Bellman equation of our control problem is a second order degenerate elliptic integro-differential variational inequality of the form

$$F(x, v, v_b, v_s, v_y, v_{ss}) := \max\Big(U(y) - \delta v + \mathcal{A}v, -v_b + \beta v_y, -(1+\lambda)v_b + v_s, (1-\mu)v_b - v_s \Big) = 0. \tag{7}$$

The point is now that value function (3) as well as the optimal control Π^* (whose existence is guaranteed by Thm. 4.5) can be found by (numerically) solving the fully nonlinear partial differential equation in (7).

The main result of our previous work [4] was a characterization of the value function (3) as the unique constrained viscosity solution [6] of the Hamilton-Jacobi-Bellman equation, which indeed constitutes a starting point for computing (numerically) the optimal value (3) as well as the optimal control Π^*.

Since we have only been able to show that the value function is continuous (see [4]), we cannot interpret the value function as a solution of (7) in the usual classical sense, but we have to resort to a weaker notion of solution that does not require differentiability of candidate solutions. The proper notion of weak solutions turns out to be that of constrained viscosity solutions as described in, e.g., Crandall, Ishii, and Lions [6].

We recall that the value function V is a *constrained viscosity solution* of (7) if it is simultaneously a viscosity subsolution in $\overline{\mathcal{D}}$ and a viscosity supersolution in $\mathcal{D}$. For example, the value function V is a *viscosity subsolution* of (7) in $\overline{\mathcal{D}}$ if $\forall \phi \in C^2(\overline{\mathcal{D}})$ (growing at most linearly as $x \to \infty$) we have:

$$\begin{cases} \text{for each } x \in \overline{\mathcal{D}} \text{ s. t. } V \le \phi \text{ and } (V - \phi)(x) = 0, \\ F(x, \phi, \phi_b, \phi_s, \phi_y, \phi(x)) \ge 0. \end{cases} \tag{8}$$

A *viscosity supersolution* is defined similarly, see [4] for details.

The purpose of this section is to give an alternative proof of the viscosity subsolution property of the value function, which is simpler than the proof in [4]. The proof below exploits that we have Thm. 4.5 at our disposal. We refer to [4] for results concerning continuity of the value function as well as uniqueness of the viscosity solution characterization.

Theorem 5.2. *The value function* (3) *is a constrained viscosity solution of* (7).

Proof. The proof of the viscosity supersolution property goes as before [4]. We therefore concentrate on the viscosity subsolution property.

Let ϕ be as in (8). Arguing by contradiction, we suppose that the subsolution inequality (8) is violated. Then, by continuity, there is a nonempty open ball $\mathcal{N}$ centered at x and $\varepsilon > 0$ such that $V \leq \phi - \varepsilon$ on $\partial\mathcal{N} \cap \mathcal{D}$ and in $\overline{\mathcal{N} \cap \mathcal{D}}$ we have

$$\beta\phi_y - \phi_b \leq 0, \quad -(1+\lambda)\phi_b + \phi_s \leq 0, \quad (1-\mu)\phi_b - \phi_s \leq 0, \tag{9}$$

as well as $U(\cdot) - \delta\phi + \mathcal{A}\phi \leq -\varepsilon\delta$.

From Thm. 4.5, there exists an optimal investment-consumption strategy $\Pi^*(t) = (L^*(t), M^*(t), C^*(t)) \in \mathcal{A}_x$. Let $X^*(t) = (B^*(t), S^*(t), Y^*(t))$ denote the corresponding optimal trajectory with $X^*(0) = x$. In Lemma 5.3 below, it is shown that $X^*(t)$ has no control-jumps P - a.s. at x. Hence P - a.s., we have

$$\tau = \inf\{t \in [0, \infty) : X^*(t) \notin \overline{\mathcal{N} \cap \mathcal{D}}\} > 0.$$

Let us introduce the short-hand notation

$$\begin{aligned}\Delta^{\Pi^*}\phi(t) := \phi\Big(&B(t-) - \Delta C^*(t) - (1+\lambda)\Delta L^*(t) + (1+\mu)\Delta M^*(t),\\ &S(t-) + S(t-)\Delta\eta(t) + \Delta L^*(t) - \Delta M^*(t),\ Y(t-) + \beta\Delta C^*(t)\Big)\\ &- \phi\Big(B(t-),\ S(t-) + S(t-)\Delta\eta(t),\ Y(t-)\Big),\end{aligned}$$

where $\Delta\eta(t)$ is a (bad!) notation for the jump-contribution due to the Lévy process. Note that by the dynamic programming principle (6), we can without loss of generality assume that $\Delta C^*(\tau) = 0$. Let $L^{*,c}$, $M^{*,c}$, and $C^{*,c}$ denote the continuous parts of L^*, M^*, and C^*, respectively.

Using Itô's formula for semimartingales together with the inequalities stated above (see 9), we get

$$\begin{aligned}V(x) &= \mathbb{E}\Big[\int_0^\tau e^{-\delta t}U(Y^*(t))\,dt + e^{-\delta\tau}V(X^*(\tau))\Big]\\ &\leq \mathbb{E}\Big[\int_0^\tau e^{-\delta t}U(Y^*(t))\,dt + e^{-\delta}\phi(X^*(\tau))\Big]\\ &\leq \mathbb{E}\Big[\phi(x) + \int_0^\tau e^{-\delta t}\Big(U(Y^*(t)) - \delta\phi(X^*(t)) + \mathcal{A}\phi(X^*(t))\Big)\,dt\Big]\\ &\quad + \mathbb{E}\Big[\int_0^\tau e^{-\delta t}\Big(-\phi_b + \beta\phi_y\Big)\,dC^{*,c}(t)\Big]\\ &\quad + \mathbb{E}\Big[\int_0^\tau e^{-\delta t}\Big(-(1+\lambda)\phi_b + \phi_s\Big)\,dL^{*,c}(t)\Big]\\ &\quad + \mathbb{E}\Big[\int_0^\tau e^{-\delta t}\Big((1-\mu)\phi_b - \phi_s\Big)\,dM^{*,c}(t)\Big]\\ &\quad + \mathbb{E}\Big[\sum_{t\in[0,1]\cap[0,\tau]} e^{-\delta t}\Delta^{\Pi^*}\phi(t)\Big] \leq \phi(x) - \varepsilon\mathbb{E}[e^{-\delta\tau}] < \phi(x),\end{aligned}$$

which is a contradiction since $(V - \phi)(x) = 0$. □

Lem. 5.3 below, which was used in the proof of Thm. 5.2, is similar to Lem. 3.5 in Davis, Panas, and Zariphopoulou [8].

Lemma 5.3. *Let $A = A(\omega)$ denote the event that the optimal trajectory $X^*(t)$ starting at $x = (b, s, y)$ has an initial control-jump of size $(\varepsilon_L, \varepsilon_M, \varepsilon_C) > 0$. Suppose that the inequalities in* (9) *hold. Then $P(A) = 0$.*

Proof. Notice that the state (after the control-jump) is

$$\hat{x}(\varepsilon_L, \varepsilon_M, \varepsilon_C) := \Big(b - (1+\lambda)\varepsilon_L + (1-\mu)\varepsilon_M - \varepsilon_C,\ s + \varepsilon_L - \varepsilon_M,\ y + \beta\varepsilon_C\Big).$$

By the dynamic programming principle (6), we can without loss of generality assume that $\hat{x}(\varepsilon_L, \varepsilon_M, \varepsilon_C) \in \overline{\mathcal{N} \cap \mathcal{D}}$. Again by (6), we have

$$V(x) = \mathbb{E}\big[V\big(\hat{x}(\varepsilon_L, \varepsilon_M, \varepsilon_C)\big)\big] = \int_{A(\omega)} V\big(\hat{x}(\varepsilon_L, \varepsilon_M, \varepsilon_C)\big)\, dP + \int_{\Omega - A(\omega)} V(x)\, dP.$$

From this equality it follows that

$$\int_{A(\omega)} \Big(V\big(\hat{x}(\varepsilon_L, \varepsilon_M, \varepsilon_C)\big) - V(x)\Big)\, dP = 0,$$

and, since $V \le \phi$ and $(V - \phi)(x) = 0$ (recall that ϕ comes from (8)),

$$\int_{A(\omega)} \Big(\phi\big(\hat{x}(\varepsilon_L, \varepsilon_M, \varepsilon_C)\big) - \phi(x)\Big)\, dP \ge 0. \tag{10}$$

From (9), we get

$$\phi\big(\hat{x}(\varepsilon_L, \varepsilon_M, \varepsilon_C)\big) \le \phi\big(\hat{x}(\varepsilon, 0, 0)\big), \qquad \forall \varepsilon \le \varepsilon_L, \tag{11}$$

$$\phi\big(\hat{x}(\varepsilon_L, \varepsilon_M, \varepsilon_C)\big) \le \phi\big(\hat{x}(0, \varepsilon, 0)\big), \qquad \forall \varepsilon \le \varepsilon_M, \tag{12}$$

$$\phi\big(\hat{x}(\varepsilon_L, \varepsilon_M, \varepsilon_C)\big) \le \phi\big(\hat{x}(0, 0, \varepsilon)\big), \qquad \forall \varepsilon \le \varepsilon_C. \tag{13}$$

Suppose $\varepsilon_L > 0$. We then claim that

$$\big[-(1+\lambda)\phi_b(x) + \phi_s(x)\big]P(A) \ge 0. \tag{14}$$

From (10) and (11) it follows that

$$\int_{A(\omega)} \Big(\phi\big(\hat{x}(\varepsilon, 0, 0)\big) - \phi(x)\Big)\, dP \ge 0, \qquad \forall \varepsilon \le \varepsilon_L,$$

and therfore by Fatou's lemma

$$\int_{A(\omega)} \limsup_{\varepsilon \to 0} \Big[\frac{\phi\big(b - (1+\lambda)\varepsilon,\ s + \varepsilon,\ y\big) - \phi(b, s, y)}{\varepsilon}\Big]\, dP \ge 0.$$

Hence, (14) follows. Similarly, if $\varepsilon_M > 0$, we can use (10) and (12) to prove

$$\big[(1-\mu)\phi_b(x) - \phi_s(x)\big]P(A) \ge 0. \tag{15}$$

Finally, if $\varepsilon_C > 0$, we can use (10) and (13) to prove

$$\big[-\phi_b(x) + \beta\phi_y(x)\big]P(A) \ge 0. \tag{16}$$

Summing up, if at least one of the jump-sizes $\varepsilon_L, \varepsilon_M, \varepsilon_C$ is greater than zero, then we can conclude (from (9), (14), (15), (16)) that $P(A) = 0$. □

References

[1] F. E. Benth, K. H. Karlsen, and K. Reikvam, *Optimal portfolio selection with consumption and nonlinear integro-differential equations with gradient constraint: A viscosity solution approach,* Preprint, MaPhySto Research Report No 21, University of Aarhus, Denmark. (1999). To appear in *Finance & Stochastics.*

[2] F. E. Benth, K. H. Karlsen, and K. Reikvam, *Optimal portfolio management rules in a non-Gaussian market with durability and intertemporal substitution,* Preprint No 12, University of Oslo, Norway. (2000). To appear in *Finance & Stochastics.*

[3] F. E. Benth, K. H. Karlsen, and K. Reikvam, *A note on portfolio management under non-Gaussian logreturns,* Preprint, MaPhySto Research Report No 5, University of Aarhus, Denmark. (2000). To appear in *Intern. J. Theor. Appl. Finance.*

[4] F. E. Benth, K. H. Karlsen, and K. Reikvam, *Portfolio optimization in a Lévy market with intertemporal substitution and transaction costs,* Preprint, MaPhySto Research Report No 15, University of Aarhus, Denmark, (2000).

[5] F. E. Benth, K. H. Karlsen, and K. Reikvam, *Finite horizon portfolio optimization in a Lévy market with intertemporal substitution and transaction costs,* manuscript in preparation (2000).

[6] M. G. Crandall, H. Ishii, and P.-L. Lions, User's guide to viscosity solutions of second order partial differential equations, *Bull. Amer. Math. Soc. (N.S.)* **27** (1992), no. 1, 1–67.

[7] J. Cvitanić and I. Karatzas, *Hedging and portfolio optimization under transaction costs: A martingale approach,* Math. Finance, **6**(2) (1996), 133-165.

[8] M. H. A. Davis, V. G. Panas, and T. Zariphopoulou, European option pricing with transaction costs, *SIAM J. Control Optim.* **31** (1993), no. 2, 470–493.

[9] A. Hindy and C. Huang, *Optimal consumption and portfolio rules with durability and local substitution,* Econometrica, **61** (1993), 85-121.

[10] I. Ekeland and R. Temam, Convex analysis and variational problems, North-Holland Publishing Co., Amsterdam (1976).

[11] I. Karatzas and S. Shreve, *Connections between optimal stopping and singular stochastic control I. Monotone follower problems,* SIAM J. Control Optim., **22**(6) (1984), 856-877.

(Benth and Reikvam)
Department of Mathematics, University of Oslo
PO Box 1053 Blindern, N-0316 Oslo, Norway
E-mail address: `fredb@math.uio.no` and `kre@math.uio.no`

(Karlsen)
Department of Mathematics, University of Bergen
Johs. Brunsgt. 12, N-5008 Bergen, Norway
E-mail address: `kennethk@mi.uib.no`

Trends in Mathematics, © 2001 Birkhäuser Verlag Basel/Switzerland

A Quadratic Approach To Interest Rates Models In Incomplete Markets

Francesca Biagini

Abstract. The aim of this paper is to apply the mean-variance hedging approach, originally formulated for risky assets, to interest rate models in presence of stochastic volatility.
In a HJM framework, we set a finite number of bonds such the volatility matrix is invertible and provide an explicit formula for the density of the variance-optimal measure which is independent by the chosen times of maturity.
Finally, we compare the mean-variance hedging approach to the local risk minimization one in the interest rate case.

1. Introduction

The aim of this paper is to extend the mean-variance hedging approach to interest rate models in presence of stochastic volatility. The interest rate case is analysed in a Heath-Jarrow-Morton framework, where the forward rate volatility is supposed to be stochastic. Here a stochastic volatility model is seen as a model with *incomplete information*, where volatility is affected by an additional source of randomness. A perfect replication of a given european option H is not possible even by using an infinite number of bonds. In order to find an approximation price and strategy, we choose the mean-variance hedging approach and consider only self-financing portfolios composed by a finite number of bonds as in the approach of [6] .
We set $T_1 < T_2 < \cdots < T_n$ times of maturity, greater than the option time of expiration T_0, such that the matrix $\int_{T_0}^{T_j} \sigma_i(t,s)ds$ is invertible P^E-almost everywhere for every t. We characterize the set of the martingale measures for $\frac{p(t,T_j)}{p(t,T_0)}$, $t \leq T_0, j = 1, \dots, n$ and compute an explicit formula for the density of the variance-optimal measure for $\frac{p(t,T_j)}{p(t,T_0)}$, $j = 1, \dots, n$, in terms of Doleans Exponential. This expression is shown to be independent of the chosen T_j.
Finally, we introduce the local risk minimization approach for interest rates and compare it with the mean-variance hedging one.

2. The Model

In the sequel, all filtrations are supposed to satisfy the so-called *"usual hypothesis"*.

Our basic model consists of two complete filtered probability spaces denoted by $(\Omega, \mathcal{F}^W, \mathcal{F}_t^W, P^W)$ and by $(E, \mathcal{E}, \mathcal{E}_t, P^E)$. We assume that W_t is a standard n-dimensional brownian motion on $\Omega = \mathcal{C}([0,T], \mathbb{R})$, P^W is the Wiener measure and $\mathcal{F}_t^W$ is the P^W-augmentation of the filtration generated by W_t. The space E represents an additional source of randomness which affects the market. The market is now incomplete as a result of *incomplete information*: if the evolution of η had been known the market would be complete.

We suppose that there exists on E a square integrable (eventually d-dimensional) martingale M_t endowed with the *predictable representation property*, i.e. for every square integrable martingale N_t there exists a predictable process H_t such that $N_t = N_0 + \int_0^t H_s dM_s$.

We analyse the mean-variance hedging criterion in the case of interest rates models. The assets to be considered on the market are zero coupon bonds with different maturities. As in [4], we represent the price at time t of a bond maturing at time T by an optional stochastic process $p(t,T)$ such that $p(t,t) = 1$ for all t.

We assume that there exists a frictionless market for T-bonds for every $T > 0$ and that for every fixed t, $p(t,T)$ is almost surely differentiable in the T-variable. The *forward rate* $f(t,T)$ is defined as $f(t,T) = -\dfrac{\partial \log p(t,T)}{\partial T}$ and the *short rate* as $r_t = f(t,t)$.

According to the Heath-Jarrow-Morton approach, we describe the forward rate dynamics. In this setting, $f(t,T)$ is represented by a process on the product probability space $(\Omega \times E, \mathcal{F}_t^W \otimes \mathcal{E}_t, P^W \otimes P^E)$ such that

$$df(t,T,\omega,\eta) = \alpha(t,T,\omega,\eta)dt + \sigma(t,T,\omega,\eta)dW_t(\omega) \tag{1}$$

with initial condition $f(0,T,\eta) = f^*(0,T)$. We make the following assumptions:

i) The equation (1) admits P^E-a.e. a unique strong solution with respect to the filtration $\mathcal{F}_t^W$. For example, it is sufficient that α and σ are P^E-a.e.-bounded.

ii) *Heath-Jarrow-Morton condition* on the drift: there exists a predictable $\mathbb{R}^n$-valued process h_t such that the integral $\int h_s dW_s$ is well defined and

$$\alpha(t,T,\omega,\eta) = \sigma(t,T,\omega,\eta) \int_t^T \sigma(t,s,\eta)ds - \sigma(t,T,\omega,\eta)h_t(\omega,\eta) \tag{HJM}$$

for every $T \geq 0$. For the sake of simplicity, in the sequel we will omit ω in the notation.

In the complete market case, this condition guarantees the existence of the unique equivalent martingale measure for $\dfrac{p(t,T)}{B_t}$ as long as $\mathcal{E}\left(\int hdW\right)$

is a uniformly integrable martingale, while in this setting of incomplete information there exists an infinite number of them. Note that it compels to impose stronger regularity on σ to obtain global solutions for equation (2). For a further discussion on the integrability conditions to impose on h_t, see [1].

By Proposition 15.5 of [4], we obtain the bond price dynamics:

$$\frac{dp(t,T)}{p(t,T)} = (r(t,\eta) + \frac{1}{2}\|S(t,T,\eta)\|^2 + A(t,T,\eta))dt + S(t,T,\eta)dW_t$$

where

1. $S(t,T,\eta) = -\int_t^T \sigma(t,s,\eta)ds$
2. $A(t,T,\eta) = -\int_t^T \alpha(t,s,\eta)ds$

Neverthless in principle an infinite number of bonds is available for trade, we consider only portfolios composed by to an arbitrarily large, but finite number of bonds as in the approach of [6]. Since are traded bonds for every time of maturity $T \in \mathbb{R}^+$, one is induced to think that the market is complete in spite of lack of information. Unfortunately, this is not true. For example, suppose in equation (1) dim $W_t = 1$ and let the volatility have a jump at a random time. The market is incomplete since the random time of jump can not be known neither through the observation of the entire term structure. For a further discussion, we refer to Example 2.1 of [1].

3. The Variance-Optimal Measure for Interest Rates

In this framework, we study the problem of hedging a certain European option H expiring at time T_0 by using a self-financing portfolio composed by a finite number of bonds of convenient maturities and eventually by the money market account B_t. In the sequel we assume to work with the filtration $(\mathcal{F}_t)_{t\in[0,T_0]}$; for the sake of simplicity we will write $\frac{dQ}{dP}$ instead of $\frac{dQ}{dP}|_{\mathcal{F}_{T_0}}$.
Since a perfect replication is not possible, we look for a self-financing portfolio which solves the following minimization problem:

$$\min E\left[(H - V_{T_0})^2\right] \tag{2}$$

Usually the money market account $B_t = \exp\left(\int_0^t r(s,\eta)ds\right)$ is used as discounting factor. Since the spot rate is now stochastic, the minimization problem (2) is equivalent to

$$\min E^B\left[(\frac{H}{B_{T_0}} - \frac{V_{T_0}}{B_{T_0}})^2\right]$$

where E^B is the expectation under the equivalent probability P^B with density

$$\frac{dP^B}{dP} = \frac{B_{T_0}^2}{E\left[B_{T_0}^2\right]}$$

The computation of the new bond dynamics under P^B can be quite complicated even in very simple cases, as shown in further details in Remark 3.9 of [1]. In order to avoid it, we can choose as numéraire the bond $p(t, T_0)$ expiring at time T_0 of maturity of H. We immediately have

$$\frac{dP^{T_0}}{dP} = \frac{p(T_0, T_0)^2}{E\left[p(T_0, T_0)^2\right]} = 1$$

or in other words $P^{T_0} \equiv P$.

More precisely, we are not simply interested in a self-financing portfolio whose final value has minimal quadratic distance by H, but, once fixed $(n+1)$ bonds $p(t, T_j)$, $j = 0, 1, \ldots, n$, where n is the dimension of W_t, we look for a solution to the minimization problem:

$$\min_{\substack{V_0 \in \mathbb{R} \\ \theta \in \Theta}} E\left[\left(H - V_0 - G_{T_0}(\theta)\right)^2\right] \tag{3}$$

where $G_t(\theta) = \int_0^t \theta_s dX_s$, $X_s^j = \dfrac{p(s, T_j)}{p(s, T_0)}$ and

$$\Theta = \left\{\theta \in L(X) : \int \theta dX \in \mathcal{S}^2\right\} \tag{4}$$

$L(X)$ is the set of integrable processes with respect to X_t and $\mathcal{S}^2$ is the space of square-integrable semimartingale.

We assume a sort of no-arbitrage condition on the underlying financial market:

$$\textbf{no-approximate profit condition}: \; 1 \notin \overline{G_{T_0}(\Theta)} \tag{5}$$

This conditions simply means that the riskless profit 1 can't be approximate by using self-financing portfolios with zero initial wealth.

Problem (3) admits a unique solution (V_0, θ) for all $H \in L^2$ under the hypothesis that $G_{T_0}(\Theta)$ is closed (see [9] for the proof). In this case, θ is called the *mean-variance optimal* strategy and V_0 the *approximation price*. The drawback of the nonclosedness of the space $G_{T_0}(\Theta)$ can be overcome by looking for a mean-variance optimal strategy in the space Θ_{GLP} of all predictable processes such that the stochastic integral $\int_0^t \theta_s dX_s$ is a Q-square-integrable martingale for every equivalent square integrable martingale measure Q (see [9]).

Problem (3) is strictly related to a particular martingale measure for X_t, since the approximation price and the mean-variance optimal strategy θ can be computed in terms of $\tilde{P}$, the *variance-optimal measure*. We denote as $\mathcal{M}_s^2(T_1, \ldots, T_n)$ and $\mathcal{M}_e^2(T_1, \ldots, T_n)$ respectively the set of *signed martingale measures* and the set of *equivalent martingale measures* for $\dfrac{p(t, T_j)}{p(t, T_0)}$, $j = 1, \ldots, n$.

The *variance-optimal* measure $\widetilde{P}$ is the element of $\mathcal{M}_s^2(T_1,\dots,T_n)$ of minimal norm, where for every $Q \in \mathcal{M}_s^2(T_1,\dots,T_n)$

$$\|\frac{dQ}{dP}\|^2 = E\left[(\frac{dQ}{dP})^2\right]$$

If (2) has solution, in [9] it is shown that $\tilde{V}_0 = \tilde{E}[H]$. Moreover, if $G_{T_0}(\Theta)$ is closed and there exists at least a martingale measure for X_t, the optimal strategy θ can be computed by using the density of $\widetilde{P}$, as shown in [9].
Apparently, this definition of $\widetilde{P}$ depends on the chosen maturities $T_1,\dots,T_n$. By imposing the following condition, we will show in the sequel that it is actually invariant under a change of the times of maturity.

There exist maturities $T_1,\dots,T_n$ greater than T_0 such that for every t

$$\text{the matrixes } \sigma_i(t,T_j) \text{ and } \int_{T_0}^{T_j} \sigma_i(t,s)ds \text{ are non-singular } P^E\text{-a.e} \quad \text{(H1)}$$

This assumption is motivated by Proposition 4.3 of [3] and by Proposition 5.5 and Theorem 5.6 by [5]. For a further discussion, see [1].
In order to obtain an explicit formula for the variance-optimal measure, we characterize the set of the martingale measure for $\dfrac{p(t,T)}{p(t,T_0)}$ for every $T > 0$. Note that we don't need $T \le T_0$ since time t cannot exceed T_0 by assumption.

Lemma 3.1. *Let Z_t be a local martingale with $Z_0 = 1$. The following conditions are equivalent:*

1. $Z_t \dfrac{p(t,T)}{p(t,T_0)}$ *is a local martingale for every $T > 0$*
2. $Z_t = \mathcal{E}\left(-\int_0^{\cdot}(h_s + S(s,T_0,\eta))dW_s\right)_t (1 + \int_0^t k_s dM_s)$ *for some predictable process k_s such that the integral $\int_0^t k_s dM_s$ is a local martingale.*

Proof. For the proof, see Lemma 3.4 of [1]. □

Lemma 3.1 shows that our condition on the drift guarantees the existence of an absolutely continuous (eventually signed) martingale measure for $\dfrac{p(t,T)}{p(t,T_0)}$ for every $T \ge 0$, $t \le T_0$.
Since we assume to invest in an arbitrary, but finite number of bonds, we choose for our portfolio $p(t,T_1),\dots,p(t,T_n)$ where $T_0 < T_1 < \dots < T_n$ are maturities such that $\int_{T_0}^{T_j} \sigma_i(t,s)ds$ is invertible for P^E-almost every η. By the following lemma, we obtain that the set of martingale measures for $\dfrac{p(t,T_j)}{p(t,T_0)}$, $j = 1,\dots,n$, coincides with the set of martingales measures for $\dfrac{p(t,T)}{p(t,T_0)}$, $T \ge 0$.

Lemma 3.2. *Let Z_t be a local martingale with $Z_0 = 1$. The following conditions are equivalent:*

1. $Z_t \frac{p(t,T_j)}{p(t,T_0)}$ *is a local martingale for every* $j = 1, \dots, n$
2. $Z_t = \mathcal{E}\left(-\int_0^\cdot (h_s + S(s,T_0,\eta))dW_s\right)_t (1 + \int_0^t k_s dM_s)$ *for some predictable process* k_s *such that the integral* $\int_0^t k_s dM_s$ *is a local martingale.*

We remark that this result is independent from the chosen maturities unless for the fact that $\int_{T_0}^{T_j} \sigma_i(t,s)ds$ must be invertible.

Proposition 3.3. 1. *If $Q \in \mathcal{M}_s^2(T_1, \dots, T_n)$, then*

$$\frac{dQ}{dP} = \mathcal{E}\left(-\int_0^\cdot (h_s(\eta) + S(s,T_0,\eta))dW_s\right)_{T_0} (1 + \int_0^{T_0} k_s dM_s)$$

for some predictable process k_t such that the above expression is square integrable.

2. *If $Q \in \mathcal{M}_e^2(T_1, \dots, T_n)$, then*

$$\frac{dQ}{dP} = \mathcal{E}\left(-\int_0^\cdot (h_s(\eta) + S(s,T_0,\eta))dW_s\right)_{T_0} \mathcal{E}\left(\int_0^\cdot k_s dM_s\right)_{T_0}$$

for some predictable process k_t such that the Doleans Expontial

$$\mathcal{E}\left(-\int_0^\cdot (h_s(\eta) + S(s,T_0,\eta))dW_s + \int_0^\cdot k_s dM_s\right)_t$$

is a square-integrable martingale and $k_t \cdot \Delta M_t > -1$.

Proof. This proposition directly follows by Lemma 3.2. □

The following Lemma is quite technical, but it allows us to write an explicit expression for the density of the variance-optimal measure.

Lemma 3.4. *Let H, K be two predictable stochastic processes whose stochastic integrals $\int_0^t H_s dW_s^*$ and $\int_0^t K_s dM_s$ are defined. The following conditions are equivalent:*

$$\exp\left(\int_0^T \|(h_s(\eta) + S(s,T_0,\eta))\|^2 ds\right) = c\,\frac{\mathcal{E}\left(\int_0^\cdot H_s dW_s^*\right)_T}{\mathcal{E}\left(\int_0^\cdot K_s dM_s\right)_T} \tag{6}$$

$$\mathcal{E}\left(-\int_0^\cdot (h_s(\eta) + S(s,T_0,\eta))dW_s + \int_0^\cdot K_s dM_s\right)_T =$$

$$= c\,\mathcal{E}\left(\int_0^\cdot (-h_s(\eta) - S(s,T_0,\eta) + H_s)d\widehat{W}_s\right)_T \tag{7}$$

where c is the same constant in both equations.

Proof. For the proof, see Lemma 3.7 of [1]. □

We recall that $X_t^j = \dfrac{p(t,T_j)}{p(t,T_0)}$ and denote by A_t the matrix whose ji-th element is given by $[A_t]_{ji} = \int_{T_0}^{T_j} \sigma_i(t,s)ds$. By exploiting Lemma 3.4, we obtain the following explicit formula for the variance-optimal measure.

Theorem 3.5. *Let H, K be two predictable processes such that the exponential martingale $\mathcal{E}\left(\int_0^\cdot H_s dW_s + \int_0^\cdot K_s dM_s\right)$ is square-integrable. Then H, K are solutions of the equation (7) of Lemma 3.4 if and only if*

$$\frac{d\widetilde{P}}{dP} = \mathcal{E}\left(-\int_0^\cdot (h_s(\eta) + S(s,T_0,\eta))dW_s + \int_0^\cdot K_s dM_s\right)_{T_0}$$

or equivalently

$$\frac{d\widetilde{P}}{dP} = \frac{\mathcal{E}\left(-\int_0^\cdot \beta_s dX_s\right)_{T_0}}{E\left[\mathcal{E}\left(-\int_0^\cdot \beta_s dX_s\right)_{T_0}\right]}$$

where $\beta_s^j = \dfrac{p(s,T_0)}{p(s,T_j)} \sum_i (h_s^i(\eta) + S^i(s,T_0,\eta)) - H_s^i)[A_s^{-1}]_{ij}$.

In particular, if $\sigma(t,T,\eta,\omega) = \sigma(t,T,\eta)$, by [1] we obtain that the density of $\widetilde{P}$ has the form

$$\frac{d\widetilde{P}}{dP} = \mathcal{E}\left(-\int_0^\cdot \lambda_s dW_s\right)_{T_0} \frac{\exp\left(-\int_0^T \|\lambda_s\|^2 ds\right)}{E\left[\exp\left(-\int_0^T \|\lambda_s\|^2 ds\right)\right]} \tag{8}$$

where $\lambda_t = h_t(\eta) + S(t,T_0,\eta)$.

We stress that the characterization of $\widetilde{P}$ provided by Theorem 3.5 is independent of the chosen maturities $T_1, \ldots, T_n$ unless for the fact that matrix A_t must be invertible.

4. Examples

The Heath-Jarrow-Morton condition on the drift allows us to modelize only the forward rate volatility $\sigma(t,T,\eta)$.

Example 4.1. *First we consider the case when* $\dim W_t = 1$ *and*

$$\sigma(t,T) = \sigma_0 I_{\{t<\eta, t\le T\}} + \sigma_1 I_{\{t\ge\eta, t\le T\}}$$

where $\sigma_0, \sigma_1 \in \mathbb{R}^+$ and η is a stopping time with a diffuse law on $\mathbb{R}^+$. Here we set $E = \mathbb{R}^+$, $\mathcal{E}_t = \mathcal{B}([0,t]) \vee (t,+\infty]$ and a fundamental martingale is given by $M_t = I_{\{t\ge\eta\}} - a_t$, where a_t is the compensator of the process $I_{\{t\ge\eta\}}$ associated to η.

Example 4.2. *More generically, the volatility can be given by a Markov process in continuous time with a finite set of states I. By following the approach of* [8], *we choose E as the space of all right-continuous, left-limited functions from* $[0,\infty)$ *to I endowed with the filtration $\mathcal{E}$ generated by η_t. By Theorem IV.20.6 of* [8], *we obtain a set of martingales on E with the predictable representation property in the following way. Let a, b be states in I such that $a \neq b$ and define $M_t^b = I_b(\eta_t) - I_b(\eta_0) - \int_0^t QI_b(\eta_s)ds$ and $H_t^a = I_a(\eta_{t-})$. The process $U_{ab}(t) = \int_0^t H_s^a dM_s^b$ is a martingale by Lemma IV.21.12 of* [8] *and the family $(U_{ab})_{a,b\in I, a\neq b}$ has the predictable representation property.*

Example 4.3. *If the volatility is given by a multivariate point process η_t, there exist no finite set of martingales with the predictable representation property. By* [7], *we obtain that the compensated integer-valued random measure $\mu - \nu$ associated to η_t has the predictable representation property on E endowed with the smallest filtration under which μ is optional.*

Example 4.4. *Finally η can be given by a diffusion process*

$$\begin{aligned} df(t,T) &= \alpha(t,T,\eta_t)dt + \sigma(t,T,\eta_t)dW_t^1 \\ d\eta_t &= F(t,T,\eta_t)dt + G(t,T,\eta_t)dW_t^2 \end{aligned}$$

where W_t^1 can be eventually correlated with W_t^2.

5. A comparison with the local risk minimizing approach

An alternative approach for pricing and hedging contingent claims in incomplete markets is the local risk minimization one. The main difference with respect to mean-variance hedging is the fact that a local risk minimizing strategy perfectly replicates the value of a given option, but it is not self-financing. More precisely, suppose we want to hedge a T_0-option H. As in the previous sections, we choose $T_1 < \cdots < T_n$ satisfying (H1) holds and consider $X_t^j = \dfrac{p(t,T_j)}{p(t,T_0)}$, $j = 1,\ldots,n$. By exploiting the approach of [2], we have the following

Definition 5.1. *An L^2-*strategy *is a pair (θ,θ^0) such that $\theta \in \Theta$ and θ^0 is a real predictable process such that the* value process *left limit $V_{t-} = \theta_t \cdot X_t + \theta_t^0$ is square integrable for $0 \le t \le T_0$.*
The (cumulative) cost process *is defined by $C_t = V_t - \int_0^t \theta_s dX_s$, $0 \le t \le T_0$.*

By Definition 5.1, we get that the portfolio's jumps coincide with the jumps in the cost process.

Definition 5.2. *Let $H \in L^2(\mathcal{F}_{T_0}, P)$ be a contingent claim. An L^2-strategy (θ,θ^0) with $V_{T_0} = H$ $P-a.s.$ is called* pseudo-locally risk-minimizing *or* pseudo-optimal *for H if the cost process C_t is a P-martingale and is strongly orthogonal to the martingale part of X.*

By Definition 5.2 follows immediately that a contingent claim $H \in L^2(\mathcal{F}_{T_0}, P)$ admits a pseudo-optimal strategy if and only if H can be written as

$$H = H_0 + \int_0^{T_0} \xi_u dX_u + L_{T_0} \tag{9}$$

where $H_0 \in L^2(\mathcal{F}_{T_0}, P)$, $\xi \in \Theta$ and L is a square integrable martingale strongly P-orthogonal to the martingale part of X. Equation (9) is usually addressed in literature as the *Föllmer-Schweizer decomposition* of H. This is connected to a suitably chosen martingale measure, the so-called *minimal martingale measure.*

Definition 5.3. *$\widehat{P}^0 \in \mathcal{M}_e^2(T_1, \ldots, T_n)$ is the* minimal measure *(with respect to $p(t, T_0)$ as numéraire) if any locally square integrable local martingale which is orthogonal to the martingale part of X under P remains a local martingale under $\widehat{P}^0$.*

By Definition 5.3 follows immediately that the pseudo-optimal portfolio $\widehat{V}(\phi)$ is a local $\widehat{P}$-martingale and we get $\widehat{V}_t(\phi) = p(t, T_0)\widehat{E}^0 \left[H | \mathcal{F}_t\right]$. By Definition 5.3 and Theorem 3.5, we obtain that

$$\frac{d\widehat{P}^0}{dP} = \mathcal{E}\left(-\int_0^{\cdot} (h_s(\eta) + S(s, T_0, \eta))dW_s\right)_{T_0}$$

define the minimal measure's density as long as the Doleans Exponential $\mathcal{E}\left(-\int_0^{\cdot}(h_s(\eta) + S(s, T_0, \eta))dW_s\right)$ is a uniformly integrable martingale.
We compute now the pseudo-optimal strategy for a T_0-call option $C = (p(T_0, T_1) - K)^+$. The pseudo-optimal portfolio is given by $\widehat{V}_t(\phi) = p(t, T_0)\widehat{E}^0 \left[C | \mathcal{F}_t\right]$ and by exploiting the same argument as in Theorem 5.1 by [2], we obtain that the optimal strategy components are $\theta_t^0 = -K\widehat{E}^0 \left[1_A | \mathcal{F}_{t-}\right]$, $\theta_t^1 = \widehat{E}^1 \left[1_A | \mathcal{F}_{t-}\right]$ and $\theta_t^j = 0$ for all $j = 2, \ldots, n$. Note that in the local risk minimization case, the pseudo-optimal strategy depends only on two assets in spite of the dimension of the driving brownian motion. On the contrary, in [1] is shown that the mean-variance optimal strategy is based on $(n+1)$ bonds, where $n = \dim W_t$.
We apply these results in order to compute the local risk minimizing strategy for a caplet $H = \dfrac{R^*}{p(T_0, T_1)}(\dfrac{1}{R^*} - p(T_0, T_1))^+ = \dfrac{R^*}{p(T_0, T_1)}K$, where K is a T_0-put option on $p(t, T_1)$. The pseudo-optimal portfolio for the caplet H is given by $\widehat{V}_t = p(t, T_1)\widehat{E}^1 \left[H | \mathcal{F}_t\right]$. For $t \leq T_0$, we have $\widehat{V}_t = p(t, T_1)\widehat{E}^1 \left[\dfrac{R^*}{p(T_0, T_1)}K \middle| \mathcal{F}_t\right] = p(t, T_0)\widehat{E}^0 \left[K | \mathcal{F}_t\right]$, since by [2] we have the following change of measure's formula $\dfrac{d\widehat{P}^1}{d\widehat{P}^0} = p(T_0, T_1)\dfrac{p(0, T_0)}{p(0, T_1)}$. For $t > T_0$, $\widehat{V}_t = \widehat{E}\left[H | \mathcal{F}_t\right] = H$ since H is

$\mathcal{F}_{T_0}$-measurable. Hence, the local risk-minimization strategies for the T_1-option H and for the T_0-option K coincide up to time T_0 and we can behave exactly as in the complete market case. On the contrary, in [1] is shown that mean-variance hedging strategy for H does not coincide with the one for K. The key is that in this approach we perfectly replicate the option value in spite of approximating it as in the mean-variance hedging criterium.

References

[1] Biagini F. (2000) " Mean-variance hedging for Interest Rates Models with Stochastic Volatility", preprint

[2] Biagini F., Pratelli M. (1999) " Local Risk Minimization and Numéraire ", Journal of Applied Probability, vol.36, number 4, 1-14.

[3] Björk T.,(1997) "Interest Rate Theory" Proceedings CIME Conference, Bressanone 1996, Springer

[4] Björk T.,(1998) "Arbitrage Theory in Continuous Time" Oxford University Press

[5] Björk T., Kabanov Y., Runggaldier W.(1997) "Bond market structure in the presence of marked point processes" Mathematical Finance **7**, n.2, 211-239

[6] Musiela M., Rutkowski M.(1997) "Martingale Methods in Financial Modelling", Springer

[7] Jacod J., Shiryaev N.A.,(1987) Limit Theorems, Springer Verlag,

[8] Rogers, Williams, (1987) "Diffusions, Markov Processes and Martingales", 2, Willey Series in Probability

[9] Schweizer M. (1999) "A Guided Tour through Quadratic Hedging Approaches" Preprint.

Dipartimento di Matematica,, Università di Bologna,, P.zza Porta S. Donato,, 40127 Bologna, Italy
E-mail address: `biagini@dm.unibo.it`

Trends in Mathematics, © 2001 Birkhäuser Verlag Basel/Switzerland

Risk Sensitive Asset Management: Two Empirical Examples

T.Bielecki[1], A.Harris[2], J.Li[3], S.Pliska[4]

Abstract. This paper is concerned with applications of risk sensitive control theory in financial decision making. In earlier work a variation of Merton's continuous-time intertemporal capital asset pricing model was developed where the infinite horizon objective is to maximize the portfolio's risk adjusted growth rate. Here the model is illustrated by applying it to two portfolio management problems based upon historical data. In the first there are four assets (a bank account, the Dow Jones Industrials index, the S&P 500 index, and the NASDAQ index) and two stochastic economic factors (the short interest rate for the bank account and a long term interest rate). In the second there are six stochastic economic factors, namely, U.S. Treasury yields of various maturities; in addition to the usual bank account the other assets are rolling horizon bonds corresponding to the factors. These examples demonstrate that the risk sensitive asset management model is tractable and provides economic insight as well as useful results, although the optimal strategies sometimes involve high levels of leverage.

1. Introduction

The famous intertemporal capital asset pricing model (ICAPM) developed more than 25 years ago by Merton [9] features both assets and the exogenous factors which affect them, all of which are modeled as diffusion processes. For the problem of maximizing expected utility of consumption and/or terminal wealth, he derived and provided economic interpretations of the corresponding Hamilton-Jacobi-Bellman equation. But while this model is widely cited, it is of limited usefulness because, unfortunately, the ICAPM's HJB equation is intractable except for a very few special cases.

In a series of recent papers, Bielecki and Pliska have developed a variation of the ICAPM where explicit, general solutions can be obtained right now. Retaining the same diffusion process model for assets and factors, their approach calls for changing from Merton's finite horizon expected utility objective to the infinite horizon objective of maximizing the risk adjusted growth rate. As explained in

The research of the authors was partially supported by NSF Grants DMS-9971307 and DMS-9971424.

their earlier work, especially in Section 3 of [4], this criterion can be viewed as being analogous to the classical Markowitz single-period approach except that instead of trading off single-period criteria the investor is trading off the long run growth rate (which by itself is maximized by the growth optimal portfolio) versus the average volatility of the portfolio. The risk adjusted growth rate objective emerges naturally from the application of recent mathematical results on risk sensitive control theory. The principal benefit of this objective is that it is an infinite horizon criterion and therefore, as with most control problems in general, the ICAPM is more tractable than if a finite horizon criterion is used.

The initial work by Bielecki and Pliska featured a model with Gaussian factors and with assets having constant volatilities and appreciation rates that are affine functions of the factor levels. The theoretical foundations of this model are found in [1], while its applications to asset allocation problems are explored in [2] and [3]. But while this model fully allows for there to be correlations between asset returns and movements of the factor levels, it has a critical shortcoming: the partial correlations between asset returns and movements of the factor levels must be zero, that is, the residuals of the asset returns must be independent of the residuals of the factors. While this assumption might be reasonable for some important applications, it is unacceptable for models where there is an explicit relationship between assets and factors such as models that include both interest rate factors and fixed income assets or models that have both a stock index and the corresponding dividend yield.

Bielecki and Pliska [4] overcame this shortcoming in subsequent work. They showed that with correlated residuals one can still obtain an explicit formula for the optimal trading strategy, provided there are no special constraints on the trading strategies such as short sales restrictions. A key to their results is recognition of the fact that the solution of the Hamilton-Jacobi-Bellman equation depends, in part, upon the solution of a continuous, algebraic Riccati equation. The basic model and main theoretical results are presented in the following section. Related results can be found in Fleming and Sheu [5] as well as Kuroda and Nagai [8].

The purpose of this paper is to demonstrate the tractability and usefulness of the risk sensitive asset management model by applying it to two portfolio management problems. In both examples the model parameters are estimated from historical data. In the first example, presented in Section 3, there are four assets (a bank account, the Dow Jones Industrials index, the S&P 500 index, and the NASDAQ index) and two stochastic economic factors (the short interest rate for the bank account and a long term U.S. Treasury yield). The second example, presented in Section 4, deals with a problem in the area of fixed income management. There are six stochastic economic factors, namely, U.S. Treasury yields of various maturities. Moreover, in addition to the usual bank account, all the other assets are rolling horizon bonds corresponding to the factors. The concept of rolling horizon bonds was developed by Rutkowski [10]; these can be thought of as mutual funds, each consisting of riskless, zero coupon bonds, all of approximately the same maturity and being frequently rolled over in a self-financing manner.

These two examples demonstrate that the risk sensitive asset management model is tractable and provides helpful economic insight about how optimal investment positions should vary with respect to underlying factors that affect the mean returns of asset prices. Indeed, the results suggest that the model can be useful for practical investment problems, although the optimal strategies sometimes involve high levels of leverage.

2. Model Formulation and Main Results

In this section we formulate and provide the main results for the general optimal dynamic asset management problem that was studied by Bielecki and Pliska [4]. The interested reader is referred to that reference for additional details.

We consider a market consisting of $n \geq 1$ risky securities and $m \geq 1$, economic factors. Denoting by $S_i(t)$ the price of the i-th security and by $X_j(t)$ the level of the j-th factor at time t, we consider the following market model for the dynamics of the security prices and factors:

$$\frac{dS_i(t)}{S_i(t)} = (a + AX(t))_i dt + \sum_{k=1}^{N} \sigma_{ik} dW_k(t), \quad S_i(0) = s_i, \quad i = 1, 2, \cdots, n \tag{1}$$

$$dX(t) = (b + BX(t))dt + \Lambda dW(t), \quad X(0) = x, \tag{2}$$

where $W(t) = (W_1(t), \ldots, W_N)'$ is a R^N valued standard Brownian motion process, $X(t) = (X_1(t), \ldots, X_m)'$ is the R^m valued factor process, the market parameters a, A, $\Sigma := [\sigma_{ij}]$, b, B, $\Lambda := [\Lambda_{ij}]$ are matrices of appropriate dimensions, and $(a + Ax)_i$ denotes the $i - th$ component of the vector $a + Ax$. Also, we denote $(\Omega, \{\mathcal{F}_t\}_{t \geq 0}, \mathcal{F}, \mathbf{P})$ for the underlying probability space, the one corresponding to the underlying Brownian motions. It is well known that a unique, strong solution exists for (1), (2), and that the processes $S_i(t)$ are positive with probability 1 (see e. g. [7], chapter 5).

There is also a bank account, that is, an instantaneously risk-free asset, denoted $S_0(t)$, whose dynamics are given by

$$\frac{dS_0(t)}{S_0(t)} = (a_0 + A_0 X(t))dt, \quad S_0(0) = s_0, \tag{3}$$

where a_0 is some scalar and A_0 is a constant row vector having m components. Thus $a_0 + A_0 X(t)$ represents a short term interest rate at time t. For future purposes it is convenient to define

$$\tilde{A} := A - \mathbf{1}A_0, \quad \tilde{a} := a - \mathbf{1}a_0,$$

where $\mathbf{1}$ is a column vector of appropriate dimension whose every component equals one.

Let $\mathcal{G}_t := \sigma((S(s), X(s)), 0 \leq s \leq t)$, where $S(t) = (S_0(t), S_1(t), \ldots, S_n(t))$ is the security price process. Let $h(t)$ denote an R^{n+1} valued investment process (or strategy) whose components $h_i(t)$, $i = 0, 1, \ldots, n$, represent the time-t proportions of wealth in the corresponding assets. It will be convenient for future purposes to partition $h(t)$ as

$$h(t) = (h_0(t), \tilde{h}(t)).$$

Definition 2.1. An investment process $h(t)$ is said to be **admissible** if the following conditions are satisfied:

(i) $\sum_{i=0}^{n} h_i(t) = 1$,

(ii) $h(t)$ is measurable and $\mathcal{G}_t$-adapted,

(iii) for every $\theta > 0$ there exists a probability measure $\mathbf{P}^{h,\theta}$ on $(\Omega, \mathcal{F})$ such that

$$\frac{d\mathbf{P}^{h(\cdot),\theta}}{dP}\bigg|_{\mathcal{F}_t} = \eta_t(h(\cdot),\theta),$$

where

$$\eta_t(h(\cdot),\theta) = exp\{-\frac{\theta^2}{8}\int_0^t \|\tilde{h}'(s)\Sigma\|^2 ds - \frac{\theta}{2}\int_0^t \tilde{h}'(s)\Sigma dW_s\},$$

(iv) $\limsup_{t\to\infty} t^{-1}\ln \mathbf{E}^{h(\cdot),\theta} exp\{(\frac{\theta}{2})(X_t' K_1(\theta) X_t + K_2(\theta) X_t\} \le 0$,

where the matrix $K_1(\theta)$ and the vector $K_2(\theta)$ are defined below, and where $\mathbf{E}^{h(\cdot),\theta}$ denotes expectation under $\mathbf{P}^{h(\cdot),\theta}$.

The class of admissible investment strategies will be denoted by $\mathcal{H}$. □

Let now $h(t)$ be an admissible investment process. Then there exists a unique, strong, and almost surely positive solution $V(t)$ to the following equation:

$$dV(t) = h_0(t)V(t)[a_0 + A_0 X(t)]dt + \sum_{i=1}^{m} h_i(t)V(t)\big[\mu_i(X(t))dt + \sum_{k=0}^{N} \sigma_{ik} dW_k(t)\big], \tag{4}$$

$$V(0) = v > 0$$

where $\mu_i(x)$ is the i-th coordinate of the vector $a + Ax$ for $x \in R^{m+1}$. The process $V(t)$ represents the investor's capital at time t, and $h_i(t)$ represents the proportion of capital that is invested in security i, so that $h_i(t)V(t)/S_i(t)$ represents the number of shares invested in security i, just as in, for example, Section 3 of [6].

In this paper we consider the following family of risk sensitized optimal investment problems, labeled as $\mathcal{P}_\theta$:

for $\theta \in (0, \infty)$, maximize the risk adjusted expected growth rate

$$J_\theta(v, x; h(\cdot)) := \liminf_{t\to\infty}(-2/\theta)t^{-1} ln\ \mathbf{E}^{h(\cdot)}\ [e^{-(\theta/2) lnV(t)} | V(0) = v, X(0) = x] \tag{5}$$

over the class of all admissible investment processes $h(\cdot)$, subject to (1) and (2), where $\mathbf{E}^{h(\cdot)}$ is the expectation with respect to $\mathbf{P}$. The notation $\mathbf{E}^{h(\cdot)}$ emphasizes that the expectation is evaluated for process $V(t)$ generated by (4) under the investment strategy $h(t)$.

As indicated earlier, the risk adjusted growth rate criterion is discussed more fully in the references by Bielecki and Pliska cited above, especially [4]. Here we shall simply remark that the parameter θ should be interpreted as the investor's risk aversion parameter, and the criterion should be interpreted as the portfolio's long run (geometric) growth rate minus a penalty term which equals $\theta/4$ times the asymptotic variance of the portfolio. For instance, if V is simple geometric Brownian motion with parameters μ and σ, then the risk adjusted growth rate is precisely equal to $\mu - \sigma^2/2 - (\theta/4)\sigma^2$. Hence the bigger the value of θ, the more risk averse the investor.

We make the following standing assumptions with regard to the problem (5):

Assumption (A1) *The spectrum of the matrix B is contained in the left half plane [i.e. the matrix B is stable].*

Remark 2.1. We observe here that Assumption (A1) is not a necessary assumption for the results below to hold. Nevertheless, we make this assumption because it is almost necessary for another assumption that follows later. This assumption means that the factor process is an ergodic Gaussian process. In economic terms this means that the factor process reverts to its ergodic mean.

Assumption (A2) *(a) The matrix $\Sigma\Sigma'$ is positive definite; (b) The matrix $\Lambda\Lambda'$ is positive definite.*

Remark 2.2. It is worth emphasizing at this point that we do not assume independence of random perturbations driving the price and the factor dynamics. That is, we do not assume that $\Sigma\Lambda' = 0$. Such an assumption was made in [1].

Let us consider now the following continuous algebraic Riccati equation (CARE)

$$K'R_1(\theta)K + K'R_2(\theta) + R_2(\theta)'K + R_3(\theta) = 0, \tag{6}$$

where

$$R_1(\theta) := \frac{\theta^2}{2}\Big(\frac{\theta}{2}+1\Big)^{-1}\Lambda\Sigma'\Psi\Sigma\Lambda' - \theta\Lambda\Lambda', \quad R_2(\theta) := -\frac{\theta}{2}\Big(\frac{\theta}{2}+1\Big)^{-1}\Lambda\Sigma'\Psi\tilde{A} + B,$$

$$R_3(\theta) := (1/2)\Big(\frac{\theta}{2}+1\Big)^{-1}\tilde{A}'\Psi\tilde{A},$$

and where $\Psi := (\Sigma\Sigma')^{-1}$. Note that any solution of this CARE will be a symmetric $m \times m$ matrix.

In order to formulate our main results we need one more assumption:

Assumption (A3) *For every $\theta > 0$ the equation (6) admits a unique, positive semi-definite solution, say $K_1(\theta)$. Moreover, the matrix $G(\theta)$ defined below is stable:*

$$G(\theta) := R_2(\theta) + R_1(\theta)K_1(\theta). \tag{7}$$

Remark 2.3. Standard controllability and observability conditions imposed on the parameters of problems $\mathcal{P}_\theta$ will guarantee that Assumption (A3) is satisfied. As a matter of fact under these conditions stability of the matrix $G(\theta)$ is implied by positive semi-definiteness of the matrix $K_1(\theta)$. See Wonham [11], for example.

In view of Assumption A3 we can define the vector

$$K_2(\theta) = -\Big(G(\theta)'\Big)^{-1}\Big((\theta/2)(\theta/2+1)^{-1}[\tilde{A} - \theta\Sigma\Lambda' K_1(\theta)]'\Psi\tilde{a} + A_0' + 2K_1(\theta)b\Big). \tag{8}$$

Theorems 2.1 and 2.2 below, repeated for convenience from Bielecki and Pliska [4], provide the main results that will be needed for this paper. Theorem 2.1 characterizes optimal, risk-sensitive investment strategies. Theorem 2.2 characterizes the optimal value of the objective criterion.

Theorem 2.1. *Assume (A1), (A2) and (A3). Fix $\theta > 0$ and consider the process $h^\theta(t) = (h_0^\theta(t), \tilde{h}^\theta(t))$ defined as*

$$\tilde{h}^\theta(t) = (\theta/2+1)^{-1}\Psi\big[(\tilde{A} - \theta\Sigma\Lambda' K_1(\theta))X_t + \tilde{a} - (\theta/2)\Sigma\Lambda' K_2(\theta)\big], \tag{9}$$

and

$$h_0^\theta(t) = 1 - \mathbf{1}'\tilde{h}^\theta(t). \tag{10}$$

Then for θ sufficiently small the investment process $h^\theta(t)$ is optimal for problem (P_θ), that is,

$$J_\theta(v, x; h(\cdot)) \leq J_\theta(v, x; h^\theta(\cdot))$$

holds for all $h(\cdot) \in \mathcal{H}$, $v > 0$, $x \in R^{m+1}$.

Theorem 2.2. *Assume (A1), (A2) and (A3), fix $\theta > 0$ sufficiently small, and consider the problem P_θ. Let $h^\theta(t)$ be as in Theorem 2.1. Then*

(a) For all $v > 0$ and $x \in R^{m+1}$ we have

$$J_\theta(v, x; h^\theta(\cdot))$$

$$= \lim_{t\to\infty}(-2/\theta)t^{-1}\ln\ \mathbf{E}^{h^\theta(\cdot)}\ [e^{-(\theta/2)\ln V(t)}|V(0) = v, X(0) = x] =: \rho(\theta).$$

(b) The constant $\rho(\theta)$ in (a) is given by the formula

$$\rho(\theta) = (1/2)(\theta/2+1)^{-1}\Big(\tilde{a} - (\theta/2)\Sigma\Lambda' K_2(\theta)\Big)'\Psi\Big(\tilde{a} - (\theta/2)\Sigma\Lambda' K_2(\theta)\Big)$$

$$+(1/2)\Big[-(\theta/2)K_2(\theta)'\Lambda\Lambda' K_2(\theta) + 2tr[\Lambda\Lambda' K_1(\theta)]\Big] + b' K_2(\theta). \tag{11}$$

In the following two sections we shall apply these theorems to two concrete portfolio management problems.

3. Asset Allocation Among U.S. Equity Indexes

In this section we consider a specification of the general model of Section 2. Specifically, we consider a portfolio management problem with four assets and two stochastic economic factors. The two factors are U.S. Treasury 3-month and 10-year yields. The four assets are the bank account (with the 3-month yield for its interest rate), the Dow Jones Industrials index, the S&P 500 index, and the NASDAQ index. This system involves 5 Brownian motions, so N=5. We estimated the model parameters using weekly data from December 31, 1993, to March 26, 1999. We regressed the asset returns against the levels of the two factors to obtain (after conversion to an annual time scale, and with the interest rates expressed as decimals)

$$a = \begin{pmatrix} 0.1283 \\ 0.1231 \\ 0.7854 \end{pmatrix}, \quad A = \begin{pmatrix} 8.0313 & -4.9402 \\ 10.4009 & -6.4751 \\ 14.7373 & -17.5576 \end{pmatrix}. \tag{12}$$

Note that the more downward sloping the yield curve, the more bullish it is for the stock market.

For the second step we regressed the weekly changes in interest rates against the beginning-of-week levels of interest rates to obtain

$$b = \begin{pmatrix} -0.0058 \\ 0.1203 \end{pmatrix}, \quad B = \begin{pmatrix} -2.1401 & 1.7013 \\ -2.2466 & -0.1664 \end{pmatrix}. \tag{13}$$

Note that B is indeed a stable matrix, and the equilibrium (mean reverting) levels of the short rate and 10-year rate were computed to be 4.87% and 6.47%, respectively.

For the Σ and Λ matrices we computed the 5×5 covariance matrix using the residuals from the regressions in step 1 and step 2. We then computed the corresponding Cholesky matrix, from which the fist three rows give

$$\Sigma = \begin{pmatrix} 0.1397 & 0 & 0 & 0 & 0 \\ 0.1261 & 0.0478 & 0 & 0 & 0 \\ 0.1410 & 0.0939 & 0.1371 & 0 & 0 \end{pmatrix} \tag{14}$$

and the last two rows give

$$\Lambda = \begin{pmatrix} 0.0003 & 0.0001 & -0.0002 & 0.0065 & 0 \\ -0.0012 & -0.0028 & 0.0014 & -0.0002 & 0.0076 \end{pmatrix}. \tag{15}$$

We used MATLAB to solve the continuous algebraic Riccati equation and compute the quantities of interest. For example, for $\theta = 2$ we obtained the optimal strategy

$$\tilde{h}^\theta(t) = \begin{pmatrix} 15.9884 \\ -33.6340 \\ 17.2978 \end{pmatrix} + \begin{pmatrix} -387.3667 & 66.5719 \\ 597.2411 & 56.2440 \\ 22.4111 & -235.5583 \end{pmatrix} X_t. \tag{16}$$

In particular, with the factors at their equilibrium levels of 0.0487 and 0.0687 this calls for highly leveraged long positions of 143% of one's wealth in the Dow Jones Industrials and 315% in the NASDAQ, financed by a short position of 91% in the

S&P and borrowing 267% from the bank. These positions are consistent with an aggressive investor having a small risk aversion parameter such as $\theta = 2$. It is also interesting to note the substantial differences in the sensitivities of these optimal proportions to the levels of the interest rates.

Figure 1 shows a graph of the optimal proportions versus θ for the situation where the interest rate factors are at their equilibrium levels. It is somewhat disconcerting to see that the optimal strategy calls for extreme positions in the assets, such as taking a long position equal to six times one's wealth in the NASDAQ index. We discuss this issue in the following section.

The maximum risk adjusted growth rate ρ is easy to compute; for instance, when $\theta = 2$ one has $\rho = 0.6295$. As a benchmark, consider the NASDAQ which, during the period for which data were collected, had a long run growth rate of about 35% and a volatility of about 20%. Corresponding to $\theta = 2$ this gives a risk adjusted growth rate of 33%.

4. Asset Allocation Among U.S. Treasuries

In this section we present a fixed income management example where there are seven factors, namely, U.S. Treasury zero-coupon yields for the following maturities: three months (this is taken to be the short rate), six months, one year, two years, five years, ten years, and thirty years. There are also seven assets, namely, the bank account and six rolling horizon bonds whose maturities correspond to the factors, that is, six months and one, two, five, ten, and thirty years. There are $N = 7$ underlying Brownian motions. For an explanation of the rolling horizon bond dynamics and other theoretical aspects of this kind of model, the reader should consult Bielecki and Pliska [4]. They also presented a simpler fixed income management example where there are only two interest rate factors and one rolling horizon bond. Here our objective is to investigate a more comprehensive example.

We estimated our parameters using monthly historical U.S. Treasury zero-coupon yields from January 1970 to December 1996. We used simple interpolation to construct the monthly yield curves, from which we obtained the monthly returns for zero-coupon bonds whose maturities are six months and one, two, five, ten, or thirty years. Using these returns as proxies for the returns of the six corresponding rolling horizon bonds, we then regressed the returns against the beginning-of-month factor levels to obtain (all returns and interest rates are decimals; the time scale is annual)

$$a = \begin{pmatrix} -.0153 \\ -.0314 \\ -.0469 \\ -.1179 \\ -.2927 \\ -.7295, \end{pmatrix} \tag{17}$$

$$A = \begin{pmatrix} -1.240 & 1.066 & 3.625 & -3.569 & 0.316 & 2.562 & -1.597 \\ -1.725 & -4.015 & 13.067 & -8.330 & -0.334 & 6.537 & -3.904 \\ -4.331 & -2.397 & 10.155 & -2.315 & -4.243 & 12.060 & -7.509 \\ -14.682 & 0.306 & 30.692 & -25.903 & 5.352 & 23.284 & -17.157 \\ -36.345 & 12.313 & 51.502 & -30.692 & -27.689 & 55.873 & -21.940 \\ -153.824 & 176.941 & -89.898 & 188.783 & -204.886 & -88.158 & 175.386 \end{pmatrix} \tag{18}$$

We then regressed the monthly changes in yields against the beginning-of-month yields to obtain (after conversion to an annual time scale)

$$b = \begin{pmatrix} .00273 \\ .00299 \\ .00268 \\ .00186 \\ .00186 \\ .00227 \\ .00179 \end{pmatrix}, \tag{19}$$

$$B = \begin{pmatrix} -.4825 & .9212 & -.9615 & .7379 & -.0994 & -.4295 & .2705 \\ .1140 & .0078 & -.5887 & .6845 & -.0595 & -.5284 & .3351 \\ .1611 & .2367 & -.9815 & .7571 & .0348 & -.5909 & .3555 \\ .1872 & .0897 & -.4632 & .1675 & .1848 & -.5061 & .3228 \\ .2483 & -.0113 & -.5050 & .4111 & -.0674 & -.3805 & .2913 \\ .3057 & -.1107 & -.4144 & .2485 & .2045 & -.4328 & .1849 \\ .4103 & -.4461 & .1888 & -.4414 & .4976 & .1990 & -.4126 \end{pmatrix} \tag{20}$$

Note that the equilibrium (mean-reverting) yields are computed to be $-(B^1b)' = (.066, .069, .071, .074, .078, .081, .083)$. With interest rates at these equilibrium values, the rolling horizon bond appreciation rates are computed to be $(a - AB^1b)' = (.070, .073, .077, .085, .092, .153)$.

Using the residuals from the last seven regressions we computed the corresponding covariance matrix, and from this we computed the corresponding Cholesky matrix, thereby giving

$$\Lambda = \begin{pmatrix} .02225 & 0 & 0 & 0 & 0 & 0 & 0 \\ .02139 & .00494 & 0 & 0 & 0 & 0 & 0 \\ .01907 & .00812 & .00382 & 0 & 0 & 0 & 0 \\ .01526 & .00835 & .00584 & .00328 & 0 & 0 & 0 \\ .01029 & .00690 & .00432 & .00500 & .00340 & 0 & 0 \\ .00768 & .00591 & .00294 & .00467 & .00506 & .00244 & 0 \\ .00453 & .00385 & .00193 & .00345 & .00550 & .00465 & .00668 \end{pmatrix}. \tag{21}$$

Finally, for the volatility matrix Σ we used the relationship (see Bielecki and Pliska [4]) $\sigma_{ij} = -T_i\lambda_{i+1,j}$, where T_i denotes the maturity of the corresponding rolling horizon bond. We thereby obtained

$$\Sigma = \begin{pmatrix} -.01070 & -.00247 & 0 & 0 & 0 & 0 & 0 \\ -.01907 & -.00812 & -.00382 & 0 & 0 & 0 & 0 \\ -.03054 & -.01669 & -.01166 & -.00658 & 0 & 0 & 0 \\ -.05153 & -.03450 & -.02151 & -.02497 & -.01703 & 0 & 0 \\ -.07679 & -.05918 & -.02944 & -.04677 & -.05052 & -.02454 & 0 \\ -.13597 & -.11518 & -.05802 & -.10392 & -.16541 & -.13943 & -.20005 \end{pmatrix}. \tag{22}$$

It can be seen that the rolling horizon bond volatilities are increasing with respect to maturity.

We used MATLAB to compute optimal solutions. But for small and moderate values of θ the optimal trading strategies involved extremely leveraged positions, with very large long positions in some assets and very large short positions in others. For example, with $\theta = 2$ and with the interest rate factors at their equilibrium levels, the optimal proportions in the bank account and six rolling horizon bonds are approximately (454, -1511, 1367, -250, -168, 128, 19), respectively (and note these are decimals, not percentages!). And the corresponding risk adjusted growth rate is 2286%.

The main explanation for this extreme behavior is that our risk sensitive asset management model appears to share the same weakness of all other optimal portfolio management models (including the Markowitz model): after making your best effort to estimate the model parameters, the optimal solutions will typically call for extreme positions in the various assets unless exogenous constraints have been imposed. In support of this opinion, we computed the "Merton proportions" $(\Sigma\Sigma')^{-1}(\mu - r\mathbf{1})$ corresponding to a log-utility investor, where μ is the vector indicated above of rolling horizon bond appreciation rates corresponding to the equilibrium yields and where r is the equilibrium short rate 6.64%. These Merton proportions for the bank account and six rolling horizon bonds are approximately (-75, 169, -129, 34, 11, -10, 1), numbers of a similar order of magnitude to the risk sensitive solution indicated above.

Another contributing factor is the fact that the volatilities of the rolling horizon bonds are very small relative to their appreciation rates. For example, the six month rolling horizon bond has a volatility of only 2.2%, whereas its appreciation rate when the interest rates are at their equilibrium values is 7.03%. Hence certain linear combinations of the assets can lead to portfolios having very attractive risk-return characteristics.

5. Conclusion

With two examples we have demonstrated that the computational approach developed by Bielecki and Pliska [4] can readily provide explicit solutions to the risk sensitive variation of the famous intertemporal capital asset pricing model.

Consequently, these methods can be used to provide better understanding of financial markets and economies. These results are also of importance to investors, although the usefulness of the model is limited. The model can be used to indicate in a qualitative way how positions in various assets should be adjusted to reflect changing factors in the economy. And the model might suggest some *ad hoc* methods for determining strategies that can profitably be used for investment. However, the model's optimal solutions should probably not be literally used by the typical investor because, as with all other optimal portfolio models, they call for too much leverage unless exogenous constraints are imposed. Consequently, the incorporation of explicit constraints on trading positions would be a fruitful direction for future research.

References

[1] Bielecki, T.R., and Pliska, S.R., (1999), "Risk Sensitive Dynamic Asset Management," *Appl. Math. Optim.*, vol. 39., pp. 337-360.

[2] Bielecki, T. R., Pliska, S. R., and Sherris, M., (1998), "Risk Sensitive Asset Allocation," *Journal of Economic Dynamics and Control*, vol. 24, pp. 1145-1177.

[3] Bielecki, T. R., and Pliska, S.R., (1998), "Risk-sensitive dynamic asset allocation," in *Asset & Liability Management: A Synthesis of New Methodologies*, RISK Books, London.

[4] Bielecki, T. R., and Pliska, S.R., (2000), "Risk Sensitive Intertemporal CAPM, with Applications to Fixed Income Management," working paper, submitted for publication.

[5] Fleming, W. H., and Sheu, S. J., (2000), "Risk Sensitive Control and an Optimal Investment Model," *Mathematical Finance*, vol. 10, pp. 197-214.

[6] Karatzas I., and Kou S.G., (1996), "On the pricing of contingent claims under constraints," *Annals of Applied Probability*, vol. 6, pp. 321-369.

[7] Karatzas I., and Shreve S. E., (1988), "Brownian Motion and Stochastic Calculus," Springer-Verlag, New York.

[8] Kuroda, K., and Nagai, H., (2000), "Risk-sensitive portfolio optimization on infinite time horizon," working paper.

[9] Merton, R. C., (1973), "An intertemporal capital asset pricing model," *Econometrica* vol. 41, pp. 866-887.

[10] Rutkowski, M., (1997), "Self-financing Trading Strategies for Sliding, Rolling-horizon, and Consol Bonds," *Mathematical Finance* vol. 9, pp. 361-385.

[11] Wonham, W.M., (1979), "Linear Multivariable Control: a Geometric Approach," Springer-Verlag, New York.

[1]Department of Mathematics,, The Northeastern Illinois University,, 5500 North St. Louis Avenue,, Chicago, IL 60625-4699 USA
E-mail address: T-Bielecki@neiu.edu

[2]Department of Mathematics,, University of Illinois at Chicago,, 581 S. Morgan Street,, Chicago, IL 60607 USA

[3]Department of Mathematics,, University of Illinois at Chicago,, 581 S. Morgan Street,, Chicago, IL 60607 USA
E-mail address: jackli@math.uic.edu

[4]Department of Finance,, University of Illinois at Chicago,, 601 S. Morgan Street,, Chicago, IL 60607-7124, USA
E-mail address: srpliska@uic.edu

Trends in Mathematics, © 2001 Birkhäuser Verlag Basel/Switzerland

Bounded Variation Singular Stochastic Control and Associated Dynkin Game

Frederik Boetius

Abstract. Motivated by the observation — yet present in derivations of the Black-Scholes formula — that the value of a dynamic optimization problem is actually a stochastic process determined by the future development of the stochastic system, we reformulate a singular control problem as optimization problem for decoupled forward-backward SDE. The controlled process takes the form of a general Brownian diffusion in one dimension, with bounded variation control.

Applications are given in the field of real investment and (real) options; many other optimization problems such as hedging with transaction costs or dividend control also exhibit a singular behaviour. Our reformulation enables us to consider nonadditive or stochastic differential utilities in the control problem, and exhibits formal similarities with reflected backward SDE and g-semisolutions.

Related to the control problem we find a two-player stochastic game of optimal stopping (or Dynkin game). We establish a well-known relation between both problems for this general situation, namely that the partial derivative of the value of the control problem $\frac{\partial}{\partial x}V$ equals the value of the Dynkin game u, and a saddle point for this game can be derived from an optimal control in the control problem. The arguments are based on stochastic analysis, especially comparison theorems for the solutions of controlled forward and backward stochastic differential equations.

1. Introduction

Interest in singular stochastic control as well as its relation with optimal stopping goes back to a study of spaceship control by Bather and Chernoff [2]. Several aspects of singular control were considered in the sequel, such as questions of finite fuel, control in higher dimensions and existence of optimal controls. References can be found in [1], [4], [10].

Applications include inventory control, portfolio selection, dividend control, real options, partly reversible or irreversible investment, capacity choice and industry equilibrium. Singular control may be seen as a continuous (limiting) version of impulse controls or sequences of installment options. Apart from economic interpretations the relation between singular control and optimal stopping is of interest if one considers existence or structure of optimal strategies and the principles of smooth fit and smooth pasting respectively.

An analytical approach to singular stochastic control uses characterizations through variational inequalities and pde techniques. The probabilistic approach taken in this paper is due to Karatzas and Shreve [10]. Equivalence of monotone singular stochastic control and optimal stopping is the topic of Baldurssons and Karatzas' study of irreversible investment and industry equilibrium [1] (see also the paper by Karatzas in this volume) and the impulse control/sequential stopping approach in Boetius and Kohlmann [4].

Nonlinear backward stochastic differential equations (BSDE) were introduced by Pardoux and Peng [12] and Duffie and Epstein [6] and studied to a large extent by several authors thereafter. Applications in finance are, among others, discussed in El Karoui et al. [8], El Karoui and Quenez [9] and Ma and Yong [11]. Detailed discussion of the subject and extensive references are given in these papers.

This paper is organized as follows: first we formulate the general setting and give definitions of the control problem and associated game of optimal stopping. We then recall some preliminaries such as comparison theorems that are central to our approach. In the third section we state the main result and give an outline of the major steps of the proof. The paper closes with a few notes on related topics in the theory of BSDE.

This paper is an abbreviated version of [3].

1.1. Setting

Let $(\Omega, \mathcal{F}, \boldsymbol{P})$ be a probability space, $\bar{\imath} := [0,T]$ a time interval, and W_t a standard d-dimensional Brownian motion adapted to a filtration $\{\mathcal{F}_t\}_{t\in\bar{\imath}}$ satisfying the usual conditions. For $(t_0, x) \in \bar{\imath} \times \mathbb{R}$ and standard data (b, σ, C, h, g, a) as defined below we consider the controlled FBSDE on $[t_0, T]$

$$\begin{aligned} dX_u &= b(u, X_u)\,du - dC_u^U + dC_u^L + \sigma(u, X_u)^\top\,dW_u,\\ dY_u &= -g(u, X_u, Y_u, Z_u)\,du - a_u^U\,dC_u^U + a_u^L\,dC_u^L + Z_u^\top\,dW_u,\\ X_{t_0} &= x,\\ Y_T &= \xi = h(X_T). \end{aligned} \tag{1}$$

We may write $dC = (1,1)\,dC = dC^U - dC^L$ if no confusion is possible.

Standard data (b, σ, C) for the forward and (h, g, a, C) for the backward equation in (1) are defined as follows, where the notation is taken from [11]:

- $b \in L^2_{\mathcal{F}}(0,T; W^{1,\infty}(\mathbb{R}, \mathbb{R}))$ and $\sigma \in L^2_{\mathcal{F}}(0,T; W^{1,\infty}(\mathbb{R}, \mathbb{R}^d))$.
- $C \in \mathcal{A}$ and $a = (a^U, a^L) : \bar{\imath} \times \Omega \to \mathbb{R}^2$ progressively measurable such that for all $C' \in \mathcal{A}$ the process $\int_{[0,t)} a_s\,dC'_s$ is in $L^2_{\mathcal{F}}(0,T;\mathbb{R})$ and has LCRL paths $\boldsymbol{P}$-a.s.
- $g \in L^2_{\mathcal{F}}(0,T; W^{1,\infty}(\mathbb{R} \times \mathbb{R} \times \mathbb{R}^d, \mathbb{R}))$ and $h \in L^2_{\mathcal{F}_T}(\Omega; W^{1,\infty}(\mathbb{R}, \mathbb{R}))$.

Here $\mathcal{A}$ consists of all processes $C := (C^U, -C^L)^\top$, where the components C^U, C^L are $\{\mathcal{F}_t\}$-adapted, its paths are a.s. left continuous with right hand limits (LCRL), increasing; further their absolute variation satisfies $E[|C|_T] < \infty$. $\mathcal{A}$ will be the class of admissible controls for the control problem.

So apart from sufficient measurability we require uniform Lipschitz continuity of the data (b, σ, g, h) and square integrability, and we consider bounded variation control in one dimension. The processes involved need not be Markoff.

For standard data (1) has a unique solution in $L^2_{\mathcal{F}}(0,T;\mathbb{R}\times\mathbb{R}\times\mathbb{R}^d)$ which we denote by $(X_t^{t_0,x,C}, Y_t^{t_0,x,C}, Z_t^{t_0,x,C}) = (X,Y,Z)^{t_0,x,C} = (X,Y,Z)^{t_0,x,C^U,C^L}$. To see this one may consider the FBSDE for the transformed variables $(\tilde{X}_t, \tilde{Y}_t, \tilde{Z}_t) := (X_t + C_t, Y_t + \int_{[0,t)} a_s\, dC_s, Z_t)$. These satisfy a "conventional" FBSDE and all standard results can be used.

For simplicity we only consider $t_0 = 0$ and omit dependence on t_0 and ω in our notation. Sometimes we will refer to the solution of the BSDE only with data (ξ, g, a, C), without reference to a forward equation.

We make the following additional assumptions on the data:

$$b \text{ is convex and differentiable in the space variable } x \tag{2a}$$

$$\sigma \text{ is linear in the space variable } x \tag{2b}$$

$$h \text{ is convex, differentiable and increasing in the space variable } x \tag{2c}$$

$$\begin{array}{l} g \text{ is convex and partially differentiable in the variables } (x,y,z) \\ \text{and increasing in } x;\ g_y \text{ is increasing in } x \text{ and } g_z \text{ is independent of} \\ x,y,z \end{array} \tag{2d}$$

$$\text{Both components of } a \text{ are nonnegative and continuous.} \tag{2e}$$

Because of (2d) we will drop dependence of Dg on z in the notation.

1.2. Control problem

For an interpretation of (1) write Y in the standard representation

$$Y_t = E\left[h(X_T) + \int_t^T g(s, X_s, Y_s, Z_s)\, ds + \int_{[t,T)} a_s^U\, dC_s^U - \int_{[t,T)} a_s^L\, dC_s^L \,\middle|\, \mathcal{F}_t\right].$$

Viewing h, g and a as the terminal, running and control cost Y can be seen as the dynamic version of some cost functional dependent on X and the control C exercised. If g is linear in y then Y_t is obtained by integrating over the appropriately discounted costs after time t.

Definition 1.1. *The value of the control problem with dynamics (b,σ) and cost structure (h, g, a) is defined as*

$$V_0(x) := \operatorname{ess\,inf}_{C\in\mathcal{A}} Y_0^{x,C}. \tag{3}$$

The control problem comprises determining the value V and finding an optimal control $C^\star \in \mathcal{A}$ with the property $V_0(x) = Y_0^{x,C^\star}$ $\boldsymbol{P}$-a.s.

We assume $-\infty < V_0(x)$ for all x. Further we will also assume

$$\text{There exists an optimal control } C^\star \text{ for the control problem in } x. \tag{4}$$

Observe that due to the backward formulation a straight forward extension of formula (3) for $t > 0$ leads to senseless results. To avoid a cumbersome definition we set $V_t(x) := Y_t^{x,C^\star}$ for all $t \in \bar{\iota}$, which is meaningful at least if the optimal control is unique. However if we consider V_0 this is just notation and we do not need uniqueness.

1.3. Game of optimal stopping

Before writing down the associated game we introduce some deflator processes: The biased deflator process Γ^{x,C^U,C^L} for a control $(C^U, -C^L) \in \mathcal{A}$ is defined as solution to the FSDE

$$\begin{aligned} d\Gamma_t^{x,C^U,C^L} &= \Gamma_t^{x,C^U,C^L}\left(b_x(t, X_t^{x,C^U,C^L})\,dt + \sigma_x(t)^\top\,dW_t\right), \\ \Gamma_0^{x,C^U,C^L} &= 1. \end{aligned} \tag{5}$$

Its uncontrolled version is abbreviated as $\Gamma^x := \Gamma^{x,0,0}$, which serves as deflator in the Dynkin game.

$\mathcal{T}_t$ denotes the class of $\mathcal{F}_s$-stopping times with values a.s. in $[t,T]$. For $\sigma, \tau \in \mathcal{T}_t$ define the payoff as $R_t^x(\sigma,\tau) := R_t$, where (R,Q) is the solution of the BSDE

$$\begin{aligned} dR_u &= -\langle Dg(t, X_t^{x,0,0}, V_t(x)), (\Gamma_t^x, R_t, Q_t)\rangle \chi_{t\le\sigma\wedge\tau}\,dt + Q_t^\top\,dW_t, \\ R_T &= \Big(h_x(X_T^{x,0,0})\chi_{\sigma\wedge\tau=T} + a_\tau^L \chi_{\substack{\tau<\sigma \\ \tau<T}} + a_\sigma^U \chi_{\substack{\sigma\le\tau \\ \sigma<T}}\Big)\Gamma_{\sigma\wedge\tau\wedge T}^x, \end{aligned} \tag{6}$$

and consider the upper and lower values

$$u_t^+(x) := \operatorname{ess\,inf}_{\sigma\in\mathcal{T}_t} \operatorname{ess\,sup}_{\tau\in\mathcal{T}_t} R_t^x(\sigma,\tau) \tag{7a}$$

$$u_t^-(x) := \operatorname{ess\,sup}_{\tau\in\mathcal{T}_t} \operatorname{ess\,inf}_{\sigma\in\mathcal{T}_t} R_t^x(\sigma,\tau) \tag{7b}$$

We assume $-\infty < u^-(x)$. By the definition of u^+ and u^-, $u^- \le u^+$ holds $\boldsymbol{P}$-a.s.

Definition 1.2. *The associated stochastic game of optimal stopping or Dynkin game consists of determining the upper and lower values u^+ and u^- and the value process u as solution to the Isaac's equation*

$$u_t(x) := u_t^+(x) = u_t^-(x), \tag{8}$$

and of finding a saddle point $(\sigma^\star, \tau^\star) \in \mathcal{T}_t \times \mathcal{T}_t$, i.e. stopping times such that

$$\forall t \in \bar{\iota} \quad \begin{aligned} u_t(x) &= R_t^x(\sigma^\star, \tau^\star) \\ &= \operatorname{ess\,inf}_{\sigma\in\mathcal{T}_t} R_t^x(\sigma,\tau^\star) = \operatorname{ess\,sup}_{\tau\in\mathcal{T}_t} R_t^x(\sigma^\star,\tau) \end{aligned} \quad [\boldsymbol{P}]. \tag{9}$$

INTERPRETATION R^x defined through (6) has the representation

$$\begin{aligned} R_t^x(\sigma,\tau) = E\Bigg[\int_t^{\sigma\wedge\tau} &\langle Dg(s, X_s^{x,0,0}, V_s(x)), (\Gamma_s^x, R_s^x, Q_s)\rangle\,ds \\ &+ h_x(X_T^{x,0,0})\Gamma_T^x \chi_{\sigma\wedge\tau=T} + a_\sigma^U \Gamma_\sigma^x \chi_{\substack{\sigma\le\tau \\ \sigma<T}} + a_\tau^L \Gamma_\tau^x \chi_{\substack{\tau<\sigma \\ \tau<T}} \,\Bigg|\, \mathcal{F}_t\Bigg]. \end{aligned} \tag{10}$$

We interpret this as a game for two players MIN, MAX: MIN pays MAX at rate $\langle Dg, \cdot\rangle$ as long as the game continues, and $h_x\Gamma_T$ upon reaching the time horizon

T. If one of the players chooses to terminate early, MIN pays MAX $a_\sigma^U \Gamma_\sigma$ or $a_\tau^L \Gamma_\tau$, depending on whether MIN or MAX stopped the game. So MIN seeks to minimize $R^x(\sigma,\tau)$ by choice of σ, whereas MAX wishes to maximize the payoff by choice of τ. A solution consists of a pair of stopping times such that neither of the two has an incentive to deviate from his strategy. The optimization of the opponent is anticipated and the right of choosing ones strategy first gives no advantage, i.e. has no additional value, which is the meaning of the Isaac's equation (8).

2. Preliminaries

In the proof we make use of a priori estimates and comparison theorems for solutions to the forward and backward equation in (1). Their proofs are similar to those given in e.g. [8], [9]. For the convenience of the reader we state the

Theorem 2.1. *(Comparison Theorem for backward equation) Let, for $i \in \{1,2\}$, (g^i, ξ^i, a^i, C^i) standard data and (Y^i, Z^i) the corresponding solutions of the BSDE. Assume further that $\boldsymbol{P}$-a.s.*

$$g^1(t,y,z) - g^2(t,y,z) \geq 0 \qquad\qquad \xi^1 - \xi^2 \geq 0 \tag{11a}$$

$$a_t^1\, d(C_t^1 - C_t^2) \geq 0 \qquad\qquad (a_t^1 - a_t^2)\, dC_t^2 \geq 0 \tag{11b}$$

Then the difference $Y_t^1 - Y_t^2$ is nonnegative almost surely. Let $A := \{\omega \,|\, Y_{t_0}^1(\omega) = Y_{t_0}^2(\omega)\}$. If $\boldsymbol{P}(A) > 0$, then equality holds in (11) for $t_0 \leq t \leq T$ and $\omega \in A$ $\boldsymbol{P}$-a.s.

Actually one of the weaker conditions on g^i

$$g^1(t,Y_t^1,Z_t^1) - g^2(t,Y_t^1,Z_t^1) \geq 0, \quad g^1(t,Y_t^2,Z_t^2) - g^2(t,Y_t^2,Z_t^2) \geq 0 \tag{11c}$$

is sufficient. This is useful when making use of the convexity of the driver. If e.g. g^1 does not depend on (Y^1, Z^1) we conclude that $g_t^1 \geq g^2(t,Y_t^1,Z_t^1)$ is also sufficient.

3. Main Result

Theorem 3.1. *Assume that (2) and (4) hold and denote the optimal control by $C^\star = (C^U, -C^L)$. Then the value function V of the control problem 1.1 is (partially) differentiable at x (with respect to the space variable). The associated Dynkin game 1.2 has a solution, and its value is the (partial) derivative of the value of the control problem:*

$$\frac{\partial}{\partial x} V_t(x) = u_t(x) \qquad \text{for } 0 \leq t \leq \sigma^0 \wedge \tau^0 \ \boldsymbol{P}\text{-a.s.}. \tag{12}$$

Here $\sigma^0 := \inf\{t \geq 0 \,|\, C_t^U > C_0^U\}$ and $\tau^0 := \inf\{t \geq 0 \,|\, C_t^L > C_0^L\}$ are the first action times of the controls C^U and C^L. They form a saddle point for the Dynkin game which has value

$$u_t(x) = R_t^x(\sigma^0, \tau^0) \qquad \boldsymbol{P}\text{-a.s.} \tag{13}$$

INTERPRETATION (12) indicates that the problem of an "incremental increase" of capacity is merely a question of optimal stopping; alternatively, the optimal capacity is obtained as continuous limit of sequential stopping problems ([4]). Optimal stopping times take the form of first action times of the controls. If the latter act on the boundary of a "no-control-region" in $\bar{\imath} \times \mathbb{R}$, the optimal stopping times are first exit times of the uncontrolled process from that region.

SKETCH OF PROOF The proof essentially rests on the four relations

$$\Delta^+ V_t(x) \le u_t^-(x) \le u_t^+(x) \le \Delta_- V_t(x) \le \Delta^+ V_t(x). \tag{14}$$

Here $\Delta^+ V$ and $\Delta_- V$ denote the upper right and lower left Dini derivatives of the value function V (cf. [10]).

The second inequality follows from the definitions in (7). $\Delta_- V(x) \le \Delta^+ V(x)$ is a consequence of the convex structure of the control problem, linearity of σ and monotonicity of costs in the space variable as formulated in conditions (2). It can be verified by applying a comparison theorem first to the forward equation in (1) and then to the backward equation. The argument then follows that in e.g. [10].

The first and third inequality are almost parallel. We give an outline of the proof for the first in the sequel.

Consider $x \in \mathbb{R}$ and a given control $C \in \mathcal{A}$. Then for τ^0 the first action time of C^L, arbitrary $\sigma \in \mathcal{T}$ and $\delta > 0$ we construct a sequence of controls C^δ and approximating solutions $(X^\delta, Y^\delta, Z^\delta)$ to (1) with initial value $x + \delta$ such that

$$\lim_{\delta \searrow 0} \frac{1}{\delta}\big(Y_t^\delta - Y_t^{x,C^U,C^L}\big) \le R_t^x(\sigma, \tau^0) \qquad \boldsymbol{P}\text{-a.s. for all } t \le \sigma \wedge \tau^0. \tag{15}$$

Then $\limsup_{\delta \searrow 0} \frac{1}{\delta}\big(V_t(x+\delta) - V_t(x)\big) \le \limsup_{\delta \searrow 0} \frac{1}{\delta}\big(Y_t^\delta - Y_t^\star\big) \le R_t^x(\sigma, \tau^0)$ if C is optimal. Therefore, as $\sigma \in \mathcal{T}$ is arbitrary $\Delta^+ V_t(x) \le \operatorname{ess\,inf}_{\sigma \in \mathcal{T}_t} R_t(\sigma, \tau^0) \le u_t^-(x)$ holds $\boldsymbol{P}$-a.s., which gives the third inequality in (14).

CONSTRUCTION OF APPROXIMATING SEQUENCE As above assume $\sigma \in \mathcal{T}$, $x \in \mathbb{R}$ and $C = (C^U, -C^L) \in \mathcal{A}$ fixed. To obtain the process Y^δ in (15), choose the control C^δ such that the state process X^δ tracks X^{x,C^U,C^L} as follows: It starts in $x + \delta$ and experiences the same upper control C^U as X^{x,C^U,C^L} and now lower control until either σ occurs or the trajectories of X^{x,C^U,C^L} and X^δ cross. Then X^δ jumps onto X^{x,C^U,C^L}. Hence let $\tau^\delta := \inf\{t \in \bar{\imath} \,|\, X_t^{x+\delta,C^U,0} \le X_t^{x,C^U,C^L}\}$ the crossing time and define

$$C_t^{\delta,U} := \begin{cases} C_t^U & t \le \sigma \\ C_t^U + X_\sigma^{x+\delta,C^U,0} - X_\sigma^{x,C^U,C^L} & t > \sigma, \sigma \le \tau^\delta \\ C_t^U & t > \sigma, \sigma > \tau^\delta \end{cases} \tag{16a}$$

$$C_t^{\delta,L} := \begin{cases} 0 & t \le \sigma \wedge \tau^\delta \\ C_t^L - C_\sigma^\lambda & t > \sigma, \sigma \le \tau^\delta \\ C_t^L - C_{\tau^\delta}^L - (X_{\tau^\delta}^{x+\delta,C^U,0} - X_{\tau^\delta}^{x,C^U,C^L}) & t > \tau^\delta, \sigma > \tau^\delta \end{cases} \tag{16b}$$

Let $(X^\delta, Y^\delta, Z^\delta) := (X,Y,Z)^{x+\delta, C^{\delta,U}, C^{\delta,L}}$ and $(X^\star, Y^\star, Z^\star) := (X,Y,Z)^{x,C^U,C^L}$. Indeed one can verify that $X_t^\delta = X_t^{x+\delta, C^U, 0}$ for $t \le \sigma \wedge \tau^\delta$ and, for $t > \sigma \wedge \tau^\delta$, $X_t^\delta = X_t^\star$ and $Y_t^\delta = Y_t^\star$. Further, $Y^\delta \ge Y^\star$ and Y^δ decreases with δ.

For $t \le \sigma \wedge \tau^\delta$ we can write $Y_t^\delta - Y_t^\star$ as

$$E\left[\int_t^{\sigma\wedge\tau^\delta} g(s, X_s^\delta, Y_s^\delta, Z_s^\delta) - g(s, X_s^\star, Y_s^\star, Z_s^\star)\, ds + \left(h(X_T^\delta) - h(X_T^\star)\right)\chi_{\sigma\wedge\tau^\delta = T}\right.$$
$$\left. + a_\sigma^U (X_\sigma^\delta - X_\sigma^\star)\chi_{\substack{\sigma\le\tau^\delta\\ \sigma<T}} + a_{\tau^\delta}^L (X_{\tau^\delta}^\delta - X_{\tau^\delta}^\star)\chi_{\substack{\tau^\delta<\sigma\\ \tau^\delta<T}} + \int_{[t,\sigma\wedge\tau^\delta)} a_s^L\, dC_s^L \,\middle|\, \mathcal{F}_t\right].$$

Although there are obvious similarities with $R_t^x(\sigma, \tau^0)$ in (10) we cannot apply the comparison theorem 2.1 directly. Instead, we consider

$$\tilde{Y}_t^\delta := \frac{1}{\delta}\Big((Y_t^\delta - Y_t^\star) + \int_{[0,\sigma\wedge\tau^\delta\wedge t)} a_s^L\, dC_s^L\Big) \qquad \text{and} \qquad \tilde{Z}_t^\delta := \frac{1}{\delta}(Z_t^\delta - Z_t^\star).$$

For $t \le \sigma \wedge \tau^0$ this transformation has no effect as $\int_{[0,\sigma\wedge\tau^0)} a_s^L\, dC_s^L = 0$. We then prove (15) by constructing processes $\tilde{R}^{up,\delta}$ and $\tilde{R}^{lo,\delta}$ with the following properties:

$$\tilde{R}_t^{lo,\delta} \le \tilde{Y}_t^\delta \le \tilde{R}_t^{up,\delta} \qquad \forall \delta > 0\ \forall t \in \bar{\imath}\ \boldsymbol{P}\text{-a.s.}, \tag{17a}$$
$$\exists \tilde{R} \in L_{\mathcal{F}}^2(0,T;\mathbb{R}) \ :\ \tilde{R}^{up,\delta} \searrow \tilde{R}, \quad \tilde{R}^{lo,\delta} \nearrow \tilde{R} \qquad \text{in } L_{\mathcal{F}}^2(0,T;\mathbb{R}) \text{ as } \delta \searrow 0, \tag{17b}$$
$$\tilde{R}_t \le R_t^x(\sigma, \tau^0) \qquad \forall t \le \sigma\wedge\tau^0\ \boldsymbol{P}\text{-a.s.}. \tag{17c}$$

To apply theorem 2.1 write $(\tilde{Y}^\delta, \tilde{Z}^\delta)$ as solution to a BSDE with data

$$\xi^{\tilde{Y}} := \frac{1}{\delta}\left(h(X_T^\delta) - h(X_T^\star)\right)\chi_{\sigma\wedge\tau^\delta=T} + a_\sigma^U \frac{1}{\delta}(X_\sigma^\delta - X_\sigma^\star)\chi_{\substack{\sigma\le\tau^\delta\\ \sigma<T}}$$
$$+ a_{\tau^\delta}^L \frac{1}{\delta}(X_{\tau^\delta}^\delta - X_{\tau^\delta}^\star)\chi_{\substack{\tau^\delta<\sigma\\ \tau^\delta<T}} + \frac{1}{\delta}\int_{[0,\sigma\wedge\tau^\delta)} a_s^L\, dC_s^L$$
$$g^{\tilde{Y}}(t,y,z) := \Big(\Delta_x\tilde{g}_t \frac{1}{\delta}(X_t^\delta - X_t^\star) + \Delta_y\tilde{g}_t\chi_{y\ge 0}\cdot y$$
$$- \Delta_y\tilde{g}_t \frac{1}{\delta}\int_{[0,t)} a_s^L\, dC_s^L\Big)\chi_{t\le\sigma\wedge\tau^\delta} + \Delta_z\tilde{g}_t\cdot z\chi_{t\le\sigma}$$

and $C_t^{\tilde{Y}} := (0,0)$. Here $\Delta_x\tilde{g}_t$, $\Delta_y\tilde{g}_t$ and $\Delta_z\tilde{g}_t$ are defined as:

$$\Delta_x\tilde{g}_t := \left(g(t, X_t^\delta, Y_t^\star, Z_t^\star) - g(t, X_t^\star, Y_t^\star, Z_t^\star)\right)\frac{1}{X_t^\delta - X_t^\star}\chi_{X_t^\delta\ne X_t^\star},$$
$$\Delta_y\tilde{g}_t := \left(g(t, X_t^\delta, Y_t^\delta, Z_t^\star) - g(t, X_t^\delta, Y_t^\star, Z_t^\star)\right)\frac{1}{Y_t^\delta - Y_t^\star}\chi_{Y_t^\delta\ne Y_t^\star},$$
$$\Delta_z\tilde{g}_t := \left(g(t, X_t^\delta, Y_t^\delta, Z_t^\delta) - g(t, X_t^\delta, Y_t^\delta, Z_t^\star)\right)\frac{1}{Z_t^\delta - Z_t^\star}\chi_{Z_t^\delta\ne Z_t^\star}.$$

Observe that $\frac{1}{\delta}\int_{[t,\sigma\wedge\tau^\delta)} a_s^L\, dC_s^L$ need not converge to zero as $\delta \searrow 0$. We derive an estimate for C_t^L if $t \le \tau^\delta$ from the definition of τ^δ in the next lemma.

Let $\tilde{\Theta}_t(\omega) \in [X_t^{x,C^U,0}(\omega), X_t^\star(\omega)]$ such that $\tilde{\Theta}$ is progressively measurable and

$$b_x(t,\tilde{\Theta}_t)\big(X_t^\star - X_t^{x,C^U,0}\big) = b(t,X_t^\star) - b(t,X_t^{x,C^U,0}) \qquad \boldsymbol{P}\text{-a.s.}$$

Denote by $\Gamma_t^{\tilde{\Theta}}$, Γ^{up} and Γ^{lo} the solutions of the linear SDE

$$\begin{aligned} d\Gamma_t^{\tilde{\Theta}} &= b_x(t,\tilde{\Theta}_t)\Gamma_t^{\tilde{\Theta}}\,dt + \sigma(t,\Gamma_t^{\tilde{\Theta}})^\top\,dW_t, & \Gamma_0^{\tilde{\Theta}} &= 1 \\ d\Gamma_t^{up} &= \Gamma_t^{up} L\,dt + \Gamma_t^{up}\sigma_x(t)^\top\,dW_t, & \Gamma_0^{up} &= 1 \\ d\Gamma_t^{lo} &= -\Gamma_t^{lo} L\,dt + \Gamma_t^{lo}\sigma_x(t)^\top\,dW_t, & \Gamma_0^{lo} &= 1, \end{aligned}$$

where L is a Lipschitz-constant for b. Apply the the Itô formula to $\big(\Gamma_t^{\tilde{\Theta}}\big)^{-1}\big(X_t^\star - X_t^{x,C^U,0}\big)$ to obtain the

Lemma 3.2. *C_t^L has $\boldsymbol{P}$-a.s. upper and lower bounds for $\tau^0 < t \leq \tau^\delta$:*

$$\big(X_t^\star - X_t^{x,C^U,0}\big)\frac{1}{\Gamma_t^{\tilde{\Theta}}}\Big(\inf_{\tau^0<s\leq t}\Gamma_s^{\tilde{\Theta}}\Big) \leq C_t^L \leq \big(X_t^\star - X_t^{x,C^U,0}\big)\frac{1}{\Gamma_t^{\tilde{\Theta}}}\Big(\sup_{\tau^0<s\leq t}\Gamma_s^{\tilde{\Theta}}\Big).$$

Thus C^L satisfies for $t \leq \tau^\delta$ $\boldsymbol{P}$-a.s. $C_t^L \leq \delta\Gamma_t^{up}\big(\Gamma_t^{lo}\big)^{-1}\Big(\sup_{\tau^0\leq s\leq t}\Gamma_s^{up}\Big) =: \delta\cdot\overline{C_t^{L,\delta}}$.

Bounds for the approximation $\tilde{Y}^\delta$ Let $(\tilde{R}^{up,\delta},\tilde{Q}^{up,\delta})$ and $(\tilde{R}^{lo,\delta},\tilde{Q}^{lo,\delta})$ the solutions to the BSDE with data $(\xi^{up,\delta}, g^{up,\delta}, 0)$ and $(\xi^{lo,\delta}, g^{lo,\delta}, 0)$, where

$$\begin{aligned} \xi^{up,\delta} &:= h_x(X_T^{x+\delta,C^U,0})\Gamma_T^{x+\delta,C^U,0}\chi_{\sigma\wedge\tau^\delta=T} + a_\sigma^U\Gamma_\sigma^{x+\delta,C^U,0}\chi_{\substack{\sigma\leq\tau^\delta\\ \sigma<T}} \\ &\quad + \Big(\max_{\tau^0\leq s\leq\tau^\delta} a_s^L\Big)\Big(\sup_{\tau^0\leq s\leq\tau^\delta}\Gamma_s^{x+\delta,C^U,0}\Big)^2\Big(\inf_{\tau^0\leq s\leq\tau^\delta}\Gamma_s^{x,C^U,0}\Big)^{-1}\chi_{\substack{\tau^0<\sigma\\ \tau^0<T}} \\ g^{up,\delta}(t,y,z) &:= \Big(g_x(t,X_t^\delta,Y_t^\star)\Gamma_t^{x+\delta,C^U,0} + g_y(t,X_t^\delta,Y_t^\delta)\chi_{y\geq0}\chi_{A_{g,\delta}^+}\cdot y \\ &\quad + L\Big(\max_{\tau^0\leq s\leq\tau^\delta\wedge t} a_s^L\Big)\overline{C_t^{L,\delta}}\Big)\chi_{t\leq\sigma\wedge\tau^\delta} + g_z(t)\cdot z\chi_{t\leq\sigma}, \\ \xi^{lo,\delta} &:= h_x(X_T^{x,C^U,0})\Gamma_T^{x,C^U,0}\chi_{\sigma\wedge\tau^0=T} + a_\sigma^U\Gamma_\sigma^{x,C^U,0}\chi_{\substack{\sigma\leq\tau^0\\ \sigma<T}} \\ &\quad + \Big(\min_{\tau^0\leq s\leq\tau^\delta} a_s^L\Big)\Big(\inf_{\tau^0\leq s\leq\tau^\delta}\Gamma_s^{x,C^U,0}\Big)^2\Big(\sup_{\tau^0\leq s\leq\tau^\delta}\Gamma_s^{x+\delta,C^U,0}\Big)^{-1}\chi_{\substack{\tau^\delta<\sigma\\ \tau^\delta<T}} \\ g^{lo,\delta}(t,y,z) &:= g_x(t,X_t^\star,Y_t^\star)\Gamma_t^{x,C^U,0}\chi_{t\leq\sigma\wedge\tau^0} + \Big(g_y(t,X_t^\star,Y_t^\star)\chi_{y\geq0}\chi_{A_{g,\delta}^-}\cdot y \\ &\quad - L\Big(\max_{\tau^0\leq s\leq\tau^\delta\wedge t} a_s^L\Big)\overline{C_t^{L,\delta}}\Big)\chi_{t\leq\sigma\wedge\tau^\delta} + g_z(t)\cdot z\chi_{t\leq\sigma}. \end{aligned}$$

L denotes a Lipschitz constant of g. We also use the non-standard convention $\max_{\tau^0\leq s\leq\tau^\delta\wedge t} a_s^L = 0$ for $t < \tau^0$. The sets $A_{g,\delta}^+$ and $A_{g,\delta}^-$ ensure appropriate monotonicity of the data for $\tilde{R}^{up,\delta}$ and $\tilde{R}^{lo,\delta}$ and are specified as follows:

$$\begin{aligned} A_{g,\delta}^+ &:= \{t\leq\sigma\wedge\tau^0\} \cup \{g_y(t,X_t^\delta,Y_t^\delta) \geq 0\} \\ A_{g,\delta}^- &:= \{t\leq\sigma\wedge\tau^0\} \cup \{g_y(t,X_t^\star,Y_t^\star) \leq 0\}. \end{aligned}$$

Finally we define the limiting process $\tilde{R}$ through the solution $(\tilde{R}, \tilde{Q})$ of the BSDE with data $(\tilde{\xi}, \tilde{g}, 0)$, where

$$\tilde{\xi} := h_x(X_T^{x,C^U,0})\Gamma_T^{x,C^U,0}\chi_{\sigma\wedge\tau^0=T} + a_\sigma^U\Gamma_\sigma^{x,C^U,0}\chi_{\substack{\sigma\leq\tau^0\\ \sigma<T}} + a_{\tau^0}^L\Gamma_{\tau^0}^{x,C^U,0}\chi_{\substack{\tau^0<\sigma\\ \tau^0<T}}$$

$$\tilde{g}(t,y,z) := \Big(g_x(t,X_t^\star,Y_t^\star)\Gamma_t^{x,C^U,0} + g_y(t,X_t^\star,Y_t^\star)\chi_{y\geq 0}\cdot y\Big)\chi_{t\leq\sigma\wedge\tau^0} + g_z(t)\cdot z\chi_{t\leq\sigma}.$$

After some lengthy technical computations involving comparison theorems and a priori estimates for forward and backward equations we arrive at the following

Lemma 3.3. *Assume* (2) *and let* $\tilde{R}^{up,\delta}$, $\tilde{R}^{lo,\delta}$ *and* $\tilde{R}$ *as above. Then* (17) *holds.*

This proves (15) and hence the first inequality in (14). By a similar argument one obtains $\operatorname{ess\,sup}_{\tau\in\mathcal{T}_t} R_t(\sigma^0,\tau) \leq \Delta_- V_t(x)$; one concludes that the third inequality in (14) holds and that (σ^0,τ^0) is an optimal pair.

4. Related topics

Consider once more Y_t as a dynamic cost functional. It is recursive in that Y_t explicitly takes into account its future values through g. If g is nonlinear in y this cost functional is not additive with respect to costs incurred at different instants. This corresponds to the properties of stochastic differential utility introduced in Duffie and Epstein [6].

One may also view Y as a g-semisolution for ξ (or g-semimartingale) in the sense of Peng [13], where g varies with C. Theorem 3.1 then relates the g-semimartingale $Y^\star$ to a Dg-martingale.

Hedging and pricing of american contingent claims was investigated through reflected BSDE by Cvitanić and Karatzas [5], El Karoui et al. [7] and Ma and Yong [11]. In the context of this paper, consider the linear reflected BSDE

$$\begin{aligned} A_t &= h_x(X_T^{x,0,0})\Gamma_T^x + (K_T^U - K_t^U) - (K_T^L - K_t^L) \\ &\quad + \int_t^T \langle Dg(s,X_s^{x,0,0},V_s(x)),(\Gamma_s^x,A_s,B_s)\rangle\,dt - \int_t^T B_s^\top\,dW_s \\ a_t^L\Gamma_t^x &\leq A_t \leq a_t^U\Gamma_t^x \\ 0 &= \int_0^T (a_s^U\Gamma_s^x - A_s)^+\,dK_s^U = \int_0^T (A_s - a_s^L\Gamma_s^x)^+\,dK_s^L. \end{aligned} \tag{18}$$

It is known that if there exist (A,B,K) satisfying (18) then the Dynkin game (1.2) has a solution. Its value is given as $u_t(x) = A_t^x$, and the pair of stopping times $\big(\hat{\sigma}_t := \inf\{s\in[t,T)\,|\,A_s^x = a_s^U\Gamma_s^x\},\ \hat{\tau}_t := \inf\{s\in[t,T)\,|\,A_s^x = a_s^L\Gamma_s^x\}\big)$ forms a saddle point for the Dynkin game .

We conclude that, if (A^x,B,K) is a solution to (18) then $A_t^x = \frac{\partial}{\partial x}V_t(x)$. Observe however that the processes C and K are somewhat different in nature: If the optimal stopping times in the stopping problem take the form of exit times from a region $\mathcal{C} \subset \bar{\imath} \times \mathbb{R}$ with a smooth boundary, then both the optimal state process $X^{x,C}$ and A^x are diffusions in $\mathcal{C}$ with reflection on $\partial\mathcal{C}$ caused by C and K respectively. But an optimal control C is singular with respect to to Lebesgue measure, whereas K is absolutely continuous (see [7]) as the strong irregularity of Brownian motion is absorbed by B.

References

[1] F. Baldursson and I. Karatzas, *Irreversible investment and industry equilibrium*, Finance Stochast., **1** (1997), 69 –89.

[2] J. Bather and H. Chernoff, *Sequential decisions in the control of a spaceship*, in Proc. Fifth Berkeley Symposium on Mathematical Statistics and Probability, vol. 3, 1966, 181 – 207.

[3] F. Boetius, *Bounded variation singular stochastic control and associated Dynkin game*, Discussion Paper Series, CoFE, Universität Konstanz, **00/12** (2000).

[4] F. Boetius and M. Kohlmann, *Connections between optimal stopping and singular stochastic control*, Stoch. Proc. Appl., **77** (1998), 253 – 281.

[5] J. Cvitanić and I. Karatzas, *Backward stochastic differential equations with reflection and Dynkin games*, Annals of Prob., **24** (1996), 2024 – 2056.

[6] D. Duffie and L. G. Epstein, *Stochastic differential utility (Appendix C with Costis Skiadas)*, Econometrica, **60** (1992), 353 – 394.

[7] N. El Karoui, C. Kapoudjian, E. Pardoux, S. Peng, and M.-C. Quenez, *Reflected solutions of backward SDE's, and related obstacle problems for PDE's*, Annals of Prob., **25** (1997), 702 – 737.

[8] N. El Karoui, S. Peng, and M.-C. Quenez, *Backward stochastic differential equations in finance*, Mathematical Finance, **7** (1997), 1 – 71.

[9] N. El Karoui and M.-C. Quenez, *Non-linear pricing theory and backward stochastic differential equations*, in Financial Mathematics, Lectures given at the 3rd Session of the C.I.M.E. in July 1996, W. Runggaldier, ed., Springer, 1997, 191 – 246.

[10] I. Karatzas and S. Shreve, *Connections between optimal stopping and singular stochastic control I. Monotone follower problems*, SIAM J. Control Optim., **22** (1984), 856 – 877.

[11] J. Ma and J. Yong, *Forward-Backward Stochastic Differential Equations and Their Applications*, Springer, Berlin, Heidelberg, New York, 1999.

[12] E. Pardoux and S. Peng, *Adapted solution of a backward stochastic differential equation*, Systems and Control Letters, **14** (1990), 55–61.

[13] S. Peng, *Monotonic limit theorem of BSDE and nonlinear decomposition theorem of Doob-Meyer's type*, Probab. Theory Related Fields, **113** (1999), 473 – 499.

Universität Konstanz, Konstanz, Germany
E-mail address: `frederik.boetius@uni-konstanz.de`

Trends in Mathematics, © 2001 Birkhäuser Verlag Basel/Switzerland

Option Pricing and Hedging Under Regular Lévy Processes of Exponential Type

Svetlana I. Boyarchenko and Sergei Z. Levendorskiǐ

Abstract. We define a class of regular Lévy processes of exponential type, which contains hyperbolic processes, normal inverse gaussian processes and Lévy processes of Koponen's family. We show that infinitesimal generators of these processes and corresponding generalized Black-Scholes equations enjoy common fairly favorable features from the point of view of the theory of pseudo-differential operators, and these properties are independent of a choice of an equivalent martingale measure (EMM) from a wide class of EMM. We discuss possible ways of fitting parameters of regular Lévy processes of exponential type and EMM to data, derive an explicit formula for a locally risk-minimizing hedging ratio and show that it is Hölder continuous for options at the money, at expiry date. We derive pricing formulas for some American and barrier options, show that in some cases the principle of the smooth fit fails, and suggest a substitute for it.

1. Introduction

By now, the drawbacks of the gaussian modeling in financial markets are well-known. In particular, gaussian models cannot produce so-called fat tails of observed probability distribution functions (PDF). Lévy stable processes advocated by Mandelbrot [20] give fat tails, but the tails are too fat (polynomially decaying); even worse, PDF of non-gaussian stable distributions have infinite second moments, which contradicts the observed convergence of empirical PDF to gaussian distributions over longer time scale. During 90th, several families of Lévy processes with PDF having semi-heavy (i.e. exponentially decaying) tails have been used by several groups of researchers to model stock returns and price options: Hyperbolic Processes (HP); Variance Gamma Processes (VGP); Truncated Lévy Processes (TLP); Normal Inverse Gaussian Processes (NIG) and Generalized Hyperbolic Processes, for which HP and NIG are special examples (see, respectively, [21]; [14]; [9, 10]; [1, 2] and [15] and the bibliography there).

The gaussian world is very convenient since in simple cases, there is only one equivalent martingale measure (EMM) in the market but the theory implies that no-one should trade in such a market, which undermines the theory. The

Corresponding author. Phone 7-8632-341292; fax 7-9632-654521 pricing under non-gaussian Lévy processes is free from the last drawback, but at a price: according to the modern theory [11], arbitrage-free prices can be obtain for any EMM, which is absolutely continuous w.r.t. to the historic measure, as expectations under a chosen EMM, and for a non-gaussian process, EMM is not unique (unless it is a Poisson process). In [13], for a wide class of Lévy processes, it is shown that prices of a European option, which can be obtained in this way, span the whole possible non-arbitrage interval, and a convenient characterization of an admissible EMM is provided.

A choice of EMM is a non-trivial task, and for more than a decade, a lot of effort had been made to obtain a prescription for the derivation of a formula for EMM from a historic measure (see e.g. [16, 19] and the bibliography there). Since a historic measure is a much easier object to study than EMM implied in option prices, and the Black-Scholes model with its possibility of deriving an EMM from a historic measure has set a sort of a benchmark, one can understand these efforts. Still, the authors do believe that there is no real reason to expect that it is possible to derive a formula for the EMM from the historic prices by using formal argument of any sort – be it based on Mathematics or Economics. We think that Mathematical Finance should be more of Applied Mathematics than pure exercises in Mathematical Economics, and hence, we believe that EMM should be inferred from the data on both option prices and stock prices – possibly, even more from the former than the latter. Certainly, we are not the first to suggest using option prices in the first place: this is the very thing practitioners do, only they use them in the language of the Black-Scholes model, as a simple technical tool to reformulate the data ("implied volatilities"), and Bouchaud, Eberlein and Madan with their co-authors use option prices to fit EMM in non-gaussian situations, which makes much more sense (see [9, 14, 21] and the bibliography there).

To sum up, we start with an assumption that an EMM has been somehow inferred from the data (in Section 2, we discuss a possible form of an EMM and suggest some ways of adjusting an EMM to data), and we are mainly concerned with the next step, which has been essentially neglected in the literature on option pricing under non-gaussian processes.

In the gaussian case, there exists not only the martingale approach to the option pricing but the partial differential equations (PDE) approach as well, which was the first in Mathematical Finance. In the gaussian modeling, the latter has been overshadowed by the former, but the martingale approach encounters many problems in the non-gaussian case, and hence, the former has some chance of a rebound. For non-gaussian Lévy processes with jumps an appropriate theory is not a theory of (local) differential operators but (non-local) pseudo-differential operators (PDO). The usage of PDO for studying of stochastic processes has been strongly advocated in [17] but the stochastic models in op.cit. were essentially modifications of Lévy stable processes. This is unrealistic from the point of view of applications (see the discussion about [20] above), and at the same time, this leads to significant complications from the point of view of the general theory of PDO. One of starting points in [17] was that the existing general theories of PDO were inadequate (which certainly was true as far as processes similar to non-gaussian Lévy stable processes were concerned), but our claim is that processes observed in Financial Markets belong to a very nice class of processes, with infinitesimal generators enjoying almost all desired properties possible but one: the transmission property [8] or even weaker property [12].

In [4], the main properties these processes enjoy had been singled out, and these processes were called Generalized Truncated Lévy Processes; the secong author is grateful to A.N.Shiryaev for pointing out that the name was inadequate. From now on, we use the name "regular Lévy processes of exponential type", suggested in [3], where a generalization of this family for Feller processes is constructed.

"Of exponential type" indicates exponential estimates for tails of PDF, and "regular" means that among all non-gaussian processes, they are simplest from the point of view of the general theory of PDO so that one can use the huge amount of available techniques. We illustrate these techniques with the problems of

(i) hedging–we produce an explicit formula for the locally risk-minimizing portfolio (for the definition, see e.g., [16]) assuming EMM given, without discussing a choice of an EMM based on general principles as in e.g. [16, 19];

(ii) pricing of perpetual American options. Recently our results [5, 6] for regular Lévy processes of exponential type has been obtained for any Lévy process [22], but without explicit formulas relevant to processes observed in Financial Markets, which we do provide; in addition, our technique allows one to demonstrate that the smooth fit principle can fail for processes observed in the financial markets, and suggest a substitute for it. Finally, the standard stochastic technique as in [22] does not allow one to obtain a discrete-time analog as in [7], where an investment problem in the discrete time was considered;

(iii) pricing of some path-dependent claims, as barrier options and touch-and-out options.

To sum up: we demonstrate which general properties infinitesimal generators of processes in many empirical studies of Financial Markets enjoy, and which results can be obtained by using the general technique of PDO.

The plan of the paper is as follows: in Section 2, we give the definition of regular Lévy processes of exponential type, discuss the properties of their infinitesimal generators and the generalized Black-Scholes equation, the problem of parameter fitting and construct locally risk-minimizing portfolio. In Section 3 we discuss perpetual American options and the principle of the smooth fit, and finally, in Section 4, barrier options and touch-and-out options.

2. Definition, basic properties and locally risk-minimizing hedging

2.1. Definition and properties of generators

In [4] and [6], two definitions are given: in terms of the Lévy measure $F(dx) = p(x)dx$: p has a singularity at the origin $p(x) \sim c|x|^{-\nu-1}$, as $x \to 0$, where $\nu \geq 0$, and bounded by decaying exponentials as $x \to \pm\infty$, and in terms of the characteristic exponent ψ, defined from $E[e^{-ikX_t}] = e^{-t\psi(k)}$. The first variant includes VGP into the family, the second one does not, which is due to the fact that properties of generators of VGP essentially differ from the ones of generators of other processes of the family. Here we use the second definition.

Definition 2.1. *Let there exist constants $c > 0$, $\nu \in (0, 2]$, $\nu' < \nu$, $\mu \in \mathbb{R}$, $\lambda_- < \lambda_+$, and C such that*

$$\psi(k) = i\mu k + \phi(k),$$

where ϕ admits the analytic continuation into a strip $\Im k \in (\lambda_-, \lambda_+)$, and the continuous extension up to the boundary of the strip, and satisfies the following two estimates: for all k in a strip $\Im k \in [\lambda_-, \lambda_+]$

$$|\phi(k) - c|k|^\nu| \leq C\langle k \rangle^{\nu'},$$

where $\langle k \rangle = (1 + |k|^2)^{1/2}$, and for any $[\lambda'_-, \lambda'_+] \subset (\lambda_-, \lambda_+)$ and all k in a strip $\Im k \in [\lambda'_-, \lambda'_+]$

$$|\phi'(k)| \leq C\langle k \rangle^{\nu-1},$$

where C depends on $[\lambda'_-, \lambda'_+]$ but not on k.

We say that X is a regular Lévy process of the order ν and the exponential type $[-\lambda_+, -\lambda_-]$.

It follows from the integro-differential representation of L, the infinitesimal generator of X, that it acts on oscillating exponents as follows: $Le^{ixk} = -\psi(-k)e^{ixk}$. By using the Fourier inversion formula, one finds that the action of L on sufficiently regular functions can be defined as

$$-Lu(x) = (2\pi)^{-1} \int_{-\infty}^{+\infty} e^{ixk}\psi(-k)\hat{u}(k)dk,$$

where $\hat{u}$ is the Fourier image of u: $u(x) = \int_{-\infty}^{+\infty} e^{-ixk}u(x)dx$. In the language of the theory of PDO, this means that $-L = \psi(-D)$ is a PDO with the symbol $\psi(-k)$. If $\nu \geq 1$ or $\mu = 0$, its order is ν, and the principal symbol is $c|k|^\nu$, so that $-L$ is a strongly elliptic PDO; if $\nu \in (0, 1)$ and $\mu \neq 0$, the order is 1, and $-L$ is not strongly elliptic. In addition, the symbol is holomorphic in a strip $\Im k \in (-\lambda_+, -\lambda_-)$ and admits a continuous extension into the closed strip.

HP and NIG are processes of order 1, the Brownian Motion – of order 2, the Koponen's family and its generalization introduced in [4, 5] as processes with the Lévy measures

$$F(dx) = c_+ x_+^{-\nu-1} e^{-\lambda_+ x} + c_- x_-^{-\nu-1} e^{-\lambda_- x},$$

and the characteristic exponents

$$\psi(k) = i\mu k + c_+\Gamma(-\nu)[\lambda_+^\nu - (\lambda_+ + ik)^\nu] + c_-\Gamma(-\nu)[(-\lambda_-)^\nu - (-\lambda_- - ik)^\nu],$$

where $c_\pm > 0, \lambda_- < 0 < \lambda_+, \nu \in (0, 2), \nu \neq 1$, are processes of order ν. When $c = c_+ = c_-$, this model produces PDF almost symmetric in the central part, and the asymmetry of tails can be achieved by choosing $-\lambda_- < \lambda_+$.

Modulo the drift term and the conjugation with an exponential, generators of NIG are of an especially nice form $-L = \delta[(\alpha^2 - \Delta)^{1/2} - \alpha]$ so the theory of fractional powers of differential operators (DO) applies; for more general processes obtained from diffusions via subordination (see e.g. [23]), more general functions of differential operators appear as generators. This example illustrates typical properties of generators of regular Lévy processes of exponential type: being of a good class of PDO, they are among nicest operators on the whole space (here, line) but being fractional powers of DO or functions of more general sort, they do not possess the transmission property [8], which makes the study of boundary problems for such operators much more difficult. Notice that the generator of a VGP is, essentially, $-\ln(1-\Delta)$, and hence, not a standard object in the theory of PDO (though the latter is applicable to such operators, too).

2.2. Choice of EMM and properties of the generalized Black-Scholes equation

Let r denote the interest rate, which we assume constant, and let $S_t = \exp X_t$ be the price process of a stock, where X_t is a purely discontinuous Lévy process under a historic measure $\mathbb{P}$. In [13], a convenient characterization of admissible EMM $\mathbb{Q}$ is provided. Let $F'(dx)$ be the Lévy measure of X_t under $\mathbb{Q}$. From Proposition 1 in [13], it follows that one may look for $F'(dx)$ in the form $F'(dx) = y(x)F(dx)$, where e.g.

$$y(x) = e^{\theta} + dx_{-}e^{\mu_{-}x},$$

where $d \geq 0, \mu_{-} < 0$. With $d = 0$, the definition of EMM leads to a unique θ, and one recovers the Esscher transform used for option pricing in [21, 14] (see also the bibliography there), and the second term can be used to take into account rare large negative jumps. If the parameters for the historic measure are not specified in advance, one may fit parameters of the historic process and θ simultaneously, by choosing them on a hyper-surface in the (parameters, θ)-space, where the EMM-requirement is satisfied (see e.g. [14]). Similarly one can fit a model with more parameters, as above, or even add more terms (for a rather general equation for admissible hyper-surfaces, see [4]). We believe, though, that it may be more reasonable to fit parameters λ_{-}, λ_{+} in a generalization of Koponen's model or similar parameters in NIG and HP models by looking at the behavior of $\ln p_t(x)/x$ as $x \to \mp\infty$, and use the maximum likelihood estimates for the set of other parameters after.

As far as the PDO-approach is concerned, the important thing is that if one chooses EMM in this way, the process under EMM is also a regular Lévy process of the same order but typically of a different exponential type, and main properties of the infinitesimal generator remain the same: the principal symbol remains unchanged, and so properties of Generalized Black-Scholes equation (the backward Kolmogorov equation) are essentially independent of the choice of EMM from a wide class. In variables $\tau = T - t$, x, the equation is of the form

$$\partial_\tau f + (r + \psi^{\mathbb{Q}}(-D_x))f = 0,$$

where $\psi^{\mathbb{Q}}$ is the characteristic exponent of X_t under a chosen EMM $\mathbb{Q}$; in order that the bond and stock can be priced, the exponential type must satisfy $\lambda_- \le 0 < 1 \le \lambda_+$. If $\nu \ge 1$ or $\mu = 0$, the generalized Black-Scholes equation is a parabolic equation, with an elliptic PDO having a positive-definite real part instead of a differential operator of second order in the Black-Scholes model. If $\nu \in (0,1)$ but $\mu \neq 0$, the operator is not parabolic since the stationary part is not elliptic. By changing variables $x \mapsto x + \mu\tau$, one can obtain a parabolic equation with the stationary part being elliptic of order ν but it spoils the boundary if one considers, say, barrier options with time-independent barriers or touch-and-out options. This is the reason why we study these options below only when the stationary part is elliptic in the initial coordinates.

Notice an evident formula for the price of an European option with the payoff g at the expiration date T:

$$f(\tau, x) = \exp[-\tau(r + \psi^{\mathbb{Q}}(-D_x))]g(x),$$

which can be made more explicit by using the Fourier transform. For instance, for a European call with the strike price normalized to 1,

$$f(\tau, x) = (2\pi)^{-1} \int_{-\infty+i\sigma}^{+\infty+i\sigma} \frac{\exp[ixk - \tau(r + \psi^{\mathbb{Q}}(-k))]}{k(k+i)} dk,$$

for any $\sigma \in (-\lambda_+, -1)$ (here we assume for simplicity that $\lambda_+ > 1$).

2.3. Locally risk-minimizing hedging ratio

Consider a portfolio short one option, and long θ_t shares of the stock. The price of the portfolio is $W_t = -f_t + \theta_t S_t$, where f_t is the price of the option. Suppose, that at each date t, an investor chooses θ_t to minimize the variance of the portfolio at $t + \Delta t$, conditionally on information available at date t.

We calculate the conditional variance of wealth under $\mathbb{P}$:

$$E_t[(W_{t+\Delta t} - E_t[W_{t+\Delta t}])^2] = E_t[(f_{t+\Delta t} - E_t[f_{t+\Delta t}])^2] -$$

$$-2\theta_t E_t[(f_{t+\Delta t} - E_t[f_{t+\Delta t}])(S_{t+\Delta t} - E_t[S_{t+\Delta t}])] + \theta_t^2 E_t[(S_{t+\Delta t} - E_t[S_{t+\Delta t}])^2].$$

Minimizing w.r.t. θ_t, we obtain the following first order condition:

$$\theta_t = \theta_{t,\Delta t} = \frac{E_t[f_{t+\Delta t} S_{t+\Delta t}] - E_t[f_{t+\Delta t}] E_t[S_{t+\Delta t}]}{E_t[S_{t+\Delta t}^2] - E_t[S_{t+\Delta t}]^2},$$

and by expanding into a series w.r.t. Δt as $\Delta t \to 0$, we obtain the following theorem (for details, see [4]).

Theorem 2.2. *Let X be a regular Lévy process: under $\mathbb{P}$, of the order ν and the exponential type $[-\lambda_+, -\lambda_-]$, where $\lambda_- < 0 < 2 < \lambda_+$, and under $\mathbb{Q}$, of the exponential type $[-\mu_+, -\mu_-]$, where $\mu_- < 0 < 1 < \mu_+$.*

Then for European options with payoffs not growing as the stock price $S \to 0$ and growing not faster than S as $S \to +\infty$, in the limit $\Delta t \to +0$,

$$\theta_t = (B(D_x)f_t)S_t^{-1},$$

where

$$B(k) = \frac{-\psi(-k+i) + \psi(-k) + \psi(i)}{-\psi(2i) + 2\psi(i)}.$$

In particular, for the European call with the strike price K,

$$\theta_t(\ln S) = \frac{K}{2\pi S} \int_{-\infty+i\sigma}^{+\infty+i\sigma} \frac{\exp[ik\ln(S/K) - \tau(r + \psi^{\mathbb{Q}}(-k))]B(k)}{k(k+i)} dk.$$

When X_t is the Brownian Motion, we recover the standard Δ-hedging result. It is straightforward to see that if $\nu < 2$ and $\nu' < 1$, the integrand decays as $|k|^{-2+\omega}$ as $|k| \to \infty$, where $\omega = \max\{\nu - 1, \nu'\} < 1$, hence θ_t is Hölder continuous at the expiry, at the money, whereas in the gaussian case, it is discontinuous. This observation shows possible advantages of non-gaussian hedging.

3. Perpetual American options and the principle of the smooth fit

By using the Dynkin formula (see e.g. eq. (41.3) in [23]), it is possible to reduce an optimal stopping problem, related to perpetual American option with a non-negative continuous payoff g, to a non-local free boundary problem of the form: find an open set J and a function V such that

$$(q - L^{\mathbb{Q}})V(x) = 0, \quad x \in J; \tag{1}$$

$$V(x) = g(x), \quad x \notin J; \tag{2}$$

$$V(x) \geq g(x), \quad x \in J; \tag{3}$$

$$(q - L^{\mathbb{Q}})V(x) \geq 0, \quad x \notin \bar{J}. \tag{4}$$

Here $q = r+\lambda$, where $\lambda \geq 0$ is the constant dividend rate. Eq. (1)–(2) are equivalent to the Wiener-Hopf equation for $u = V - g$ on J, and can be easily solved by the Wiener-Hopf factorization method if J is a half-axis as it is the case with European calls and puts. Take puts, for instance. For any $J = (h, +\infty)$, we obtain an explicit formula for u:

$$u(x) = A_-(D)^{-1}\mathbf{1}_{[h,+\infty)}(x)(A_-(-i)e^x - A_-(0)K), \tag{5}$$

where $A_-(k)$ is one of the factors in the Wiener-Hopf factorization formula for $A(k) = q + \psi^{\mathbb{Q}}(-k)$: $A(k) = A_-(k)A_+(k)$. Here $A_\pm(k)$ is holomorphic in a half-plane $\pm\Im k > 0$ and continuous up to the boundary. The factorization exists for any Lévy process [23], Theorem 45.2, but explicit formulas can be obtained only for special cases. Luckily, regular Lévy processes are good enough (here "the exponential type" condition, i.e. the analyticity of $\psi^{\mathbb{Q}}$ is not needed: smoothness suffices), so that an explicit formulas for $A_\pm(k)$ can be obtained (see [5, 6]). By using (5), we relatively easily verify that for any choice of h but one, (3) or (4) fails. If $\nu < 2$, the unique candidate for the free boundary can be characterized as h such that $u'(x)$ is bounded as $x \to h$. This is a substitute for the smooth fit principle for regular Lévy processes with sufficiently smooth characteristic exponents; it can be shown [6], that for processes of order $\nu \in (0, 1)$, the standard smooth fit principle fails iff $\mu > 0$ (for calls, iff $\mu < 0$).

The unique candidate is determined from an equation $A_-(-i)e^x - A_-(0)K = 0$, and we obtain the optimal exercise price $H = KA_-(0)/A_-(-i)$. After that, we verify that (3)–(4) holds. In [6], we conjectured that this formula (and its analog for calls) is valid for any Lévy process; in [22] it has been proved, without an explicit formula for $A_-(0)/A_-(-i)$ for processes in real Financial Markets, though.

4. Pricing of barrier options and touch-and-out options

These problems can be reduced to non-local boundary problems for a parabolic pseudo-differential equation (notice that the rebate for a barrier option must be specified not only at the barrier as in the gaussian case but for all values of x the other side of the barrier), and solved by using the representation theorem for analytic semigroups [24]. To apply this theorem, the resolvent $(\lambda + \tilde{A})^{-1}$ of the operator $\tilde{A}$ of an appropriate boundary problem for an elliptic operator $r+\psi^{\mathbb{Q}}(-D)$ must be constructed, for $\lambda \in \mathbb{C}$ satisfying $\arg \lambda \in (-\theta, \theta)$ for some $\theta \in (\pi/2, \pi)$, and an estimate $||\lambda(\lambda+\tilde{A})^{-1}|| \le C$, uniform w.r.t. λ must be proved. The Wiener-Hopf factorization allows us to do that (for some $\theta \in (\pi/2, \pi)$) but only if either $\nu \ge 1$ or $\mu = 0$, i.e. exactly in cases when the stationary part of the generalized Black-Scholes equation is elliptic. So, the pricing formulas, which results from Section 10 in [24], and are formulated below, are valid only in this case.

Consider a down-and-out call option. Let $H = e^h$ be the barrier, and $g_0 e^{\beta x}$, $g_0 \ge 0$, $\beta \ge 0$ be the rebate paid when $S = e^x$ reaches H or falls below it (more general rebates and other types of barrier options can also be considered). Let $\lambda + r + \psi^{\mathbb{Q}}(-k) = A_+(\lambda, k)A_-(\lambda, k)$ be the Wiener-Hopf factorization, and $\tilde{L}_\theta = \{\lambda(\sigma) \mid -\infty < \sigma < +\infty\}$ be the contour defined by $\lambda(\sigma) = -\theta, \sigma < 0$, and $\lambda(\sigma) = \theta, \sigma > 0$. Then for the case of H normalized to 1, $f(\tau, x)$, the price of the barrier option, is determined by the following set of formulas:

$$f(\tau, x) = e^x + (2\pi)^{-1} \int_{-\infty - i\omega}^{+\infty - i\omega} e^{ikx} \hat{v}_1(k) dk + v_2(\tau, x),$$

where $\omega < \min\{1, \beta\}$, and

$$\hat{v}_1(k) = A_-(0,k)^{-1}[A_-(0,-i)/(ik-1) - g_0 A_-(0,-i\beta)/(ik-\beta)];$$

$$v_2(\tau, x) = -(2\pi i)^{-3} \int_{\tilde{L}_\theta} \int_{-\infty - i\epsilon_1}^{+\infty - i\epsilon_1} \int_{-\infty - i\epsilon}^{+\infty - i\epsilon} \frac{e^{\lambda\tau + ixk}(\hat{G}(k') - \hat{v}_1(k'))dk'dkd\lambda}{A_-(\lambda,k)(k-k')A_+(\lambda,k')},$$

where $0 < \epsilon < \epsilon_1 < \lambda_+$ are arbitrary, and $G(x) = \mathbf{1}_{x>0}(x)\max\{-K, -e^x\}$. If $\nu \in (1, 2]$ or $\mu = 0$, then $A_-(\lambda, k) = ((1+\lambda^2)^{1/2} + ik)^{\nu/2} \exp b_-(\lambda, k)$, $\Im k \le 0$, where

$$b_-(\lambda, k) = -\frac{i}{2\pi} \int_{-\infty + i\omega}^{+\infty + i\omega} \frac{\ln[c^{-1}(1+\lambda^2+l^2)^{-\nu/2}(r + \psi^{\mathbb{Q}}(-l))]}{k-l} dl,$$

and $\omega > 0$ is sufficiently small so that ln above is well-defined. (By the Cauchy theorem, the result is independent of a choice of such ω). In the case $\nu = 1, \mu \ne 0$, the formula is more complicated.

Now we consider a touch-and-out option ("American digital put"), which gives the right to get \$1 the first time the price of the stock crosses the level K from above. If K is normalized to 1, the price $f(\tau, x)$ is given by

$$f(\tau, x) = v_1(x) + v_2(\tau, x);$$

$$v_1(x) = -\frac{A_-(0,0)}{2\pi i} \int_{-\infty-i\omega}^{+\infty-i\omega} \frac{e^{ixk}dk}{A_-(0,k)k},$$

where $\omega > 0$ is arbitrary;

$$v_2(\tau, x) = (2\pi)^{-3} \int_{\tilde{L}_\theta} \int_{-\infty-i\epsilon_1}^{+\infty-i\epsilon_1} \int_{-\infty-i\epsilon}^{+\infty-i\epsilon} \frac{A_-(0,0)e^{\lambda\tau+ixk}dk'dkd\lambda}{A_-(\lambda,k)(k-k')A_+(\lambda,k')A_-(0,k')k'},$$

where $0 < \epsilon < \epsilon_1 < \lambda_+$ are arbitrary.

References

[1] O.E. Barndorff-Nielsen, *Processes of Normal Inverse Gaussian Type*, Finance and Stochastics, **2** (1998), 41–68.

[2] O.E. Barndorff-Nielsen and W. Jiang, *An initial analysis of some German stock price series*, Working Paper Series, Aarhus: CAF Univ. of Aarhus/Aarhus School of Business **15** (1998).

[3] O.E. Barndorff-Nielsen and S.Z.Levendorskiǐ, *Feller processes of normal inverse gaussian type*, Mimeo (2000)

[4] S.I. Boyarchenko and S.Z. Levendorskiǐ, *Generalizations of the Black-Scholes equation for truncated Lévy processes*, Mimeo (1999).

[5] S.I. Boyarchenko and S.Z. Levendorskiǐ, *Option pricing for truncated Lévy processes*, Intern. Journ. Theor. and Appl. Finance (2000).

[6] S.I. Boyarchenko and S.Z. Levendorskiǐ, *Perpetual American options under Lévy processes*, Submitted to SIAM J.Control and Optimization (2000).

[7] S.I. Boyarchenko and S.Z. Levendorskiǐ, *A discrete time model of investment under non-gaussian shocks*, submitted to Journ. of Math. Economics (2000)

[8] L. Boutet de Monvel, *Boundary problems for pseudo-differential operators*, Acta Math. **126** (1971), 11–51.

[9] J-P.Bouchaud and M.Potters, *Theory of Financial Risk*, Paris: Aléa-Saclay, Eurolles 1997

[10] R. Cont, M. Potters and J.-P. Bouchaud, *Scaling in stock market data: stable laws and beyond*, in: B. Dubrulle, F. Graner, and D. Sornette, Eds., Scale Invariance and beyond (Proceedings of the CNRS Workshop on Scale Invariance, Les Houches, March 1997) (Springer, Berlin) (1997), 75–85.

[11] F.Delbaen and W. Schachermayer, *A general version of the fundamental theorem of asset pricing*, Mathematische Annalen **300** (1994), 463–520.

[12] G.I.Eskin, *Boundary Problems for Elliptic Pseudo-Differential Equations*, Moscow: Nauka, 1973 (Transl. of Mathematical Monographs **52**, Providence, Rhode Island: Amer. Math. Soc., 1980).

[13] E. Eberlein and J. Jacod, *On the range of option prices*, Finance and Stochastics **1** (1997), 131–140.

[14] E. Eberlein, U. Keller and K. Prause, *New insights into smile, mispricing and value at risk: The hyperbolic model*, Journ. of Business **71** (1998), 371–406.

[15] E.Eberlein and K.Prause, *The Generalized Hyperbolic Model: Financial Derivatives and Risk Measures*, Preprint Universität Freiburg i. Br. N 56, November 1998.

[16] H.Föllmer and M.Schweizer, *Hedging of contingent claims under incomplete information*, in: M.H.A.Davies, R.J.Elliott, Eds., Contributions to Mathematical Economics (North Holland, Amsterdam) (1991), 389–414.

[17] N.Jacob, *Pseudo-differential operators and Markov processes*, Berlin: Akademie Verlag, 1996

[18] I. Koponen, *Analytic approach to the problem of convergence of truncated Lévy flights towards the Gaussian stochastic process*, Phys. Rev. E **52** (1995), 1197–1199.

[19] J.P.Laurent and Huyen Pham, *Dynamic programming and mean-variance hedging*, Finance and Stochastics, **3** (1999), 83–110.

[20] B.Mandelbrot, *The variation of certain speculative prices*, Journal of Business **36** (1963), 392–417.

[21] D.B. Madan, P.Carr and E.C.Chang, *The variance Gamma process and option pricing*, European Finance Review **2** (1998), 79–105.

[22] E.Mordecki, *Optimal stopping and perpetual options for Lévy processes*, Prepublicaciones de Matemática de la Universidad de la Republica, 2000/35.

[23] K.Sato, *Lévy processes and infinitely divisible distributions*, Cambridge: Cambridge University Press, 1999

[24] K.Yosida, *Functional Analysis*, Berlin Heidelberg New-York: Springer-Verlag, 1964

Department of Economics,
University of Pennsylvania,
3718, Locust Walk
PA 19104 Philadelphia, USA
E-mail address: `sboyarch@ssc.upenn.edu`

Department of Mathematics,
Rostov State University of Economics,
69, B.Sadovaya
344007 Rostov-on-Don, Russia
E-mail address: `leven@ns.rnd.runnet.ru`

Trends in Mathematics, © 2001 Birkhäuser Verlag Basel/Switzerland

Installment Options and Static Hedging

Mark H. A. Davis[1], Walter Schachermayer[2] and Robert G. Tompkins[2]

Abstract. An installment option is a European option in which the premium, instead of being paid up-front, is paid in a series of installments. If all installments are paid the holder receives the exercise value, but the holder has the right to terminate payments on any payment date, in which case the option lapses with no further payments on either side. We discuss pricing and risk management for these options, in particular the use of static hedges to obtain both no-arbitrage pricing bounds and very effective hedging strategies with almost no vega risk.

1. Introduction

In a conventional option contract the buyer pays the premium up front and acquires the right, but not the obligation, to exercise the option at a fixed time T in the future (for European-style exercise) or at any time at or before T (for American-style exercise). In this paper we consider an alternative form of contract in which the buyer pays a smaller up-front premium and then a sequence of "installments", i.e. further premium payments at equally spaced time intervals before the maturity time T. If all installments are paid the buyer can exercise the option, European-style, at time T. Crucially, though, the buyer has the right to 'walk away': if any installment is not paid then the contract terminates with no further payments on either side. This provides useful extra optionality to the buyer while, as we shall see, the seller can hedge the option using simple static hedges that largely eliminate model risk.

The case of two installments is equivalent to a compound option, previously considered by Geske [7] and Selby and Hodges [8]. Let $C(t, T, S, K)$ denote the Black-Scholes value at time t of a European call option with strike K maturing at time T when the current underlying price is S (all other model parameters are constant). Installments p_0, p_1 are paid at times t_0, t_1 and final exercise is at time $T > t_1$. At time t_1 the holder can either pay the premium p_1 and continue to hold the option, or walk away, so the value at t_1 is $\max(C(t_1, T, S(t_1), K) - p_1, 0)$. The holder will pay the premium p_1 if this is less than the value of the call option. The value of this contract at t_0 is thus the value of a call on C with 'strike' p_1.

Another way of looking at it, that will be useful later, is this: the holder buys the underlying call at time t_0 for a premium $p = p_0 + e^{-r(t_1-t_0)}p_1$ (the NPV of the two premium payments where r denotes the riskless interest rate) but has the right to sell the option at time t_1 for price p_1. The compound call is thus equivalent to the underlying call option plus a put on the call with exercise at time t_1 and strike price p_1. The value p is thus greater than the Black-Scholes value $C(t_0, T, S(t_0), K)$, the difference being the value of the put on the call.

A similar analysis applies to installment options with premium payments $p_0, p_1, \dots, p_k$ at times $t_0, \dots, t_k < T$. The NPV of the premium payments is $p = \sum_{i=0}^{k} p_i e^{-r(t_i-t_0)}$ and the installment option is equivalent to paying p at time t_0 and acquiring the underlying option plus the right to sell it at time t_j, $1 \le j \le k$ at a price $q_j = \sum_{i=j}^{k} p_i e^{-r(t_i-t_j)}$ (all subsequent premiums are 'refunded' when the right to sell is exercised). The installment option is thus equivalent to the underlying option plus a Bermuda put on the underlying option with time-varying strike q_i.

In the next section we discuss pricing in the Black-Scholes framework. As for American options a finite-difference algorithm must be used. In section 3, simply-stated 'no-arbitrage' bounds on the price are derived valid for very general price process models. As will be seen, these depend on comparison with other options and suggest possible classes of hedging strategies. In section 4 we introduce and analyse static hedges for installment options, concentrating on a specific example to illustrate our point. Concluding remarks are given in Section 5.

This paper is largely a summary of our companion paper [5], which also contains a discussion of 'continuous-installment options', not considered here.

2. Pricing in the Black-Scholes framework

Consider an asset whose price process S_t is the conventional log-normal diffusion

$$dS_t = rS_t\,dt + \sigma S_t\,dw_t, \tag{1}$$

where r is the riskless rate and w_t a standard Brownian motion; thus (1) is the price process in the risk-neutral measure. We consider a European call option on S_t with exercise time T and payoff

$$[S_T - K]^+ = \max(S_T - K, 0). \tag{2}$$

The Black-Scholes value of this option at time 0 is of course

$$p_{\mathrm{BS}} = Ee^{-rT}[S_T - K]^+. \tag{3}$$

p_{BS} is the unique arbitrage-free price for the option, to be paid at time 0. As an illustrative example we will take $T = 1$ year, $r = 0$, $K = 100$, $S_0 = 100$ and $\sigma = 25.132\%$, giving $p_{\mathrm{BS}} = 10.00$.

In an installment option we choose times $0 = t_0 < t_1 < \cdots < t_n = T$ (generally $t_i = iT/n$ to a close approximation). We pay an upfront premium p_0 at t_0 and pay an 'installment' of p_1 at each of the $n-1$ times $t_1, \dots, t_{n-1}$. We also have the right to walk away from the deal at each time t_i: if the installment due at t_i is not paid then the deal is terminated with no further payments on either side. The pricing problem is to determine what is the no arbitrage value of the premium p_1 for a given value of p_0. The present value of premium payments – assuming they are all paid – is

$$p_0 + p_1 \sum_{i=1}^{n-1} e^{-rt_i}, \tag{4}$$

and this must exceed the Black-Scholes value in view of the extra optionality.

Computing the exact value is straightforward in principle. Let $V_i(S)$ denote the net value of the deal to the holder at time t_i when the asset price is $S_{t_i} = S$. In particular

$$V_n(S) = [S - K]^+. \tag{5}$$

At time t_i we can either walk away, or pay p_1 to continue, the continuation value being

$$E_{t_i,S(t_i)}[e^{-r(t_{i+1}-t_i)}V_{i+1}(S_{t_{i+1}})]. \tag{6}$$

Thus

$$V_i(S) = \max(0, E_{t_i,S}[e^{-r(t_{i+1}-t_i)}V_{i+1}(S_{t_{i+1}})] - p_1). \tag{7}$$

In particular, V_{n-1} is just the maximum of 0 and BS $- p_1$, where BS denotes the Black-Scholes value of the option at time t_{n-1}. The unique arbitrage free value of the initial premium is then

$$p_0 = V_0^+(S_0, p_1) := E_{t_0,S_0}\left[e^{-r(t_1-t_0)}V_1(S_{t_1})\right].$$

For fixed p_1, $V_0^+(S_0, p_1)$ is easily evaluated using a binomial or trinomial tree and this determines the up-front payment p_0. If we want to go the other way round, pre-specifying p_0, then we need a simple one-dimensional search to solve the equation $p_0 = V_0^+(S_0, p_1)$ for p_1. A similar search solves the equation $\widehat{p} = V_0^+(S_0, \widehat{p})$ giving the installment value $\widehat{p}$ when all installments, including the initial one, are the same.

Figure 1 shows the price $\widehat{p}$ at time 0 for our standard example with 4 equal installments. For comparison, one quarter of the Black-Scholes value is also shown. At $S_0 = 100, \widehat{p} = 3.284$, which is 31% greater than one quarter of the Black-Scholes value. Figure 2 shows the value at time t_1; when $S(t_1) \leq 98.28$ it is optimal not to continue and the option has value zero.

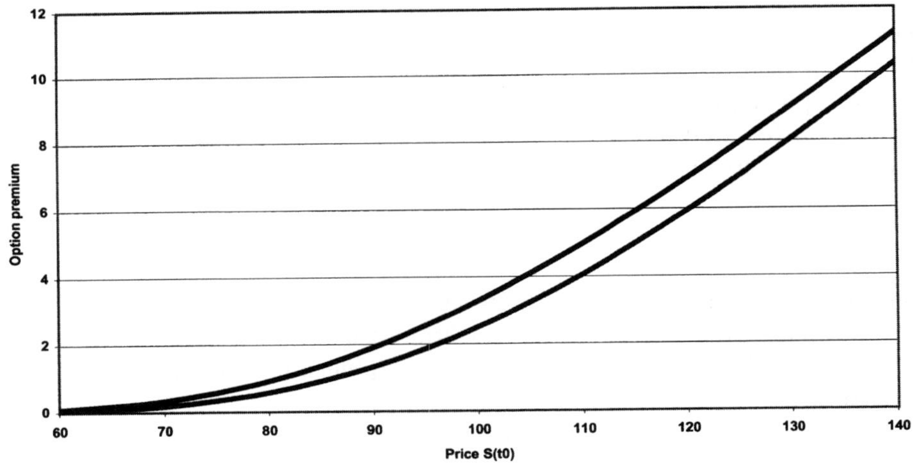

FIGURE 1. *Fair installment value and Black-Scholes value*

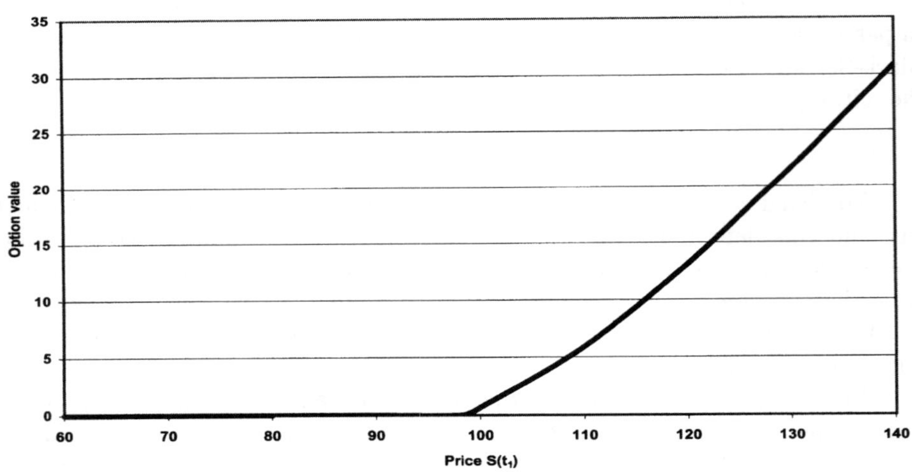

FIGURE 2. *Value of installment option at time t_1 as function of price $S(t_1)$*

3. No-arbitrage bounds derived from static hedges

The pricing model of the previous section makes the standard Black-Scholes assumptions: log-normal price process, constant volatility. By considering static super-replicating portfolios, however, we can determine easily computable bounds on the price valid for essentially arbitrary price models. We need only assume that for any $s \in [t_0, T]$ there is a liquid market for European calls with maturities $t \in [s, T]$, the price being given by

$$C(s,t,K) = \mathbb{E}_Q \left[e^{-r(t-s)}(S_t - K)^+ \middle| \mathcal{F}_s \right] \tag{9}$$

where Q is a martingale measure for the process S and $\mathcal{F}_s$ denotes the information available at time s. By put-call parity this also determines the value of put options $P(s,t,K)$. We know today's prices $C(t_0,t,K)$ and that is *all* we know about the process S and the measure Q. We ignore interest rate volatility, assuming for notational convenience that the riskless rate is a constant, r, in continuously compounding terms. We also assume that no dividends are paid.

Let us first consider a 2-installment, i.e. compound, option, with premiums p_0, p_1 paid at t_0, t_1 for an underlying option with strike K maturing at $T = t_2$.

The subsequent result provides no-arbitrage bounds on the prices p_0, p_1 which are independent of the special choice of the model S and the equivalent martingale measure Q.

Proposition 3.1. *For the compound option described above, there is an arbitrage opportunity if p_0, p_1 do not satisfy the inequalities*

$$C\left(t_0, T, K + e^{r(T-t_1)}p_1\right) > p_0 > C(t_0,T,K) - e^{-r(t_1-t_0)}p_1 + P(t_0,t_1,p_1). \tag{10}$$

Proof. Denote $K' = K + e^{r(T-t_1)}p_1$ and suppose we sell the compound option with agreed premium payments p_0, p_1 such that $p_0 \geq C(t_0,T,K')$. We then buy the call with strike K' and place $x = p_0 - C(t_0,T,K') \geq 0$ in the riskless account. If the second installment is not paid, the value of our position at time t_1 is $xe^{r(t_1-t_0)} + C(t_1,T,K') \geq 0$, whereas if the second installment is paid we add it to the cash account, and the value at time T is then $xe^{r(T-t_0)} + p_1 e^{r(T-t_1)} + C(T,T,K') - C(T,T,K) \geq 0$. This is an arbitrage opportunity, giving the left-hand inequality in (10).

Now suppose the compound option is available at p_0, p_1 satisfying

$$p_0 + e^{-r(t_1-t_0)}p_1 \leq C(t_0,T,K) + P(t_0,t_1,p_1). \tag{11}$$

We buy it, i.e. pay p_0, and sell the two options on the right (call them $\widehat{C}$, $\widehat{P}$), so our cash position is $\widehat{C} + \widehat{P} - p_0 \geq e^{-r(t_1-t_0)}p_1$. At time t_1 the cash position is therefore $x \geq p_1$, and we have the right to pay p_1 and receive the call option. We exercise this right if $C(t_1,T,K) \geq p_1$. Then our cash position is $x - p_1 \geq 0$, $\widehat{C}$ is covered and $\widehat{P}$ will not be exercised because $p_1 < C(t_1,T,K) \leq S(t_1)$ (the call option value is never greater than the value of the underlying asset). On the other hand, $\max(p_1 - C(t_1,T,K), 0) \geq \max(p_1 - S(t_1), 0)$, so if $p_1 > C(t_1,T,K)$ we do not pay the second installment and still have enough cash to cover $\widehat{C}$ and $\widehat{P}$. Thus there is an arbitrage opportunity when the right-hand inequality in (10) is violated. □

Of course, for practical purpose $P(t_0, t_1, p_1)$ will be a negligible quantity (and there will be no liquid market as typically $p_1 \ll S_0$) but for obtaining theoretically sharp bounds one must not forget this term.

In fact, the above inequalities are sharp: it is not hard to construct examples of arbitrage-free markets such that the differences in the left (resp. right) inequality in (10) become arbitrarily small.

Finally let us interpret the right hand side of inequality (10) by using the interpretation of the compound option given in the introductory section: the net present value $p_0 + e^{-r(t_1-t_0)}p_1$ of the payment for the compound must equal — by no-arbitrage — the price $C(t_0, T, K)$ of the corresponding European option plus a put option to sell this call option at time t_1 at price p_1. Denoting the latter security by Put(Call) we obtain the no-arbitrage equality.

$$p_0 = C(t_0, T, K) - e^{-r(t_1-t_0)}p_1 + \mathrm{Put}(\mathrm{Call}). \tag{12}$$

In the proof of proposition 3.1 we have (trivially) estimated this Put on the Call from below by the corresponding Put $P(t_0, t_1, p_1)$ on the underlying S. We now see that the difference in the right hand inequality of (10) is precisely equal to the difference $\mathrm{Put}(\mathrm{Call}) - P(t_0, t_1, p_1)$ in this estimation.

Similar arguments apply for n installments, where holding the installment option is equivalent to holding the underlying option plus the right to sell this option at any installment date at a price equal to the NPV of all future installments. The value of the Bermuda option on the option is greater than the equivalent option on the stock. This gives us the following result.

Proposition 3.2. *For the n-installment call option with premium payment p_0 at time t_0 and p_1 at times $t_1, \dots, t_{n-1}$ there is an arbitrage opportunity if p_0, p_1 do not satisfy*

$$C(t_0, T, K + \widehat{p}_1) > p_0 > \left[C(t_0, T, K) - e^{r(T-t_0)}\widehat{p}_1 + P_{\mathrm{Berm}}(t_0)\right]_+ . \tag{13}$$

Here

$$\widehat{p}_1 = p_1 \sum_{i=1}^{n-1} e^{r(T-t_i)} \tag{14}$$

and $P_{\mathrm{Berm}}(t_0)$ denotes the price at time t_0 of a Bermuda put option on the underlying $S(t)$ with exercise times $t_1, \dots, t_{n-1}$ and strike price K_i at time t_i, where

$$K_i = p_1 \sum_{j=i}^{n-1} e^{-r(t_j-t_i)}. \tag{15}$$

4. Static Hedging

Static hedging is a technique that has come increasingly into favour in recent years because of its robustness to market friction and model error. See [1],[2],[3],[6] and [9], for example.

In Section 2 the pricing formula for installment options was derived under the perfect market assumptions made by Black and Scholes. The price is intimately associated with the construction of a riskless dynamic hedging portfolio. In Section 3 we saw that – without making any assumptions about the price process model – arbitrage opportunities exist if the installment option price fails to satisfy certain bounds, derived by comparison with the values of static portfolios of plain vanilla options. In this section our objective is to show that if the price does lie within these bounds then a static hedging strategy based on these portfolios nevertheless provides an excellent hedge. Typically, we find that with these hedges the installment option writer faces a maximum loss that is bounded, with probability 1, by a number that is a modest fraction – say 20% – of the original option premium.

In this paper we shall just analyse one example, in order to convey the ideas and convince the reader that static hedges are potentially very effective. The companion paper [5] gives much more evidence, based on extensive simulation under alternative assumptions about the price process, including stochastic volatility and market friction in the form of transaction costs.

Our test case will be the standard example of Section 2 with two equal installments, paid at times $t_0 = 0, t_1 = 0.5$. (This is simply a compound option). Recall that the Black-Scholes value of the underlying plain vanilla option is $p_{\mathrm{BS}} = 10.00$. The Black-Scholes value of equal installment premiums is $p_0 = p_1 = 5.855$. We noted in Section 3 that there would be an arbitrage opportunity if the first installment p_0 were enough to buy a 1-year call option with strike $K' = K + p_1 = 105.855$. In fact the value of this option is $p' = 7.627$, so there is – of course – no arbitrage. To construct our hedge, we receive p_0 as premium, borrow the difference $p' - p_0 = 1.772$ and buy the K'-call. The second installment will not be paid if $S(t_1) \leq 97.59$, and in this case we close out the position at t_1; otherwise the hedge is held to maturity. The curve labelled 'Hedge 1' in Figure 3 shows the P&L of the hedged position at t_1 as a function of the price $S(t_1)$. The P&L is equal to $\min\{p_0 - p' + C', p_0 - p' + C' + p_1 - C\}$, where $C' = C(t_1, T, S(t_1), K'), C = C(t_1, T, S(t_1), K)$. The maximum loss is $p' - p_0$, which is 17.72% of the Black-Scholes premium for the underlying option, and the maximum gain is 20.3% of this premium, realized when the price is on the continuation boundary. Figure 4 shows the distribution of P&L under the risk-neutral measure; this turns out to be close to the uniform distribution. Of course this distribution would be different under different price modelling assumptions, but the main point is that under any reasonable model the expected P&L is close to zero and the maximum loss is confined to the amount borrowed, which is not very much.

We can do better. Looking at the P&L profile for Hedge 1 in Figure 3 we note that it is very similar to the value of a *calendar spread* (the difference between two call options with the same strike but different maturities). This suggests that instead of borrowing 1.772 from the bank to finance purchase of the K' call we should raise the money by selling a calendar spread. After some experimentation we find that a suitable calendar spread has strike 97.59 (equal to the price at which the P&L of Hedge 1 is maximum) and maturity times $T_1 = 0.5, T_2 = 1.0$.

The Black-Scholes value of this calendar spread at time 0 is 2.865, so we need to sell $0.619 = 1.772/2.865$ of them to finance purchase of the K' call. We thus create a hedge whose value at time 0 is p_0 consisting of the following plain vanilla call options ((K,T) denotes strike and maturity): long $(97.59, 0.5)$ and $(105.86, 1.0)$, short $(97.59, 1.0)$. The P&L profile of this hedge at t_1 is 'Hedge 2' in Figure 3. It is a considerable improvement: the maximum loss has been reduced to 5.1% of the underlying Black-Scholes premium.

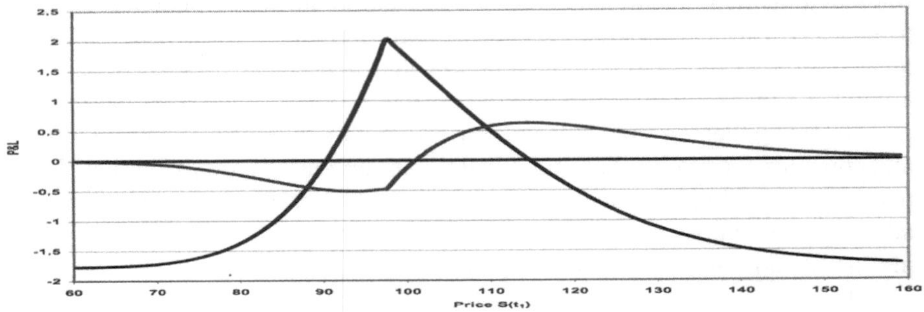

FIGURE 3. *Profit/Loss profiles of compound option static hedges as functions of the underlying asset price* $S(t_1)$

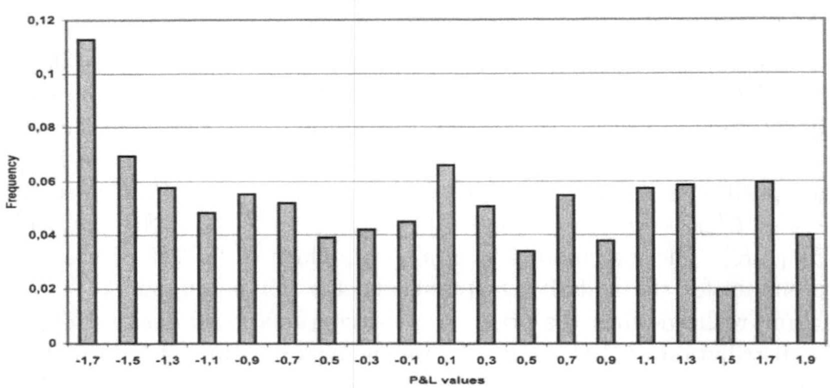

FIGURE 4. *Distribution of Static Hedge P&L under the risk neutral measure*

5. Concluding Remarks

The reader will find in the companion paper [5] a much more complete analysis of static hedges for installment options – including ones with multiple installments – examining examining the performance relative to dynamic hedging under realistic market scenarios and including the effects of transaction costs and stochastic volatility. This study confirms what is suggested by the example presented above, namely that static hedging is a highly effective technique that decouples the hedger almost entirely from the malign effects of market friction and vega risk.

The construction of Hedge 2 in Section 4 – which was frankly done by trial and error – suggests that we should take a more systematic approach. Effectively, plain vanilla option payoffs are being used as basis functions, in terms of which we seek to approximate the payoff profile of an exotic option. While expansion of option payoffs in Fourier series has been investigated by Desrolles and Laurent [4], we are not aware of any studies in which other option payoffs are used as basis functions, although such an approach would not be hard to implement and was in fact suggested to the authors years ago by Michael Selby. We plan to return to this topic.

References

[1] P. Carr, K. Ellis and V. Gupta, (1998), *Static Hedging of Exotic Options.* Journal of Finance, Vol. 53, No. 3, pp. 1165–1190.

[2] P. Carr and J.-F. Picron, (1999), *Static Hedging of Timing Risk.* The Journal of Derivatives, Vol. 6, No. 3, pp. 57–70.

[3] A. Chou and G. Georgiev, (1998), *A Uniform Approach to Static Replication.* The Journal of Risk, Vol. 1, No. 1, pp. 73–87.

[4] S. Darolles and J.-P. Laurent, (2000), *Approximating payoffs and pricing formulas.* Journal of Economic Dynamics and Control, Vol. 24, pp. 1721-1746.

[5] M. Davis, W. Schachermayer and R. Tompkins,(2000),*Pricing, no-arbitrage bounds and robust hedging of installment options.* Preprint, Technische Universität Wien.

[6] E. Derman, D. Ergener and I. Kani, (1995), *Static Options Replication.* The Journal of Derivatives, Vol. 2, No. 4, pp. 78–95.

[7] R. Geske, (1979), *The valuation of compound options.* Journal of Financial Economics, Vol. 7, pp. 63–81.

[8] M.J.P. Selby, S.D. Hodges, (1987), *On the evaluation of compound options.* Management Science, Vol. 33, pp. 347–355.

[9] R. Tompkins, (1997), *Static versus dynamic hedging of exotic options: an evaluation of hedge performance via simulation.* Netexposure, Vol. 1, No. 2, pp. 1–28.

[1]Department of Mathematics,
Imperial College,
London SW7 2BZ, England

[2]Financial and Actuarial Mathematics Group
Technische Universität Wien
Wiener Hauptstrasse 8
A-1040 Wien, Austria

E-mail address: `mark.davis@ic.ac.uk`

Trends in Mathematics, © 2001 Birkhäuser Verlag Basel/Switzerland

Fractional Brownian Motion And Financial Modelling

R.J. Elliott[1)], J. van der Hoek[2)]

Abstract. We present a new framework for fractional Brownian motion in which processes with all indices can be considered under the same probability measure. Our results extend recent contributions by Hu, Øksendal, Duncan, Pasik-Duncan and others. As an application we develop option pricing in a fractional Black-Scholas market with a noise process driven by a sum of fractional Brownian motions with various Hurst indices.

1. Introduction

There have been several recent contributions to white noise theory. Those most related to the current work include the book by Holden et al. [5], and the papers by Aase, Øksendal, Privault and Ubøe [1], Hu and Øksendal [6], and Duncan, Hu and Pasik-Duncan [3].

A natural framework for this theory is to take the underlying probability space to be $(\mathcal{S}'(R), \mathcal{F})$ the space of tempered distributions, with $\mathcal{F}$ the Borel field. A probability measure P is given on $(\mathcal{S}'(R), \mathcal{F})$ by the Bochner-Minlos theorem. Fractional Brownian motion B^H is defined in [6], for a Hurst parameter H with a value in $(1/2, 1)$ by introducing a measure P_H on $(\mathcal{S}'(R), \mathcal{F})$, again given by the Bochner-Minlos theorem (see [8]).

The present paper defines fractional Brownian motions on $(\mathcal{S}'(R), \mathcal{F})$ using one probability measure P. Advantages of our approach are that Brownian motions for all Hurst indices H can be discussed simultaneously, and their Hurst parameters can take any value in $(0, 1)$.

Our definition and development of multiple fractional Brownian motions appears natural and new. Modifying results and proofs from [1] and [6] we derive option pricing formulae in a multiple fractional Brownian Black-Scholes market.

The support of SSHRC is gratefully acknowledged.

2. Fractional Brownian Motion

Definition 2.1. *For* $0 < H < 1$ *fractional Brownian motion (FBM)* $B^H = \{B^H(t),\, t \in R\}$, *for Hurst parameter* $H \in (0,1)$, *is the Gaussian process with mean* 0 *for all* t:

$$E[B^H(t)] = 0$$

and co-variance

$$E[B^H(t)B^H(s)] = \frac{1}{2}\{\, |t|^{2H} + |s|^{2H} - |t-s|^{2H}\,\}. \tag{1}$$

We take $B^H(0) = 0$. *For* $H = 1/2$ $B^{1/2}$ *is standard Brownian motion.*

We first develop fractional Brownian motion for a fixed $H \in (0,1)$. Take the underlying probability space to be $\mathcal{S}'(R)$ with the Borel sigma field $\mathcal{F}$.

For $f \in \mathcal{S}(R)$ and $\|f\|^2 = \int_R f^2(x)dx$ the Bochner-Minlos theorem gives a probability measure P on $(\mathcal{S}'(R), \mathcal{F})$ such that

$$E[\exp i\,\langle f, \xi\rangle\,] = e^{-\frac{1}{2}\|f\|^2}. \tag{2}$$

Here $\langle f, \xi\rangle$ denotes the action of $\xi \in \mathcal{S}'(R)$ on $f \in \mathcal{S}(R)$. Note

$$E[\,\langle f,\xi\rangle\,] = 0 \quad\text{and}\quad E[\,\langle f,\xi\rangle^2] = \|f\|^2 \quad\text{for}\quad f \in \mathcal{S}(R), \tag{3}$$

and the duality can be extended to $f \in L^2(R) = \{f : \|f\|^2 < \infty\}$.

Consider the indicator function

$$I(0,t)(s) = \begin{cases} 1 & \text{if} \quad 0 \le s \le t \\ -1 & \text{if} \quad -t \le s \le 0 \\ 0 & \text{otherwise} \end{cases} \tag{4}$$

For $0 < H < 1/2$ and $f \in \mathcal{S}(R)$ define the operator M_H by:

$$M_H f(x) := \Big(2\Gamma\big(H - \frac{1}{2}\big)\cos\big(\frac{\pi}{2}\,(H - \frac{1}{2})\big)\Big)^{-1} \int_R \frac{f(x-t) - f(x)}{|t|^{\frac{3}{2}-H}}\,dt. \tag{5}$$

For $1/2 < H < 1$ and $f \in \mathcal{S}(R)$

$$M_H f(x) := \Big(2\Gamma\big(H - \frac{1}{2}\big)\cos\big(\frac{\pi}{2}\,(H - \frac{1}{2})\big)\Big)^{-1} \int \frac{f(t)}{|t-x|^{\frac{3}{2}-H}}\,dt. \tag{6}$$

For $H = 1/2$ we put $(M_H f)(x) = f(x)$, the identity map.

Even more basically, $M_H f$ can be defined as a constant multiple of

$$-\frac{d}{dx}\int_R (t-x)|t-x|^{H-\frac{3}{2}} f(t)dt$$

for each $0 < H < 1$. The expressions above are obtained using integration by parts. The constant is chosen so that (8) holds.

With the hat denoting the Fourier transform

$$\widehat{f}(\xi) = \int_R e^{-ix\xi} f(x)dx \tag{7}$$

we have for $0 < H < 1$

$$\widehat{M_H f}(\xi) = |\xi|^{\frac{1}{2}-H}\, \widehat{f}(\xi). \tag{8}$$

When $f(x) = I(0,t)(x)$ write

$$M_H f(x) = M_H(0,t)(x). \tag{9}$$

Define

$$L^2_H(R) = \{f : M_H f \in L^2(R)\} \tag{10}$$

$$= \{f : |\xi|^{\frac{1}{2}-H}\, \widehat{f}(\xi) \in L^2(R)\}. \tag{11}$$

For $H \in (0,1)$ define

$$\widetilde{B}_H(t)(\xi) = \langle M_H(0,t), \xi\rangle.$$

Again $\widetilde{B}_H(t)$ is a Gaussian random variable with

$$E[\widetilde{B}_H(t)] = 0$$

$$E[\widetilde{B}_H(t)\widetilde{B}_H(s)] = \int_R M_H(0,t)M_H(0,s)dx \tag{12}$$

$$= \frac{1}{2\sin(\pi H)\Gamma(2H+1)} \{|t|^{2H} + |s|^{2H} - |t-s|^{2H}\} \tag{13}$$

The process $t \to \widetilde{B}_H(f)$ has a continuous version which we denote by

$$B_H = \{B_H(t),\, t \in R\}.$$

Note that when $H = 1/2$

$$\frac{1}{2}\{|t|^{2H} + |s|^{2H} - |t-s|^{2H}\} = \min(t,s).$$

The process B_H, defined under P, is, therefore, a fractional Brownian motion with Hurst parameter H. If $f \in L^2_H(R)$, approximating by step functions we see that

$$\langle f, \xi\rangle = \int_R f(t)dB_H(t)(\xi). \tag{14}$$

Note that with this definition all fractional Brownian motions B_H for all Hurst parameters $H \in (0,1)$ can be considered simultaneously. This is in contrast to [3], [5] where a single Hurst parameter $H \in (1/2,1)$ is considered with an associated probability measure P_H. Of course, multiple FBM can be introduced in basically two ways: as above, where they are all related to the same Brownian motion or, in terms of different independent underlying Brownian motions.

Indeed consider fixed values $H_1, \dots, H_m$, $0 < H_k < 1$, $k = 1, 2, \dots, m$, and constants $\sigma_k > 0$, $k = 1, 2, \dots, m$. Then we can define the operator

$$M := \sigma_1 M_{H_1} + \cdots + \sigma_m M_{H_m} \tag{15}$$

and the process

$$B_M := \sigma_1 B_{H_1} + \cdots + \sigma_m B_{H_m}. \tag{16}$$

In fact $B_M = \{B_M(t),\, t \in R\}$ is a continuous version of $\widetilde{B}_M$ where

$$\widetilde{B}_M(t)(\xi) = \langle M(0,t), \varphi\rangle \tag{17}$$

$$= \sum_{i=1}^{m} \sigma_i \langle M_{H_i}(0,f), \xi\rangle. \tag{18}$$

Clearly, if $\sigma_1 = 1$ and $\sigma_i = 0$ for $i > 1$ then $B_M = B_{H_1}$, as above. For $0 < H,\, L < 1$,

$$\int_R M_H(0,t) M_L(0,s) dx \tag{19}$$

$$= \frac{1}{2\sin\left(\frac{\pi}{2}(H+L)\right)\Gamma(H+L+1)} \left[|t|^{H+L} + |s|^{H+L} - |s-t|^{H+L}\right]. \tag{20}$$

In particular

$$\int_R M_H(0,t)^2 dx = \frac{|t|^{2H}}{\sin(\pi H)\Gamma(1+2H)}\,. \tag{21}$$

Write

$$\sigma_M(t)^2 = \sum_{i,j=1}^{m} \left(\sin\left(\frac{\pi}{2}(H_i + H_j)\right)\Gamma(H_i + H_j + 1)\right)^{-1} |t|^{H_i + H_j}. \tag{22}$$

Then for each t $B_M(t)(\xi)$ is a Gaussian random variable with zero mean and variance $\sigma_M(t)^2$.
For $f \in \mathcal{S}(R)$, taking Fourier transforms

$$\widehat{Mf}(\xi) = \left(\sum_{i=1}^{m} \sigma_i |\xi|^{\frac{1}{2} - H_i}\right) \widehat{f}(\xi). \tag{23}$$

Define $L^2_M(R) = \{f : Mf \in L^2(R)\} = \{f : \widehat{Mf} \in L^2(R)\}$. The inner product for $f, g \in L^2_M(R)$ is defined as $\langle f, g\rangle_M = \int_R (Mf)(Mg) dx$. Note that M^{-1} is defined in terms of Fourier transforms by setting

$$\widehat{M^{-1}f}(\xi) = \left(\sum_{i=1}^{m} \sigma_i |\xi|^{\frac{1}{2} - H_i}\right)^{-1} \widehat{f}(\xi). \tag{24}$$

2.1. Hermite functions

Recall the Hermite polynomials

$$h_n(x) = (-1)^n e^{\frac{x^2}{2}} \frac{d^n}{dx^n} e^{-\frac{x^2}{2}}, \quad n = 0, 1, 2, \dots \tag{25}$$

The Hermite functions are then defined as

$$\widetilde{h}_n(x) = \pi^{-\frac{1}{4}} \left((n-1)!\right)^{-\frac{1}{2}} h_{n-1}(\sqrt{2}\,x) e^{-\frac{x^2}{2}}, \quad n = 1, 2, \dots \tag{26}$$

Then $\{\widetilde{h}_n,\ n = 1, 2, \dots\}$ is an orthonormal basis of $L^2(R)$ and

$$Cn^{-\frac{1}{12}} \text{ if } \ |x| \leq 2\sqrt{n} \tag{27}$$

$$|\widetilde{h}_n(x)| \leq \tag{28}$$

$$Ce^{-\gamma x^2} \text{ if } \ |x| > 2\sqrt{n} \tag{29}$$

where C and γ are constants independent of n.
Define

$$e_i(x) := M^{-1}\,\widetilde{h}_i(x), \quad i = 1, 2, \dots . \tag{30}$$

Then $\{e_i,\ i = 1, 2, \dots\}$ is an orthonormal basis of $L^2_M(R)$.

3. Fractional White Noise

Recall

$$B_M(t)(\xi) = \langle M(0,t), \xi\rangle \tag{31}$$

$$= \langle I(0,t), M\xi\rangle \tag{32}$$

$$= \sum_{i=1}^{\infty} \langle Me_i, \xi\rangle \int_0^t M\widetilde{h}_i(x)dx\,. \tag{33}$$

The M-fractional white noise is the derivative in $(S)^*$ of $B_M(t)$, so

$$W_M(t)(\xi) = \sum_{i=1}^{\infty} M\widetilde{h}_i(t)\langle Me_i, \xi\rangle. \tag{34}$$

We can also define geometric fractional Brownian motion related to B_M. Consider the stochastic differential equation:

$$dX(t) = \mu X(t)dt + \sigma X(t)dB_M(t) \tag{35}$$

$$\text{i.e.} \quad \frac{dX(t)}{dt} = \mu X(t) + \sigma X(t) \diamond W_M(t) \tag{36}$$

$$= \big(\mu + \sigma W_M(t)\big) \diamond X(t). \tag{37}$$

From the Wick calculus this has a solution

$$X(t) = x \exp^{\diamond}\left(\mu t + \sigma \int_0^t W_M(s)ds\right) \tag{38}$$

$$= x \exp^{\diamond}\left(\mu t + \sigma B_M(t)\right) \tag{39}$$

$$= x \exp \mu t \cdot \exp\left(\sigma B_M(t) - \frac{\sigma^2}{2}\sigma_M(t)^2\right). \tag{40}$$

Note $E[X(t)] = x \exp \mu t$. We now give a version of Girsanov's Theorem. Our setting and result differ from that given in [5].

We have been working with the probability measure P on $(\mathcal{S}'(R), \mathcal{F})$. For $\phi \in L^2(R)$ we can define a new probability measure $\widehat{P}$ by setting

$$\frac{d\widehat{P}}{dP}(\xi) = \exp\left[\langle \phi, \xi \rangle - \frac{1}{2}\|\phi\|^2\right].$$

The classical Girsanov Theorem then states that under $\widehat{P}$ the process $\widehat{B}$ is a standard Brownian motion where

$$\widehat{B}(t) := B(t) - \int_0^t \phi(s)ds.$$

Consequently,

$$\widehat{B}_M(t) = \int_R M(0,t)(s)d\widehat{B}(s) \tag{41}$$

$$= \int_R M(0,t)(s)dB(s) - \int_R M(0,t)(s)\phi(s)ds \tag{42}$$

is a fractional Brownian motion under $\widehat{P}$.

To eliminate a drift in a fractional Brownian motion we shall need to solve equations of the form

$$\int_R M(0,t)(s)\phi(s)ds = g(t).$$

That is

$$\int_R I(0,t)(s)M\phi(s)ds = g(t)$$

or

$$\int_0^t M\phi(s)ds = g(t).$$

Consequently, $(M\phi)(t) = g'(t)$ and

$$\phi(t) = (M^{-1}g')(t). \tag{43}$$

In particular, if $M = M_H$ and $g'(t) = A$ on $[0,T]$ then

$$\phi(t) = AM^{-1}(0,T)(t) \tag{44}$$

$$= AM_{1-H}(0,T)(t) \tag{45}$$

$$= A\Big[\frac{T-t}{(T-t)^{\frac{3}{2}-H}} + \frac{t}{|t|^{\frac{3}{2}-H}}\Big] \tag{46}$$

and for $0 \le t \le T$ this is

$$= A[(T-t)^{H-\frac{1}{2}} + t^{H-\frac{1}{2}}]. \tag{47}$$

4. Differentiation

For fractional Brownian motion B_M the definitions of [1] and [5] must be modified.

Definition 4.1. *Suppose* $F : \mathcal{S}'(R) \to R$ *and suppose* $\gamma \in \mathcal{S}'(R)$. *We say* F *has a directional* M*-derivative in the direction* γ *if*

$$D_\gamma^M F(\xi) := \underset{\varepsilon \to 0}{\to} \lim \frac{F(\xi + \varepsilon M\gamma) - F(\xi)}{\varepsilon}$$

exists in $(S)^*$. *We then call* $D_\gamma^M F$ *the directional* M*-derivative of* F *in the direction* γ.

Example 4.2. *Suppose*

$$F(\xi) = \langle Mf, \xi\rangle = \int_R f dB_M$$

for some $f \in \mathcal{S}(R)$, *and* $\gamma \in L^2_M(R)$. *Then*

$$D_\gamma^M F(\xi) = \underset{\varepsilon \to 0}{\to} \lim \frac{1}{\varepsilon}[\langle Mf, \xi + \varepsilon M\gamma\rangle - \langle Mf, \xi\rangle] \tag{48}$$

$$= \langle Mf, M\gamma\rangle = \int_R Mf \cdot M\gamma = \langle f, \gamma\rangle_M . \tag{49}$$

Definition 4.3. *We say* $F : \mathcal{S}'(R) \to R$ *is* M*-differentiable if there is a map*

$$\Psi : R \to (S)^*$$

such that $\big(M\Psi(t)\big) \cdot (M\gamma)(t)$ *is* $(S)^*$ *integrable and*

$$D_\gamma^M F(\xi) = \int_R \big(M\Psi(t,\xi)\big)(M\gamma)(t)dt = \langle \Psi(t), \gamma\rangle_M \tag{50}$$

for all $\gamma \in L^2_M(R)$. *We then define*

$$D_t^M F(\xi) = \Psi(t,\xi) \tag{51}$$

and call $D_t^M F(\xi)$ *the stochastic* M*-gradient of* F *at* t.

Our fractional Clarke-Ocone Theorem now follows as in [1] and [5].

Theorem 4.4. a) *Suppose* $G \in \mathcal{G}^*(M)$ *is* $\mathcal{F}_T^M$ *measurable. Then* $D_t^M G \in \mathcal{G}^*(M)$ *and* $\widetilde{E}_M[D_t^M G|\mathcal{F}_t^M] \in \mathcal{G}^*(M)$ *for almost all* t.

$$\widetilde{E}_M[D_t^M G|\mathcal{F}_t^M]\diamond W_M(t)$$

is integrable in S^* *and*

$$G(\xi) = E[G] + \int_0^T \widetilde{E}_M[D_t^M G|\mathcal{F}_t^M]\diamond W_M(t)dt. \tag{52}$$

b) *Suppose* $G \in \mathcal{L}_M^2$ *is* $\mathcal{F}_T^M$ *measurable. Then* $(t,\xi) \to \widetilde{E}_M[D_t^M G|\mathcal{F}_t^M](\xi)$ *is in* $L_M^2(\lambda \times P)$, *where* λ *is Lebesgue measure in* $[0,T]$. *Further,*

$$G(\xi) = E[G] + \int_0^T \widetilde{E}_M[D_t^M G|\mathcal{F}_t^M]dB_M(t). \tag{53}$$

Notation : Write $\nabla_t^M G := \widetilde{E}_M[D_t^M G|\mathcal{F}_t^M]$.

Theorem 4.5. *Suppose* f *is such that*

$$E\big[f(B_M(T))\big] < \infty.$$

Then

$$\nabla_t^M f(B_M(T)) = \int_R q_{t,T}(B_M(T) - y) f(y)dy \tag{54}$$

where

$$q_{t,T}(x) = \frac{d}{dx} P_{t,T}(x)$$

and

$$P_{t,T}(x) = \frac{1}{\sqrt{2\pi(\sigma_M(T)^2 - \sigma_M(t)^2)}} \exp\Big[\frac{-x^2}{2(\sigma_M(T)^2 - \sigma_M(t)^2)}\Big]. \tag{55}$$

5. A Financial Application

We consider a fractional Black-Scholes market. This contains a bank account or bond, $S_0(t)$ with dynamics

$$dS_0(t) = rS_0(t)dt \tag{56}$$

$$S_0(0) = 1, \quad 0 \le t \le T, \tag{57}$$

so that $S_0(t) = e^{rt}$, and a stock whose price $S_1(t)$ has fractional log normal dynamics:

$$dS_1(t) = \mu S_1(t)dt + \sigma S_1(t)dB_M(t) \tag{58}$$

$$S_1(0) = x > 0. \tag{59}$$

We have seen in (2.4) that in this case

$$S_1(t) = x \exp\big(\mu t + \sigma B_M(t) - \frac{\sigma^2}{2}\sigma_M(t)^2\big).$$

A portfolio $\theta(t) = (u(t), v(t))$ is a pair of $\mathcal{F}_t^M$ adapted processes which give the number of units of S_0 and S_1, respectively, held at time t.
The corresponding wealth process is

$$Z^\theta(t) = u(t)S_0(t) + v(t)S_1(t). \tag{60}$$

θ is called *admissible* if $Z^\theta(t)$ is bounded below for all $t \in [0,T]$, and θ is *self-financing*, that is

$$dZ^\theta(t) = u(t)dS_0(t) + v(t)dS_1(t). \tag{61}$$

This means the only changes to the wealth process come from changes in the values of the assets S_0 and S_1.

Note

$$u(t) = \frac{Z^\theta(t) - v(t)S_1(t)}{S_0(t)}$$

so

$$dZ^\theta(t) = rZ^\theta(t)dt + \sigma v(t)S_1(t)\Big[\frac{\mu - r}{\sigma}\,dt + dB_M(t)\Big]. \tag{62}$$

From the form of Girsanov's Theorem, given in $(58), (59)$

$$\widehat{B}_M(t) := \frac{\mu - r}{\sigma}\,t + B_M(t) \tag{63}$$

is a fractional Brownian motion with respect to the measure $\widehat{P}$ defined on $\mathcal{F}_T^M$ by:

$$\frac{d\widetilde{P}}{dP}\Big|_{\mathcal{F}_T^M} = \exp\left[\langle\phi,\xi\rangle - \frac{1}{2}\,\|\phi\|^2\right]$$

where $\phi(t) = \left(\frac{\mu-r}{\sigma}\right)M^{-1}\big(I(0,T)\big)$. In particular, we have seen that if $M = M_H$ then

$$\phi(t) = \left(\frac{\mu - r}{\sigma}\right)\left[\frac{(T-t)}{|T-t|^{\frac{3}{2}-H}} + \frac{t}{|t|^{\frac{3}{2}-H}}\right]. \tag{64}$$

Consequently, under $\widehat{P}$, Z^θ has dynamics

$$dZ^\theta(t) = rZ^\theta(t)dt + \sigma v(t)S_1(t)d\widehat{B}_M(t). \tag{65}$$

The usual argument now shows our market is arbitrage free:

$$e^{-rt}Z^\theta(t) = z + \int_0^t e^{-rs}\sigma v(s)S_1(s)d\widehat{B}_M(s) \tag{66}$$

$$\text{and} \qquad \widehat{E}[e^{rT}Z^\theta(T) = Z^\theta(0) = z \tag{67}$$

Therefore, there cannot be a portfolio $\theta(t) = (u(t), v(t))$ for which $Z^\theta(0) \le 0$ and $Z^\theta(T) \ge 0$ with $P\big(Z^\theta(T) > 0\big) > 0$.

Furthermore, our market is complete: Suppose $F(\xi)$ is an $\mathcal{F}_T^M$-measurable random variable in $\mathcal{L}_M^2$. The market is complete if there is a portfolio $\theta = (u, v)$ and in initial investment such that

$$F(\xi) = Z^{\theta,z}(T,\xi) \quad \text{a.s.}$$

This is the same as requiring

$$e^{-rT}F(\xi) = z + \int_0^T e^{-rs}\sigma v(s)S_1(s)d\widehat{B}_M(s). \tag{68}$$

However, applying the fractional representation result to $e^{-rT}F(\xi)$, with B_M replaced by $\widehat{B}_M$, we have

$$e^{-rT}F(\xi) = \widehat{E}[e^{-rT}F] + \int_0^T \widetilde{E}_M[e^{-rT}\widehat{D}_t^M f|\mathcal{F}_t^M]d\widehat{B}_M(t). \tag{69}$$

Here , $\widehat{D}_t^M$ is the stochastic M-gradient under measure $\widehat{P}$. Therefore,

$$z = Z^\theta(0) = \widehat{E}[e^{-rT}F]$$

is the price of the claim and

$$v(t) = S_1(t)^{-1}e^{-r(T-t)}\sigma^{-1}\widehat{E}_M[e^{-rT}\widehat{D}_t^M F|\mathcal{F}_t^M] \tag{70}$$

determines the portfolio.

A Black-Scholes Formula. Suppose F is a European call:

$$F(\xi) = \big(S_1(T,\xi) - K\big)^+, \tag{71}$$

with strike price K. Applying the above results, we can show as in [5] the price for the call at time 0 is

$$x\Phi(y_1) - Ke^{-rT}\Phi(y_2)$$

where $S_1(0) = x$

$$y_1 = \Big[\log\frac{x}{K} + rT + \frac{\sigma^2}{2}\sigma_M(T)^2\Big]/\sigma\cdot\sigma_M(T) \tag{72}$$

$$y_2 = \Big[\log\frac{x}{K} + rT - \frac{\sigma^2}{2}\sigma_M(T)^2\Big]/\sigma\cdot\sigma_M(T). \tag{73}$$

Here $\Phi(y) = \frac{1}{\sqrt{2\pi}}\int_{-\infty}^y e^{-x^2/2}dx$ is the cumulative Gaussian distribution. The delta hedge portfolio is given by

$$v(t) = \sigma^{-1}e^{-r(T-t)}S_1(t)^{-1}\widehat{\nabla}_t^M F(\xi) \tag{74}$$

$$\text{where} \qquad \widehat{\nabla}_t^M F = \widehat{E}_M[\widehat{D}_t^M F|\mathcal{F}_t^M] \tag{75}$$

is the fractional Clark-Ocone M-derivative of F when F is considered a functional of

$$\widehat{B}_M(t) = B_M(t) + \int_0^t \frac{\mu - r}{\sigma}\,ds.$$

Now suppose

$$F(\xi) = \big(S(T,\xi) - K\big)^+ \tag{76}$$

$$= \big(x\ \exp[\sigma\widehat{B}_M(T) + rT - \frac{\sigma^2}{2}\,\sigma_M(T)^2] - K\big)^+ \tag{77}$$

$$= f(\widehat{B}_M(t)) \tag{78}$$

where

$$f(z) = \Big(x\ \exp\ \big(\sigma z + rT - \frac{\sigma^2}{2}\,\sigma_M(T)^2\big) - K\Big)^+. \tag{79}$$

Then, from equation (75)

$$\widehat{\nabla}_t^M F = \int_R q_{t,T}\big(\widehat{B}_M(t) - y\big)g(y)dy \tag{80}$$

where

$$q_{t,T}(x) = \frac{d}{dx}\,P_{t,T}(x) \tag{81}$$

$$\text{and}\quad P_{t,T}(x) = \frac{1}{\sqrt{2\pi(\sigma_M(T)^2 - \sigma_M(t)^2)}}\ e^{\frac{-x^2}{2(\sigma_M(T)^2 - \sigma_M(t)^2)}}\ . \tag{82}$$

As in the classical case it follows that

$$v(t) = \Phi\big(y_1(T,t)\big) \tag{83}$$

where

$$y_1(T,t) = \frac{\log\ \big(\frac{S(t)}{K}\big) + r(T-t) + \frac{\sigma^2}{2}\,\big(\sigma_M(T)^2 - \sigma_M(t)^2\big)}{\sigma\sqrt{\sigma_M(T)^2 - \sigma_M(t)^2}}\ .$$

References

[1] K. Aase, B. Øksendal, N. Privault and J. Ubøe: White noise generalizations of the Clark-Haussmann-Ocone theorem with application to mathematical finance, Preprint, University of Oslo 1999

[2] L. Decreusefond and A.S. Üstünel: Stochastic calculus for fractional Brownian motion, Potential Analysis 10, 1999, 177-214

[3] T.E. Duncan, Y. Hu and B. Pasik-Duncan: Stochastic calculus for fractional Brownian motion. I. Theory, SIAM Jour. Control and Optimization 38, 2000, 582-612

[4] T. Hida, H.H. Kuo, J. Potthoff and L. Streit: White Noise Analysis, Kluwer 1993

[5] H. Holden, B. Øksendal, J. Ubøe and T. Zhang: Stochastic Partial Differential Equations, Birkhauser 1996

[6] Y. Hu and B. Øksendal: Fractional white noise analysis and applications to finance, Preprint, University of Oslo, 1999

[7] K. Ito: Multiple Wiener integral, J. Math. Soc. Japan 3, 1951, 157-169

[8] H.-H. Kuo: White Noise Distribution Theory. CRC Press, 1996

[9] T. Lindstøm: Fractional Brownian fields and integrals of white noise, Bull. London Math. Soc. 25, 1993, 83-88

[10] G. Samorodnitsky and M.S. Taqqu: Stable Non-Gaussian Random Processes, Chapman and Hall, London 1994

[11] S.G. Samko, A.A. Kilbas and O.I. Marichev: Fractional Integrals and Derivatives: Theory and Applications, Gordon and Breach Scientific Publisher, London 1987

1) Department of Mathematical Sciences, University of Alberta, Edmonton, Alberta, Canada T6G 2G1

2) Department of Applied Mathematics, University of Adelaide, Adelaide, South Australia 5005

E-mail address: 1) relliott@ualberta.ca

E-mail address: 2) jvanderh@maths.adelaide.edu.au

Trends in Mathematics, © 2001 Birkhäuser Verlag Basel/Switzerland

Stochastic Volatility and Epsilon-Martingale Decomposition

Jean-Pierre Fouque, George Papanicolaou and Ronnie Sircar

Abstract. We address the problems of pricing and hedging derivative securities in an environment of uncertain and changing market volatility. We show that when volatility is stochastic but fast mean reverting Black-Scholes pricing theory can be corrected. The correction accounts for the effect of stochastic volatility and the associated market price of risk. For European derivatives it is given by explicit formulas which involve parsimonous parameters directly calibrated from the implied volatility surface. The method presented here is based on a martingale decomposition result which enables us to treat non-Markovian models as well.

1. Stochastic Volatility Models

We consider stochastic volatility models where the asset price $(X_t)_{t\geq 0}$ satisfies the stochastic differential equation

$$dX_t = \mu X_t dt + \sigma_t X_t dW_t,$$

and $(\sigma_t)_{t\geq 0}$ is called the volatility process. It must satisfy some regularity conditions for the model to be well-defined, but it does not have to be an Itô process: it can be a jump process, a Markov chain, etc. In order for it to be a volatility, it should be positive. Unlike the implied deterministic volatility models for which the volatility is a deterministic function $\sigma(t, X_t)$ of time and price, the volatility process is *not* perfectly correlated with the Brownian motion (W_t). Therefore, volatility is modeled to have an independent random component and since σ_t is not the price of a traded asset, the market is incomplete and there is no longer a unique equivalent martingale measure. We refer to [6], [5] or [3](Ch.2) for reviews of stochastic volatility models.

1.1. Mean-Reverting Stochastic Volatility Models

We consider first volatility processes which are Itô processes satisfying stochastic differential equations driven by a second Brownian motion. This is a convenient way to incorporate correlation with stock price changes.

One feature that most volatility models seem to like is *mean-reversion.* The term mean-reverting refers to the characteristic (typical) time it takes for a process to get back to the mean-level of its invariant distribution (the long-run distribution of the process). In other words we assume that σ_t is *ergodic* with additional mixing properties. From a financial modeling perspective, mean-reverting refers to a linear pull-back term in the drift of the volatility process itself, or in the drift of some (underlying) process of which volatility is a function. Let us denote $\sigma_t = f(Y_t)$ where f is some positive function. Then mean-reverting stochastic volatility means that the stochastic differential equation for (Y_t) looks like

$$dY_t = \alpha(m - Y_t)dt + \beta d\hat{Z}_t,$$

where $(\hat{Z}_t)_{t\geq 0}$ is a Brownian motion correlated with (W_t). Here α is called the *rate of mean-reversion* and m is the long-run mean-level of Y. The drift term pulls Y towards m and consequently we would expect that σ_t is pulled towards the mean value of $f(Y)$, with respect to the long-run distribution of Y.

Choosing $\beta > 0$ constant corresponds to the Ornstein-Uhlenbeck process which is a Gaussian process with the normal invariant distribution $\mathcal{N}(m, \beta^2/2\alpha)$. This choice, though not necessary, is particularly convenient to explain the concept of fast mean-reversion and to show through relatively explicit computations how to exploit this property in pricing and hedging problems. It is still very flexible since the function f is unspecified.

The second Brownian motion $(\hat{Z}_t)$ is correlated with the Brownian motion (W_t) driving the asset price equation. We denote by $\rho \in [-1, 1]$ the instantaneous correlation coefficient defined by

$$d\langle W, \hat{Z}\rangle_t = \rho dt.$$

It is also convenient to write

$$\hat{Z}_t = \rho W_t + \sqrt{1-\rho^2} Z_t,$$

where (Z_t) is a standard Brownian motion independent of (W_t). It is often found from financial data that $\rho < 0$, and there are economic arguments for a negative correlation or *leverage effect* between stock price and volatility shocks. From common experience and empirical studies, when volatility goes up, asset prices tend to go down. In general, the correlation may depend on time $\rho(t) \in [-1, 1]$, but we shall assume it a constant for notational simplicity and because, in most practical situations, it is taken to be such.

1.2. Pricing with Equivalent Martingale Measures

Because of the additional source of randomness in the volatility process, contingent claims cannot, in general, be replicated by self-financing portfolios made of stocks and riskfree bonds for which we assume a constant interest rate r for simplicity. There is no uniqueness of no-arbitrage prices. We take the point of view that the market is choosing one equivalent martingale measure to determine prices of derivatives. This translates into the introduction of a *market price of volatility risk*

γ_t, an adapted process such that

$$\frac{dI\!P^{\star(\gamma)}}{dI\!P} = \exp\left(-\frac{1}{2}\int_0^T (\theta_s^2+\gamma_s^2)ds - \int_0^T \theta_s dW_s - \int_0^T \gamma_s dZ_s\right),$$

defines an equivalent probability $I\!P^{\star(\gamma)}$ with $\theta_t = (\mu - r)/f(Y_t)$. By Girsanov's theorem

$$W_t^\star = W_t + \int_0^t \frac{(\mu-r)}{f(Y_s)}ds \ , \ Z_t^\star = Z_t + \int_0^t \gamma_s ds,$$

are two independent Brownian motions under $I\!P^{\star(\gamma)}$. We assume that $\gamma_t = \gamma(Y_t)$ is a function of Y_t only and our model under the equivalent martingale measure $I\!P^{\star(\gamma)}$ becomes

$$\begin{aligned} dX_t &= rX_t dt + f(Y_t)X_t dW_t^\star, \\ dY_t &= [\alpha(m-Y_t) - \beta\Lambda(Y_t)]\,dt + \beta d\hat{Z}_t^\star, \end{aligned}$$

where $\hat{Z}_t^\star = \rho W_t^\star + \sqrt{1-\rho^2}\,Z_t^\star$ and $\Lambda(y) = \rho\frac{(\mu-r)}{f(y)} + \gamma(y)\sqrt{1-\rho^2}$ is a combined market price of risk.

Denoting by $I\!E^{\star(\gamma)}$ the expectation with respect to $I\!P^{\star(\gamma)}$, derivatives with time T payoff H are then priced by using the formula

$$V_t = I\!E^{\star(\gamma)}\{e^{-r(T-t)}H|\mathcal{F}_t\},$$

for all $t \le T$, excluding arbitrage opportunities. The filtration $(\mathcal{F}_t)$ is generated by (W, Z) or $(W^\star, Z^\star)$. Even though, in these Markovian models, derivatives prices can be obtained as solutions of PDE's through the Feynman-Kac formula, we will not use this point of view. The martingale approach developed here has the advantage to generalize naturally to non-Markovian models.

1.3. Fast Mean-Reversion

We consider the "regime" where the rate of mean-reversion α is large while the variance of the invariant distribution of Y_t remains of order one ($\nu^2 = \beta^2/2\alpha$ in the OU case). In other words the volatility clock is running faster than typical maturities of order one unit of time (the year for instance). Empirical evidence for this regime in the S&P500 are given in [3](Ch.4) and in the forthcoming detailed analysis [4]. It is mathematically convenient to introduce the small parameter

$$\varepsilon = 1/\alpha,$$

and to rescale β accordingly to $\beta = \sqrt{2}\nu/\sqrt{\varepsilon}$ where ν does not vary with ε. With this notation our model under the pricing equivalent martingale measure becomes

$$\begin{aligned} dX_t^\varepsilon &= rX_t^\varepsilon dt + f(Y_t^\varepsilon)X_t^\varepsilon dW_t^\star, \\ dY_t^\varepsilon &= \left[\frac{1}{\varepsilon}(m - Y_t^\varepsilon) - \frac{\nu\sqrt{2}}{\sqrt{\varepsilon}}\Lambda(Y_t^\varepsilon)\right]dt + \frac{\nu\sqrt{2}}{\sqrt{\varepsilon}}d\hat{Z}_t^\star, \end{aligned}$$

where we write $(X^\varepsilon, Y^\varepsilon)$ to indicate explicitly the dependence upon ε.

2. Asymptotic Pricing

We consider a European derivative given by its nonnegative payoff function $h(x)$ and its maturity time T. We assume that h is a smooth function in order to avoid technicalities in explaining the principle of our asymptotic method first introduced in [3](Ch.10). The nonsmooth case is more technical and not presented here. The price $P^\varepsilon(t)$ of this derivative at time $t < T$ is given by

$$P^\varepsilon(t) = \mathbb{E}^{\star(\gamma)}\{e^{-r(T-t)}h(X_T^\varepsilon)|\mathcal{F}_t\},$$

where the conditional expectation is with respect to the filtration $(\mathcal{F}_t)$ of the two Brownian motions. Instead of characterizing this price as a function of the current values (x,y) of $(X_t^\varepsilon, Y_t^\varepsilon)$ satisfying a two-dimensional PDE, we characterize $P^\varepsilon(t)$ by the fact that the process M^ε defined by

$$M_t^\varepsilon = e^{-rt}P^\varepsilon(t) = \mathbb{E}^{\star(\gamma)}\{e^{-rT}h(X_T^\varepsilon)|\mathcal{F}_t\},$$

is a martingale with a terminal value given by

$$M_T^\varepsilon = e^{-rT}h(X_T^\varepsilon).$$

Our goal is to show that $P^\varepsilon(t)$ can be approximated up to the order $\mathcal{O}(\varepsilon)$ by $Q^\varepsilon(t, X_t^\varepsilon)$ where $Q^\varepsilon(t,x)$ is a function, independent of y, to be determined.

2.1. The Epsilon-Martingale Decomposition Argument

For a function $Q^\varepsilon(t,x)$ to be determined which may depend on ε but not on y, we consider the process N^ε defined by

$$N_t^\varepsilon = e^{-rt}Q^\varepsilon(t, X_t^\varepsilon).$$

Requiring that the function Q^ε satisfies $Q^\varepsilon(T,x) = h(x)$ at the final time T, we have

$$M_T^\varepsilon = N_T^\varepsilon.$$

The method consists in finding $Q^\varepsilon(t,x)$ such that N^ε can be **decomposed** as

$$N_t^\varepsilon = \widetilde{M_t^\varepsilon} + R_t^\varepsilon,$$

where $\widetilde{M^\varepsilon}$ is a martingale and R_t^ε is of order ε. Observe that in this **decomposition** the terms of order $\sqrt{\varepsilon}$ are absorbed in the martingale part.

Supposing that this has been established, by taking a conditional expectation with respect to $\mathcal{F}_t$ on both sides of the equality

$$N_T^\varepsilon = \widetilde{M_T^\varepsilon} + R_T^\varepsilon,$$

and using the martingale property of $\widetilde{M^\varepsilon}$, one obtains

$$\mathbb{E}^\star\{N_T^\varepsilon|\mathcal{F}_t\} = \widetilde{M_t^\varepsilon} + \mathbb{E}^\star\{R_T^\varepsilon|\mathcal{F}_t\}.$$

From the terminal condition $M_T^\varepsilon = N_T^\varepsilon$ and the martingale property of M^ε we deduce that the left hand side is also equal to M_t^ε. From the decomposition of N_t^ε we have $\widetilde{M_t^\varepsilon} = N_t^\varepsilon - R_t^\varepsilon$ and therefore

$$M_t^\varepsilon = N_t^\varepsilon + \mathbb{E}^\star\{R_T^\varepsilon|\mathcal{F}_t\} - R_t^\varepsilon.$$

Multiplying by e^{rt} and using the definitions of M_t^ε and N_t^ε one deduces

$$P^\varepsilon(t) = Q^\varepsilon(t, X_t^\varepsilon) + \mathcal{O}(\varepsilon),$$

which is the desired **approximation result**. Indeed it remains to determine Q^ε leading to the epsilon-decomposition of N^ε.

2.2. Decomposition Result

Assuming a priori sufficient smoothness of the function Q^ε we write

$$\begin{aligned} dN_t^\varepsilon &= d\left(e^{-rt} Q^\varepsilon(t, X_t^\varepsilon)\right) \\ &= e^{-rt}\left(\frac{\partial}{\partial t} + \frac{1}{2} f(Y_t^\varepsilon)^2 (X_t^\varepsilon)^2 \frac{\partial^2}{\partial x^2} + r X_t^\varepsilon \frac{\partial}{\partial x} - r\right) Q^\varepsilon(t, X_t^\varepsilon) dt \\ &+ e^{-rt}\left(\frac{\partial Q^\varepsilon}{\partial x}(t, X_t^\varepsilon)\right) f(Y_t^\varepsilon) X_t^\varepsilon dW_t^\star . \end{aligned}$$

The method consists in cancelling the bounded variation terms as much as possible. The first obvious step in that direction is to replace the volatility $f(Y_t^\varepsilon)$ by a constant volatility and to consider $Q^\varepsilon(t,x)$ as a perturbation of the corresponding Black-Scholes pricing function. A natural choice of a constant volatility is given by averaging f^2 with respect to the invariant distribution of Y^ε. Denoting this averaging by $\langle\cdot\rangle_\varepsilon$ we define

$$\bar{\sigma}_\varepsilon^2 = \langle f^2 \rangle_\varepsilon .$$

In our example, Y^ε is a perturbed Ornstein-Uhlenbeck process and its invariant distribution admits the density

$$J_\varepsilon \exp\left\{-\frac{(y-m)^2}{2\nu^2} - \frac{\sqrt{2\varepsilon}}{\nu}\widetilde{\Lambda}(y)\right\},$$

where $\widetilde{\Lambda}$ is an antiderivative of the market price of risk Λ and J_ε is the appropriate normalizing constant. Introducing the usual Black-Scholes operator

$$\mathcal{L}_{BS}(\sigma) = \frac{\partial}{\partial t} + \frac{1}{2}\sigma^2 x^2 \frac{\partial^2}{\partial x^2} + r\left(x\frac{\partial}{\partial x} - \cdot\right),$$

we have

$$\begin{aligned} dN_t^\varepsilon &= e^{-rt}\left(\mathcal{L}_{BS}(\bar{\sigma}_\varepsilon) + \frac{1}{2}\left(f(Y_t^\varepsilon)^2 - \bar{\sigma}_\varepsilon^2\right)(X_t^\varepsilon)^2 \frac{\partial^2}{\partial x^2}\right) Q^\varepsilon(t, X_t^\varepsilon) dt \\ &+ e^{-rt}\left(\frac{\partial Q^\varepsilon}{\partial x}(t, X_t^\varepsilon)\right) f(Y_t^\varepsilon) X_t^\varepsilon dW_t^\star . \end{aligned}$$

Setting $Q^\varepsilon = P_0^\varepsilon + \widetilde{Q_1^\varepsilon}$ where P_0^ε is the the solution of the Black-Scholes equation $\mathcal{L}_{BS}(\bar{\sigma}_\varepsilon) P_0^\varepsilon = 0$ with the terminal condition $P_0^\varepsilon(T,x) = h(x)$, we deduce

$$\begin{aligned} dN_t^\varepsilon &= e^{-rt}\left(\mathcal{L}_{BS}(\bar{\sigma}_\varepsilon)\widetilde{Q_1^\varepsilon}(t, X_t^\varepsilon) + \frac{1}{2}\left(f(Y_t^\varepsilon)^2 - \bar{\sigma}_\varepsilon^2\right)(X_t^\varepsilon)^2 \frac{\partial^2 Q^\varepsilon}{\partial x^2}(t, X_t^\varepsilon)\right) dt \\ &+ e^{-rt}\left(\frac{\partial Q^\varepsilon}{\partial x}(t, X_t^\varepsilon)\right) f(Y_t^\varepsilon) X_t^\varepsilon dW_t^\star . \end{aligned}$$

The second term will be small of order $\mathcal{O}(\sqrt{\varepsilon})$ by a central limit type argument and $\widetilde{Q_1^\varepsilon}$ will be chosen to combine with it into a term of order $\mathcal{O}(\varepsilon)$ and a martingale term. For clarity we do that first by using the Markov property of Y^ε.

2.2.1. MARKOVIAN CASE. We denote by $\varepsilon^{-1}\mathcal{L}_0$ (resp. $\varepsilon^{-1}\mathcal{L}_0^\varepsilon$) the infinitesimal generator of the unperturbed (resp. perturbed) Markov process Y^ε. In the particular case of an OU process one has:

$$\begin{aligned}
\mathcal{L}_0 &= (m-y)\frac{\partial}{\partial y} + \nu^2\frac{\partial^2}{\partial y^2}, \\
\mathcal{L}_0^\varepsilon &= \mathcal{L}_0 - \nu\sqrt{2\varepsilon}\,\Lambda(y)\frac{\partial}{\partial y} \\
&= (m - y - \nu\sqrt{2\varepsilon}\,\Lambda(y))\frac{\partial}{\partial y} + \nu^2\frac{\partial^2}{\partial y^2}.
\end{aligned}$$

We then consider a solution ϕ_ε of the Poisson equation

$$\mathcal{L}_0^\varepsilon\phi_\varepsilon(y) = f(y)^2 - \bar{\sigma}_\varepsilon^2,$$

which we assume to be well defined (up to an additive constant) and to have a bounded derivative as in the OU case with a bounded function f.

We then deduce from Itô's formula that

$$\left(f(Y_t^\varepsilon)^2 - \bar{\sigma}_\varepsilon^2\right)dt = \varepsilon\left\{d(\phi_\varepsilon(Y_t^\varepsilon)) - \frac{\nu\sqrt{2}}{\sqrt{\varepsilon}}\,\phi_\varepsilon'(Y_t^\varepsilon)d\hat{Z}_t^\star\right\},$$

leading to

$$\begin{aligned}
dN_t^\varepsilon &= e^{-rt}\left(\mathcal{L}_{BS}(\bar{\sigma}_\varepsilon)\widetilde{Q_1^\varepsilon}(t,X_t^\varepsilon)dt + \frac{1}{2}\varepsilon(X_t^\varepsilon)^2\frac{\partial^2 Q^\varepsilon}{\partial x^2}(t,X_t^\varepsilon)d(\phi_\varepsilon(Y_t^\varepsilon))\right) \\
&+ e^{-rt}\left(\frac{\partial Q^\varepsilon}{\partial x}(t,X_t^\varepsilon)\right)f(Y_t^\varepsilon)X_t^\varepsilon dW_t^\star \\
&- \frac{\nu\sqrt{\varepsilon}}{\sqrt{2}}e^{-rt}\left((X_t^\varepsilon)^2\frac{\partial^2 Q^\varepsilon}{\partial x^2}(t,X_t^\varepsilon)\right)\phi_\varepsilon'(Y_t^\varepsilon)d\hat{Z}_t^\star.
\end{aligned}$$

The second term is computed by using the integration by parts formula:

$$\begin{aligned}
(X_t^\varepsilon)^2\frac{\partial^2 Q^\varepsilon}{\partial x^2}d\left(\phi_\varepsilon(Y_t^\varepsilon)\right) &= d\left((X_t^\varepsilon)^2\frac{\partial^2 Q^\varepsilon}{\partial x^2}\phi_\varepsilon(Y_t^\varepsilon)\right) - \phi_\varepsilon(Y_t^\varepsilon)d\left((X_t^\varepsilon)^2\frac{\partial^2 Q^\varepsilon}{\partial x^2}\right) \\
&- d\left\langle (X^\varepsilon)^2\frac{\partial^2 Q^\varepsilon}{\partial x^2}, \phi_\varepsilon(Y^\varepsilon)\right\rangle_t,
\end{aligned}$$

where the covariation term is given by

$$\begin{aligned}
&d\left\langle (X^\varepsilon)^2\frac{\partial^2 Q^\varepsilon}{\partial x^2}, \phi_\varepsilon(Y^\varepsilon)\right\rangle_t \\
&= \frac{\nu\rho\sqrt{2}}{\sqrt{\varepsilon}}\,\phi_\varepsilon'(Y_t^\varepsilon)\frac{\partial}{\partial x}\left(x^2\frac{\partial^2 Q^\varepsilon}{\partial x^2}\right)(t,X_t^\varepsilon)f(Y_t^\varepsilon)X_t^\varepsilon dt.
\end{aligned}$$

Collecting the martingale terms and the bounded variation terms of order $\mathcal{O}(\varepsilon)$, we get

$$\begin{aligned} N_t^\varepsilon &= N_0^\varepsilon + \text{Martingale} \\ &+ \int_0^t e^{-rs} \\ &\times \left(\mathcal{L}_{BS}(\bar{\sigma}_\varepsilon)\widetilde{Q_1^\varepsilon}(s, X_s^\varepsilon) - \frac{\nu\rho\sqrt{\varepsilon}}{\sqrt{2}} \phi_\varepsilon'(Y_s^\varepsilon) f(Y_s^\varepsilon) X_s^\varepsilon \frac{\partial}{\partial x}\left(x^2 \frac{\partial^2 P_0^\varepsilon}{\partial x^2}\right)(t, X_s^\varepsilon)\right) ds \\ &+ \mathcal{O}(\varepsilon)\,. \end{aligned}$$

We choose now $\widetilde{Q_1^\varepsilon}$ such that the quantity to be integrated in time is centered with respect to the invariant distribution of Y^ε. Repeating the argument above, one can show that this integral can be replaced by the sum of a martingale and a term of order $\mathcal{O}(\varepsilon)$. The centering involves the third derivative of the Black-Scholes price P_0^ε. Introducing the small constant

$$V_3^\varepsilon = \frac{\rho\nu\sqrt{\varepsilon}}{\sqrt{2}} \langle f\phi_\varepsilon' \rangle_\varepsilon \,,$$

and defining the source function

$$H^\varepsilon(t,x) = V_3^\varepsilon\, x \frac{\partial}{\partial x}\left(x^2 \frac{\partial^2 P_0^\varepsilon}{\partial x^2}\right) = V_3^\varepsilon \left(2x^2 \frac{\partial^2 P_0^\varepsilon}{\partial x^2} + x^3 \frac{\partial^3 P_0^\varepsilon}{\partial x^3}\right),$$

we choose $\widetilde{Q_1^\varepsilon}(t,x)$ to be the solution of the Black-Scholes equation

$$\mathcal{L}_{BS}(\bar{\sigma}_\varepsilon)\widetilde{Q_1^\varepsilon} = H^\varepsilon \,,$$

with the zero terminal condition $\widetilde{Q_1^\varepsilon}(T,x) = 0$. The desired decomposition follows

$$N_t^\varepsilon = \widetilde{M_t^\varepsilon} + R_t^\varepsilon \,,$$

where $\widetilde{M^\varepsilon}$ is a martingale and R^ε is small of order $\mathcal{O}(\varepsilon)$. Note that $Q^\varepsilon = P_0^\varepsilon + \widetilde{Q_1^\varepsilon}$ do not depend on y and satisfies $Q^\varepsilon(T,x) = h(x)$. The correction $\widetilde{Q_1^\varepsilon}$ is of order $\mathcal{O}(\sqrt{\varepsilon})$ and $P^\varepsilon(t) = Q^\varepsilon(t,x) + \mathcal{O}(\varepsilon)$. Note also that in the less interesting uncorrelated case $\rho = 0$, the correction $\widetilde{Q_1^\varepsilon}$ is simply zero and the approximated price is the Black-Scholes price P_0^ε with the modified volatility $\bar{\sigma}_\varepsilon$.

2.2.2. NON-MARKOVIAN CASE. We briefly indicate how to obtain the decomposition of N^ε in a non-Markovian stochastic volatility environment. We assume that Y^ε has an invariant distribution and is strongly mixing with a fast exponential rate of mixing denoted by α. We again set $\varepsilon = 1/\alpha$ and we assume that the variance of the invariant distribution of the unperturbed process Y denoted by ν^2 is independent of ε. We also suppose that Y has a diffusion part driven by a Brownian motion correlated to W through the constant ρ as in the Markovian case.

The decomposition result is then obtained as in the Markovian case except that we cannot define the "corrector" ϕ_ε by using the Poisson equation involving an infinitesimal generator. Instead $\phi_\varepsilon(Y_t^\varepsilon)$ is replaced by the random quantity $\widetilde{\phi_\varepsilon}(t)$ defined by the *conditional shift*

$$\widetilde{\phi_\varepsilon}(t) = -\frac{1}{\varepsilon} I\!E^\star \left\{ \int_t^T \left(f(Y_s^\varepsilon)^2 - \bar{\sigma}_\varepsilon^2 \right) ds | \mathcal{F}_t^\varepsilon \right\}.$$

It is very similar to the way the solutions of the Poisson equations are constructed in the Markovian case since the simple change of variable $u = s/\varepsilon$ shows that it is comparable to

$$\phi_\varepsilon(y) = -\int_0^{+\infty} I\!E^\star \left\{ \left(f(Y_t)^2 - \bar{\sigma}_\varepsilon^2 \right) | Y_0 = y \right\} dt,$$

used in the Markovian case. The method of conditional shift (or pseudo-generator) is treated in [7] for instance. The rest of the proof is very similar to the Markovian case once we observe that

$$\mathcal{M}_t^\varepsilon = \widetilde{\phi_\varepsilon}(t) - \frac{1}{\varepsilon}\int_0^t \left(f(Y_s^\varepsilon)^2 - \bar{\sigma}_\varepsilon^2 \right) ds,$$

is a martingale. Defining $\psi_\varepsilon(t)$ by

$$\frac{\sqrt{\varepsilon}}{2} d\left\langle W^\star, \mathcal{M}^\varepsilon \right\rangle_t = \psi_\varepsilon(t) dt,$$

so that it plays the role of $\frac{\nu\rho}{\sqrt{2}}\phi'(Y_t^\varepsilon)$ in the Markovian case, the small constant V_3^ε is given by the averaging

$$V_3^\varepsilon = \langle f \psi_\varepsilon \rangle_\varepsilon .$$

The source function $H^\varepsilon(t,x)$ and the correction $\widetilde{Q_1^\varepsilon}(t,x)$ are then defined exactly as in the Markovian case and the same conlusion follows: $\widetilde{Q_1^\varepsilon}$ is of order $\mathcal{O}(\sqrt{\varepsilon})$ and $P^\varepsilon(t) = P_0^\varepsilon(t,x) + \widetilde{Q_1^\varepsilon}(t,x) + \mathcal{O}(\varepsilon)$.

3. Practical Form of the Approximated Price

The computation of the approximated price $P_0^\varepsilon + \widetilde{Q_1^\varepsilon}$ requires the two parameters $\bar{\sigma}_\varepsilon$ and V_3^ε in order to compute first the Black-Scholes price P_0^ε and then deduce the correction $\widetilde{Q_1^\varepsilon}$. In fact it is easily shown, [3](Ch.5), that

$$\widetilde{Q_1^\varepsilon} = -(T-t)H^\varepsilon = -(T-t)V_3^\varepsilon \, x \frac{\partial}{\partial x}\left(x^2 \frac{\partial^2 P_0^\varepsilon}{\partial x^2} \right).$$

The mean square volatility $\bar{\sigma}_\varepsilon^2$ would be easy to estimate except for the fact that we do not observe stock returns under the risk neutral probability $I\!P^{\star(\gamma)}$ but rather under the subjective probability $I\!P$ which does not involve the market price of volatility risk γ. For practical purpose we need to rewrite our approximation formula with a different set of parameters.

3.1. Perturbed Invariant Distribution

Denoting by $\langle\cdot\rangle$ the average with respect to the invariant distribution of the unperturbed Y process, we have the following expansion of the invariant distribution of the perturbed Y^ε process

$$\langle\cdot\rangle_\varepsilon = \langle\cdot\rangle - \frac{\sqrt{2\varepsilon}}{\nu}\left\langle\widetilde{\Lambda}(\cdot - \langle\cdot\rangle)\right\rangle + \mathcal{O}(\varepsilon),$$

where $\widetilde{\Lambda}$ is an antiderivative of the combined market price of risk Λ. Defining $\bar{\sigma}^2 = \langle f^2\rangle$, it follows that

$$\begin{aligned}\mathcal{L}_{BS}(\bar{\sigma}_\varepsilon) &= \mathcal{L}_{BS}(\bar{\sigma}) + \frac{1}{2}\left(\langle f^2\rangle_\varepsilon - \langle f^2\rangle\right)x^2\frac{\partial^2}{\partial x^2}\\ &= \mathcal{L}_{BS}(\bar{\sigma}) - \frac{\sqrt{\varepsilon}}{\nu\sqrt{2}}\langle\widetilde{\Lambda}(f^2 - \langle f^2\rangle)\rangle x^2\frac{\partial^2}{\partial x^2} + \mathcal{O}(\varepsilon).\end{aligned}$$

Introducing $P_0(t,x)$, the Black-Scholes solution with constant voaltility $\bar{\sigma}$, one can easily write

$$\widetilde{P_1} = -(T-t)\left(V_2 x^2\frac{\partial^2 P_0}{\partial x^2} + V_3 x^3\frac{\partial^3 P_0}{\partial x^3}\right),$$

for two parameters V_2 and V_3, small of order $\mathcal{O}(\sqrt{\varepsilon})$, and such that the approximation $P^\varepsilon(t) = P_0(t,x) + \widetilde{P_1}(t,x) + \mathcal{O}(\varepsilon)$ holds. This is the form which arises naturally when performing the asymptotics on the pricing PDE in the Markovian case as described first in [1] and detailed in [3](Ch.5).

3.2. Calibration

Without going into details we see that, in order to use this formula, we need the three parameters $(\bar{\sigma}, V_2, V_3)$. The first one is easily estimated on historical returns data. The two others, the V's, can be obtained by using observed call option prices or equivalently by fitting the term structure of implied volatility. Explicit formulas are given in [2] and, in [3](Ch.6) with a stability analysis.

4. Hedging

To stay within a reasonable length we simply recall that, in presence of stochastic volatility, a perfect self-financing hedge of a derivative, with stocks and bonds only, is not possible in general. When volatility is fast mean-reverting as described in this paper, it is possible to correct a pure Black-Scholes hedge ratio in order to reduce the bias introduced in the cost of the hedging strategy. This hedging correction gives a *mean self-financing strategy* up to order $\mathcal{O}(\varepsilon)$. It is obtained by the method described in Section 2.2 but performed under the objective probability $\mathbb{P}$. We refer to [3](Ch.7) for details and formulas.

References

[1] J.P. Fouque, G. Papanicolaou and R. Sircar, *Asymptotics of a Two-Scale Stochastic Volatility Model,* Equations aux dérivées partielles et applications, Articles dédiés à Jacques-Louis Lions, Gauthier-Villars, Paris (1998) 517–526.

[2] J.P. Fouque, G. Papanicolaou and R. Sircar, *Mean-Reverting Stochastic Volatility,* International Journal of Theoretical and Applied Finance, **3**(1) (2000), 101–142.

[3] J.P. Fouque, G. Papanicolaou and R. Sircar, *Derivatives in Financial Markets with Stochastic Volatility,* Cambridge University Press, 2000.

[4] J.P. Fouque, G. Papanicolaou, R. Sircar and K. Solna, *Mean Reversion of S&P* 500 *Volatility,* preprint (2000).

[5] R. Frey, *Derivative Asset Analysis in Models with Level-Dependent and Stochastic Volatility,* CWI Quarterly, **10**(1) (1996), 1–34.

[6] E. Ghysels, A. Harvey and E. Renault, *Stochastic Volatility,* Statistical Methods in Finance. Handbook of Statistics, **14**(Ch.5) (1996) 119–191.

[7] A. Kushner, *Approximation and Weak Convergence Methods for random Processes,* MIT Press, 1984.

Jean-Pierre Fouque,
Department of Mathematics,
North Carolina State University,
Raleigh NC 27695-8205, USA
Partially supported by NSF/DMS-0071744
E-mail address: fouque@math.ncsu.edu

George Papanicolaou,
Department of Mathematics,
Stanford University,
Stanford CA 94305, USA
E-mail address: papanico@math.stanford.edu

Ronnie Sircar,
Department of Operations Research & Financial Engineering,
Princeton University, E-Quad
Princeton NJ 08544, USA
Partially supported by NSF/DMS-0090067
E-mail address: sircar@princeton.edu

Trends in Mathematics, © 2001 Birkhäuser Verlag Basel/Switzerland

Mutual Debts Compensation as Graph Theory Problem

Vladimir Gazda

Abstract. The article deals with a serious problem of many postcommunist countries i.e. firms' financial insolvency as a result of their mutual debts. Some authors proposed graph theory application in solution of the problem. Creditor - debtor relationship is modelled by a digraf of unpayable debts $G = (V, E, y)$, where the set of vertices V represents the firms and the set of edges E represents the creditor–debtor relationship. Function y assigns positive value of debt to each of the edges. The role of ministry of finance in mutual debts compensation process is discussed in the article.

1. Introduction

There have been made some attempts to build a theory of mutual debts minimising in Slovakia. Fecenko [3] in the first article proposed solving of this problem as circulation maximising. In the same article he was looking for appropriate method of firms debts relations minimisation a discovered influence capital subsidiaries on mutual debts compensation. As his article is in Slovak his approach of circulation maximisation is described in the second and in the third section. Gazda [4] discovered alternative method of mutual debts minimisation using subsidiaries and he investigated possibilities of mutual debts compensation with possibility of debts structure changing. Gazda and Paluch [5] reformulated circulation maximising problem as flow problem from source to sink in a network. They solved problem of controlled bancrupcy of the firm with regards to mutual debts minimising. The mutual debts minimisation organised by Ministry of Finance is considered in this article. The ministry may solve another target function, i.e. minimising of its unpayeble assets, or minimising of its unpayable debts. This approach is considered in the fourth section.

2. Digraph of debts

Let $V = \{1, 2, \ldots, n\}$ be a set of firms. Then y_{ij} is a debt which firm i owes towards firm j. We assume that $0 \leq y_{ij}$ and at least one of equations $y_{ij} = 0$, $y_{ji} = 0$ is fulfiled. Let us have a set of creditor–debtor relations defined as follows

$$E = \{(i,j) \mid y_{ij} > 0 \quad \text{and} \quad \mathrm{i} \neq j\} \tag{1}$$

Then, function $y : E \to R^+$ assigns a positive real value of the debt's level to each creditor – debtor relation. That means following

$$\forall e \in E, \quad e = (i,j) \qquad y(e) = y_{ij} \tag{2}$$

Then,

$$G = (V, E, y) \tag{3}$$

is weighted directed graph called debt digraph. Balance of each firm $i \in V$ can be written as follows

$$b(i) = \sum_{ji \in I_i^+} y_{ji} - \sum_{ij \in I_i^-} y_{ij} \tag{4}$$

where $I_i^+ = \{ji \mid \exists(j,i)\}$ is a set of indeces pairs representing edges entering the i-th vertex and $I_i^- = \{ji \mid \exists(i,j)\}$ is a set of indeces pairs representing edges leaving the i-th vertex in graph G. The first sum in (4) are receivable assets of the i-th firm and the (4) are debts of this firm. Have a look at

$$\sum_{i=1}^{n} b(i) = 0. \tag{5}$$

It is fulfiled, because:

$$\sum_{i=1}^{n} b(i) = \sum_{i=1}^{n}((\sum_{ji \in I+} y_{ji} - \sum_{ij \in I-} y_{ij}))) = \sum_{i=1}^{n} \sum_{ji \in I+} y_{ji} - \sum_{i=1}^{n} \sum_{ij \in I-} y_{ij} = 0 \tag{6}$$

Let

$$c = v_1, (v_1, v_2), v_2, \ldots, v_{k-1}, (v_{k-1}, v_k), v_k, \tag{7}$$

be a cycle in digraph $G = (V, H, y)$ where $v_1 = v_k$. Capacity of cycle c is defined as follows

$$y(c) = \min_{e \in c}\{y(e)\}. \tag{8}$$

If there is a cycle (7) in graph (3) with capacity (8), it means presence of the cyclical mutual debts in the firms structure. The situation can be simplified by subtracting the capacity of the cycle $y(c)$ from each weight $y(e)$ of the all edges $e \in c$:

$$y(e) := y(e) - y(c) \tag{9}$$

It means reduction of total debts in digraph and in such a way at least one of the edges from cycle c gets new value $y(e)$ equal to zero. That means, the creditor–debtor relation represented by the edge with zero weight is removed from the digraph G. In such a way we can remove all cycles from G and the graph becomes acyclical.

3. Mutual debts' elimination as problem of circulation maximisation in the graph.

Let us define x_{ij} as a portion of debt which was eliminated between the i-th firm and j-th one. Then, we can express the total amount of eliminated debts as

$$\sum_{i=1}^{n} \sum_{ij \in I_i^-} x_{ij}. \tag{10}$$

Because x_{ij} represents an eliminated debt, the following relation must be fulfiled:

$$0 \leq x_{ij} \leq y_{ij}. \tag{11}$$

Mutual debts compensation must not worsen nor improve balance of payments of each firm in V. The equation

$$\sum_{ji \in I_i^+} x_{ji} = \sum_{ij \in I_i^-} x_{ij} \tag{12}$$

is fulfiled for each $i \in V$. Balance of payments of each vertex before the elimination can be expressed as

$$\sum_{ji \in I_i^+} y_{ji} - \sum_{ij \in I_i^-} y_{ij} = b(i) \tag{13}$$

and must not be changed by the elimination. That means, that

$$\sum_{ji \in I_i^+} (y_{ji} - x_{ji}) - \sum_{ij \in I_i^-} (y_{ij} - x_{ij}) = b(i) \tag{14}$$

is fulfiled for each $i \in V$ only if condition (12) is satisfied for each $i \in V$.

The total debts minimising problem can be formulated as minimisation of (10) with regards to (11) and (12).

Note: Based on (9) we can conclude that $y_{ij} = 0$ implies $x_{ij} = 0$. Debts minimisation does not cause creation of new creditor–debtor relations what is considered as good characteristic of proposed mathematical model.

4. Ministry of Finance and problem of its receivable assets minimising

We assume that ministry of finance receives its income from firms' taxes and spends its outcome by paying it to firms. We do not take into consideration ministry relations to another subjects. Then a set $\bar{V}$ envolves all subjects included to mutual debts compensation, i.e.

$$\bar{V} = V \cup \{n+1\} \tag{15}$$

where V is a set of n firms and vertex $n+1$ represents ministry of finance. The debts relation are represented by set $\bar{E}$ where

$$\bar{E} = E \cup \{(n+1, i) \mid i \in V\} \cup \{(i, n+1) \mid i \in V\}. \tag{16}$$

Function $\bar{y}$ is defined as follows

$$\bar{y} : \bar{E} \to R^+. \tag{17}$$

That means that a real positive value is assigned to each of the debt relations and the assigned value represents an amount of the debt which owes the i-th subject towards the j-th subject. More concretely

$$\bar{y}(e) = y(e) \quad \text{for} \quad e \in E \tag{18}$$

$$\bar{y}(e) = y_{i\ n+1} \quad \text{for} \quad e \in \{(i, n+1) | i \in V\} \tag{19}$$

$$\bar{y}(e) = y_{n+1\ i} \quad \text{for} \quad e \in \{(n+1, i) | i \in V\} \tag{20}$$

Then digraph $\bar{G}(\bar{V}, \bar{E}, \bar{y})$ represents mutual debts structure among the firms and ministry of finance, too. The debts balance of ministry of finance is represented as

$$\sum_{j\ n+1 \in I^+_{n+1}} y_{j\ n+1} - \sum_{n+1\ j \in I^-_{n+1}} y_{n+1\ j} = b_{n+1} \tag{21}$$

The balance of payments of each subject must stay unchanged after the cycle elimination. The same characteristic has also Ministry of Finance.
If balance of payments of n subjects in digraph $\bar{G}(\bar{V}, \bar{E}, \bar{y})$ stays unchanged after the debts elimination, balance of debts of the $n+1$-st subject must stay unchanged, too. It prevents the ministry to improve its balance of debts by organising the mutual debts compensation by worsening of balance of payment of other subjects.
Let us prove the assertion mentioned above. If the debts sum of firms and ministry of finance before elimination is

$$\sum_{i=1}^{n+1} \sum_{ij \in I_i^-} y_{ij} - \sum_{i=1}^{n+1} \sum_{ji \in I_i^+} y_{ji} = 0 \tag{22}$$

as it was proved in the second section. The same sum must stay unchanged after elimination, i.e.

$$\sum_{i=1}^{n} \sum_{ij \in I_i^+} (y_{ij} - x_{ij}) - \sum_{i=1}^{n} \sum_{ji \in I_i^-} (y_{ji} - x_{ji}) + \tag{23}$$

$$\sum_{i=1}^{n} (y_{i\ n+1} - x_{i\ n+1}) + \sum_{j=1}^{n} (y_{n+1\ j} - x_{n+1\ j}) = 0$$

and if condition of compensated debts

$$\sum_{ij \in I_i^-} x_{ij} = \sum_{ji \in I_i^+} x_{ji} \tag{24}$$

is satisfied for n firms, then sum of compensated debts and assets of the ministry must be also equal

$$\sum_{n+1\ j \in I^+_{n+1}} x_{n+1\ j} = \sum_{j\ n+1 \in I^-_{n+1}} x_{j\ n+1}. \tag{25}$$

Then, the problem of ministry's debts minimisation achieved due to debts compensation in the whole graph

$$min \sum_{j\ n+1 \in I^-_{n+1}} x_{j\ n+1} \tag{26}$$

$$\sum_{ij \in I^+_i} x_{ij} = \sum_{ji \in I^-_i} x_{ji} \quad \text{for} \quad \mathrm{i} \in \bar{V} \tag{27}$$

$$\sum_{ij \in I^+_i} x_{ij} = \sum_{ji \in I^-_i} x_{ji} \quad \text{for} \quad \mathrm{i} \in \bar{V} \tag{28}$$

$$0 \leq x_{ij} \geq y_{ij} \quad \text{for} \quad (i,j) \in \bar{E} \tag{29}$$

The proposed optimisation problem is based on linear programming. Using this method we can have problems with large scale models. Because of that we propose solution of the same problem as problem of flow maximising in network with source 0 and sink $n+1$. Vertex 0 is virtual one that will be created respecting the following rules: Let us define a new set of vertices

$$\bar{\bar{V}} = \{0\} \cup \bar{V}. \tag{30}$$

All edges entering vertex $n+1$ are redirect to virtual vertex 0, i.e.

$$\{(i, n+1) \mid i \in V\} \rightarrow \{(i, 0) \mid i \in V\} \tag{31}$$

and a new edge $(0, n+1)$ is also created. Then a new set of edges is defined as

$$\bar{\bar{E}} = E \cup \{(i,0) \mid i \in V\} \cup (0, n+1) \cup \{(n+1, i) \mid i \in V\} \tag{32}$$

Function $\bar{\bar{y}}$ assigns a real positive value of debt to each $\bar{\bar{e}} \in \bar{\bar{E}}$ as follows

$$\bar{\bar{y}}(i,j) = y(i,j) \quad \text{for} \quad \mathrm{i}, j \in V \tag{33}$$

$$\bar{\bar{y}}(i,0) = y(i, \bar{n}+1) \quad \text{for} \quad \mathrm{i} \in V \tag{34}$$

$$\bar{\bar{y}}(0, n+1) = \sum_{i \in I^+_{n+1}} \bar{y}_{i_n+1} \tag{35}$$

$$\bar{\bar{y}}(n+1, i) = \bar{y}(n+1, i) \quad \text{for} \quad \mathrm{i} \in V. \tag{36}$$

Then graph $\bar{\bar{G}}(\bar{\bar{V}}, \bar{\bar{E}}, \bar{\bar{y}})$ contains virtual vertex and ministry's unpayable assets minimising problem is equivalent with a problem of flow maximisation thru the edge $(0, n+1)$. Because of equality of input and output flows in each vertex we can eliminate edge $(0, n+1)$. In such a way vertex $(n+1)$ becomes source and vertex 0 becomes sink of the network. Then, we can apply usual methods of network flow maximisation to minimise amount of unpayable assets of the ministry of finance.

5. Discussion

Mutual debts compensation using graph theory is not closed topic. The proposed method can be developed by including of import-export vertices and impact of revaluation or devaluation of domestic currency can be investigated. New fields can be discovered using dynamic approach by providing real money to some subjects in the graph. Process of random walk in graph can be simulated in the same case. Another field of graph theory application in finance is investing via investment funds where complicated owners' structure can be found.

References

[1] G.Chartrand,-O.R.Oellermann, *Applied and Algorithmic Graph Theory,* McGraw-Hill, Inc., **1993**

[2] R.Diestel, *Graph Theory,* Springer Verlag, New York, **1997**

[3] J. Fecenko, *About Optimisation of Receivables and Payables Compensation,* Ekonomicky casopis, **42** 1994, p.360 – 374. (in Slovak)

[4] V.Gazda, *Mutual Debts compensation as Graph Theory Application,* Proceedigs of the Conference in Canakale held in April 2000, CD - ROM.

[5] V.Gazda and S.Paluch, *One Method of Mutual Debts Compensation* yet unpublished

[6] Harary,F.,*Graph Theory,* Adison-Wesley Publishing Company, **1968**

Department of Economics,
University of Economics,
Tajovskeho 11
040 01 Kosice, Slovakia
E-mail address: gazda@economy.euke.sk

Trends in Mathematics, © 2001 Birkhäuser Verlag Basel/Switzerland

First Steps to Stochastic Finance

Hans-Joachim Girlich

Abstract. The first World congress of the Bachelier Finance Society has celebrated the centenary of Louis Bachelier's thesis "Théorie de la spéculation". A brief account of this dissertation together with life and work of the author is given on the nice paper [1]. The aim of the present note is to embedding this landmark event [2] into the development of probability and finance, a relationship which is ignored in the literature to a great extend comparing with the intimate relation between probability and physics (see [3]).

1. Towards Stochastic Finance

Under Bachelier's predecessors who have prepared tools and first steps to stochastic finance Carl Friedrich Gauss should be mentioned first. Bachelier cited "la loi de Gauss" which was introduced as an error distribution by Gauss in 1809 (see [2], [4]). We enlighten a fewer known side of the "mathematicorum princeps", his work as a successful investor. Since the 1830s, in defiance (or because) of his modest salary, Gauss invested his savings in bank stocks and obligations and leaved an estate of more than 150000 thalers (see [5], [6]). His great experience with financial operations on a bourse took effect also in public affairs. He reorganized the fund for professor's widows at the university of Göttingen during 1845 - 1851. For this purpose he calculated special life-insurance tables (see [7]).

In the 1860s Henry Dunning Macleod showed in his main work the "principles of Currency and Banking, their progressive development in practice, and the laws at present affecting them" (see [8]).
Francis Ysidro Edgeworth found a connection between finance and probability. He investigated the problem of bankruptcy and postulated the law of error as the foundation of banking. Edgeworth tested his hypothesis with real data, the amount of returns of Bank of England notes in the hand of the public in the period of 1833 to 1844. Upon this foundation he computed the probability that the demand will not exceed any proposed limit. Furthermore, he presented a complex model of banking as a new game of chance formulated as a certain portfolio problem with three possibilities to handle a disposable fund (see [9], [10]).

The main exchange in Europe in this time was the Paris Bourse. Ambroise Buchère wrote an important handbook on financial operations in the exchange (see [11]).

There were a few dissertation about the markets of the exchange, but all presented to a faculty of law. The first thesis on speculation from a stochastic point of view was written by Louis Bachelier. It has been seen and allowed for publication in January 1900 by Jean Darboux the dean of the Paris Faculty of Science and was published inmediately in February (see [12], [20], [2]).

Henry Poincaré suggested in his report on Bachelier's thesis to study into the details of Fourier's analysis the relationship of stochastic processes with partial differential equations. Bachelier realized this hint. In the year 1906 three attemps were published to handle dependent random phenomena. They were given in the case of discrete time by Andrei Markov in St. Petersburg and by Heinrich Bruns in Leipzig, but in continuous time by Louis Bachelier in Paris (see [13], [14], [15]). Bachelier presented a framework which covered not only the Brownian motion as in [2] but also the Ornstein-Uhlenbeck process. Furthermore, he disclosed a connection with the work of Pierre Simon Laplace (see [15], p. 275, [16] II 14, §17). Unfortunately, Bachelier's approach failed the necessary rigour for a general acceptance. In the special case of the Ornstein-Uhlenbeck process Markov's method of moments was successful yet. The general classical case in Poincaré's sense has been solved by Andrei Kolmogorov (see [17], [18]).

A long time was necessary for establishing a mathematical theory of stochastic processes. In Dynkin's monograph on Markov processes published in 1963 Bachelier's contributions are not more mentioned. At the same time Paul Cootner rediscovered Bachelier's dissertation now for finance and published an English translation together with important remarks and further papers in [20] opening the door to stochastic finance.

References

[1] J.-M. Courtault, Y. Kabanov, B. Bru, P. Crépel, I. Lebon and A. Le Marchand, *Louis Bachelier - On the Centenary of Théorie de la Spéculation*, Mathematical Finance 10 (2000), 339–353.

[2] L. Bachelier, *Théorie de la spéculation*, Annales Scientifiques de l' École Normale Supérieure, 3e Series, 17 (1900), 21 - 86.

[3] D. Hilbert, *Mathematische Probleme*, Nachrichten der Königl. Ges. d. Wiss. zu Göttingen, math.-phys. Klasse, 1900, Heft 3, 253 - 297.

[4] C.F. Gauss, *Theoria Motus Corporum Coelestium*, in: Carl Friedrich Gauss Werke, Vol. 7, 1 - 261, Teubner, Leipig, 1906.

[5] *Gauss an Olbers, 1838 Maerz 4*, In C. Schilling (ed.): Wilhelm Olbers, sein Leben und seine Werke, Vol. 2, 666 - 668, Berlin, 1909.

[6] *Wertpapier-Verzeichnisse*, Nachlass von Carl Friedrich Gauss, G IX 21: 7, Stadtarchiv Braunschweig.

[7] C.F. Gauss, Nachlass. *Anwendung der Wahrscheinlichkeitsrechnung auf die Bestimmung der Bilanz für Witwenkassen*, in: Carl Friedrich Gauss Werke, Vol. 4, 119 - 188, Göttingen, 1880.

[8] H. D. Macleod, *The Theory and Practice of Banking*, Longmans, Green, Reader, & Dyer, London, 1866.

[9] F. Y. Edgeworth, *The Mathematical Theory of Banking*, Journal of the Royal Statistical Society of London 51 (1888), 113 - 127.

[10] F. Y. Edgeworth, *Problems in Probability*, London Philosophical Magazine, Series 5, 22 (1886), 371 - 384.

[11] A. Buchère, *Traité theéorique et pratique des Opérations de la Bourse*, A. Marescq Ainé, Paris, 1877.

[12] J. F. Jeanneret, *Le Jeu, le Pari et les Marchés de Bourse*, Thèse de doctorat, Faculté de Droit de l'Université de Berne, Neuchatel, 1892.

[13] A. Markov, *Extension of the Law of Large Numbers to the Dependent Case*, Notices of the Physico Mathematical Society on the University of Kazan, 2nd Series, 15 (1906) 135 - 156 (in Russian).

[14] H. Bruns, *Das Gruppenschema für zufällige Ereignisse*, Leipziger Abhandlungen 51 (1906), 575 - 628.

[15] L. Bachelier, *Théorie des probabilités continues*, Journal de Mathématiques Pures et Appliquées, Sixième Série, 12 (1906), 259 - 327.

[16] P. S. Laplace, *Théorie analytique des Probabilités*, Courcier, Paris, 1820.

[17] A. Markov, *On a problem of Laplace*, Bulletin de l'Académie Impériale des Sciences, Petrograd, VI Série, 9 (1915), 87 - 104 (in Russian).

[18] A. Kolmogoroff, *Über die analytischen Methoden in der Wahrscheinlichkeitsrechnung*, Mathematische Annalen 104 (1931), 415 - 458.

[19] E. B. Dynkin, *Markov Processes*, Grundlehren, Vol. 121/122, Springer, Berlin, 1965 (Moscow, 1963).

[20] P. Cootner (ed.), *The Random Character of Stock Marked Prices*, The M.I.T. Press, Cambridge, 1964.

Department of Mathematics,
University of Dresden,
Dresden, Germany
E-mail address: girlich@uni-dresden.de

Trends in Mathematics, © 2001 Birkhäuser Verlag Basel/Switzerland

Fractional Calculus and Continuous-Time Finance III: the Diffusion Limit

Rudolf Gorenflo[1], Francesco Mainardi[2], Enrico Scalas[3] and Marco Raberto[4]

Abstract. A proper transition to the so-called diffusion or hydrodynamic limit is discussed for continuous time random walks. It turns out that the probability density function for the limit process obeys a fractional diffusion equation. The relevance of these results for financial applications is briefly discussed.

1. Introduction

A theory of tick-by-tick dynamics in financial markets, based on a continuous-time random walk (CTRW) model, was recently proposed by Scalas et al [1]. In a less general way, a similar approach had been developed by Parkinson [2]. This leads to a time-fractional master equation whose empirical usefulness has been studied by Mainardi et al [3] and Raberto et al [4], by analysing the behaviour observed in the waiting-time distribution for some future prices traded at the London International Financial Futures and Options Exchange (LIFFE). Here, we refine the theory by investigating the so-called *diffusion* or *hydrodynamic* limit, which allows the transition from the time-fractional master equation to a space-time fractional diffusion equation. The correct transition to this limit does need a special care and involves the introduction of a vanishing length scale h and the adaption of the Gnedenko limit theorem of probability theory (a generalization of the classical central limit theorem for *i.i.d.* random variables with heavy-tailed probability density functions (*pdf's*)).

For readers' convenience, in Section 2, we outline the theory leading to the time-fractional master equation. The core of the paper is in Section 3, where we discuss the transition to the space-time fractional diffusion equation. In Appendix A and Appendix B, the correct time and space fractional derivatives are introduced, which enter the fractional diffusion equation. Finally, the main conclusions are drawn in Section 4.

2. The time-fractional master equation

In financial markets, not only prices and returns can be modelled as random variables, but also the waiting times between two consecutive transactions vary in a stochastic way.

Let us denote by x the logarithm of an asset price. We assume that both the log-price jumps $\xi_i = x(t_i) - x(t_{i-1})$ and the waiting times between two consecutive trades $\tau_i = t_i - t_{i-1}$ are i.i.d. stochastic variables. We also assume that these variables are characterized by the two probability density functions: $w(\xi)$ and $\psi(\tau)$. The physicist can think of x as the position of a random walker performing discrete jumps in one dimension at randomly distributed instants, according to the model of continuous-time random-walk (CTRW), introduced in Statistical Mechanics by Montroll and Weiss [5]. In the following, we shall often use the physicist's CTRW language.

The evolution equation for $p(x,t)$, the probability of occurrence of the log-price x at time t, or of finding the random walker at position x at time instant t, can be written as follows, assuming the initial condition $p(x,0) = \delta(x)$ (i.e. the walker is initially at the position $x = 0$) [3]:

$$p(x,t) = \delta(x)\,\Psi(t) + \int_0^t \psi(t-t')\left[\int_{-\infty}^{+\infty} w(x-x')\,p(x',t')\,dx'\right] dt'\,, \qquad (2.1)$$

where

$$\Psi(t) = \int_t^\infty \psi(t')\,dt' = 1 - \int_0^t \psi(t')\,dt'\,, \quad \psi(t) = -\frac{d}{dt}\Psi(t)\,. \qquad (2.2)$$

This is called the *master equation* of the CTRW. It is usually derived in the Fourier-Laplace domain.

We note that $\int_0^\tau \psi(t')\,dt'$ represents the probability that at least one step is taken at some instant within the interval $[0,\tau)$, hence $\Psi(\tau)$ is the probability that the diffusing quantity x does not change value during a time interval of duration τ after a jump.

In [3], it was convenient to present an alternative form of (2.2) involving the first time derivative of $p(x,t)$ (along with an additional auxiliary function) so that the resulting equation can be interpreted as an *evolution* equation of generalized *Fokker-Planck-Kolmogorov* type. Then the master equation reads:

$$\int_0^t \Phi(t-t')\,\frac{\partial}{\partial t'}p(x,t')\,dt' = -p(x,t) + \int_{-\infty}^{+\infty} w(x-x')\,p(x',t)\,dx'\,, \qquad (2.3)$$

where the auxiliary function $\Phi(t)$ is such that $\Psi(t) = \int_0^t \Phi(t-t')\,\psi(t')\,dt'$.

From (2.3) we recognize the role of memory function played by $\Phi(t)$. As a consequence, in general, the CTRW turns out to be a non-Markovian process. The process is *memoryless*, namely Markovian if (and only if) the above memory function degenerates into a delta function, multiplied by a certain positive constant, so that $\Psi(t)$ and $\psi(t)$ may differ only by a multiplying positive constant. By an appropriate choice of the unit of time we assume $\Phi(t) = \delta(t)$, $t \geq 0$. In this case Eq. (2.3) reduces to

$$\frac{\partial}{\partial t} p(x,t) = -p(x,t) + \int_{-\infty}^{+\infty} w(x-x')\, p(x',t)\, dx' , \quad p(x,0) = \delta(x) . \tag{2.4}$$

This is, up to a change of the time unit (which means multiplication of the R.H.S by a positive constant), the most general *master equation* for a *Markovian CTRW*; it is called the *Kolmogorov-Feller equation* in [6].

We note that, by exhibiting a weighted first-time derivative, the form (2.3) is original as far as we know; it allows a natural characterization of a peculiar class of *long-memory* processes, namely non-Markovian processes characterized by a memory function $\Phi(t)$ with power-law time decay. Within this class, a possible and interesting choice is

$$\Phi(t) = \frac{t^{-\beta}}{\Gamma(1-\beta)} , \quad t \geq 0 , \quad 0 < \beta < 1 . \tag{2.5}$$

Thus, $\Phi(t)$ becomes a function weakly singular at $t = 0$ that, in the limiting case $\beta = 1$, reduces to $\Phi(t) = \delta(t)$, according to the formal representation of the Dirac generalized function, $\delta(t) = t^{-1}/\Gamma(0)$, $t \geq 0$ (see e.g. [7]). As a consequence of the choice (2.5), as shown in [3], the resulting Eq. (2.4) can be written as

$$\frac{\partial^\beta}{\partial t^\beta} p(x,t) = -p(x,t) + \int_{-\infty}^{+\infty} w(x-x')\, p(x',t)\, dx' , \quad p(x,0) = \delta(x) , \tag{2.6}$$

where $\partial^\beta / \partial t^\beta$ is the pseudo-differential operator explicitly defined in the Appendix A, that we call the *Caputo* fractional derivative of order β. Eq. (2.6) is a time-fractional generalization of Eq. (2.4) and can consequently be called the *time-fractional Kolmogorov-Feller equation*.

The choice of $\Phi(t)$ yields peculiar forms for the functions $\Psi(t)$ and $\psi(t)$ generalizing the exponential behaviour of the waiting time *pdf* of the Markovian case. In fact, as shown in [3], we get for $t \geq 0$,

$$\Psi(t) = E_\beta(-t^\beta) , \quad \psi(t) = -\frac{d}{dt} E_\beta(-t^\beta) , \quad 0 < \beta < 1 , \tag{2.7}$$

where E_β is the Mittag-Leffler function of order β, an entire transcendental function defined in the complex plane by the power series

$$E_\beta(z) := \sum_{n=0}^{\infty} \frac{z^n}{\Gamma(\beta n + 1)} , \quad \beta > 0 , \quad z \in \mathbf{C} . \tag{2.8}$$

For detailed information on the Mittag-Leffler-type functions and their Laplace transforms the reader can consult e.g. [8, 9, 10] and references therein.

Hereafter, it is convenient to summarize the main features of the functions $\Psi(t)$ and $\psi(t)$ relevant for our purposes. We begin to quote their series expansions (for $t \geq 0$) and asymptotic representations (for $t \to \infty$):

$$\Psi(t) = \sum_{n=0}^{\infty} (-1)^n \frac{t^{\beta n}}{\Gamma(\beta n+1)} \sim \frac{\sin(\beta\pi)}{\pi} \frac{\Gamma(\beta)}{t^{\beta}}, \tag{2.9}$$

and

$$\psi(t) = \frac{1}{t^{1-\beta}} \sum_{n=0}^{\infty} (-1)^n \frac{t^{\beta n}}{\Gamma(\beta n+\beta)} \sim \frac{\sin(\beta\pi)}{\pi} \frac{\Gamma(\beta+1)}{t^{\beta+1}}. \tag{2.10}$$

In the limit $\beta \to 1$, we recover the exponential functions of the Markovian case. We note that for $0 < \beta < 1$ both functions $\psi(t)$, $\Psi(t)$, even if losing their exponential decay by exhibiting power-law tails for large times, keep their completely monotonic character. Complete monotonicity of the functions $\psi(t)$, $\Psi(t)$, $t > 0$, means:

$$(-1)^n \frac{d^n}{dt^n} \Psi(t) \geq 0, \quad (-1)^n \frac{d^n}{dt^n} \psi(t) \geq 0, \quad n = 0, 1, 2, \ldots \tag{2.11}$$

or, equivalently, their representability as (real) Laplace transforms of non-negative functions [3]. It may be useful to note that, for sufficiently small times, $\Psi(t)$ has the same behaviour as a stretched exponential; in fact we have

$$E_\beta(-t^\beta) \simeq 1 - \frac{t^\beta}{\Gamma(\beta+1)} \simeq \exp\{-t^\beta/\Gamma(1+\beta)\}, \quad 0 \leq t \ll 1. \tag{2.12}$$

From the above considerations, it can be seen that the choice leading to our time-fractional master equation yields a peculiar time behaviour for the survival probability and waiting-time *pdf*, interpolating between a stretched exponential onset and a power-law decay. This is an interesting effect, especially in the cases where β is close to one; in these cases we get a quasi-markovian behaviour coupled to a power-law decay emerging for sufficiently long waiting times. Such an effect has been just pointed out in [3, 4], where the function $\Psi(\tau)$ has been estimated from empirical data concerning the waiting-time distribution of future prices on German and Italian Government bonds with fixed interest rates traded at LIFFE.

3. The transition to space-time fractional diffusion

Let us now consider the so called *diffusion* or *hydrodynamic limit*. In this limit one gets the transition from the time-fractional master equation (2.6) to a space-time fractional diffusion equation. Our aim is to prove this transition in a correct and transparent way by using proper scaling arguments. In other words, we would like to show which scaling assumptions ensure the validity of the fractional diffusion limit. Indeed, the correct transition to this limit needs a special care and involves the introduction of a vanishing length scale h (a limit of infinitely fine discretization in the CTRW).

In our opinion and to our knowledge, the approaches to the diffusion limit appearing in the literature, even if usually correct for practical purposes, may present flaws, from the mathematical point of view.

Usually (see [1] and references therein), the transition to the diffusion limit is obtained by approximating the Fourier transform of the jump *pdf* and the Laplace transform of the waiting-time *pdf* as follows,

$$\widetilde{\psi}(s) \sim 1 - s^{\beta}\,, \quad s \to 0\,, \quad 0 < \beta \le 1\,, \tag{3.1}$$

$$\widehat{w}(\kappa) \sim 1 - |\kappa|^{\alpha}\,, \quad \kappa \to 0\,, \quad 0 < \alpha \le 2\,. \tag{3.2}$$

In order to properly carry out the transition, we start from Eqs. (2.4) and (2.6), and pass through their Fourier-Laplace counterparts. The process described by the equations originates via a sequence of jumps, each jump being a sample of the real random variable Y. The particle is at position $Y_1 + Y_2 + \ldots Y_n$ during the time interval $t_n \le t < t_{n+1}$. The Y_k are *i.i.d.* random variables all having, like Y, the *pdf* $w(x)$, and consequently the characteristic function $\widehat{w}(\kappa)$. To be precise, we require that the *pdf* $w(x)$ is such that, if $\alpha = 2$:

$$\sigma^2 = \int_{-\infty}^{+\infty} x^2\, w(x)\, dx < \infty\,, \tag{3.3}$$

whereas if $0 < \alpha < 2$:

$$w(x) = (b + \epsilon(|x|))\, |x|^{-(\alpha+1)}\,, \quad b > 0\,, \quad \epsilon(|x|) \to 0 \text{ as } |x| \to \infty\,, \tag{3.4}$$

where $\epsilon(|x|)$ is assumed to be bounded and $O\left(|x|^{-\eta}\right)$ with $\eta > 0$ as $|x| \to \infty$.

We now consider a sequence of random processes *pdf*'s $p_h(x,t)$ resulting from jumps of size $h\,Y_k$ instead of Y_k, and from an acceleration of the process by a factor (*the scaling factor*) $\mu^{-1/\beta}\, h^{-\alpha/\beta}$, where μ must fulfill some conditions to be stated later. The *pdf* of the jump size is $w_h(x) = w(x/h)/h$, so that its characteristic function is $\widehat{w_h}(\kappa) = \widehat{w}(\kappa h)$. For $0 < \alpha \le 2$ and $0 < \beta \le 1$ we replace Eq. (2.6) (including (2.4) in the special case $\beta = 1$) by the sequence of equations

$$\mu h^{\alpha}\, \frac{\partial^{\beta}}{\partial t^{\beta}} p_h(x,t) = -p_h(x,t) + \int_{-\infty}^{+\infty} w_h(x - x')\, p_h(x',t)\, dx'\,. \tag{3.5}$$

Applying the Fourier-Laplace transform and recalling the Laplace transform of the Caputo time-fractional derivative, see (A.1), we get

$$\mu h^{\alpha} \left\{ s^{\beta}\, \widehat{\widetilde{p}}_h(\kappa,s) - s^{\beta-1} \right\} = \left[\widehat{w}_h(\kappa) - 1\right] \widehat{\widetilde{p}}_h(\kappa,s)\,. \tag{3.6}$$

Let us now be guided by the ***classical central limit theorem*** or by the ***Gnedenko limit theorem***, see Gnedenko & Kolmogorov [11], both expressed in terms of characteristic functions. We recall that the Gnedenko limit theorem provides a suitable generalization of the classical central limit theorem for space *pdf*'s with infinite variance, decaying according to condition (3.4).

The transition to the *diffusion limit* is based on the following Lemma due to Gorenflo (for the proof see [12]):
With the scaling parameter

$$\mu = \frac{\sigma^2}{2}\,, \quad \text{if} \quad \alpha = 2\,, \tag{3.7}$$

$$\mu = \frac{b\,\pi}{\Gamma(\alpha+1)\,\sin(\alpha\pi/2)}\,, \quad \text{if} \quad 0<\alpha<2\,, \tag{3.8}$$

we have

$$\lim_{h\to 0} \frac{\widehat{w}(\kappa h)-1}{\mu\, h^{\alpha}} = -|\kappa|^{\alpha}\,, \quad 0<\alpha\le 2\,, \quad \kappa\in\mathbf{R}\,. \tag{3.9}$$

Setting now

$$\rho_h(\kappa) = \frac{\widehat{w}(\kappa h)-1}{\mu h^{\alpha}}\,, \tag{3.10}$$

the sequence of equations (3.6) reads

$$s^{\beta}\,\widehat{\widetilde{p_h}}(\kappa,s) - s^{\beta-1} = \rho_h(\kappa)\,\widehat{\widetilde{p_h}}(\kappa,s)\,, \tag{3.11}$$

and, in the limit $h\to 0$, thanks to (3.8),

$$s^{\beta}\,\widehat{\widetilde{p_0}}(\kappa,s) - s^{\beta-1} = -|\kappa|^{\alpha}\,\widehat{\widetilde{p_0}}(\kappa,s)\,, \quad 0<\alpha\le 2\,, \quad 0<\beta\le 1\,. \tag{3.12}$$

By inversion and recalling the Fourier transform of the Riesz space-fractional derivative, see (B.2), we finally obtain the equation

$$\frac{\partial^{\beta}}{\partial t^{\beta}} p_0(x,t) = \frac{\partial^{\alpha}}{\partial |x|^{\alpha}} p_0(x,t)\,, \quad p_0(x,0)=\delta(x)\,, \tag{3.13}$$

which is a space-time fractional diffusion equation. For $\beta=1$ and $\alpha=2$, it reduces to the standard diffusion equation.

In summary, we have shown a formally correct transition to the diffusion limit starting from the general master equation of the CTRW, namely Eq. (2.1) or Eq. (2.3). Indeed, by invoking the continuity theorem of probability theory, see e.g. Lukacs [13], we can convince ourselves that the random variable whose density is $p_h(x,t)$ converges in distribution (*weakly* or *in law*) to the random variable whose density is $p_0(x,t)$.

Solving Eq. (3.11) for $\widehat{\widetilde{p_h}}(\kappa,s)$, Eq. (3.12) for $\widehat{\widetilde{p_0}}(\kappa,s)$, gives

$$\widehat{\widetilde{p_h}}(\kappa,s) = \frac{s^{\beta-1}}{s^{\beta}-\rho_h(\kappa)}\,, \quad \widehat{\widetilde{p_0}}(\kappa,s) = \frac{s^{\beta-1}}{s^{\beta}+|\kappa|^{\alpha}}\,, \tag{3.14}$$

which implies

$$\widehat{p_h}(\kappa,t) = E_{\beta}\left(\rho_h(\kappa)t^{\beta}\right)\,, \quad \widehat{p_0}(\kappa,t) = E_{\beta}\left(-|\kappa|^{\alpha}t^{\beta}\right)\,. \tag{3.15}$$

By (3.9) $\rho_h(\kappa)\to -|\kappa|^{\alpha}$ as $h\to 0$, hence

$$p_h(x,t)\to p_0(x,t)\,, \quad \text{for} \quad t>0\,, \quad h\to 0\,. \tag{3.16}$$

To get the fundamental solutions for the space-time fractional diffusion equation (3.13) in the general case $\{0<\alpha\le 2\,, 0<\beta\le 1\}$ we refer the reader to [14] and [15]. We also refer to the above references for the particular cases $\{0<\alpha\le 2\,, \beta=1\}$ (space fractional diffusion) and $\{\alpha=2\,, 0<\beta\le 1\}$ (time fractional diffusion), already discussed in the literature. The scaling properties of the fundamental solutions of (3.13) have been outlined in [1].

4. Discussion and Conclusions

In this paper we have proposed a scaling method to get a correct transition to the diffusion limit from the CTRW master equation governing a stochastic process. It may be useful to add some comments on the meaning of Eq. (3.5).

The factor μh^α can be viewed as causing the jump process to run faster and faster (the waiting times becoming shorter and shorter) as h becomes smaller and smaller. The replacement of the density $w(x)$ by the density $w_h(x) = w(x/h)/h$, and, accordingly, of the jumps Y by hY, means that the jump size becomes smaller and smaller as the scaling length h tends to zero.

An alternative interpretation is that we look at the same process with a discrete number of jumps occurring after finite times, from far away and after long time, so that spatial distances and time intervals of normal size appear very small, being x replaced by x/h, t replaced by $t/(\mu^{1/\beta}\, h^{\alpha/\beta})$.

The stochastic processes described by Eq. (3.13) exhibit infinite variance if $\alpha < 2$. This has limited their applications in finance. In the case of option pricing, for instance, it is not possible to apply the usual martingale methods, as one cannot define the analogous of the exponential Wiener process. On the other hand, if $\beta < 1$ the stochastic processes are non-Markovian with memory effects that can be responsible of long-range time correlations. If $\alpha = 2$ and $\beta < 1$ the processes exhibit a finite variance with sub-diffusive effects. Thus, the general space-time fractional diffusion equation governs a large class of stochastic processes which, through their simulation by random-walks, are expected to be useful both for modeling the dynamics of financial markets and for risk management. The mathematical aspects of the random-walks are under our intensive investigation and the financial applications will be the subject of future papers of ours.

Appendix A: The Caputo time-fractional derivative

For readers' convenience, here, we present an introduction to the *Caputo* fractional derivative starting from its representation in the Laplace domain and pointing out its difference from the standard *Riemann-Liouville* fractional derivative. In so doing, we avoid the subtleties lying in the inversion of fractional integrals. If $f(t)$ is a (sufficiently well-behaving) function with Laplace transform $\mathbf{L}\{f(t); s\} = \widetilde{f}(s) = \int_0^\infty \mathrm{e}^{-st} f(t)\, dt$, we have

$$\mathbf{L}\left\{\frac{d^\beta}{dt^\beta} f(t); s\right\} = s^\beta\, \widetilde{f}(s) - s^{\beta-1}\, f(0^+)\,, \quad 0 < \beta < 1\,, \tag{A.1}$$

if we define

$$\frac{d^\beta}{dt^\beta}\, f(t) := \frac{1}{\Gamma(1-\beta)} \int_0^t \frac{df(\tau)}{d\tau} \frac{d\tau}{(t-\tau)^\beta}\,. \tag{A.2}$$

We can also write

$$\frac{d^\beta}{dt^\beta} f(t) = \frac{1}{\Gamma(1-\beta)} \frac{d}{dt} \left\{ \int_0^t [f(\tau) - f(0^+)]\, \frac{d\tau}{(t-\tau)^\beta} \right\}, \tag{A.3}$$

$$\frac{d^\beta}{dt^\beta} f(t) = \frac{1}{\Gamma(1-\beta)} \frac{d}{dt} \left\{ \int_0^t \frac{f(\tau)}{(t-\tau)^\beta} d\tau \right\} - \frac{t^{-\beta}}{\Gamma(1-\beta)} f(0^+) . \qquad (A.4)$$

The Eqs. (A.1-4) can be extended to any non integer $\beta > 1$, see e.g. the survey by Gorenflo & Mainardi [9]. We refer to the fractional derivative defined by (A.2) as the *Caputo* fractional derivative, since it was formerly applied by Caputo in the late sixties for modelling dissipation effects in *linear viscoelasticity*, see e.g. [16, 17] and [18].

The reader should observe that this definition differs from the usual one named after Riemann and Liouville, which is given by the first term in the R.H.S. of (A.4), see e.g. the treatise on Fractional Calculus by Samko, Kilbas & Marichev[19].

Gorenflo & Mainardi have pointed out the major usefulness of the Caputo fractional derivative in the treatment of differential equations of fractional order for *physical applications*. In fact, in physical problems, the initial conditions are usually expressed in terms of a given number of bounded values assumed by the field variable and its derivatives of integer order, despite the fact that the governing evolution equation may be a generic integro-differential equation and therefore, in particular, a fractional differential equation.

Appendix B: The Riesz space-fractional derivative

If $f(x)$ is a (sufficiently well-behaving) function with Fourier transform

$$\mathbf{F}\,\{f(x);\kappa\} = \hat{f}(\kappa) = \int_{-\infty}^{+\infty} \mathrm{e}^{i\kappa x}\, f(x)\, dx\,, \quad \kappa \in \mathbf{R}\,,$$

we have

$$\mathbf{F}\left\{ \frac{d^\alpha}{d|x|^\alpha} f(x);\kappa \right\} = -|\kappa|^\alpha\, \hat{f}(\kappa)\,, \quad 0 < \alpha < 2\,, \qquad (B.1)$$

if we define

$$\frac{d^\alpha}{d|x|^\alpha} f(x) = \Gamma(1+\alpha)\, \frac{\sin(\alpha\pi/2)}{\pi} \int_0^\infty \frac{f(x+\xi) - 2f(x) + f(x-\xi)}{\xi^{1+\alpha}}\, d\xi\,. \qquad (B.2)$$

The fractional derivative defined by (B.2) can be referred to as the *Riesz fractional derivative* since it is obtained from the inversion of the fractional integral originally introduced by Marcel Riesz, known as the *Riesz potential*, see e.g. [19]. Our representation (B.2), formerly used by us in [12], is more explicit and convenient than other ones found in the literature, see e.g. [6], [19]. It is based on a suitable regularization of a hyper-singular integral.

For $\alpha = 2$ the Riesz derivatives reduces to the standard derivative of order 2 since $-|\kappa|^2 = -\kappa^2$.

For $\alpha = 1$ the Riesz derivative is related to the Hilbert transform, resulting in the formula

$$\frac{d}{d|x|} f(x) = -\frac{1}{\pi} \frac{d}{dx} \int_{-\infty}^{+\infty} \frac{f(\xi)}{x-\xi}\, d\xi\,. \qquad (B.3)$$

We note, by writing $-|\kappa|^\alpha = -(\kappa^2)^{\alpha/2}$, that the Riesz derivative of order α can be interpreted as the opposite of the $\alpha/2$ power of the (positive definite) operator $-D^2 = -\frac{d^2}{dx^2}$, namely

$$\frac{d^\alpha}{d|x|^\alpha} = -\left(-\frac{d^2}{dx^2}\right)^{\alpha/2} . \qquad (B.4)$$

The notation used above is due to Saichev & Zaslavsky, see e.g. [6]. A different notation which takes into account asymmetries is due to Gorenflo & Mainardi, see e.g. [20], [21].

References

[1] E. Scalas, R. Gorenflo and F. Mainardi, *Fractional calculus and continuous-time finance*, Physica A **284** (2000), 376-384.

[2] M. Parkinson, *Option pricing: the American put*, J. Business **50** (1977), 21-36.

[3] F. Mainardi, M. Raberto, R. Gorenflo and E. Scalas, *Fractional calculus and continuous-time finance II: the waiting-time distribution*, Physica A **287** (2000), 468-481.

[4] M. Raberto, E. Scalas, R. Gorenflo and F. Mainardi, *The waiting-time distribution of LIFFE bond futures*, J. Quantitative Finance (2001), to appear.

[5] E.W. Montroll and G.H. Weiss, *Random walks on lattices, II*, J. Math. Phys. **6** (1965), 167–181.

[6] A.I. Saichev and G.M. Zaslavsky, *Fractional kinetic equations: solutions and applications*, Chaos **7** (1997), 753–764.

[7] I.M. Gel'fand and G.E. Shilov, *Generalized Functions*, Vol. 1, Academic Press, New York, 1964.

[8] A. Erdélyi, W. Magnus, F. Oberhettinger, F.G. Tricomi, *Higher Transcendental Functions*, Bateman Project, Vol. 3, McGraw-Hill, New York, 1955. [Ch. 18, pp. 206-227.]

[9] R. Gorenflo and F. Mainardi, *Fractional calculus, integral and differential equations of fractional order*, in: A. Carpinteri and F. Mainardi (Editors), *Fractals and Fractional Calculus in Continuum Mechanics*, Springer Verlag, Wien and New York, 1997, pp. 223–276.

[10] F. Mainardi and R. Gorenflo, *On Mittag-Leffler type functions in fractional evolution processes*, J. Comput. & Appl. Mathematics **118** (2000), 283-299.

[11] B.V. Gnedenko and A.N. Kolmogorov, *Limit Distributions for Sums of Independent Random Variables*, Addison-Wesley, Cambridge, Mass., 1954. [Translated from the Russian edition of 1949]

[12] R. Gorenflo and F. Mainardi, *Random walk models approximating symmetric space-fractional diffusion processes*, in: J. Elschner, I. Gohberg and B. Silbermann (Editors), *Problems in Mathematical Physics* (Siegfried Prössdorf Memorial Volume), Proceedings of the 11-th TMP Conference. Birkhäuser Verlag, Boston-Basel-Berlin, 2000, pp. 120-145.

[13] E. Lukacs, *Characteristic Functions*, Griffin, London, 1960.

[14] R. Gorenflo, A. Iskenderov and Yu. Luchko, *Mapping between solutions of fractional diffusion-wave equations*, Fractional Calculus and Applied Analysis **3** (2000), 75–86.

[15] F. Mainardi, Yu Luchko and G. Pagnini, *The fundamental solutions of the space-time fractional diffusion equation*, Fractional Calculus and Applied Analysis **4** (2001), to appear

[16] M. Caputo, *Linear models of dissipation whose Q is almost frequency independent, Part II*, Geophys. J. R. Astr. Soc. **13** (1967), 529–539.

[17] M. Caputo, *Elasticità e Dissipazione*, Zanichelli, Bologna, 1969. [in Italian]

[18] M. Caputo and F. Mainardi, *Linear models of dissipation in anelastic solids*, Riv. Nuovo Cimento (Ser. II) **1** (1971), 161–198.

[19] S.G. Samko, A.A. Kilbas and O.I. Marichev, *Fractional Integrals and Derivatives, Theory and Applications*, Gordon and Breach, Amsterdam, 1993. [Original edition in Russian, Nauka i Tekhnika, Minsk, 1987]

[20] R. Gorenflo and F. Mainardi, *Random walk models for space-fractional diffusion processes*, Fractional Calculus and Applied Analysis **1** (1998), 167–191.

[21] R. Gorenflo and F. Mainardi, *Approximation of Lévy-Feller diffusion by random walk*, J. Analysis and its Applications (ZAA) **18** (1999), 231–246.

[1]Erstes Mathematisches Institut,
Freie Universität Berlin,
Arnimallee 3, D-14195 Berlin, Germany
E-mail address: gorenflo@math.fu-berlin.de

[2]Dipartimento di Fisica,
Università di Bologna,
via Irnerio 46, I–40126 Bologna, Italy
E-mail address: mainardi@bo.infn.it

[3]Dipartimento di Scienze e Tecnologie Avanzate,
Università del Piemonte Orientale,
via Cavour 84, I–15100 Alessandria, Italy
E-mail address: scalas@unipmn.it

[4]Dipartimento di Ingegneria Biofisica ed Elettronica,
Università di Genova,
via dell'Opera Pia 11a, I–16145 Genova, Italy
E-mail address: raberto@dibe.unige.it

Trends in Mathematics, © 2001 Birkhäuser Verlag Basel/Switzerland

Passport Options Outside the Black Scholes World

Vicky Henderson

Abstract. Passport option pricing is well established for Black Scholes style models where the underlying follows exponential Brownian motion, and recently results have been proved for diffusion models. In this paper, we extend passport option theory to two areas where traditional constant volatility models are not sufficient. We first consider equity passport options using a stochastic volatility model and second, provide a new class of hedging strategies for commodity producers, using passport options on forward contracts. Both cases are illustrated with numerical results and these are compared to the Black Scholes style world.

1. Introduction

Passport options are a type of exotic option which give the holder a zero-strike call option on the value of a trading account. The holder of the option trades dynamically on an underlying. At exercise, if the trades have realised a profit, the holder receives the proceeds of the trading account; if the account is in deficit then the losses are bourne by the option seller. In return for the option the holder makes an upfront payment, the option premium, to the seller. Since the holder undertakes trades over the life of the option, pricing becomes complex and must involve identifying the "best" trading strategy for the holder.

We can classify the literature on passport options according to the methods used in the analysis and by the models used for the undelying asset. Of course, there are further differences based on the many types of contracts, for more details see Henderson [10]. The two main methods used to treat passport option problems are partial differential equations and using probabilistic techniques. The other division is based on models and early work on these options focused entirely on Black Scholes style models, establishing the basic features of pricing under exponential Brownian motion. Passport options were introduced in the paper of Hyer et al [18] and have since been studied by Andersen et al [2], Ahn et al [1] and Lipton [20] using pde methods. These methods use a scaling which relies on the fact that the underlying is modelled by an exponential Brownian motion.

More recently Henderson and Hobson [12], Henderson [10], Shreve and Večeř [22] and Delbaen and Yor [5] have used probabilistic methods to study the passport problem. Shreve and Večeř remain in the exponential Brownian motion framework whilst the other authors prove the results for more general models.

One of the interesting features of the solution is the choice of trading strategy made by the option holder. It is known in the case of exponential Brownian motion that the optimal strategy is to be long if the trading account is below zero and short if the account is above zero. In Henderson and Hobson [12], this strategy was shown to remain optimal for a wide class of diffusion models. This means the pricing is robust to model misspecification, see also El Karoui et al [6], Hobson [16] and Henderson [11].

In this paper we give two cases of passport options outside the familiar Black Scholes setting, and develop the pricing for each, using probabilistic methods. We show the robustness of the optimal strategy illustrated in Henderson and Hobson [12] can be extended to include stochastic volatility models. This is our first case, which is still in the equity passport option setting. A second case centres on commodity hedging using the Gibson-Schwartz [9] model applied to forward prices. This is a new application outside the traditional realm of equities and currencies for the passport option. For an extension using futures as the hedging instrument, see Henderson, Hobson and Kentwell [14]. They also propose a second hedging tool, and compare the two hedges.

Both stochastic volatility and the Gibson-Schwartz models have been introduced to overcome some of the deficiencies of the Black Scholes style model in the equity and commodity markets, respectively. Namely, implied volatility displays a 'smile' and convenience yields exhibit mean reversion.

I would like to acknowledge the helpful comments from Rajna Gibson concerning the commodity example and David Hobson for useful suggestions. The financial support of RiskLab, ETH Zurich, where I was a postdoctoral researcher when this paper was written, was appreciated.

2. The Passport Option

Beginning with the dynamics for the asset price P, under the real world measure, $\mathbb{P}$, we take a general specification which becomes useful in the later models. A secons process ξ may affect the drift and volatility of the price P.

$$\begin{aligned} \frac{dP_u}{P_u} &= a_P(u,\xi_u,P_u)du + \sigma_P(u,\xi_u,P_u)dB_u \\ d\xi_u &= b(u,\xi_u)du + \eta(u,\xi_u)dW_u \end{aligned} \tag{1}$$

where we set $a_P(u,\xi_u,P_u) = \mu_P(u,\xi_u,P_u)\sigma_P(u,\xi_u,P_u)$ and $dB_u dW_u = \rho du$.

We will use the diffusion model of Henderson and Hobson [12] in this section, to describe the options' characteristics. Take $a_P(u,\xi,P) = r$ and σ_P independent of ξ. The price process under the pricing measure $\tilde{\mathbb{P}}$, with $\tilde{B}$, Brownian motion is then $\frac{dP_u}{P_u} = rdu + \sigma_P(u, P_u)d\tilde{B}_u$. Discounting by the interest rate r we have

$$\frac{dS_u}{S_u} = \sigma(u, S_u)d\tilde{B}_u \tag{2}$$

for the dynamics of the discounted price, $S_t = e^{-rt}P_t$. The martingale measure $\tilde{\mathbb{P}}$ is the measure under which S is a martingale, so for the moment, is unique. We also assume $S \in H^1$, however, we will not dwell on technicalities and refer the reader to Henderson [10] for more details. Later, in §4 we obtain a class of equivalent martingle measures for S.

If the buyer holds q_u units of the underlying at time u then the dynamics of the trading account are

$$d\psi_u(q) = r(\psi_u(q) - q_u P_u)du + q_u dP_u$$

where rate r is earnt on money not invested in the risky asset. We assume that ψ_0, the initial wealth, can be non-zero. The discounted gains $G_u := e^{-ru}\psi_u$ follow $dG_u(q) = q_u dS_u$. Note $G(q)$ is a martingale under the measure $\tilde{\mathbb{P}}$, see Chapter 2, Henderson [10]. This specification is known as the 'symmetric' passport option.

Given that the investor follows the admissible strategy $(q_u)_{\{0 \le u \le T\}}$, bounded by a constant D, the passport option pricing problem is to find the strategy q_u^* such that the price (at time 0) is maximised:

$$\sup_{|q_u| \le D} \tilde{\mathbb{E}} G_T^+(q). \tag{3}$$

The problem is linear in the constraint so henceforth we shall take $D = 1$.

For the Black Scholes model, with r and σ constants, an explicit price is given by Hyer et al [18], Andersen et al [2], Ahn et al [1], Shreve and Večeř [22] and Henderson [10]. They also show that the optimal holdings are $q_u = 1$ if $G_u \le 0$ and $q_u = -1$ if $G_u > 0$. The last two authors also treat the 'asymmetric' case, where the strategy may lie in $[\kappa_1, \kappa_2]$, not necessarily equal in absolute value.

3. Beyond Black Scholes

The first work to extend the model beyond the Black Scholes framework was Henderson and Hobson [12], who dealt with the diffusion case with a_P, σ_P independent of ξ. The key result in the pricing of passport options is the fact that the price can be identified with the price of a lookback option. In particular, in Henderson and Hobson [12] (see also Delbaen and Yor [5]) it is shown that the price of the passport option reduces to

$$\tfrac{1}{2} \sup_{|v_u| \le 1} \tilde{\mathbb{E}}(\overline{M}_T(v) - |G_0|)^+ + G_0^+ \tag{4}$$

where $M_p(v) = \int_0^p v_r dS_r$ and $\overline{M}_z(v) = \sup_{0 \le r \le z} M_r(v)$. Further, v in the supremum of (4) is related to the strategy q in (3) by $v_u = -q_u \mathrm{sgn}(G_u)$. This is proved in Henderson and Hobson [12], Henderson [10] and Delbaen and Yor [5] and uses Tanaka's formula, local time and the Skorokhod lemma.

Interestingly Henderson and Hobson [12] also prove that the optimal strategy (stated above for exponential Brownian motion) is the same for a class of diffusion models where the diffusion coefficient, $\sigma(u, S_u)S_u$ is non-decreasing in S, see their Theorem 3.4. This was achieved using coupling of stochastic processes. In the next two sections, we consider our two examples of other models where a similar analysis can be used.

4. Stochastic Volatility models

Stochastic volatility models have evolved in response to empirical evidence that call prices are inconsistent with the constant volatility assumption. See the survey articles Frey [8] and Hobson [15]. These models are incomplete, hence there are many equivalent martingale measures which can be used for pricing. This choice of measure does not affect the optimal strategy for a passport option, although it will affect the price.

When there is no correlation between the asset price process and the volatility process we express the price of the passport option as an integral of a lookback option over the variance of the log-price. By using power series methods, we price the passport option under the Hull and White model [17]. We discover that the Black Scholes passport option price does not necessarily underestimate the passport price with uncertain volatility. Instead the relationship between the price in the Black-Scholes model and a stochastic volatility alternative depends on the convexity of the lookback option price and the parameter values considered. For more details, proofs and other models and examples, see Henderson and Hobson [13].

We assume the price dynamics in (1), with independent Brownian motions, hence $\rho = 0$. Thus ξ represents an independent process impacting on the price volatility. Denote the class of equivalent martingale measures by $\tilde{\mathbb{P}} \in P$. Discounting again by r, we see

$$\begin{aligned} dS_u &= S_u\sigma(u,\xi_u,S_u)d\tilde{B}_u \\ d\xi_u &= [b(u,\xi_u) + \eta(u,\xi_u)\lambda_u]du + \eta(u,\xi_u)d\tilde{W}_u \end{aligned} \tag{5}$$

where we introduce a market price of risk parameter, λ, see Frey [8].

This specification covers many of the stochastic volatility models studied in the literature. Hull and White's (1987) model [17] assumes that the square of volatility follows exponential Brownian motion and $\sigma_P(u,\xi_u) = \sqrt{\xi_u}$. Wiggins [24] models the log of volatility as an arithmetic Ornstein Uhlenbeck process. Another model of this type is Stein and Stein [23].

The passport option price in this stochastic volatility set-up is again given by (3) where the gains follow the same SDE as in Section 2. It is proved in Henderson and Hobson [13] that the price reduces again to a lookback style price, as in (4). Thus we need to obtain the optimal strategy v, or equivalently q. The result is as follows, see the appendix of the aforementioned paper for a proof.

Theorem 4.1. *Let $M_p(v) = \int_0^p v_r dS_r$ and let $\overline{M}_z(v) = \sup_{0\leq r\leq z} M_r(v)$. For the class of stochastic volatility models in (5), if $s\sigma(u,\xi,s)$ is non-decreasing in s (for all u,ξ) then, for any k, $\sup_{|v_u|\leq 1} \tilde{\mathbb{E}}(\overline{M}_T(v) - k)^+$ is attained by $v_u \equiv 1$ or $q_u = -sgn(G_u)$.*

Thus we have the robustness result that the strategy which held for diffusions (and Black-Scholes), in §3, also holds for this class of stochastic volatility models. An additional feature is that Theorem 4.1 holds for *any* $\tilde{\mathbb{P}} \in P$.

The problem of pricing remains. For *any* fixed $\tilde{\mathbb{P}} \in P$, $v = 1$ and the price is

$$\tfrac{1}{2}\tilde{\mathbb{E}}(\overline{S}_T - S_0 - |G_0|)^+ + G_0^+. \tag{6}$$

If the asset price P and volatility ξ are uncorrelated, and the drift and diffusion coefficient of ξ do not depend on P then ξ and B are independent. This is the case for the three models mentioned above. We assume that a and σ_P do not depend on P, for pricing purposes.

Putting $K = S_0 + |G_0|$ and total variance V_T with

$$V_u = \int_0^u \sigma^2(s,\xi_s)ds, \tag{7}$$

we may condition on the path $(\xi_s)_{\{0\leq s\leq u\}}$ giving

$$\begin{aligned}
\tilde{\mathbb{E}}(\overline{S}_T - K)^+ &= \tilde{\mathbb{E}}\left[\tilde{\mathbb{E}}\left((\overline{S}_T - K)^+ \mid \int_0^T \sigma^2(s,\xi_s)ds\right)\right] \\
&= \tilde{\mathbb{E}}[LB(S_0;V_T;K)] = \int_{V_T=v} LB(S_0;v;K)\tilde{f}(dv) \qquad (8)
\end{aligned}$$

where $\tilde{f}$ is the density function for the distribution of the total variance and $LB(S_0;v;K)$ is the time 0 price of a fixed strike lookback call option with strike K, expiry T and volatility $\sqrt{v/T}$ under exponential Brownian motion. The lookback price is calculated by standard arguments to be

$$LB(S_0;v;K) = S_0[N(d) + \sqrt{v}(N'(d) + dN(d))] - KN(d - \sqrt{v}) \tag{9}$$

where $d = \frac{-\ln(K/S_0)+\frac{1}{2}v}{\sqrt{v}}$. We use power series methods to characterise the law of V_T, as used by Hull and White [17]. Other techniques have been proposed in the literature, each involve computing the moment generating function of V_T.

To calculate prices, we must choose our market price of risk, λ_u, or martingale measure $\tilde{\mathbb{P}}$. We take the "minimal martingale measure" of Follmer and Schweizer [7], corresponding to $\lambda_u = 0$. Henceforth this particular pricing measure will be denoted $\tilde{\mathbb{P}}^0$ with expectation operator $\tilde{\mathbb{E}}^0$. Other measures could be chosen here, but this is left for future study.

Since the option price is a nonlinear function of the total variance $V \equiv V_T$ we expand $LB(S_0; v; K)$ in a Taylor series about the expected value $\tilde{\mathbb{E}}^0 V$ of V:

$$\int_v LB(S_0; v; K)\tilde{f}_0(v)dv = LB(S_0; \tilde{\mathbb{E}}^0 V; K) + \tfrac{1}{2}\frac{\partial^2 LB(.)}{\partial v^2}\bigg|_{\tilde{\mathbb{E}}^0(V)} \mathrm{Var}(V)$$

$$+ \frac{1}{6}\frac{\partial^3 LB(.)}{\partial v^3}\bigg|_{\tilde{\mathbb{E}}^0(V)} \mathrm{Skew}(V) + \textit{higher order terms}\ldots \tag{10}$$

Explicit expressions for the derivatives of the lookback price and the moments can be calculated for many stochastic volatility models. The exponential Brownian motion prices are calculated using (6).

4.1. Numerical results for the Hull White model

Under the pricing measure $\tilde{\mathbb{P}}^0$, the Hull and White model is

$$\begin{aligned} \frac{dS_u}{S_u} &= \sqrt{\xi_u} d\tilde{B}^0_u \\ d\xi_u &= \bar{b}\xi_u du + \delta\xi_u d\tilde{W}^0_u \end{aligned}$$

where $\tilde{B}^0, \tilde{W}^0$ are independent $\mathbb{Q}_0$-Brownian motions. Figure 1 shows the relationship between the stochastic volatility option price and the price under an exponential Brownian motion model for a range of values of the parameter δ and $G_0 = 50$. The difference refers to the Hull and White price of the passport option minus the price for the simple exponential Brownian motion model expressed as a percentage of the Black Scholes price. As $\delta \to 0$ the Hull White price approaches the exponential Brownian motion price, as is to be expected. We see for low values of ξ_0, the constant volatility model underestimates the passport price in a stochastic volatility regime. For large ξ_0, this relationship is reversed. This is unusual in that often the introduction of stochastic volatility increases the price of an option. This means that the standard practice of adjusting prices upwards to allow for the uncertainty of volatility is inappropriate for the passport option. This can be explained by looking at the convexity of the lookback pricing function in (9).

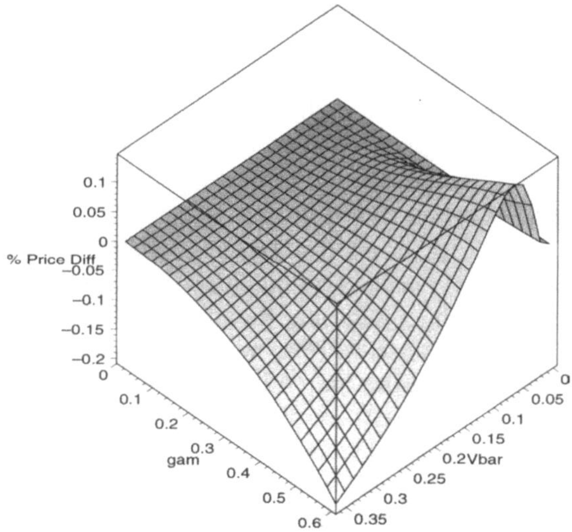

FIGURE 1. *Percentage difference in the passport option prices using the Hull and White model and exponential Brownian motion with* $S_0 = 100$, $T = 1$, $G_0 = 50$, $0 \leq \xi_0 \leq 0.36$ *and* $0 \leq \delta \leq 0.6$.

Recall the lookback option price under stochastic volatility in (8) is

$$\tilde{\mathbb{E}}^0 [LB(S_0, V, K)]$$

and we multiply by a half and add G_0^+ to get the passport option price. If the function $LB(.)$ is locally concave in v then $\tilde{\mathbb{E}}^0 LB(S_0, V, K) < LB(S_0, \tilde{\mathbb{E}}^0 V, K)$ and so the stochastic volatility price is lower than the Black Scholes price. Conversely, if $LB(.)$ is convex in v then the inequality is reversed.

In this example, for large ξ_0, $\tilde{\mathbb{E}}^0 V$ is large and stochastic volatility decreases the price. Indeed, this statement will always be true: for any value of G_0, a sufficiently large value of the initial volatility ξ_0 will lead to the passport option price being overestimated by the Black Scholes model.

5. Commodity Hedging

In our second example of a non Black-Scholes passport option model, we focus on commodity hedging. A commodity producer typically hedges the commodity price risk at a specific future date by taking positions in a sold forward contract or in a futures contract. Hedging exposures at a series of dates can be achieved by strips of sold forward contracts which result in an equivalent swap rate. The effectiveness of this approach depends on whether the market is in contango (gold, silver) or backwardation (oil, base metals). Other derivative structures can be embedded in these forward contracts which potentially result in a higher forward selling price. This is achieved by the producer selling options and thus taking views on the direction of the commodity price. Whilst it is possible to embed upside views on the price movements, these payoffs are typically European and cannot necessarily capture daily or weekly moves in the futures curve.

It seems reasonable to develop a hedging strategy whereby the potential profits of correctly revising a hedge are compared to the potential profits of a static futures hedge, with the hedger choosing the greater of the two. In the case when a producer must hedge to lock in a revenue stream, then this new structure will protect losses but keep the upside potential. The producer enters into a contract with a bank which entitles him to receive the returns from the standard hedge, plus the gains from his additional trades provided these trades have made a profit. If the additional trades have made a loss, then the loss is borne by the financial institution. The only constraint on the additional trades is that there is a limit on the number of forwards that the producer may hold at any one time. In effect, this is simply a passport option on a forward contract. A rationale for using this type of product may be that producers who have access to the underlying commodity may be in a better, more informed position to ascertain whether movements in the futures curve have been influenced by fundamental supply and demand effects. In addition, producers are able to use this product to speculate on the upside.

We use a forward contract as the hedging instrument, this is extended in Henderson, Hobson and Kentwell [14] to futures contracts. In the forwards case, we can calculate the explicit cost analytically. When the hedging is done with futures, the option fits into the case of equity passport option with dividends in Henderson [10, Chapter 5], Delbaen and Yor [5] and Lipton [20] and a pde must be solved numerically to obtain prices. Henderson, Hobson and Kentwell [14] examine another strategy, they call the 'unhedged' case, whereby the commodity producer remains unhedged, but, subject to position limits trades actively in the hedge instrument.

The traditional Black-Scholes assumptions are violated in these markets and a more appropriate model is needed. Our choice of forward price model, the Gibson-Schwartz model [9] has the important feature of mean-reversion for the asset price and reasonable probabilities for movements of futures curves from contango into backwardation and vice-versa, see Gibson and Schwartz [9], Schwartz [21] and Bjerksund [3].

5.1. The Black and Gibson-Schwartz models

Gibson and Schwartz [9] introduced a two factor model which allows for a stochastic convenience yield. The commodity price and the convenience yield are assumed to follow the model (1), with $a_P(u,\xi,P) = \mu, \sigma_P(u,\xi,P) = \sigma_1, b(u,\xi) = \kappa(\alpha - \delta_u), \eta(u,\xi) = \sigma_2$ and (the relabelling) $\delta = \xi$. Under the risk neutral measure $\tilde{\mathbb{P}}$,

$$
\begin{aligned}
dP &= (r - \delta_t)Pdt + \sigma_1 Pd\tilde{B}, \\
d\delta &= \kappa\left(\left(\alpha - \frac{\sigma_2\lambda_t}{\kappa}\right) - \delta\right) dt + \sigma_2 d\tilde{W}.
\end{aligned} \tag{11}
$$

As before, $d\tilde{B}d\tilde{W} = \rho dt$ and λ now represents the market price of convenience yield risk, assumed constant.

Bjerksund [3] and Jamshidian and Fein [19] solve for the forward price to get

$$F_t = P_t e^{-\{(1-e^{-\kappa(T-t)})/\kappa\}\delta_t + A(T-t)}$$

where A is a deterministic function. Using Ito and Levy's theorem it can be shown that F_t satisfies the SDE

$$\frac{dF_t}{F_t} = \gamma(T-t)d\beta \tag{12}$$

where β_t is a Brownian motion under the risk-neutral measure and γ is a time dependent function

$$\{\gamma(u)\}^2 = \sigma_1^2 + \frac{\sigma_2^2}{\kappa^2}(1-e^{-\kappa u})^2 - 2\frac{\rho\sigma_1\sigma_2}{\kappa}(1-e^{-\kappa u}). \tag{13}$$

It follows that the forward price has zero drift, and the log-forward price is Gaussian.

The constant volatility model, the Black [4] model assumes P is exponential Brownian motion and F solves

$$\frac{dF_t}{F_t} = \sigma d\tilde{B}_t. \tag{14}$$

With a deterministic convenience yield $\frac{dP_t}{P_t} = (r_t - \delta_t)dt + \sigma d\tilde{B}_t$ and again (14) holds.

5.2. The Commodity Passport Option

Consider a producer who has a fixed quantity of copper to sell at some future date T and would usually hedge by selling forward. Now suppose the producer makes additional trades in forwards creating an extra trading profit or loss at time T of

$$\psi_t = \int_0^t q_u dF_u. \tag{15}$$

where the number of forwards held at time t is denoted q_t. As in the passport option framework in §2, the position q_t is bounded by some constant D, which we take to be 1. In forward markets any gains (or losses) are realised at maturity so there is no need to account for interest earned. Since we are concerned now with gains resulting from trading in forwards, the forward price F, takes the place of the discounted asset price S, in §2.

Let P_T denote the spot price at maturity and F_0 the forward price at $t = 0$. Consider a cost of product K with total production A_T. The unhedged revenue is $R_T = A_T(P_T - K)$, but (fully hedged) with a sold forward is then

$$R_T = A_T(P_T - K) - A_T(P_T - F_0) = A_T(F_0 - K). \tag{16}$$

Thus when $F_0 > K$ the producer may hedge and receive a fixed positive revenue. This simple forward selling strategy serves as a benchmark. We assume $A_T = 1$ since the problem is linear in this variable. We allow the producer to hedge with a forward at time 0, and undertake any trading strategy subject to the position limit $q_t \leq 1$. At maturity he has the option to either receive the terminal wealth which results from following his chosen strategy, or to forget the additional trades and to receive the monies due from a forward hedge. The payoff to the gold producer at time T is then

$$\max[\psi_T + (F_0 - K), F_0 - K] = \max[\psi_T, 0] + (F_0 - K). \tag{17}$$

5.3. Pricing the Commodity options

As in §2, the bank assumes the producer behaves optimally and the price is

$$\sup_{|q|\leq 1} e^{-rT}\tilde{\mathbb{E}}\left[\psi_T^+ + (F_0 - K)\right] \tag{18}$$

From the work of Henderson and Hobson [12] discussed in §3, we can deduce the optimal q in (18) and see that it is unchanged. Then using (12) and (13) we have that $F_T = F_0 e^{\int_0^T \gamma_u d\tilde{B}_u - \frac{1}{2}\int_0^T \gamma_u^2 du}$. If we set $\xi = \int_0^T \gamma_u^2 du$, we deduce

Lemma 5.1. *For the option with payoff as in (18) and for the Gibson-Schwartz model with stochastic convenience yield, the price is*

$$\begin{aligned} & \sup_{|q|\leq 1} e^{-rT}\tilde{\mathbb{E}}\left(\int_0^T q dF\right)^+ + e^{-rT}(F_0 - K) \\ = & \ \tfrac{1}{2}e^{-rT}F_0\left\{(2N(d) - 1) + \sqrt{\xi}(N'(d) + dN(d))\right\} + e^{-rT}(F_0 - K) \end{aligned}$$

where $d = \frac{1}{2}\sqrt{\xi}$.

For the Black model, we know $q_t^* = -\text{sgn}(\psi_t)$ from §2.1, and the price is a simplification of the Gibson-Schwartz price with $\sqrt{\xi} = \sigma\sqrt{T}$.

Gibson and Schwartz empirically tested their model using weekly oil futures prices from 1984-88. We will use their estimated parameters for an example, comparing the Black price and the Gibson-Schwartz price in Lemmas 5.1 on a graph, see Figure 2. On the graph, the solid line represents the Black model price whilst the Gibson-Schwartz model price is given by the dotted line. It can be seen that the Gibson Schwartz price is higher than the Black price.

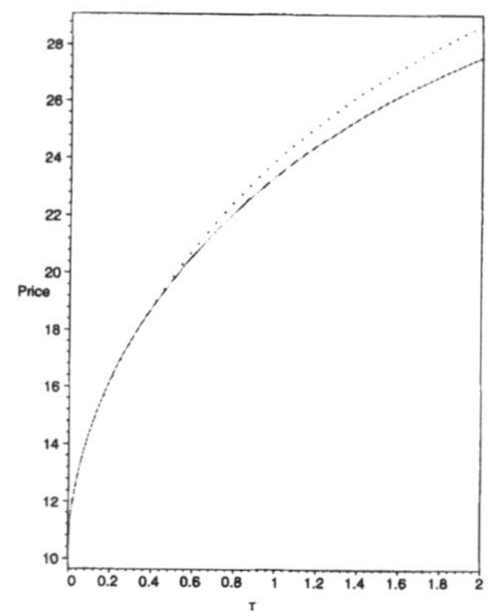

FIGURE 2. *Prices of the commodity option using the Black and Gibson-Schwartz models with* $F_0 = 100, K = 90$. *In both models* $r = 0.09$. *In the Black model we have* $\sigma = 0.35$ *and the parameter values for the Gibson-Schwartz model are* $\kappa = 4, \sigma_1 = 0.35, \sigma_2 = 0.4, \rho = 0.32$.

6. Conclusion

The literature on passport option theory so far has required the underlying to follow exponential Brownian motion, or in the case of Henderson and Hobson [12] a diffusion. Developments outside the Black Scholes framework may allow passport options to be written on other assets for which the exponential Brownian motion assumption is not a good one.

This paper introduces two new areas outside the Black Scholes model where passport options could be used. We demonstrate the results concerning the optimal strategy hold for many stochastic volatility models, regardless of the EMM chosen. It is shown that the convexity of the lookback function determines whether the Black Scholes passport option price is over or underestimated using stochastic volatility.

Commodity hedging using forward contracts provides a second example of a non Black Scholes model. A new downside hedge is developed using a forward contract and a trading component where the producer may buy and sell forward contracts until the maturity date. Since the positive part of the trading component is received, the producer cannot do worse than the forward hedge. Results using the Gibson-Schwartz [9] and Black [4] models were given.

References

[1] H. Ahn, A. Penaud and P. Wilmott, *Various passport options and their valuation*, to appear in Applied Mathematical Finance.

[2] L. Andersen, J. Andreasen and R. Brotherton-Ratcliffe, *The Passport Option*, Journal of Computational Finance, **1** (**3**),(1998), 15–36.

[3] P. Bjerksund, *Contingent Claims Evaluation when the Convenience Yield is Stochastic: Analytical Results*, Working paper, No.1, Norwegian School of Economics and Business Administration, 1991.

[4] F. Black, *The Pricing of Commodity Contracts*, Journal of Financial Economics, **3**, (1976) 167–179.

[5] F. Delbaen and M. Yor, *Passport options*, Preprint, ETH, April 1999.

[6] El Karoui N, Jeanblanc Picque M. and S.E. Shreve; Robustness of the Black and Scholes formula, *Mathematical Finance*, Vol 8, No.2, p93-126, 1998.

[7] H. Follmer and M. Schweizer; Hedging of contingent claims under incomplete information, *in Applied Stochastic Analysis*, Gordon and Breach, London, (1991), 205–223.

[8] R. Frey; *Derivative asset analysis in models with level-dependent and stochastic volatility*, CWI Quaterly 10, **1** (special issue on the Mathematics of Finance) (1997), 1–34.

[9] R. Gibson and E.S. Schwartz, *Stochastic Convenience Yield and the Pricing of Oil Contingent Claims*, Journal of Finance, **XLV** (**3**), (1990).

[10] V. Henderson *A Probabilistic Approach to Passport Options*, PhD Thesis, University of Bath (1999).

[11] V. Henderson, *Price comparison results and super-replication: an application to Passport options*, to appear in Applied Stochastic Models in Business and Industry, (2000).

[12] V. Henderson and D. Hobson, *Local time, coupling and the Passport option*, Finance and Stochastics, **4** (**1**), (2000), 69–80.

[13] V. Henderson and D. Hobson, *Passport Options with Stochastic Volatility*, Preprint, submitted, March, 2000.

[14] V. Henderson, D. Hobson and G. Kentwell, *A New Class of Commodity Hedging Strategies: A Passport Option Approach*, Preprint, July, 2000.

[15] D. Hobson, *Stochastic Volatility*, in *Statistics and Finance*. Editors Hand D.J. and S.D. Jacka; Arnold, (1998), 283–306.

[16] Hobson D.G.; Volatility mis-specification, option pricing and super-replication via coupling, *Annals of Applied Probability*, 8, No.1, p193-205, 1998.

[17] J. Hull and A. White, *The pricing of options on assets with stochastic volatilities*, Journal of Finance, **42**, (1987), 281–300.

[18] T. Hyer, A. Lipton-Lifschitz and D. Pugachevsky, *Passport to Success*, RISK, **10** (**9**), (1997), 127–131.

[19] Jamshidian F. and M. Fein; Closed form solutions for oil futures and European options in the Gibson Schwartz model: A comment, *unpublished, Merrill Lynch Capital Markets*, 1990.

[20] A. Lipton, *Similarities via self-similarities*, RISK, (1999), 101–105.

[21] E.S. Schwartz, *The Stochastic Behaviour of Commodity Prices: Implications for Valuation and Hedging*, Journal of Finance, **LII (3)**,(1997) 923–973.

[22] Shreve S.E. and J. Večeř, *Options on a Traded Account: Vacation Calls, Vacation Puts and Passport Options*, Finance and Stochastics, **4, 3**, (2000), 255–274.

[23] E.M. Stein and J.C. Stein, *Stock price distributions with stochastic volatility: an analytic approach*, Review of Financial Studies,**4**, (1991), 727–752.

[24] J.B. Wiggins, *Option values under stochastic volatility, theory and empirical estimates*, Journal of Financial Economics, **19**, (1987) 351–372.

Senior Research Fellow,
Financial Options Research Centre,
Warwick Business School,
University of Warwick,
Coventry, CV4 7AL, UK
E-mail address: FORCVH@rapier.wbs.warwick.ac.uk

Trends in Mathematics, © 2001 Birkhäuser Verlag Basel/Switzerland

New Developments in Backward Stochastic Riccati Equations and Their Applications

Michael Kohlmann[1] and Shanjian Tang[2]

Abstract. The following backward stochastic Riccati differential equation (BSRDE in short)

$$\begin{aligned} dK_t &= -[A_t'K_t + K_tA_t + \sum_{i=1}^d C_{it}'K_tC_{it} + Q_t + \sum_{i=1}^d (C_{it}'L_{it} + L_{it}C_{it}) \\ &\quad -(K_tB_t + \sum_{i=1}^d C_{it}'K_tD_{it} + \sum_{i=1}^d L_{it}D_{it})(N_t + \sum_{i=1}^d D_{it}'K_tD_{it})^{-1} \\ &\quad \times(K_tB_t + \sum_{i=1}^d C_{it}'K_tD_{it} + \sum_{i=1}^d L_{it}D_{it})']\,dt + \sum_{i=1}^d L_{it}\,dw_{it}, \\ &\quad 0 \le t < T, \\ K_T &= M. \end{aligned}$$

is motivated, and is then studied. Various properties are presented. The existence and uniqueness of a global adapted solution to a BSRDE has been open for the case $D_i \neq 0$ for more than two decades. Recently, we have made a breakthrough on this problem. On this topic, the literature is reviewed, and our recent results are surveyed. Finally, applications are outlined, both in finance and control, which contain another unique feature of our recent works — exposing the novel crucial role of *stochastic maximum principle* and the *BMO-property* of the solutions for BSRDEs in obtaining the necessary regularities, so as to overcome the new difficulty which arises from applications of BSRDEs and which roots in a general combination of both the random nature of the coefficients and the control-dependent nature of the system noise.

1. Introduction

Let $(\Omega, \mathcal{F}, P, \{\mathcal{F}_t\}_{t\ge 0})$ be a fixed complete probability space on which is defined a standard d-dimensional $\mathcal{F}_t$-adapted Brownian motion $w_t \equiv (w_{1t}, \cdots, w_{dt})'$. Denote by $\mathcal{F}_t$ the completion, by the totality $\mathcal{N}$ of all null sets of $\mathcal{F}$, of the natural filtration $\{\mathcal{F}_t^w\}$ generated by w.

Both authors gratefully acknowledge the support by the Center of Finance and Econometrics, University of Konstanz. The second author is supported by a Research Fellowship from the Alexander von Humboldt Foundation and by the National Natural Science Foundation of China under Grant No. 79790130.

Consider the following BSRDE:

$$\begin{cases} dK_t = -G(t, K_t, L_t)\,dt + \sum_{i=1}^{d} L_{it}\,dw_{it}, & 0 \le t < T, \\ K_T = M \end{cases} \tag{1}$$

where the generator G is given by

$$\begin{aligned} G(t, K, L) := \; & A_t'K + KA_t + \sum_{i=1}^{d} C_{it}'KC_{it} + Q_t \\ & + \sum_{i=1}^{d}(C_{it}'L_i + L_iC_{it}) + F(t, K, L) \end{aligned} \tag{2}$$

with

$$\begin{aligned} F(t, K, L) := \; & -(KB_t + \sum_{i=1}^{d} C_{it}'KD_{it} + \sum_{i=1}^{d} L_iD_{it})(N_t + \sum_{i=1}^{d} D_{it}'KD_{it})^{-1} \\ & \times(KB_t + \sum_{i=1}^{d} C_{it}'KD_{it} + \sum_{i=1}^{d} L_iD_{it})', \\ & \forall (t, K, L) \in [0, T] \times \mathcal{S}_+^n \times (\mathcal{S}^n)^d. \end{aligned} \tag{3}$$

It will be called the BSRDE $(A, B; C_i, D_i, i = 1, \ldots, d; Q, N, M)$ in the following for convenience of indicating the associated coefficients. The coefficients appearing here will be defined in Section 2.

BSRDEs have at least the following two motivations.

(1) The control-theoretic motivation. The BSRDE (1) arises when solving the optimal control problem

$$\inf_{u \in \mathcal{L}_{\mathcal{F}}^2(0,T;R^m)} J(u; 0, x) \tag{4}$$

where for $t \in [0, T]$ and $x \in R^n$,

$$J(u; t, x) := E^{\mathcal{F}_t}[\int_t^T [(N_su_s, u_s) + (Q_sX_s^{t,x;u}, X_s^{t,x;u})]\,ds + (MX_T^{t,x;u}, X_T^{t,x;u})] \tag{5}$$

and $X^{t,x;u}$ solves the following *stochastic differential equation* (SDE in short)

$$\begin{cases} dX_s = (A_sX_s + B_su_s)\,ds + \sum_{i=1}^{d}(C_{is}X_s + D_{is}u_s)\,dw_{is}, & t \le s \le T, \\ X_t = x. \end{cases} \tag{6}$$

We have the following connection: if the BSRDE (1) has a solution (K, L), the solution for the above *linear-quadratic optimal control problem* (LQ problem in short) has the following closed form (also called the feedback form)

$$u_s = -(N_s + \sum_{i=1}^{d} D_{is}'K_sD_{is})^{-1}(K_sB_s + \sum_{i=1}^{d} C_{is}'K_sD_{is} + \sum_{i=1}^{d} L_{is}D_{is})'X_s \tag{7}$$

and the associated value function V is the following quadratic form

$$V(t,x) := \inf_{u\in\mathcal{L}^2_{\mathcal{F}}(t,T;R^m)} J(u;t,x) = (K_t x, x), \qquad 0 \le t \le T, x \in R^n. \tag{8}$$

In this way, the problem of solving an LQ problem is reduced to that of solving the associated BSRDE (1). The LQ problem with a terminal expected constraint ($EX(T) = x_T$ for some fixed $x_T \in R^n$, for example), is also reduced to solving a BSRDE.

(2) The financial motivation. A mean-variance hedging problem is a one-dimensional, nonhomogeneous, singular stochastic LQ problem. A mean-variance portfolio selection problem is a one-dimensional, nonhomogeneous, singular stochastic LQ problem with an expected terminal state constraint. Solution of these two classes of mathematical financial problems is reduced to solving the associated one-dimensional BSRDEs.

The rest of the report is organized as follows. Preliminaries are given in Section 2 where the notation is listed, basic assumptions are made, and a solution of a BSRDE is defined. In Section 3, various properties of BSRDEs are collected. In Section 4, a historical review is given on BSRDEs, and the known existence and uniqueness result due to Bismut [5] and Peng [27] is stated. Section 5 surveys our recent results on the existence and uniqueness of a global adapted solution of BSRDE (1). Finally in section 6, BSRDEs are applied to control and finance where a new difficulty is exposed and the solution is outlined.

Note that the following exposition is only a short version of our forthcoming review paper [17].

2. Preliminaries

Notation. Throughout this paper, the following notation will be used.

M'	:	the transpose of any vector or matrix M;
$\lvert M\rvert$	:	$= \sqrt{\sum_{ij} m_{ij}^2}$ for any vector or matrix $M = (m_{ij})$;
(M_1, M_2)	:	the inner product of the two vectors M_1 and M_2;
R^n	:	the n-dimensional Euclidean space;
R_+	:	the set of all nonnegative real numbers;
$\mathcal{S}^n$	:	the Euclidean space of all $n \times n$ symmetric matrices;
$\mathcal{S}^n_+$	:	the set of all $n \times n$ nonnegative definite matrices;
$C([0,T];H)$	:	the Banach space of H-valued continuous functions on $[0,T]$, endowed with the maximum norm for a given Hilbert space H;
$\mathcal{L}^2_{\mathcal{F}}(0,T;H)$	:	the Banach space of H-valued $\mathcal{F}_t$-adapted square-integrable stochastic processes f on $[0,T]$, endowed with the norm $(E\int_0^T \lvert f_t\rvert^2\,dt)^{1/2}$ for a given Euclidean space H;

$\mathcal{L}^\infty_{\mathcal{F}}(0,T;H)$	:	the Banach space of H-valued, $\mathcal{F}_t$-adapted, essentially bounded stochastic processes f on $[0,T]$, endowed with the norm $\operatorname{ess\,sup}_{t,\omega}\|f_t\|$ for a given Euclidean space H;
$L^2(\Omega,\mathcal{F},P;H)$	:	the Banach space of H-valued norm-square-integrable random variables on the probability space $(\Omega,\mathcal{F},P)$ for a given Banach space H;

and $L^\infty(\Omega,\mathcal{F},P;C([0,T];R^n))$ is the Banach space of $C([0,T];R^n)$-valued, essentially maximum-norm-bounded random variables f on the probability space $(\Omega,\mathcal{F},P)$, endowed with the norm $\operatorname{ess\,sup}_{\omega\in\Omega}\max_{0\le t\le T}|f_t|$.

We make the following two *basic assumptions:*

($\mathcal{A}1$) *The coefficients A, B, C_i, and D_i are $\mathcal{F}_t$-progressively measurable bounded matrix-valued processes, defined on $\Omega\times[0,T]$, of dimensions $n\times n, n\times m, n\times n, n\times m$ respectively. M is an $\mathcal{F}_T$-measurable, nonnegative, and bounded $n\times n$ random matrix, and Q and N are $\mathcal{F}_t$-progressively measurable, bounded, and nonnegative $n\times n$ and $m\times m$ matrix processes, respectively.*

($\mathcal{A}2$) *N_t is uniformly positive,* or

($\mathcal{A}3$) *M and $\sum_{i=1}^d D'_{it}D_{it}$ are uniformly positive.*

Definition 2.1. A solution of the BSRDE (1) is a pair (K,L) of processes such that

(i) $K\in\mathcal{L}^\infty_{\mathcal{F}}(0,T;\mathcal{S}^n)\cap L^\infty(\Omega,\mathcal{F}_T,P;C([0,T];\mathcal{S}^n)),\ L\in\left(\mathcal{L}^2_{\mathcal{F}}(0,T;\mathcal{S}^n)\right)^d,$

(ii) $N_t+\sum_{i=1}^d D'_{it}K_tD_{it}$ is uniformly positive with respect to (t,ω),

(iii) $K_t=M+\int_t^T G(s,K_s,L_s)\,ds-\int_t^T L_s\,dw_s,\ 0\le t\le T.$

When the pair (K,L) is a solution of the BSRDE (1), we also say that it solves the BSRDE (1).

3. Fundamentals of BSRDEs

This section collects various properties of BSRDEs. They are consequences of the special structure of BSRDE (1) from different points of view.

Define $\Gamma:[0,T]\times\mathcal{S}^n\times R^{n\times d}\to R^{m\times n}$ by

$$\Gamma(t,K,L)=-(N_t+\sum_{i=1}^d D'_{it}KD_{it})^{-1}(KB_t+\sum_{i=1}^d C'_{it}KD_{it}+\sum_{i=1}^d L_iD_{it})' \tag{9}$$

and

$$\widehat{A}_t:=A_t+B_t\Gamma(t,K_t,L_t),\widehat{C}_{it}:=C_{it}+D_{it}\Gamma(t,K_t,L_t),\quad i=1,\dots,d. \tag{10}$$

Consider the following SDE

$$\left\{\begin{array}{rcl} dY_s & = & \widehat{A}_s Y_s\, ds + \sum_{i=1}^{d} \widehat{C}_{is} Y_s\, dw_{is}, \quad t \le s \le T, \\ Y_t & = & I_{n\times n}. \end{array}\right. \tag{11}$$

In view of Gal'chuk [10], this equation has a unique strong solution (denoted by $\Phi(\cdot,t)$) when $L \in (\mathcal{L}^2_{\mathcal{F}}(0,T;\mathcal{S}^n))^d$.

3.1. Regularity

Theorem 3.1. *Let (K,L) solve the BSRDE* (1). *Then,*

$$\Gamma(\cdot,K,L)\Phi(\cdot,t) \in \mathcal{L}^2_{\mathcal{F}}(t,T;R^{m\times n}),\ \Phi^* := \max_{t\le s\le T} |\Phi(s,t)| \in L^2(\Omega,\mathcal{F},P). \tag{12}$$

Consider the optimal control problem (4). Using the same methods of functional analysis as in Bismut [5] for the regular case and in Kohlmann and Tang [14] for the singular case, we show that the assumption ($\mathcal{A}2$) or ($\mathcal{A}3$) implies the existence and uniqueness of the optimal control $\widehat{u} \in \mathcal{L}^2_{\mathcal{F}}(0,T;R^m)$. From the optimality conditions (see Peng [31], and Tang and Li [37]), we have

$$\widehat{u}_s = \Gamma(s,K_s,L_s)\Phi(s,t)x,\ X_s^{t,x;\widehat{u}} = \Phi(s,t)x.$$

Theorem 3.1 then follows.

3.2. The Feynman-Kac Representation

Theorem 3.2. *Let (K,L) solve the BSRDE* (1). *Then,*

$$(K_t x, x) = V(t,x) := \inf_{u\in L^2_{\mathcal{F}}(t,T;R^m)} J(u;t,x),\ \forall t \in [0,T],\ x \in R^n. \tag{13}$$

This theorem plays a crucial role in the proof of uniqueness: it immediately gives the uniqueness of K as the first part of the solution. It also plays an important role in the proof of the closed property of the solutions (see Theorem 3.9).

3.3. Nonnegativity

Theorem 3.3. *If (K,L) solves the BSRDE* (1), *then $K_t \ge 0,\ 0 \le t \le T$.*

3.4. Monotonicity

Theorem 3.4. *Let (K,L) and $(\widetilde{K},\widetilde{L})$ be the solutions to the BSRDEs $(A,B;C_i,D_i,i=1,\cdots,d;Q,N,M)$ and $(A,B;C_i,D_i,i=1,\cdots,d;\widetilde{Q},\widetilde{N},\widetilde{M})$, respectively. If*

$$\widetilde{Q} \ge Q \ge 0, \widetilde{N} \ge N > 0, \widetilde{M} \ge M \ge 0,$$

then $\widetilde{K} \ge K$, a.s.a.e..

3.5. Minimality

Theorem 3.5. *Let $\widetilde{\Gamma} \in \mathcal{L}^{\infty}_{\mathcal{F}}(0,T;R^{m\times n})$ and let $(\widetilde{K},\widetilde{L})$ solve the linear BSDE:*

$$\begin{cases} dK_t &= -[\widetilde{A}_t'K_t + K_t\widetilde{A}_t + \sum_{i=1}^{d} \widetilde{C}_{it}'K_t\widetilde{C}_{it} + Q_t + \sum_{i=1}^{d}(\widetilde{C}_{it}'L_{it} + L_{it}\widetilde{C}_{it}) \\ & \quad +\widetilde{\Gamma}_t N_t \widetilde{\Gamma}_t]\, dt + \sum_{i=1}^{d} L_{it}\, dw_{it}, \qquad 0 \le t < T, \\ K_T &= M \end{cases} \tag{14}$$

where

$$\widetilde{A} := A + B\widetilde{\Gamma}, \quad \widetilde{C}_i := C_i + D_i\widetilde{\Gamma},\ I = 1,\dots,d. \tag{15}$$

If (K,L) solves BSRDE (1), then $K \le \widetilde{K}$, a.s.a.e..

Note that a deterministic version of Theorem 3.5 was given by Wonham [38].

3.6. ***A priori*** **estimates and the BMO-property.**

Theorem 3.6. (Kohlmann and Tang [16]) *Let (K,L) solve the BSRDE $(A,B;C_i,D_i,$ $i=1,\dots,d;Q,N,M)$. Then, there is a deterministic positive constant ε_{01} such that the following estimates hold:*

$$0 \le K_t \le \varepsilon_{01} I_{n\times n}, \quad \text{and} \quad E^{\mathcal{F}_t}\left(\int_t^T |L_s|^2\, ds\right)^p \le \varepsilon_{01}, \quad \forall p \ge 1. \tag{16}$$

Here the universal constant ε_{01} depends on the uniform upper bound of all the coefficients and the uniform positive lower bound for eigen-values of positive matrices M and $\sum_{i=1}^d D_{it}'D_{it}$.

Theorem 3.7. *Let (K,L) solve the BSRDE (1). Then, $\int_0^{\cdot} L_s\, dw_s$ is a* BMO(P) *-martingale.*

Note that a simple case (where the generator of BSRDE (1) depends on the martingale term L in a linear way) of Theorem 3.7 has been obtained by Bismut [5] by a quite different argument. This theorem plays an important role in solving a general non-homogeneous stochastic LQ problem and in solving a general mean-variance hedging problem.

Theorem 3.8. *Let the two assumptions $(\mathcal{A}1)$ and $(\mathcal{A}3)$ be satisfied, and (K,L) solve the BSRDE (1). Then, there is a deterministic positive constant ε_{02} such that*

$$K_t \ge \varepsilon_{02} I_{n\times n}. \tag{17}$$

The proof is an immediate consequence of a combination of Theorem 3.2 and the following estimate.

Lemma 3.1. (Kohlmann and Tang [14]) *Let the two assumptions $(\mathcal{A}1)$ and $(\mathcal{A}3)$ be satisfied. Let $X^{t,x;u}$ solve the SDE (6). Then, there is a deterministic positive constant ε_{03}, which is independent of the control u, such that*

$$|x|^2 + \frac{1}{2}E^{\mathcal{F}_t}\int_t^T |u_s|^2\, ds \le \exp\left(\varepsilon_{03}(T-t)\right)E^{\mathcal{F}_t}|X_T^{t,x;u}|^2, \ \forall u \in \mathcal{L}^2_{\mathcal{F}}(t,T;R^m). \tag{18}$$

3.7. Closedness of solution.

Theorem 3.9. (Kohlmann and Tang [16]) *Assume that* $\forall\gamma \geq 0$ *the coefficients* $A^\gamma, B^\gamma, C_i^\gamma, D_i^\gamma, Q^\gamma$, *and* N^γ *are* $\mathcal{F}_t$*-progressively measurable matrix-valued processes, defined on* $\Omega \times [0,T]$, *of dimensions* $n\times n, n\times m, n\times n, n\times m, n\times n$, *and* $m\times m$, *respectively. Assume that* M^γ *is an* $\mathcal{F}_T$*-measurable and nonnegative* $n\times n$ *random matrix. Assume that* Q^γ *is a.s.a.e. nonnegative. Assume that there are two deterministic positive constants* ε_1 *and* ε_2 *which are independent of the parameter* γ, *such that*

$$|A_t^\gamma|, |B_t^\gamma|, |C_{it}^\gamma|, |D_{it}^\gamma|, |Q_t^\gamma|, |N_t^\gamma|, |M^\gamma| \leq \varepsilon_1$$

and

$$N_t^\gamma \geq \varepsilon_2 I_{m\times m}.$$

Assume that as $\gamma\to 0$, $A_t^\gamma, B_t^\gamma, C_{it}^\gamma, D_{it}^\gamma, Q_t^\gamma$, *and* N_t^γ *converge uniformly in* (t,ω) *to* $A_t^0, B_t^0, C_{it}^0, D_{it}^0, Q_t^0$ *and* N_t^0, *respectively. Assume that* M^γ *uniformly converges to* M^0 *as* $\gamma\to 0$. *Assume that* $\forall\gamma > 0$ *the BSRDE* $(A^\gamma, B^\gamma; C_i^\gamma, D_i^\gamma, i = 1,\ldots,d; Q^\gamma, N^\gamma, M^\gamma)$ *has a unique solution* (K^γ, L^γ) *with* $K_t^\gamma \geq 0, t\in[0,T]$. *Then, there is a pair of processes* (K, L) *with*

$$K \in \mathcal{L}_{\mathcal{F}}^\infty(0,T;\mathcal{S}_+^n) \cap L^\infty(\Omega,\mathcal{F}_T,P;C([0,T];\mathcal{S}_+^n)), \quad L \in (\mathcal{L}_{\mathcal{F}}^2(0,T;\mathcal{S}^n))^d,$$

such that

$$\begin{aligned}
&\lim_{\gamma\to 0} K^\gamma = K \quad \text{strongly in } \mathcal{L}_{\mathcal{F}}^\infty(0,T;\mathcal{S}_+^n) \cap L^\infty(\Omega,\mathcal{F}_T,P;C([0,T];\mathcal{S}_+^n)),\\
&\lim_{\gamma\to 0} L^\gamma = L \quad \text{strongly in } (\mathcal{L}_{\mathcal{F}}^2(0,T;\mathcal{S}^n))^d,
\end{aligned} \tag{19}$$

and such that (K, L) *solves the BSRDE* $(A^0, B^0; C^0, D^0; Q^0, N^0, M^0)$.

If the above assumption of uniform convergence of $(A^\gamma, C^\gamma, Q^\gamma)$ *is replaced by the following one:*

$$\lim_{\gamma\to 0} \operatorname*{esssup}_{\omega\in\Omega} \int_0^T (|A_s^\gamma - A_s^0| + |C_s^\gamma - C_s^0|^2 + |Q_s^\gamma - Q_s^0|)\, ds \to 0. \tag{20}$$

then the above assertions still hold.

Remark 3.1. *When the assumption of uniform positivity on the control weight matrix* N_t *is relaxed to nonnegativity, Theorem 3.9 still holds under the additional assumption that there is a deterministic positive constant* ε_3 *such that*

$$\sum_{i=1}^d (D_{it}^\gamma)' D_{it}^\gamma \geq \varepsilon_3 I_{m\times m}, \quad M^\gamma \geq \varepsilon_3 I_{n\times n}.$$

3.8. Transformation of BSRDEs

Let Φ solve the differential equation:

$$\begin{cases} \dfrac{d\Phi}{dt}(t) &= A_t\Phi(t), \quad t\in(0,T],\\ \Phi(0) &= I_{n\times n}. \end{cases}$$

Consider the following transformation

$$\widetilde{K} := \Phi' K \Phi, \quad \widetilde{L} := \Phi' L \Phi \tag{21}$$

of the BSRDE (1).

Using Itô's formula, we verify that the BSRDE $(A, B; C_i, D_i, i = 1, \ldots, d; Q, N, M)$ becomes the new BSRDE $(0, \widetilde{B}; \widetilde{C}_i, \widetilde{D}_i, i = 1, \cdots, d; \widetilde{Q}, N, \widetilde{M})$, which is satisfied by $(\widetilde{K}, \widetilde{L})$. Here,

$$\begin{array}{lll} \widetilde{B} := \Phi^{-1}B, & \widetilde{C} := \Phi^{-1}C\Phi, & \widetilde{D} := \Phi^{-1}D \\ \widetilde{Q} := \Phi' Q \Phi, & \widetilde{M} := \Phi(T)' M \Phi(T). & \end{array} \tag{22}$$

The advantage of doing the transformation (21) is that it makes A disappear. Since A is bounded, the resulting transformation Φ is also bouned.

It is interesting to discuss the more general transformation Ψ which solves the SDE:

$$\left\{ \begin{array}{lll} d\Psi(t) & = & A_t \Psi(t)\, dt + \sum_{i=1}^{d} C_{it} \Psi(t)\, dw_{it}, \quad t \in (0, T], \\ \Psi(0) & = & I_{n \times n}. \end{array} \right. \tag{23}$$

This transformation can anneal both A and C in the BSRDE (1). Note that Ψ is not necessarily bounded though A and $C = (C_1, \ldots, C_d)$ are assumed to be bounded.

3.9. The Markov case : a PDE characterization

Since optimal controls or optimal hedging/investing strategies are characterized in terms of the solutions of associated BSRDEs, it is important to characterize the solutions of BSRDEs. When the coefficients of BSRDEs are Markovian functions of an Itô's process, it is natural to connect the solution of a BSRDE with a system of parabolic PDEs. Note that in general the associated system of parabolic PDEs contains a quadratic term of the gradient. We are not going into the details due to the limitation of space. The reader is referred to among others, Peng [28, 29], Pardoux and Peng [25], and Pardoux and Tang [26] for this direction.

4. A Historical Review

First, consider the regular case, i.e., N is assumed to be uniformly positive. When the coefficients A, B, C_i, D_i, Q, N, M are all deterministic, then $L_1 = \cdots = L_d = 0$ and the BSRDE (1) is reduced to the following nonlinear matrix ordinary differential equation

$$\left\{ \begin{array}{lll} dK_t & = & -[A_t' K_t + K_t A_t + \sum_{i=1}^{d} C_{it}' K_t C_{it} + Q_t - (K_t B_t + \sum_{i=1}^{d} C_{it}' K_t D_{it}) \\ & & \times (N_t + \sum_{i=1}^{d} D_{it}' K_t D_{it})^{-1} (K_t B_t + \sum_{i=1}^{d} C_{it}' K_t D_{it})']\, dt, \ 0 \le t < T, \\ K_T & = & M, \end{array} \right. \tag{24}$$

which was completely solved by Wonham [38, 39] by applying Bellman's principle of quasi-linearization (see Bellman [2]) and a monotone convergence result of symmetric matrices.

The attention to the randomness of the coefficients A, B, C, D, Q, N, M is dated back at least to Bismut [4, 5]. Bismut [4, 5] studied the case of random coefficients, but he could solve only some special simple cases at that time. Let the integer $d_0 \geq 0$, and denote by $\{\mathcal{F}_t^2, 0 \leq t \leq T\}$ the P-augmented natural filtration generated by the $(d-d_0)$-dimensional Brownian motion $(w_{d_0+1}, \ldots, w_d)$. Bismut [4, 5] assumed that the randomness of the coefficients only comes from the smaller filtration $\{\mathcal{F}_t^2\}$, which leads to $L_1 = \cdots = L_{d_0} = 0$. He further assumed in the paper [4] that

$$C_{d_0+1} = \cdots = C_d = 0, \quad D_{d_0+1} = \cdots = D_d = 0, \tag{25}$$

under which the BSRDE (1) is reduced to the one:

$$\left\{\begin{array}{rcl} dK_t &=& -[A_t'K_t + K_tA_t + \sum_{i=1}^{d_0} C_{it}'K_tC_{it} + Q_t - (K_tB_t + \sum_{i=1}^{d_0} C_{it}'K_tD_{it}) \\ && \times(N_t + \sum_{i=1}^{d_0} D_{it}'K_tD_{it})^{-1}(K_tB_t + \sum_{i=1}^{d_0} C_{it}'K_tD_{it})']\,dt \\ && + \sum_{i=d_0+1}^{d} L_{it}\,dw_{it}, \\ K_T &=& M, \end{array}\right. \tag{26}$$

and the generator does not involve L at all. In the work [5], Bismut assumed that

$$D_{d_0+1} = \cdots = D_d = 0, \tag{27}$$

under which the BSRDE (1) becomes the following one

$$\left\{\begin{array}{rcl} dK_t &=& -[A_t'K_t + K_tA_t + \sum_{i=1}^{d} C_{it}'K_tC_{it} + Q_t + \sum_{i=d_0+1}^{d} (C_{it}'L_{it} + L_{it}C_{it}) \\ && -(K_tB_t + \sum_{i=1}^{d_0} C_{it}'K_tD_{it}) \\ && \times(N_t + \sum_{i=1}^{d_0} D_{it}'K_tD_{it})^{-1}(K_tB_t + \sum_{i=1}^{d_0} C_{it}'K_tD_{it})']\,dt \\ && + \sum_{i=d_0+1}^{d} L_{it}\,dw_{it}, \quad 0 \leq t < T, \\ K_T &=& M, \end{array}\right. \tag{28}$$

and the generator depends on the second unknown variable $(L_{d_0+1}, \ldots, L_d)'$ in a linear way. Moreover, his method consists of constructing a contraction mapping and then using a fixed point theorem.

Later, Peng [27] gave a nice treatment on the proof of existence and uniqueness for the BSRDE (28), by using Bellman's principle of quasilinearization and a monotone convergence result of symmetric matrices—a generalization of Wonham's approach to the random situation.

The following proposition states the above-mentioned result on existence and uniqueness of a global adapted solution to the BSRDE (1), due to Bismut [5] and Peng [27].

Theorem 4.1. *Let assumptions* $(\mathcal{A}1)$ *and* $(\mathcal{A}2)$ *be satisfied. Assume that* $D_{d_0+1} = \cdots = D_d = 0$, *all the coefficients* $A, B, \{C_i\}_{i=1}^d, \{D_i\}_{i=1}^{d_0}, Q$ *and* N *are* $\mathcal{F}_t^2$*-progressively measurable, and that* M *is* $\mathcal{F}_T^2$*-measurable. Then, the* BSRDE (1) *has a unique* $\mathcal{F}_t^2$*-adapted global solution* (K, L) *with* $K_t \geq 0, t \in [0,T]$, *and* $L_1 = \cdots = L_{d_0} = 0$.

Solution of the general BSRDE (1), whose generator is allowed to contain a quadratic term of L, turns out to become a long-standing problem. As early as in 1978, Bismut [5] commented on page 220 that:"Nous ne pourrons pas démontrer l'existence de solution pour l'équation (2.49) dans le cas général." (We could not prove the existence of solution for equation (2.49) for the general case.) On page 238, he pointed out that the essential difficulty for solution of the general BSRDE (1) lies in the integrand of the martingale term which appears in the generator in a quadratic way. Two decades later in 1998, Peng [30] included the above problem in his list of open problems on *backward stochastic differential equations* (BSDEs, in short).

Second, consider the singular case $N = 0$. Kohlmann and Zhou [18] studied the following case:

$$C = 0, M = I_{n\times n}, D_i u = u_i, \ i = 1, \cdots, d.$$

They made the two rather restrictive assumptions: (a) all the coefficients are deterministic and (b) $A + A' \geq BB'$. Under the above-described framework, they obtain the existence and uniqueness of a global solution. Their method is based on an existence criterion of Chen, Li and Zhou [6].

5. Recent Advances on Existence and Uniqueness.

There are *two essential* difficulties in the proof of existence and uniqueness of a global adapted solution to a BSRDE. One, resulting from a singular BSRDE, is how to formulate the conditions on the coefficients such that the solution K is *a priori* guaranteed to be *uniformly* (the uniform requirement is obviously not necessary but convenient) positive and such that singularity is therefore *actually* excluded in the BSRDE even if $R = 0$. The other essential difficulty, is the existence of the quadratic term of L in the generator, as we have argued in the last section. Recently, we have made a breakthrough in overcoming the two difficulties, and have obtained the global existence and uniqueness result for many BSRDEs.

Kohlmann and Tang [14] proved Theorem 3.8 with a *simple yet nice* application of the *a priori estimate* result Lemma 3.1, and therefore were able to extend Theorem 4.1 to the singular case under assumption ($\mathcal{A}3$), via a perturbation technique.

Theorem 5.1. *Let assumption* ($\mathcal{A}3$) *be satisfied. Then, Theorem* 3.1 *still holds even if* $N = 0$. *Moreover,* K_t *is uniformly positive.*

Kohlmann and Tang [15] completely solved the *one-dimensional* case of the above Bismut-Peng's problem.

Theorem 5.2. *Let the assumptions* ($\mathcal{A}1$) *and* ($\mathcal{A}2$) *be satisfied, and* $n = 1$. *Then, BSRDE* (1) *has a unique solution. Further, assume that* ($\mathcal{A}3$) *is satisfied. Then, BSRDE* (1) *still has a unique solution* (K, L) *even if* $N = 0$, *and moreover,* K_t *is uniformly positive.*

This theorem meets the current needs for BSRDEs to be applied to solution of the mean-variance problems in finance. An approximation technique is used, which is motivated by the works of Kobylansky [13], and Lepeltier and San Martin [20, 21].

For the *multi-dimensional* case, Kohlmann and Tang [16] proved the global existence and uniqueness result for some classes of BSRDE (1): These BSRDEs are typical, for the generators contain a quadratic term of L. The results are stated by the following two theorems.

Theorem 5.3. (the singular case) *Let assumption* ($\mathcal{A}1$) *be satisfied,* $n = m$, *and* $d = 1$. *Assume that there is a deterministic positive constant* ε *such that*

$$M \geq \varepsilon I_{n\times n} \tag{29}$$

and

$$D_t' D_t \geq \varepsilon I_{m\times m}. \tag{30}$$

Then, the BSRDE:

$$\begin{cases} dK_t &= -[A_t'K_t + K_tA_t + C_t'K_tC_t + Q_t + C_t'L_t + L_tC_t \\ &\quad -(K_tB_t + C_t'K_tD_t + L_tD_t)(D_t'K_tD_t)^{-1} \\ &\quad \times(K_tB_t + C_t'K_tD_t + L_tD_t)']\,dt \\ &\quad +L_t\,dw_t, \ \ 0 \leq t < T, \\ K_T &= M. \end{cases} \tag{31}$$

has a unique solution (K, L) *with* K_t *being uniformly positive w.r.t.* (t, ω).

Theorem 5.4. (the regular case) *Let assumptions* ($\mathcal{A}1$) *and* ($\mathcal{A}2$) *be satisfied. Further assume that* $d = 1, B = C = 0$, *and* D *and* N *satisfy the following*

$$\begin{aligned} \lim_{h\to 0+} \operatorname*{esssup}_{\omega\in\Omega} \max_{t_1,t_2\in[0,T];\,|t_1-t_2|\leq h} |D_{t_1} - D_{t_2}| &= 0, \\ \lim_{h\to 0+} \operatorname*{esssup}_{\omega\in\Omega} \max_{t_1,t_2\in[0,T];\,|t_1-t_2|\leq h} |N_{t_1} - N_{t_2}| &= 0. \end{aligned} \tag{32}$$

Then, the BSRDE:

$$\begin{cases} dK_t &= -(A_t'K_t + K_tA_t + Q_t - L_tD_t(N_t + D_t'K_tD_t)^{-1}D_t'L_t)\,dt \\ &\quad +L_t\,dw_t, \qquad 0 \leq t < T, \\ K_T &= M. \end{cases} \tag{33}$$

has a unique solution (K, L) *with* $K_t \geq 0, \forall t \in [0, T]$.

The main point for the proof of Theorem 5.3 is to consider the inverse transformation: $\widetilde{K} := K^{-1}$. $\widetilde{K}$ turns out to satisfy a Riccati equation whose generator depends on the martingale term in a linear way.

Remark 5.1 *It is interesting to note that the technique of doing the inverse transformation is also applicable to the proof of the second assertion of Theorem 5.2, but only for the special case of $m = d$ — not for the case of $m < d$.*

The proof of Theorem 5.4 is a little complex: the inverse transformation used above can not eliminate the quadratic term of the second unknown variable. However, the proof still consists of doing some appropriate transformations and some additional arguments of approximating. The whole proof is divided into several propositions.

Proposition 5.1. *Assume that $Q \geq A'(D^{-1})'ND^{-1} + (D^{-1})'ND^{-1}A, m = n$, and D and N are nonsingular constant matrices. Then, Theorem 5.4 holds.*

Proposition 5.2. *Assume that $A = 0$ and D and N are constant matrices. Then, Theorem 5.4 holds.*

Proposition 5.3. *Assume that $A = 0$, and D and N are piece-wisely constant, bounded, and $\mathcal{F}_t$-adapted matrix processes. Then, Theorem 5.4 holds.*

Proposition 5.4. *Assume that $A = 0$. Then, Theorem 5.4 holds.*

At this stage, the proof of Theorem 5.4 is finished by an immediate application of the transformation technique presented in Subsection 3.8.

We close this section with the following remark.

Remark 5.2. *Theorem 3.9 on closedness of solution can be improved using Theorem 3.6 on a priori estimate. Therefore, the condition (32) of Theorem 5.4 can be relaxed accordingly. The relevant details will be given elsewhere.*

6. Applications in Control and Finance

6.1. Control-theoretic application

Assume that

$$\xi \in L^2(\Omega, \mathcal{F}_T, P; R^n), \quad q, f, g_i \in \mathcal{L}^2_{\mathcal{F}}(0,T; R^n). \tag{34}$$

Consider the following optimal control problem (denoted by $\mathcal{P}_0$):

$$\min_{u \in \mathcal{L}^2_{\mathcal{F}}(0,T;R^m)} J(u; 0, x) \tag{35}$$

with

$$\begin{aligned} J(u; t, x) = \quad & E^{\mathcal{F}_t}(M(X_T^{t,x;u} - \xi), X_T^{t,x;u} - \xi) \\ & + E^{\mathcal{F}_t} \int_t^T [(Q_s(X_s^{t,x;u} - q_s), X_s^{t,x;u} - q_s) + (N_s u_s, u_s)]\, ds \end{aligned} \tag{36}$$

and $X^{t,x;u}$ solving the following linear SDE

$$\begin{cases} dX_s &= (A_sX_s + B_su_s + f_s)\,ds + \sum_{i=1}^{d}(C_{is}X_s + D_{is}u_s + g_{is})\,dw_{is}, \\ & \qquad t < s \le T, \\ X_t &= x, \quad u \in \mathcal{L}^2_{\mathcal{F}}(t,T;R^m). \end{cases} \tag{37}$$

The value function V is defined as

$$V(t,x) := \min_{u \in \mathcal{L}^2_{\mathcal{F}}(t,T;R^m)} J(u;t,x), \quad (t,x) \in [0,T] \times R^n. \tag{38}$$

Theorem 6.1 *Let the two assumptions $(\mathcal{A}1)$ and $(\mathcal{A}2)$, or $(\mathcal{A}1)$ and $(\mathcal{A}3)$ be satisfied. Let (K,L) solve the BSRDE (1). Then, the BSDE*

$$\begin{cases} d\psi_t &= -[\widehat{A}'_t\psi_t + \sum_{i=1}^{d}\widehat{C}'_{it}(\phi_{it} - K_tg_{it}) - K_tf_t - \sum_{i=1}^{d}L_{it}g_{it} + Q_tq_t]\,dt \\ & \qquad + \sum_{i=1}^{d}\phi_{it}\,dw_{it}, \\ \psi_T &= M\xi \end{cases} \tag{39}$$

where $\widehat{A}$ and $\widehat{C}_i$ are defined by (10), has a unique $\mathcal{F}_t$-adapted solution (ψ,ϕ) with

$$\psi \in \mathcal{L}^2_{\mathcal{F}}(0,T;R^n) \cap L^2(\Omega,\mathcal{F}_T,P;C([0,T];R^n)),\ \phi \in \left(\mathcal{L}^2_{\mathcal{F}}(0,T;R^n)\right)^d. \tag{40}$$

Moreover, the optimal control $\widehat{u}$ for the non-homogeneous stochastic LQ problem $\mathcal{P}_0$ exists uniquely and has the following feedback law

$$\begin{aligned} \widehat{u}_t \;=\; & -(N_t + \sum_{i=1}^{d}D'_{it}K_tD_{it})^{-1}[(B'_tK_t + \sum_{i=1}^{d}D'_{it}K_tC_{it} + \sum_{i=1}^{d}D'_{it}L_{it})\widehat{X}_t \\ & -B'_t\psi_t + \sum_{i=1}^{d}D'_{it}(K_tg_{it} - \phi_{it})] \end{aligned} \tag{41}$$

where $\widehat{X} := X^{0,x;\widehat{u}}$.

Remark 6.1. *Note that $\widehat{A}$ and $\widehat{C}_i$ depend on L in general, and thus they might not be uniformly bounded. In this case, we have no available—to the authors' best knowledge—theorem to guarantee the existence and the uniqueness of a global adapted solution, though the BSDE (39) is linear.*

Proof of Theorem 6.1. The assumptions guarantee that there is a unique optimal control $\widehat{u} \in \mathcal{L}^2_{\mathcal{F}}(0,T;R^m)$. The optimality condition (see Peng [31], and Tang and Li [37]) implies

$$B'\widetilde{p} + \sum_{i=1} D'_i\widetilde{q}_i + N\widehat{u} = 0$$

where $(\widetilde{p}, \widetilde{q})$ solves the BSDE (called the adjoint equation)

$$\begin{cases} d\widetilde{p}_t &= -[A_t'\widetilde{p}_t + \sum_{i=1}^d C_{it}'\widetilde{q}_{it} + Q_t(\widehat{X}_t - q_t)]\, dt + \sum_{i=1}^d \widetilde{q}_{it}\, dw_{it}, \\ \widetilde{p}_T &= M(\widehat{X}_T - \xi) \end{cases} \tag{42}$$

with

$$\widetilde{p} \in \mathcal{L}^2_{\mathcal{F}}(0,T;R^n) \cap L^2(\Omega, \mathcal{F}_T, P; C([0,T];R^n)), \ \widetilde{q} \in \left(\mathcal{L}^2_{\mathcal{F}}(0,T;R^n)\right)^d. \tag{43}$$

Via Itô's formula, we check out that the pair (ψ, ϕ) defined by the following

$$\psi_t := K_t\widehat{X}_t - \widetilde{p}_t, \ \phi_{it} := K_t(C_i\widehat{X}_t + D_i\widehat{u}_t + g_{it}) + L_{it}\widehat{X}_t - \widetilde{q}_{it}$$

solves the BSDE (39). It is obvious that

$$\psi \in \mathcal{L}^2_{\mathcal{F}}(0,T;R^n) \cap L^2(\Omega, \mathcal{F}_T, P; C([0,T];R^n)).$$

Since $\int_0^\cdot L_s\, dw_s$ is a BMO(P)-martingale, it follows from Theorem 1.1 (i) and (iii) of Bañuelos and Bennett [1] that $\int_0^T L_{is}\widehat{X}_s\, dw_{is}$ is square-integrable. Therefore, $L_i\widehat{X} \in \mathcal{L}^2_{\mathcal{F}}(0,T;R^n)$, and $\phi_i \in \mathcal{L}^2_{\mathcal{F}}(0,T;R^n)$.

It is standard to get the explicit formula (41) of $\widehat{u}$ from the optimality condition. The uniqueness proof is omitted. The proof is complete.

Theorem 6.2 *Suppose that the two assumptions $(\mathcal{A}1)$ and $(\mathcal{A}2)$, or $(\mathcal{A}1)$ and $(\mathcal{A}3)$ are satisfied. Let (K, L) solve BSRDE (1). Then, the value function $V(t,x), (t,x) \in [0,T] \times R^n$ has the following explicit formula*

$$V(t,x) = (K_t x, x) - 2(\psi_t, x) + V^0(t), \quad (t,x) \in [0,T] \times R^n \tag{44}$$

with

$$\begin{aligned} V^0(t) := \ & E^{\mathcal{F}_t}(M\xi, \xi) + E^{\mathcal{F}_t}\int_t^T (Q_s q_s, q_s)\, ds - 2E^{\mathcal{F}_t}\int_t^T (\psi_s, f_s)\, ds \\ & + E^{\mathcal{F}_t}\int_t^T \sum_{i=1}^d [(K_s g_{is}, g_{is}) - 2(\phi_{is} g_{is})]\, ds \\ & - E^{\mathcal{F}_t}\int_t^T ((N_s + \sum_{i=1}^d D_{is}'K_s D_{is})u_s^0, u_s^0)\, ds \end{aligned} \tag{45}$$

and

$$u_s^0 := (N_s + \sum_{i=1}^d D_{is}'K_s D_{is})^{-1}[B_s'\psi_s + \sum_{i=1}^d D_{is}'(\phi_{is} - K_s g_{is})], \quad t \le s \le T. \tag{46}$$

6.2. Financial application.

As an application of Theorems 6.1 and 6.2, the mean-variance hedging problem with random market conditions is considered. The mean-variance hedging problem was initially introduced by Föllmer and Sondermann [8], and later widely studied by Duffie and Richardson [7], Föllmer and Schweizer [9], Schweizer [34, 35, 36], Hipp [12], Monat and Stricker [23], Pham, Rheinländer and Schweizer [32], Gourieroux, Laurent and Pham [11], and Laurent and Pham [19]. All of these works are based on a projection argument. Recently, Kohlmann and Zhou [18] used a natural LQ theory approach to solve the case of deterministic market conditions. Kohlmann and Tang [14] used a natural LQ theory approach to solve the case of stochastic market conditions, but the market conditions are only allowed to come from a smaller filtration $\{\mathcal{F}_t^2\}$. Kohlmann and Tang [15] completely solved the case of random market conditions by using Theorems 6.1 and 6.2, and the optimal hedging portfolio and the variance-optimal martingale measure are characterized by the solution of the associated BSRDE.

Consider a financial market in which there are $m+1$ primitive assets: one nonrisky asset (the bond) of price process

$$S_0(t) = \exp\Big(\int_0^t r_s\, ds\Big), \quad 0 \le t \le T, \tag{47}$$

and m risky assets (the stocks)

$$dS(t) = \mathrm{diag}(S(t))(\mu_t\, dt + \sigma_t\, dw_t), \quad 0 \le t \le T. \tag{48}$$

Assume that the instantaneous interest rate r, the m-dimensional appreciation vector process μ, and the volatility $m \times d$ matrix process σ are progressively measurable with respect to $\{\mathcal{F}_t, 0 \le t \le T\}$, and uniformly bounded, and that there exists a positive constant ε such that

$$\sigma_t \sigma_t' \ge \varepsilon I_{m\times m}, \quad 0 \le t \le T, a.s. \tag{49}$$

The risk premium process is given by

$$\lambda_t = \sigma_t'(\sigma_t\sigma_t')^{-1}\widetilde{\mu}_t, \quad 0 \le t \le T \tag{50}$$

where $e_m = (1, \ldots, 1)' \in R^m$, and $\widetilde{\mu} := \mu - r e_m$.

For any $x \in R$ and $\pi \in \mathcal{L}^2_{\mathcal{F}}(0,T;R^m)$, define the self-financed wealth process X with initial capital x and with quantity π invested in the risky asset S by

$$\begin{cases} dX_t &= [r_t X_t + (\widetilde{\mu}_t, \pi_t)]\, dt + \pi_t'\sigma_t\, dw_t, \quad 0 < t \le T, \\ X_0 &= x, \quad \pi \in \mathcal{L}^2_{\mathcal{F}}(0,T;R^m). \end{cases} \tag{51}$$

Given $\xi \in L^2(\Omega, \mathcal{F}_T, P)$, consider the LQ problem:

$$\textbf{Problem } \mathcal{P}_{0,x}(\xi) \qquad \min_{\pi\in \mathcal{L}^2_{\mathcal{F}}(0,T;R^m)} E|X_T^{0,x;\pi} - \xi|^2 \tag{52}$$

where $X^{0,x;\pi}$ solves the wealth equation (51). The associated value function is denoted by $V(t,x), (t,x) \in [0,T] \times R$. The minimum point of $V(t,x)$ over $x \in R$ for given time t is defined to be the approximate price of the contingent claim ξ at time t.

Problem $\mathcal{P}_{0,x}(\xi)$ is the so-called mean-variance hedging problem in mathematical finance. It is a one-dimensional, singular, nonhomogeneous, stochastic LQ problem $\mathcal{P}_0$.

Denote by σ_{it} the i-th column of the volatility matrix σ_t. The associated Riccati equation is a non-linear singular BSDE:

$$\begin{cases} dK_t = & -[(2r_t - |\lambda_t|^2)K_t - 2(\lambda_t, L_t) - K_t^{-1} L_t' \sigma_t' (\sigma_t \sigma_t')^{-1} \sigma_t L_t]\, dt \\ & +(L_t,\, dw_t), \\ K_T = & 1. \end{cases} \tag{53}$$

Let (ψ, ϕ) be the $\mathcal{F}_t$-adapted solution of the BSDE

$$\begin{cases} d\psi_t = & -\{[r_t - |\lambda_t|^2 - (\lambda_t, K_t^{-1} L_t)]\psi_t \\ & \qquad -(\lambda_t + K_t^{-1}\sigma_t'(\sigma_t\sigma_t')^{-1}\sigma_t L_t, \phi_t)\}\, dt \\ & +(\phi_t,\, dw_t), \\ \psi_T = & \xi \end{cases} \tag{54}$$

Theorem 6.1 provides an explicit formula for the optimal hedging portfolio:

$$\pi_t = -(\sigma_t\sigma_t')^{-1}[(\widetilde{\mu}_t + \sigma_t K_t^{-1} L_t) X_t - \widetilde{\mu}_t K_t^{-1}\psi_t - \sigma_t K_t^{-1}\phi_t] \tag{55}$$

where (K, L) solves the equation (53). The value function V is also given by $V(t, x)$

$$= K_t x^2 - 2\psi_t x + E^{\mathcal{F}_t}\xi^2 - E^{\mathcal{F}_t}\int_t^T (\widetilde{\mu}_s\psi_s + \sigma_s\phi_s)'(\sigma_s K_s \sigma_s')(\widetilde{\mu}_s\psi_s + \sigma_s\phi_s)\, ds \tag{56}$$

where $\phi := (\phi_1, \ldots, \phi_n)'$. So, the approximate price p_t of the contingent claim ξ at time t is given by

$$p_t = K_t^{-1}\psi_t. \tag{57}$$

The above solution need not introduce the additional concepts of the so-called *hedging numeraire* and *variance-optimal martingale measure,* and therefore is simpler than that of Gourieroux et al [11], and Laurent and Pham [19]. To be connected to the latter, the optimal hedging portfolio (55) is rewritten as

$$\pi = -(\sigma\sigma')^{-1}[(\widetilde{\mu} + \sigma\widetilde{L})(X - \widetilde{\psi}) - \sigma\widetilde{\phi}]. \tag{58}$$

Here,

$$\widetilde{L} := LK^{-1}, \quad \widetilde{\psi} := \psi K^{-1}, \quad \widetilde{\phi} := \phi K^{-1} - L\psi K^{-2}. \tag{59}$$

and the pair $(\widetilde{\psi}, \widetilde{\phi})$ solves the BSDE:

$$\begin{cases} d\widetilde{\psi}_t &= \{r_t\widetilde{\psi}_t + (\widetilde{\lambda}_t, \widetilde{\phi}_t)\}\, dt + (\widetilde{\phi}_t,\, dw_t), \quad 0 \le t < T, \\ \widetilde{\psi}_T &= \xi \end{cases} \tag{60}$$

with

$$\widetilde{\lambda} := \lambda - [I - \sigma'(\sigma\sigma')^{-1}\sigma]LK^{-1}. \tag{61}$$

The process $\widetilde{\psi}$ is just the *approximate price process,* and the BSDE (60) is the *approximate pricing equation.*

In view of Theorem 6.1, it follows from Theorem 1.1 (i) and (iii) of Bañuelos and Bennett [1] that

$$\widetilde{\psi} \in \mathcal{L}^2_{\mathcal{F}}(0, T; R^n) \cap L^2(\Omega, \mathcal{F}_T, P; C([0, T]; R)), \ \widetilde{\phi} \in \mathcal{L}^2_{\mathcal{F}}(0, T; R^d). \tag{62}$$

Note that the optimal hedging portfolio (55) consists of the two parts:

$$\pi^1 := -(\sigma\sigma')^{-1}(\widetilde{\mu} + \sigma\widetilde{L})X \tag{63}$$

and

$$\pi^0 := (\sigma\sigma')^{-1}[(\widetilde{\mu} + \sigma\widetilde{L})\widetilde{\psi} + \sigma\widetilde{\phi}], \tag{64}$$

and they satisfy

$$\pi = \pi^1 + \pi^0. \tag{65}$$

The first part π^1 is the optimal solution of the *homogeneous* mean-variance hedging problem $\mathcal{P}_{0,x}(0)$ (that is the case of $\xi = 0$ for the problem $\mathcal{P}_{0,x}(\xi)$). The corresponding optimal wealth process $X^{0,1;\pi^1}$ solves the following *optimal closed system*

$$\begin{cases} dX_t &= X_t[(r_t - |\lambda_t|^2 - (\lambda_t, \widetilde{L}_t))\,dt - (\lambda_t + \sigma_t'(\sigma_t\sigma_t')^{-1}\sigma_t\widetilde{L}_t,\, dw_t)], \\ & \qquad 0 < t \le T, \\ X_0 &= 1, \end{cases} \tag{66}$$

and it is just the *hedging numéraire*. So, the *hedging numéraire* is just *the state (wealth) transition process* of the optimal closed system (66) from time 0, or *the fundamental solution* of the optimal closed system (66).

To understand the quantity $\widetilde{\lambda}$, consider the BSDE satisfied by $(\mathcal{K}, L)$

$$\begin{cases} d\mathcal{K}_t &= \{(2r_t - |\lambda_t|^2)\mathcal{K}_t + 2(\lambda_t, \mathcal{L}_t) + \mathcal{K}_t^{-1}\mathcal{L}_t'[I - \sigma_t'(\sigma_t\sigma_t')^{-1}\sigma_t]\mathcal{L}_t\}\,dt \\ & \qquad +(\mathcal{L}_t,\, dw_t), \\ \mathcal{K}_T &= 1 \end{cases} \tag{67}$$

with $\mathcal{K} := K^{-1}$ and $\mathcal{L} := -LK^{-2}$. It is the BSRDE associated with the singular stochastic LQ problem (denoted by $\mathcal{P}_{0,x}'$):

$$\textbf{Problem } \mathcal{P}_{0,x}' \qquad \min_{\theta \in \mathcal{L}^2_{\mathcal{F}}(0,T;R^d)} E|\mathcal{X}_T^{0,x;\theta}|^2 \tag{68}$$

where $\mathcal{X}^{0,x;\theta}$ solves the following SDE

$$\begin{cases} d\mathcal{X}_t &= \mathcal{X}_t[-r_t\,dt - (\lambda_t,\, dw_t)] + ([I - \sigma_t'(\sigma_t\sigma_t')^{-1}\sigma_t]\theta_t,\, dw_t), \\ & \qquad 0 \le t \le T, \\ \mathcal{X}_0 &= x, \quad \theta \in \mathcal{L}^2_{\mathcal{F}}(0,T;R^d). \end{cases} \tag{69}$$

The following feedback control $\widehat{\theta}$ is optimal:

$$\widehat{\theta} = -\mathcal{K}^{-1}\mathcal{L}\mathcal{X} = LK^{-1}\mathcal{X}. \tag{70}$$

The problem $\mathcal{P}_{0,1}'$ is just the so-called *dual problem* of the problem $\mathcal{P}_{0,1}(0)$ in [11, 19], and so the variance-optimal martingale measure is P_* defined as

$$dP_* := \exp\left\{-\int_0^T (\widetilde{\lambda}_t,\, dw_t) - \frac{1}{2}\int_0^T |\widetilde{\lambda}_t|^2\,dt\right\} dP. \tag{71}$$

P_* is an equivalent martingale measure.

Note that $\widetilde{\psi}$ has the following explicit formula:

$$\widetilde{\psi}_t = E_*^{\mathcal{F}_t} \xi \exp\left(-\int_t^T r_s\, ds\right), \quad 0 \le t \le T. \tag{72}$$

Here, the notation $E_*^{\mathcal{F}_t}$ stands for the expectation operator conditioning on the σ-algebra $\mathcal{F}_t$ with respect to the probability P_*. The discounted $\widetilde{\phi}$ is just the integrand of the stochastic-integral-representation of the P_*-martingale

$$\{E_*^{\mathcal{F}_t} \xi \exp\left(-\int_0^T r_s\, ds\right), 0 \le t \le T\}$$

(w.r.t. the P_*-martingale $W + \int_0^\cdot \widetilde{\lambda}_t\, dt$).

As in Kohlmann and Zhou [18], again, the formula (58) has an interesting interpretation in terms of mathematical finance. The optimal hedging portfolio π in (58) consists of the two components:
(a) $(\sigma\sigma')^{-1}\sigma\tilde{\phi}$ — it may be interpreted as the perfect hedging portfolio for the contingent claim ξ with the risk premium process $\tilde{\lambda}$ (that is, under the variance-optimal martingale measure),
(b) $(\sigma\sigma')^{-1}(\tilde{\mu} + \sigma\tilde{L})(\tilde{\psi} - X)$ — it is a generalized Merton-type portfolio for a terminal utility function $c(x) = x^2$ (see Merton [22]), which invests the capital $(\tilde{\psi} - X)$ left over after fulfilling the obligation from the perfect hedge under the variance-optimal martingale measure.

6.3. A closing comment

As already touched in Remark 6.1, the general application of BSRDEs to stochastic linear quadratic optimal control problems, including mean-variance hedging and portfolio selection problems in finance, gives rise to a *new essential* difficulty: due to the new feature that the optimal feedback coefficient contains L and thus might not be bounded, the coefficients of the optimal closed system and those of the associated *linear* BSDE (39) (whose solution are used in the construction of the optimal control), might *not be bounded.* Then, the *fundamental* problem of existence and uniqueness of a solution of *either* the optimal closed system *or* the linear BSDE (39), becomes both *nontrivial* and *crucial.*

To overcome the above new difficulty, we use the **optimality condition** (namely the stochastic maximum principle) (see Subsections 3.1, 6.1 and 6.2) and the **BMO-property** (see Subsections 6.1 and 6.2), to obtain the necessary regularities. All these arguments constitute another novel unique feature of our recent works [14, 15, 16, 17].

Notes added after the conference. During the conference, the authors got to know the interesting report by Andrew Lim (see the paper in this volume by A. Lim and X. Y. Zhou). Lim and Zhou used the same method of applying the inverse transformation as in our work, but only to prove the existence and uniqueness of a global solution of a *one-dimensional singular* BSRDE for the case in which $C_1 = C_2 = \cdots = C_d = 0$ and $d = m$. As already commented by the lecturer (the second author) during the conference, first, both their existence and uniqueness result on BSRDEs (compared with our Theorem 5.2) and their application of applying the inverse transformation (compared with the arguments of Section 5 after Theorem 5.3), are special cases of ours. Second, the technique of doing the inverse transformation is *essentially* difficult to be applied to the case of $d > m$, even for the *one-dimensional* BSRDEs. Nevertheless, the work by Lim and Zhou in this volume and our recent works which are reviewed in this report were independently carried out.

Acknowledgement The second author would like to thank the hospitality of Department of Mathematics and Statistics, and the Center of Finance and Econometrics, Universität Konstanz, Germany.

References

[1] Bañuelos, R. and Bennett, A., *Paraproducts and commutators of martingale transforms.* Proceedings of the American Mathematical Society, 103 (1988), 1226-1234

[2] Bellman, R., *Functional equations in the theory of dynamic programming, positivity and quasilinearity,* Proc. Nat. Acad. Sci. U.S.A., 41 (1955), 743–746

[3] Bismut, J. M., *Conjugate convex functions in optimal stochastic control,* J. Math. Anal. Appl., 44 (1973), 384–404

[4] Bismut, J. M., *Linear quadratic optimal stochastic control with random coefficients,* SIAM J. Control Optim., 14 (1976), 419-444

[5] Bismut, J. M., *Controle des systems lineares quadratiques: applications de l'integrale stochastique,* Séminaire de Probabilités XII, eds : C. Dellacherie, P. A. Meyer et M. Weil, LNM 649, Springer-Verlag, Berlin 1978

[6] Chen, S., Li, X. and Zhou, X., *Stochastic linear quadratic regulators with indefinite control weight costs*, SIAM J. Control Optim. **36 (1998)**, 1685-1702

[7] Duffie, D. and Richardson, H. R., *Mean-variance hedging in continuous time,* Ann. Appl. Prob., 1 (1991), 1-15

[8] Föllmer, H. and Sonderman, D., *Hedging of non-redundant contingent claims.* In: Mas-Colell, A., Hildebrand, W. (eds.) Contributions to Mathematical Economics. Amsterdam: North Holland 1986, pp. 205-223

[9] Föllmer, H. and Schweizer, M., *Hedging of contingent claims under incomplete information.* In: Davis, M.H.A., Elliott, R.J. (eds.) Applied Stochastic Analysis. (Stochastics Monographs vol. 5) London-New York: Gordon & Breach, 1991, pp. 389-414

[10] Gal'chuk, L. I., *Existence and uniqueness of a solution for stochastic equations with respect to semimartingales,* Theory of Probability and Its Applications, 23 (1978), 751–763

[11] Gourieroux, C., Laurent, J. P. and Pham, H., *Mean-variance hedging and numéraire,* Mathematical Finance, 8 (1998), 179-200.

[12] Hipp, C., *Hedging general claims,* Proceedings of the 3rd AFIR colloqium, Rome, Vol. 2, 603-613, 1993

[13] Kobylanski, M., *Résultats d'existence et d'unicité pour des équations différentielles stochastiques rétrogrades avec des générateurs á croissance quadratique,* C. R. Acad. Sci. Paris, 324 (1997), Série I, 81-86

[14] Kohlmann, M. and Tang, S., *Optimal control of linear stochastic systems with singular costs, and the mean-variance hedging problem with stochastic market conditions.* submitted

[15] Kohlmann, M. and Tang, S., *Global adapted solution of one-dimensional backward stochastic Riccati equations, with application to the mean-variance hedging.* submitted

[16] Kohlmann, M. and Tang, S., *Multi-dimensional backward stochastic Riccati equations and applications.* submitted

[17] Kohlmann, M. and Tang, S., *Recent advances in linear-quadratic optimal stochastic control problems.* Preprint

[18] Kohlmann, M. and Zhou, X., *Relationship between backward stochastic differential equations and stochastic controls: a linear-quadratic approach,* SIAM J. Control & Optim., 38 (2000), 1392-1407

[19] Laurent, J. P. and Pham, H., *Dynamic programming and mean-variance hedging,* Finance and Stochastics, 3 (1999), 83-110

[20] Lepeltier, J. P. and San Martin, J., *Backward stochastic differential equations with continuous coefficient,* Statistics & Probability Letters, 32 (1997), 425-430

[21] Lepeltier, J. P. and San Martin, J., *Existence for BSDE with superlinear-quadratic coefficient,* Stochastics & Stochastics Reports, 63 (1998), 227-240

[22] Merton, R., *Optimum consumption and portfolio rules in a continuous time model,* J. Econ. Theory, 3 (1971), pp. 373-413;Erratum 6 (1973), 213-214.

[23] Monat, P. and Stricker, C., *Föllmer-Schweizer decomposition and mean-variance hedging of general claims,* Ann. Probab. 23, 605–628

[24] Pardoux, E. and Peng, S., *Adapted solution of backward stochastic equation,* Systems Control Lett., 14 (1990), 55–61

[25] Pardoux, E. and Peng, S., *Backward stochastic differential equations and quasi-linear parabolic partial differential equations,* in: Rozovskii, B. L., Sowers, R. S. (eds.) Stochastic Partial Differential Equations and Their Applications, Lecture Notes in Control and Information Science 176, Springer, Berlin, Heidelberg, New York 1992, pp. 200–217

[26] Pardoux, E. and Tang, S., *Forward-backward stochastic differential equations with application to quasi-linear partial differential equations of second-order*, Probability Theory and Related Fields, 114 (1999), 123-150

[27] Peng, S., *Stochastic Hamilton-Jacobi-Bellman equations*, SIAM J. Control and Optimization **30 (1992)** , 284-304

[28] Peng, S., *Probabilistic interpretation for systems of semilinear parabolic PDEs,* Stochastics & Stochastic Reports, 37 (1991), 61-74

[29] Peng, S., *A generalized dynamic programming principle and Hamilton-Jacobi-Bellman equation,* Stochastics & Stochastics Reports, 38 (1992), 119–134

[30] Peng, S., *Some open problems on backward stochastic differential equations,* Control of distributed parameter and stochastic systems, proceedings of the IFIP WG 7.2 international conference, June 19-22, 1998, Hangzhou, China

[31] Peng, S., *A general stochastic maximum principle for optimal control problems,* SIAM J. Control Optim., 28 (1990), 966-979

[32] Pham, H., Rheinländer, T., and Schweizer, M., *Mean-variance hedging for continuous processes: New proofs and examples,* Finance and Stochastics, 2 (1998), 173-198

[33] Richardson, H, *A minimum result in continuous trading portfolio optimization.* Management Science, 35 (1989), 1045-1055

[34] Schweizer, M., *Mean-variance hedging for general claims,* Ann. Appl. Prob., 2 (1992), 171-179

[35] Schweizer, M., *Approaching random variables by stochastic integrals,* Ann. Probab., 22 (1994), 1536–1575

[36] Schweizer, M., *Approximation pricing and the variance-optimal martingale measure,* Ann. Probab., 24 (1996), 206–236

[37] Tang, S. and Li, X., *Necessary conditions for optimal control of stochastic systems with random jumps,* SIAM J. Control Optim., 32 (1994), 1447–1475

[38] Wonham, W. M., *On a matrix Riccati equation of stochastic control,* SIAM J. Control Optim., 6 (1968), 312-326

[39] Wonham, W. M., *Random differential equations in control theory,* in: Bharucha-Reid, A. T. (ed.), Probabilistic Methods in Applied Mathematics, Vol. II, Academic Press, New York, London, 1970, pp. 131–212

[40] Zhou, X. and Li, D., *Continuous time mean-variance portfolio selection: a stochastic LQ framework,* Applied Math. and Optim., 42 (2000), 19-33

[1]Michael Kohlmann, Department of Mathematics and Statistics, University of Konstanz, D-78457, Konstanz, Germany
E-mail address: `michael.kohlmann@uni-konstanz.de`

[2]Shanjian Tang, Department of Mathematics, Fudan University, Shanghai 200433, China
E-mail address: `tang@fmi.uni-konstanz.de`

Trends in Mathematics, © 2001 Birkhäuser Verlag Basel/Switzerland

Quantile hedging for a jump-diffusion financial market model

R.N.Krutchenko, A.V.Melnikov

Abstract. The paper is devoted to the problem of hedging contingent claims in the framework of a jump-diffusion model. Based on the results of H. Föllmer and P. Leukert [1]-[2] in a general semimartingale setting, we study the question how an investor can maximize the probability of a successful hedge under the constraint that he invests not more than a fixed amount of capital which is strictly less than the price of the option. We derive explicit formulas for this so-called quantile hedging strategy.

1. Introduction

One of the basic problems in *Contingent Claim Analysis* is the problem of hedging options. A number of papers, including the famous paper by Black and Scholes, are devoted to the analysis of hedges which succeed with probability one (see, for example Shiryaev[3]). The solution of such a problem in the case of a complete market yields to the so-called *fair price* as the minimal capital that is required to replicate the contingent claim. But in general the initial capital of an investor can be less this price. The natural question arises: What kind of hedging strategy should an investor pursue who is short the option in this situation? One answer to this question is the following: The investor establishes a self-financing hedging strategy that successfully replicates the option *with maximal probability* over all self-financing strategies that do not require more capital than he has at disposal. General results concerning this type of hedging (quantile hedging, hedging with a given probability) were given by H. Föllmer and P. Leukert [1]-[2] when the price process of the underlying asset is a semimartingale. We consider here the special case of a jump-diffusion market model firstly introduced by Aase[4] and derive the corresponding stochastic differential equations for hedging strategy, its value and the price of call-option.

---

[1]The work is supported by INTAS-99-0016

2. Description of the Model and Auxiliary results

Let $(\Omega, \mathcal{F}, \mathbf{F} = (\mathcal{F}_t)_{t\geqslant 0}, \mathrm{P})$ be a standard stochastic basis. We assume there are two risky assets (stocks) $S^i, i = 1, 2$, whose price-processes are described by the following stochastic differential equations

$$(1) \qquad dS_t^i = S_{t-}^i(\mu^i dt + \sigma^i dW_t - \nu^i d\Pi_t), i = 1, 2,$$

where W is a standard Wiener process, Π is a Poisson process with positive intensity λ. Suppose also that W and Π are independent and the filtration $\mathbf{F}$ is generated by W and Π, $\mu^i \in \mathbb{R}, \sigma^i > 0, \nu^i < 1$.

There is a non-risky asset B (bond or bank account) which satisfies the equation

$$(2) \qquad dB_t = rB_t dt, B_0 = 1, r \in \mathbb{R}$$

Every predictable process $\pi = (\pi_t)_{t\geqslant 0} = ((\beta_t, \gamma_t^1, \gamma_t^2))_{t\geqslant 0}$ can be regarded as *trading strategy* or portfolio. The value of such a portfolio is given by

$$(3) \qquad X_t^\pi = \beta_t B_t + \gamma_t^1 S_t^1 + \gamma_t^2 S_t^2.$$

A strategy with non-negative value is called *admissible*. If the discounted value $\frac{X^\pi}{B}$ of such a strategy π can be represented in the form

$$(4) \qquad \frac{X_t^\pi}{B_t} = \frac{X_0^\pi}{B_0} + \int_0^t \sum_{i=1}^2 \gamma_u^i d\left(\frac{S_u^i}{B_u}\right) (\mathrm{P} - \text{a.s.}),$$

then π is called by *self-financing* ($\pi \in \mathbb{SF}$).

The market (1)-(2) with the class $\mathbb{SF}$ is *complete* if the following conditions are fulfiled

$$(5) \qquad \sigma^2\nu^1 - \sigma^1\nu^2 \neq 0, \frac{(\mu^1 - r)\sigma^2 - (\mu^2 - r)\sigma^1}{\sigma^2\nu^1 - \sigma^1\nu^2} > 0.$$

Under condition (5) there exists a unique equivalent martingale measure P^* with local density

$$(6) \quad Z_t = \frac{d\mathrm{P}_t^*}{d\mathrm{P}_t} = \frac{d\mathrm{P}^*}{d\mathrm{P}}\bigg| \mathcal{F}_t = exp(\alpha^* W_t - \frac{\alpha^{*2}}{2}t + (\lambda - \lambda^*)t + (ln\lambda^* - ln\lambda)\Pi_t),$$

where the pair (α^*, λ^*) is given by the unique solution of the equation (see, for instance Melnikov and Shiryaev [8], Volkov and Kramkov [8])

$$(7) \qquad \begin{cases} \mu^1 - r = & -\sigma^1\alpha^* + \nu^1\lambda^* \\ \mu^2 - r = & -\sigma^2\alpha^* + \nu^2\lambda^* \end{cases}, \lambda^* > 0.$$

Under the measure P^*, $W_t^* = W_t - \alpha^* t$ is a Wiener process, Π is a Poisson process with intensity $\lambda^* > 0$, and W^* is independent of Π.

A non-negative $\mathcal{F}_T$-measurable function f_T is called contingent claim. For a *perfect* hedge, we have to find a self-financing strategy π that eliminates the risk completely (i.e. $\mathrm{P}\{X_T^\pi \geqslant f_T\} = 1$) and requires minimal initial capital $X_0^\pi = \mathbb{C}(T, S_0^1)$.

We consider classical options of the form $f_T = f(S^1_T)$. According to the general theory of perfect hedging (see [3]-[7]) the fair price of the option is given by

$$\mathbb{C}(T, S^1_0) = \mathbf{E}^* f_T e^{-rT}, \tag{8}$$

where $\mathbf{E}^*$ denotes expectation w.r. to P^*.

Let us note by the Ito formula that

$$\begin{aligned} S^i_t &= S^i_0 exp(\sigma^i W_t + (\mu^i - \tfrac{1}{2}(\sigma^i)^2)t)(1-\nu^i)^{\Pi_t} \\ &= S^i_0 exp(\sigma^i W^*_t + (\mu^i + \sigma^i\alpha^* - \tfrac{1}{2}(\sigma^i)^2)t)(1-\nu^i)^{\Pi_t} \\ &= S^i_0 exp(\sigma^i W^*_t + (r + \nu^i\lambda^* - \tfrac{1}{2}(\sigma^i)^2)t)(1-\nu^i)^{\Pi_t}, \end{aligned} \tag{9}$$

$$Y^i_t = \frac{S^i_t}{B_t} = Y^i_0 exp(\sigma W^*_t + (\nu^i\lambda^* - \frac{1}{2}(\sigma^i)^2)t)(1-\nu^i)^{\Pi_t}$$

Using (8)-(9) and the independence of W^* and Π yields

$$\mathbb{C}(T, S^1_0) = \sum_{n=0}^{\infty} \mathbf{E}^* \left[f\left(S^1_0 e^{\nu^1\lambda^* T} e^{(\sigma^1 W^*_T + (r - \frac{1}{2}(\sigma^1)^2)T)} (1-\nu^1)^n \right) e^{-rT} \right] e^{-\lambda^* T} \frac{(\lambda^* T)^n}{n!} \tag{10}$$

In the case of a call option we obtain from (10) formula (see [4],[7]):

$$\mathbb{C}(T, S^1_0) = e^{-\lambda^* T} \sum_{n=0}^{\infty} \mathbb{C}^{BS}(S^1_0(1-\nu^1)^n e^{\nu^1\lambda^* T}, K, T, r, \sigma^1) \frac{(\lambda^* T)^n}{n!} \tag{11}$$

where $\mathbb{C}^{BS}$ is the *Black-Scholes price*

$$\mathbb{C}^{BS}(S_0, K, T, r, \sigma) = S_0\Phi(y_+(T)) - Ke^{-rT}\Phi(y_-(T))$$

$$y_\pm(T-t) = \frac{ln\frac{S_t}{K} + (T-t)(r \pm \frac{\sigma^2}{2})}{\sigma\sqrt{T-t}}$$

and $\Phi(x) = \frac{1}{\sqrt{2\pi}} \int_{-\infty}^{x} e^{-\frac{y^2}{2}} dy$.

The value of the corresponding hedging strategy π at time t is given by

$$X^\pi_t = \mathbb{C}(T-t, S^1_t) =$$
$$e^{-\lambda^*(T-t)} \sum_{n=0}^{\infty} \mathbb{C}^{BS}(S^1_t(1-\nu^1)^n e^{\nu^1\lambda^*(T-t)}, K, T-t, r, \sigma^1) \frac{(\lambda^*(T-t))^n}{n!}.$$

The components γ^1_t, γ^2_t of this hedging strategy are solutions of the equations

$$\begin{cases} \gamma^1_t \sigma^1 S^1_{t-} + \gamma^2_t \sigma^2 S^2_{t-} = S^1_{t-}\sigma^1 \frac{\partial}{\partial x}\mathbb{C}(S^1_{t-}, t) \\ \gamma^1_t \nu^1 S^1_{t-} + \gamma^2_t \nu^2 S^2_{t-} = \mathbb{C}(S^1_{t-}, t) - \mathbb{C}(S^1_{t-}(1-\nu^1), t). \end{cases} \tag{17}$$

We note that the equation (17) has a unique solution in view (5).

The first component of the hedge can be recognized from the *balance equation*

(18) $$\beta_t = \frac{\mathbb{C}(S^1_{t-}, t) - \gamma^1_t S^1_{t-} - \gamma^2_t S^2_{t-}}{B_t}$$

The value $X^\pi_t = \mathbb{C}(S^1_t, t)$ of such a strategy satisfies to the following equation

(19)
$$\left[\mathbb{C}(S^1_{t-}(1-\nu^1), t) - \mathbb{C}(S^1_{t-}, t)\right]\lambda^* + \frac{\partial}{\partial t}\mathbb{C}(S^1_{t-}, t) + \frac{1}{2}(\sigma^1 S^1_{t-})^2 \frac{\partial^2}{\partial x^2}\mathbb{C}(S^1_{t-}, t)$$
$$+ rS^1_{t-}\frac{\partial}{\partial x}\mathbb{C}(S^1_{t-}, t) - r\mathbb{C}(S^1_{t-}, t) + \frac{\partial}{\partial x}\mathbb{C}(S^1_{t-}, t)\nu^1\lambda^* S^1_{t-} = 0.$$

We now introduce a constraint which is a crucial motivation for *quantile hedging*: the initial capital X_0 of the investor is less than $\mathbb{C}(T, S^1_0)$. Faced with the impossibility to hedge the option with certainty the investor chooses a strategy $\pi \in \mathbb{SF}$ that maximizes the success-probability $\mathrm{P}\{X^\pi_T \geqslant f_T\}$ over all self-financing strategies with initial value $X^\pi_0 \leqslant X_0 < \mathbb{C}(T, S^1_0)$. We paraphrase this as the following problem

(20) $$1 - \varepsilon = \sup_\pi P(X^\pi_T \geqslant f_T), X^\pi_0 \leqslant X_0 < \mathbb{C}(T, S^1_0)$$

where the "optimal" $\varepsilon \in [0,1]$ and π should be found.

The problem (20) was considered by H. Föllmer and P. Leukert [1]-[2] in a "semimartingale" $(1, S)$ - setting. They derived the following general methodology to solve the problem (20):

If $\tilde{\pi}$ is the minimal hedging strategy for the contingent claim $\tilde{\phi}_T f_T$, where

$$\tilde{\phi}_T = I_{\{\frac{dP}{dP^*} > \tilde{a} f_T\}} + \gamma I_{\{\frac{dP}{dP^*} = \tilde{a} f_T\}},$$

$$\gamma = \frac{X_0 - \mathbf{E}^*\left[e^{-rT} f_T I_{\{\frac{dP}{dP^*} > \tilde{a} f_T\}}\right]}{\mathbf{E}^*\left[e^{-rT} f_T I_{\{\frac{dP}{dP^*} = \tilde{a} f_T\}}\right]},$$

$$\tilde{a} = \inf\{a : \mathbf{E}^*\left[e^{-rT} f_T I_{\{\frac{dP}{dP^*} > a f_T\}}\right] \leqslant X_0\},$$

then $\tilde{\pi}$ is the solution to the quantile hedging problem (20) and

$$1 - \varepsilon = \mathbf{E}(\tilde{\phi}_T).$$

Remark 1. *The critical value $\tilde{a}$ can be determined by condition*

(21) $$\mathbf{E}^*\left[e^{-rT}\tilde{\phi}_T f_T\right] = X_0.$$

If $P\left[\frac{dP}{dP^*} = const \cdot f_T\right] = 0$ holds then we obtain $\tilde{\phi}_T = I_{\{\frac{dP}{dP^*} > \tilde{a} f_T\}}$.

3. Main results

We consider the quantile hedging problem (20) in the case of the model (1)-(2), applying the methodology described above. The specification of the jump-diffusion model by (1)-(2) allows us to give the explicit solution to problem (20).

Theorem 1. *Assume that* $f_T = f(S_T^1)$. *Then the value process* $\mathbb{C}(S_t^1, S_t^2, t) := X_t^\pi$ *and the components* $\beta, \gamma^1, \gamma^2$ *of the optimal hedge for the problem (20) satisfy the equations*

(22)

$$\left[\mathbb{C}(S_{t-}^1(1-\nu^1), S_{t-}^2(1-\nu^2), t) - \mathbb{C}(S_{t-}^1, S_{t-}^2, t)\right]\lambda^* + rS_{t-}^1 \tfrac{\partial}{\partial x}\mathbb{C}(S_{t-}^1, S_{t-}^2, t)$$
$$+ rS_{t-}^2 \tfrac{\partial}{\partial y}\mathbb{C}(S_{t-}^1, S_{t-}^2, t) + \tfrac{\partial}{\partial t}\mathbb{C}(S_{t-}^1, S_{t-}^2, t) + \tfrac{1}{2}(\sigma^1 S_{t-}^1)^2 \tfrac{\partial^2}{\partial x^2}\mathbb{C}(S_{t-}^1, S_{t-}^2, t)$$
$$+ \tfrac{1}{2}(\sigma^2 S_{t-}^2)^2 \tfrac{\partial^2}{\partial y^2}\mathbb{C}(S_{t-}^1, S_{t-}^2, t) + (\sigma^1\sigma^2 S_{t-}^1 S_{t-}^2)\tfrac{\partial^2}{\partial x\partial y}\mathbb{C}(S_{t-}^1, S_{t-}^2, t)$$
$$- r\mathbb{C}(S_{t-}^1, S_{t-}^2, t) + \tfrac{\partial}{\partial x}\mathbb{C}(S_{t-}^1, S_{t-}^2, t)\nu^1\lambda^* S_{t-}^1 + \tfrac{\partial}{\partial y}\mathbb{C}(S_{t-}^1, S_{t-}^2, t)\nu^2\lambda^* S_{t-}^2 = 0,$$

$$(23)\quad \begin{cases} \gamma_t^1\sigma^1 S_{t-}^1 + \gamma_t^2\sigma^2 S_{t-}^2 = S_{t-}^1\sigma^1 \tfrac{\partial}{\partial x}\mathbb{C}(S_{t-}^1, S_{t-}^2, t) + S_{t-}^2\sigma^2 \tfrac{\partial}{\partial y}\mathbb{C}(S_{t-}^1, S_{t-}^2, t) \\ \gamma_t^1\nu^1 S_{t-}^1 + \gamma_t^2\nu^2 S_{t-}^2 = \mathbb{C}(S_{t-}^1, S_{t-}^2, t) - \mathbb{C}(S_{t-}^1(1-\nu^1), S_{t-}^2(1-\nu^2), t), \end{cases}$$

$$(24)\qquad \beta_t = \frac{\mathbb{C}(S_{t-}^1, S_{t-}^2, t) - \gamma_t^1 S_{t-}^1 - \gamma_t^2 S_{t-}^2}{B_t}.$$

Note that the value $\mathbb{C}$ of the quantile hedging strategy depends on S_t^1 and S_t^2 whereas the value of the perfect hedging strategy depends on S_t^1 only. Nevertheless we use here the same letter for the value.

Consider $f = (S_T^1 - K)^+$. We have to distinguish two cases:
a) $-\frac{\alpha^*}{\sigma^1} \leqslant 1$ and b) $-\frac{\alpha^*}{\sigma^1} > 1$

The following theorem gives the full answer to problem (20).

Theorem 2.

Case a.

1) The maximal probability for "success hedging" equals

$$(25)\qquad 1 - \varepsilon = e^{-\lambda T}\sum_{n=0}^{\infty}\Phi\left(\frac{c_n'(\tilde{a}) + \alpha^* T}{\sqrt{T}}\right)\frac{(\lambda T)^n}{n!}$$

where $\tilde{a}$ *is the solution of the equation*

(26)

$$X_0 = e^{-\lambda^* T}\sum_{n=0}^{\infty}\Big[\mathbb{C}^{BS}(S_0^1(1-\nu^1)^n e^{\nu^1\lambda^* T}, K, T, r, \sigma^1)$$
$$-\mathbb{C}^{BS}(S_0^1(1-\nu^1)^n e^{\nu^1\lambda^* T}, c_n(a), T, r, \sigma^1) - (c_n(a) - K)e^{-rT}\Phi\left(\tfrac{-c_n'(a)}{\sqrt{T}}\right)\Big]\frac{(\lambda^* T)^n}{n!},$$

$c_n(a)$ is the unique root of $x^{-\frac{\alpha^*}{\sigma^1}} = g \cdot b^n a e^{-rT}(x-K)$,

$$c_n'(a) = \frac{1}{\sigma^1}(\ln(\frac{c_n(a)}{(1-\nu^1)^n S_0^1}) - (r+\nu^1\lambda^* - \frac{1}{2}(\sigma^1)^2)T), \tag{27}$$

and

$$g = \frac{1}{S_0^{1\,\frac{\alpha^*}{\sigma^1}}} \exp\left(-\frac{\alpha^*\mu^1}{\sigma^1}T + \frac{\sigma^1\alpha^*}{2}T - \frac{\alpha^{*2}}{2}T + (\lambda-\lambda^*)T)\right), \tag{28}$$

$$b = \frac{\lambda^*}{\lambda(1-\nu^1)^{\frac{\alpha^*}{\sigma^1}}}. \tag{29}$$

2) The value of the optimal strategy is given by

$$\begin{aligned} X_t^\pi = e^{-\lambda^*(T-t)} \textstyle\sum_{n=0}^\infty \Big[& \mathbb{C}^{BS}(S_t^1(1-\nu^1)^n e^{\nu^1\lambda^*(T-t)}, K, (T-t), r, \sigma^1) \\ & - \mathbb{C}^{BS}(S_t^1(1-\nu^1)^n e^{\nu^1\lambda^*(T-t)}, \tilde{c_n}, (T-t), r, \sigma^1) \\ & - (\tilde{c_n} - K)e^{-r(T-t)} \cdot \Phi(-\tfrac{\tilde{c}_n'}{\sqrt{T-t}})\Big] \tfrac{(\lambda^*(T-t))^n}{n!}, \end{aligned} \tag{30}$$

where $\tilde{c_n}$ *is the root of the equation:*

$$x^{-\frac{\alpha^*}{\sigma^1}} = g \cdot b^n b^{\Pi(S_t^1, S_t^2)} \tilde{a} e^{-rT}(x-K)^+, \tag{31}$$

$$\tilde{c}_n' = \frac{1}{\sigma^1}(\ln(\frac{\tilde{c_n}}{(1-\nu^1)^n S_t^1}) - (r+\nu^1\lambda^* - \frac{1}{2}(\sigma^1)^2)(T-t)), \tag{32}$$

$$\Pi(S_t^1, S_t^2) = \frac{\frac{1}{\sigma^1}\ln\frac{S_t^1}{S_0^1} - (r+\nu^1\lambda^* - \frac{(\sigma^1)^2}{2})\frac{t}{\sigma^1} - \frac{1}{\sigma^2}\ln\frac{S_t^2}{S_0^2} + (r+\nu^2\lambda^* - \frac{(\sigma^2)^2}{2})\frac{t}{\sigma^2}}{\frac{1}{\sigma^1}\ln(1-\nu^1) - \frac{1}{\sigma^2}\ln(1-\nu^2)}. \tag{33}$$

Case b.

1) The maximal probability for "success hedging" equals

$$1-\varepsilon = e^{-\lambda T}\sum_{n=0}^{\infty}\left[\Phi(\frac{{c^1}_n'(\tilde{a}) + \alpha^* T}{\sqrt{T}}) + \Phi(-\frac{{c^2}_n'(\tilde{a}) + \alpha^* T}{\sqrt{T}})\right]\frac{(\lambda T)^n}{n!}, \tag{34}$$

where $\tilde{a}$ *is the solution of*

$$\begin{aligned} X_0 = e^{-\lambda^* T}\textstyle\sum_{n=0}^\infty \Big[& \mathbb{C}^{BS}(S_0^1(1-\nu^1)^n e^{\nu^1\lambda^* T}, K, T, r, \sigma^1) \\ & - \mathbb{C}^{BS}(S_0^1(1-\nu^1)^n e^{\nu^1\lambda^* T}, c_n^1(a), T, r, \sigma^1) \\ & + \mathbb{C}^{BS}(S_0^1(1-\nu^1)^n e^{\nu^1\lambda^* T}, c_n^2(a), T, r, \sigma^1) \\ & -(c_n^1(a) - K)e^{-rT} \cdot \Phi(\tfrac{-{c^1}_n'(a)}{\sqrt{T}}) + (c_n^2(a) - K)e^{-rT} \cdot \Phi(\tfrac{-{c^2}_n'(a)}{\sqrt{T}})\Big] \tfrac{(\lambda^* T)^n}{n!}, \end{aligned} \tag{35}$$

$c_n^1(a), c_n^2(a)$ are the roots of the equation $x^{-\frac{\alpha^*}{\sigma^1}} = g \cdot b^n a e^{-rT}(x-K)$,

(37) $$c'^i_n(a) = \frac{1}{\sigma^1}(\ln(\frac{c_n^i(a)}{(1-\nu^1)^n S_0^1}) - (r + \nu^1\lambda^* - \frac{1}{2}(\sigma^1)^2)T), i = 1,2$$

g , b were defined by (28)-(29).

2) The value of the optimal strategy is given by

(38)

$$X_t^\pi = e^{-\lambda^*(T-t)} \sum_{n=0}^\infty \Big[\mathbb{C}^{BS}(S_t^1(1-\nu^1)^n e^{\nu^1\lambda^*(T-t)}, K, (T-t), r, \sigma^1)$$

$$- \mathbb{C}^{BS}(S_t^1(1-\nu^1)^n e^{\nu^1\lambda^*(T-t)}, \tilde{c}_n^1, (T-t), r, \sigma^1)$$

$$+ \mathbb{C}^{BS}(S_t^1(1-\nu^1)^n e^{\nu^1\lambda^*(T-t)}, \tilde{c}_n^2, (T-t), r, \sigma^1)$$

$$- (\tilde{c}_n^1 - K)e^{-r(T-t)} \cdot \Phi(-\frac{\tilde{c}'^1_n}{\sqrt{T-t}}) + (\tilde{c}_n^2 - K)e^{-r(T-t)} \cdot \Phi(-\frac{\tilde{c}'^2_n}{\sqrt{T-t}})\Big] \frac{(\lambda^*(T-t))^n}{n!},$$

where $\tilde{c}_n^1, \tilde{c}_n^2$ are roots of the equation:

(39) $$x^{-\frac{\alpha^*}{\sigma^1}} = g \cdot b^n b^{\Pi(S_t^1, S_t^2)} \tilde{a} e^{-rT}(x-K)^+,$$

(40) $$\tilde{c}'^i_n = \frac{1}{\sigma^1}(\ln(\frac{\tilde{c}_n^i}{(1-\nu^1)^n S_t^1}) - (r + \nu^1\lambda^* - \frac{1}{2}(\sigma^1)^2)(T-t)), i = 1,2,$$

$\Pi(S_t^1, S_t^2)$, g, b were defined by (33), (28), (29), $\tilde{a}$ was defined by (35).

Let us note that the *boundary condition* to equation (22) in case of a call option $f_T = (S_T^1 - K)^+$ is given by

(41) $$\mathbb{C}(S_T^1, S_T^2, T) = (S_T^1 - K)^+ I_{\left\{S_T^{1\,-\frac{\alpha^*}{\sigma^1}} > g \cdot b^{\Pi(S_T^1, S_T^2)} \tilde{a} e^{-rT}(S_T^1 - K)^+\right\}},$$

where g, b, $\Pi(S_T^1, S_T^2)$, $\tilde{a}$ were defined by (28), (29), (33) and (26) (resp. (35)).

The proofs of these theorems are given in the Appendix.

Appendix.

We prove Theorem 2 first.

Taking into account that in the jump-diffusion model (1)-(2) the condition $P\left[\frac{dP}{dP^*} = const \cdot f\right] = 0$ is satisfied we obtain from Remark1:

$$(A.1)\qquad \tilde{\phi_T} = I_{\{\frac{dP}{dP^*} > \tilde{a} f\}}.$$

Let us paraphrase Z_T in terms S_T^1:

$$(A.2)\qquad \begin{aligned} \frac{dP^*}{dP} &= \exp(\alpha^* W_T - \frac{\alpha^{*2}}{2}T + (\lambda - \lambda^*)T + (ln\lambda^* - ln\lambda)\Pi_T) \\ &= \left(S_0^1 \exp\left\{\sigma^1 W_T + (\mu^1 - \frac{\sigma^{12}}{2})T\right\}(1-\nu^1)^{\Pi_T}\right)^{\frac{\alpha^*}{\sigma^1}} \times \\ &\times \frac{1}{S_0^{1\,\frac{\alpha^*}{\sigma^1}}} \exp\left(-\frac{\alpha^*\mu^1}{\sigma^1}T + \frac{\sigma^1\alpha^*}{2}T - \frac{\alpha^{*2}}{2}T + (\lambda - \lambda^*)T)\right) \times \\ &\times \left(\frac{\lambda^*}{\lambda(1-\nu^1)^{\frac{\alpha^*}{\sigma^1}}}\right)^{\Pi_T} \\ &= g \cdot (S_T^1)^{\frac{\alpha^*}{\sigma^1}} \cdot b^{\Pi_T}, \end{aligned}$$

where $g = \frac{1}{S_0^{1\,\frac{\alpha^*}{\sigma^1}}} \exp\left(-\frac{\alpha^*\mu^1}{\sigma^1}T + \frac{\sigma^1\alpha^*}{2}T - \frac{\alpha^{*2}}{2}T + (\lambda - \lambda^*)T)\right)$, $b = \frac{\lambda^*}{\lambda(1-\nu^1)^{\frac{\alpha^*}{\sigma^1}}}$.

Using (A.1)-(A.2) we can represent $\tilde{\phi}$ in the form

$$(A.3)\qquad \tilde{\phi} = I_{\{\frac{dP}{dP^*} > \tilde{a} f\}} = I_{\left\{S_T^{1\,-\frac{\alpha^*}{\sigma^1}} > g \cdot b^{\Pi_T} \tilde{a} e^{-rT}(S_T - K)^+\right\}}.$$

We show how to compute $\tilde{a}$ by means of the condition $\mathbf{E}^*\left[e^{-rT}\phi f\right] = X_0$. It follows from (A.3) that

(A.4)

$$\begin{aligned} \mathbf{E}^*\left[e^{-rT}\phi f\right] &= \mathbf{E}^*\left[e^{-rT}(S_T^1 - K)^+ I_{\left\{S_T^{1\,-\frac{\alpha^*}{\sigma^1}} > g \cdot b^{\Pi_T} \tilde{a} e^{-rT}(S_T - K)^+\right\}}\right] \\ &= \sum_{n=0}^{\infty} \mathbf{E}^*\left[e^{-rT}(S_T^1 - K)^+ I_{\left\{S_T^{1\,-\frac{\alpha^*}{\sigma^1}} > g \cdot b^n \tilde{a} e^{-rT}(S_T - K)^+\right\}}\right] \cdot \frac{(\lambda^* T)^n}{n!} e^{-\lambda^* T}. \end{aligned}$$

Case a, $-\frac{\alpha^*}{\sigma^1} \leqslant 1$:
In this case, the equation

$$x^{-\frac{\alpha^*}{\sigma^1}} = g \cdot b^n a e^{-rT}(x - K)^+$$

has a unique root $c_n(a)$. Thus the inequality

$$x^{-\frac{\alpha^*}{\sigma^1}} > g \cdot b^n a e^{-rT}(x - K)^+.$$

is equivalent to $x < c_n(a)$. This implies

$$(A.5)\qquad I_{\left\{S_T^{1\,-\frac{\alpha^*}{\sigma^1}} > g \cdot b^n a e^{-rT}(S_T^1 - K)^+\right\}} = I_{\{S_T^1 < c_n(a)\}}.$$

Using (A.5) we can transform (A.4) as follows:

(A.6)

$$\mathbf{E}^*\left[e^{-rT}\phi f\right] = \sum_{n=0}^{\infty} \mathbf{E}^*\left[e^{-rT}(S_T^1-K)^+ I_{\{S_T^1<c_n(a)\}}\right]\cdot \frac{(\lambda^*T)^n}{n!}e^{-\lambda^*T}$$

$$= \sum_{n=0}^{\infty} \mathbf{E}^*\left[e^{-rT}\left((S_T^1-K)^+ - (S_T^1-c_n(a))^+ - (c_n(a)-K)I_{\{S_T^1\geqslant c_n(a)\}}\right)\right]\times$$

$$\times \frac{(\lambda^*T)^n}{n!}e^{-\lambda^*T}$$

$$= e^{-\lambda^*T}\sum_{n=0}^{\infty}\Big[\mathbb{C}^{BS}(S_0^1(1-\nu^1)^n e^{\nu^1\lambda^*T},K,T,r,\sigma^1)$$

$$-\mathbb{C}^{BS}(S_0^1(1-\nu^1)^n e^{\nu^1\lambda^*T},c_n(a),T,r,\sigma^1) - (c_n(a)-K)e^{-rT}\times$$

$$\times P^*\left(S_0^1 exp(\sigma^1W_T^* + (r+\nu^1\lambda^* - \tfrac{1}{2}(\sigma^1)^2)T)(1-\nu^1)^n \geqslant c_n(a)\right)\Big]\frac{(\lambda^*T)^n}{n!}$$

$$= e^{-\lambda^*T}\sum_{n=0}^{\infty}\Big[\mathbb{C}^{BS}(S_0^1(1-\nu^1)^n e^{\nu^1\lambda^*T},K,T,r,\sigma^1)$$

$$-\mathbb{C}^{BS}(S_0^1(1-\nu^1)^n e^{\nu^1\lambda^*T},c_n(a),T,r,\sigma^1) - (c_n(a)-K)e^{-rT}\Phi(\frac{-c_n'(a)}{\sqrt{T}})\Big]\frac{(\lambda^*T)^n}{n!}$$

where

$$c_n'(a) = \frac{1}{\sigma^1}(\ln(\frac{c_n(a)}{(1-\nu^1)^n S_0^1}) - (r+\nu^1\lambda^* - \frac{1}{2}(\sigma^1)^2)T).$$

Let $\tilde{a}$ be determined by the condition $\mathbf{E}^*\left[e^{-rT}\phi f\right] = X_0$ by means of (A.6). The maximal probability for a successful hedge is given by (cf. (A.5))

$$1-\varepsilon = P\left\{S_T^{1-\frac{\alpha^*}{\sigma^1}} > g\cdot b^{\Pi_T}\tilde{a}e^{-rT}(S_T^1-K)^+\right\}$$

$$= e^{-\lambda T}\sum_{n=0}^{\infty} P\{S_T^1 < c_n(\tilde{a})\}\frac{(\lambda T)^n}{n!} \tag{A.7}$$

$$= e^{-\lambda T}\sum_{n=0}^{\infty}\Phi(\frac{c_n'(\tilde{a})+\alpha^*T}{\sqrt{T}})\frac{(\lambda T)^n}{n!}.$$

The quantile hedging strategy is given by the perfect hedge of the modified claim $\tilde{\phi_T}f_T$. We calculate the value of this strategy:

$$X_t^{\pi} = \mathbf{E}^*\left[\tilde{\phi_T}f_T e^{-r(T-t)}|\mathcal{F}_t\right]$$

$$= \mathbf{E}^*\left[e^{-r(T-t)}(S_T^1-K)^+ I_{\left\{S_T^{1-\frac{\alpha^*}{\sigma^1}} > g\cdot b^{\Pi_T}\tilde{a}e^{-rT}(S_T^1-K)^+\right\}}|\mathcal{F}_t\right]. \tag{A.8}$$

We want to describe Π_t in terms of the observable variables $S_t^i, i=1,2$. Since

$$S_t^i = S_0^i exp(\sigma^i W_t^* + (r+\nu^i\lambda^* - \frac{1}{2}(\sigma^i)^2)t)(1-\nu^i)^{\Pi_t},$$

$$W_t^* = \frac{1}{\sigma^i}\left(\ln\frac{\frac{S_t^i}{S_0^i}}{(1-\nu^i)^{\Pi_t}} - (r+\nu^i\lambda^* - \frac{(\sigma^i)^2}{2})t\right)$$

we obtain

$$\frac{1}{\sigma^1}\left(\ln\frac{S_t^1}{S_0^1}-\Pi_t\ln(1-\nu^1)-(r+\nu^1\lambda^*-\frac{(\sigma^1)^2}{2})t\right)$$
$$=\frac{1}{\sigma^2}\left(\ln\frac{S_t^2}{S_0^2}-\Pi_t\ln(1-\nu^2)-(r+\nu^2\lambda^*-\frac{(\sigma^2)^2}{2})t\right),$$

which implies

$$(A.9)\quad \Pi_t=\frac{\frac{1}{\sigma^1}\ln\frac{S_t^1}{S_0^1}-(r+\nu^1\lambda^*-\frac{(\sigma^1)^2}{2})\frac{t}{\sigma^1}-\frac{1}{\sigma^2}\ln\frac{S_t^2}{S_0^2}+(r+\nu^2\lambda^*-\frac{(\sigma^2)^2}{2})\frac{t}{\sigma^2}}{\frac{1}{\sigma^1}\ln(1-\nu^1)-\frac{1}{\sigma^2}\ln(1-\nu^2)}.$$

Hence we arrive at $\Pi_t=\Pi(S_t^1,S_t^2)$. Taking into account (A.8) we obtain
$(A.10)$

$$X_t^\pi=\mathbf{E}^*\left[e^{-r(T-t)}(S_T^1-K)^+I_{\left\{S_T^{1\,-\frac{\alpha^*}{\sigma^1}}>g\cdot b^{\Pi_{T-t}}b^{\Pi(S_t^1,S_t^2)}\tilde{a}e^{-rT}(S_T^1-K)^+\right\}}|\mathcal{F}_t\right]$$
$$=\sum_{n=0}^\infty\mathbf{E}^*\left[e^{-r(T-t)}(S_T^1-K)^+I_{\left\{S_T^{1\,-\frac{\alpha^*}{\sigma^1}}>g\cdot b^n b^{\Pi(S_t^1,S_t^2)}\tilde{a}e^{-rT}(S_T^1-K)^+\right\}}|\mathcal{F}_t\right]\times$$
$$\times\frac{(\lambda^*(T-t))^n}{n!}e^{-\lambda^*(T-t)}.$$

As in case (A.6) the equality (A.10) can be represented in the form
$(A.11)$

$$X_t^\pi=\sum_{n=0}^\infty\mathbf{E}^*\left[e^{-r(T-t)}\left((S_T^1-K)^+-(S_T^1-\tilde{c}_n)^+-(\tilde{c}_n-K)I_{\{S_T^1\geqslant\tilde{c}_n\}}\right)|\mathcal{F}_t\right]\times$$
$$\times\frac{(\lambda^*(T-t))^n}{n!}e^{-\lambda^*(T-t)},$$

where $\tilde{c}_n$ is the unique root of

$$x^{-\frac{\alpha^*}{\sigma^1}}=g\cdot b^n b^{\Pi(S_t^1,S_t^2)}\tilde{a}e^{-rT}(x-K)^+.$$

Equation (A.11) implies the final formula
$(A.12)$

$$X_t^\pi=e^{-\lambda^*(T-t)}\sum_{n=0}^\infty\Big[\mathbb{C}^{BS}(S_t^1(1-\nu^1)^n e^{\nu^1\lambda^*(T-t)},K,(T-t),r,\sigma^1)$$
$$-\mathbb{C}^{BS}(S_t^1(1-\nu^1)^n e^{\nu^1\lambda^*(T-t)},\tilde{c}_n,(T-t),r,\sigma^1)-(\tilde{c}_n-K)e^{-r(T-t)}\times$$
$$\times P^*(S_t^1\exp(\sigma^1W^*_{(T-t)}+(r+\nu^1\lambda^*-\tfrac{1}{2}(\sigma^1)^2)(T-t))(1-\nu^1)^n\geqslant\tilde{c}_n|\mathcal{F}_t)\Big]\times$$
$$\times\frac{(\lambda^*(T-t))^n}{n!}$$
$$=e^{-\lambda^*(T-t)}\sum_{n=0}^\infty\Big[\mathbb{C}^{BS}(S_t^1(1-\nu^1)^n e^{\nu^1\lambda^*(T-t)},K,(T-t),r,\sigma^1)$$
$$-\mathbb{C}^{BS}(S_t^1(1-\nu^1)^n e^{\nu^1\lambda^*(T-t)},\tilde{c}_n,(T-t),r,\sigma^1)$$
$$-(\tilde{c}_n-K)e^{-r(T-t)}\cdot\Phi(-\frac{\tilde{c}_n'}{\sqrt{T-t}})\Big]\frac{(\lambda^*(T-t))^n}{n!},$$

where

$$\tilde{c}'_n = \frac{1}{\sigma^1}(\ln(\frac{\tilde{c}_n}{(1-\nu^1)^n S^1_t}) - (r + \nu^1\lambda^* - \frac{1}{2}(\sigma^1)^2)(T-t)).$$

Case b, $-\frac{\alpha^*}{\sigma^1} > 1$.

Now the equation

$$x^{-\frac{\alpha^*}{\sigma^1}} = g \cdot b^n a e^{-rT}(x-K)^+$$

has two roots. Thus the inequality

$$x^{-\frac{\alpha^*}{\sigma^1}} > g \cdot b^n a e^{-rT}(x-K)^+$$

is equivalent to $x < c^1_n(a)$ or $x > c^2_n(a)$, where $K < c^1_n(a) \leqslant c^2_n(a)$ are the solutions of the equation $x^{-\frac{\alpha^*}{\sigma^1}} = g \cdot b^n a e^{-rT}(x-K)$. Hence

$$(A.13) \qquad I_{\left\{S_T^{1\,-\frac{\alpha^*}{\sigma^1}} > g\cdot b^n a e^{-rT}(S^1_T - K)^+\right\}} = I_{\{S^1_T < c^1_n(a)\}} + I_{\{S^1_T > c^2_n(a)\}}.$$

Equations (A.6), (A.7), (A.13) imply

(A.14)

$$\mathbf{E}^*\left[e^{-rT}\phi f\right] =$$

$$\sum_{n=0}^{\infty} \mathbf{E}^*\left[e^{-rT}(S^1_T - K)^+ (I_{\{S^1_T < c^1_n(a)\}} + I_{\{S^1_T > c^2_n(a)\}}\right] \frac{(\lambda^* T)^n}{n!} e^{-\lambda^* T}$$

$$= \sum_{n=0}^{\infty} \mathbf{E}^*\left[e^{-rT}\left((S^1_T - K)^+ - (S^1_T - c^1_n(a))^+ + (S^1_T - c^2_n(a))^+\right.\right.$$

$$\left.\left. -(c^1_n(a) - K) I_{\{S^1_T \geqslant c^1_n(a)\}} + (c^2_n(a) - K) I_{\{S^1_T > c^2_n(a)\}}\right)\right] \cdot \frac{(\lambda^* T)^n}{n!} e^{-\lambda^* T}$$

$$= e^{-\lambda^* T} \sum_{n=0}^{\infty} \left[\mathbb{C}^{BS}(S^1_0(1-\nu^1)^n e^{\nu^1\lambda^* T}, K, T, r, \sigma^1)\right.$$

$$- \mathbb{C}^{BS}(S^1_0(1-\nu^1)^n e^{\nu^1\lambda^* T}, c^1_n(a), T, r, \sigma^1)$$

$$+ \mathbb{C}^{BS}(S^1_0(1-\nu^1)^n e^{\nu^1\lambda^* T}, c^2_n(a), T, r, \sigma^1)$$

$$\left. -(c^1_n(a) - K)e^{-rT} \cdot \Phi(\frac{-c^{1'}_n(a)}{\sqrt{T}}) + (c^2_n(a) - K)e^{-rT} \cdot \Phi(\frac{-c^{2'}_n(a)}{\sqrt{T}})\right] \frac{(\lambda^* T)^n}{n!},$$

where

$$c^{i'}_n(a) = \frac{1}{\sigma^1}(\ln(\frac{c^i_n(a)}{(1-\nu^1)^n S^1_0}) - (r + \nu^1\lambda^* - \frac{1}{2}(\sigma^1)^2)T).$$

Now we can determine $\tilde{\phi}$ because $\tilde{a}$ can be calculated by the condition $\mathbf{E}^*\left[e^{-rT}\phi f\right] = X_0$ and (A.14).

The maximal probability for a successful hedge is given by

$$(A.15) \qquad 1 - \varepsilon = e^{-\lambda T} \sum_{n=0}^{\infty} \left[\Phi(\frac{c^{1'}_n(\tilde{a}) + \alpha^* T}{\sqrt{T}}) + \Phi(-\frac{c^{2'}_n(\tilde{a}) + \alpha^* T}{\sqrt{T}})\right] \frac{(\lambda T)^n}{n!}.$$

As in (A.10)-(A.12) the value of the quantile hedging strategy is given by

$$\sum_{n=0}^{\infty} \mathbf{E}^* \Big[e^{-r(T-t)} \Big((S_T^1 - K)^+ - (S_T^1 - \tilde{c}_n^1)^+ + (S_T^1 - \tilde{c}_n^2)^+ -(\tilde{c}_n^1 - K)I_{\{S_T^1 \geqslant \tilde{c}_n^1\}} + (\tilde{c}_n^2 - K)I_{\{S_T^1 > \tilde{c}_n^2\}}\Big) |\mathcal{F}_t\Big] \cdot \frac{(\lambda^*(T-t))^n}{n!} e^{-\lambda^*(T-t)},$$

where $\tilde{c}_n^i$ are the roots of the equation

$$x^{-\frac{\alpha^*}{\sigma^1}} = g \cdot b^n b^{\Pi(S_t^1, S_t^2)} \tilde{a} e^{-rT}(x-K)^+.$$

This yields

(A.16)

$$X_t^\pi = e^{-\lambda^*(T-t)} \sum_{n=0}^{\infty} \Big[\mathbb{C}^{BS}(S_t^1(1-\nu^1)^n e^{\nu^1\lambda^*(T-t)}, K, (T-t), r, \sigma^1)$$
$$- \mathbb{C}^{BS}(S_t^1(1-\nu^1)^n e^{\nu^1\lambda^*(T-t)}, \tilde{c}_n^1, (T-t), r, \sigma^1)$$
$$+ \mathbb{C}^{BS}(S_t^1(1-\nu^1)^n e^{\nu^1\lambda^*(T-t)}, \tilde{c}_n^2, (T-t), r, \sigma^1)$$
$$- (\tilde{c}_n^1 - K)e^{-r(T-t)} \cdot \Phi(-\frac{\tilde{c^1}'_n}{\sqrt{T-t}}) + (\tilde{c}_n^2 - K)e^{-r(T-t)} \cdot \Phi(-\frac{\tilde{c^2}'_n}{\sqrt{T-t}})\Big] \frac{(\lambda^*(T-t))^n}{n!},$$

where

$$\tilde{c^i}'_n = \frac{1}{\sigma^1}(\ln(\frac{\tilde{c}_n^i}{(1-\nu^1)^n S_t^1}) - (r + \nu^1\lambda^* - \frac{1}{2}(\sigma^1)^2)(T-t)).$$

The value given by (A.12) (resp. (A.16)) depends on S_t^1 and S_t^2. We can apply Theorem 1 to determine the components $\beta, \gamma^1, \gamma^2$ of the hedging strategy (see (23)-(24)).

We now prove Theorem 1.

By equations ((A.10)-(A.12)) the value of the quantile hedging strategy for $f_T = f(S_T^1)$ is given by

(A.17)

$$X_t^\pi = \mathbf{E}^* \left[e^{-r(T-t)} f(S_T^1) I_{\left\{S_T^{1\,-\frac{\alpha^*}{\sigma^1}} > g \cdot b^{\Pi_{T-t}} b^{\Pi(S_t^1, S_t^2)} \tilde{a} e^{-rT} f(S_T^1)\right\}} |\mathcal{F}_t\right] =$$
$$\sum_{n=0}^{\infty} \mathbf{E}^* \left[e^{-r(T-t)} f(S_T^1) I_{\left\{S_T^{1\,-\frac{\alpha^*}{\sigma^1}} > g \cdot b^n b^{\Pi(S_t^1, S_t^2)} \tilde{a} e^{-rT} f(S_T^1)\right\}} |\mathcal{F}_t\right] \frac{(\lambda^*(T-t))^n e^{-\lambda^*(T-t)}}{n!}$$
$$= \sum_{n=0}^{\infty} e^{-r(T-t)} \frac{(\lambda^*(T-t))^n}{n!} e^{-\lambda^*(T-t)} \times$$
$$\times \Bigg[\int_{-\infty}^{\infty} \Bigg(\phi_{T-t}(x) f(S_t^1 \exp(\sigma^1 x + (r + \nu^1\lambda^* - \tfrac{1}{2}(\sigma^1)^2)(T-t))(1-\nu^1)^n) \times$$
$$\times I\Big\{\left[S_t^1 \exp(\sigma^1 x + (r + \nu^1\lambda^* - \tfrac{1}{2}(\sigma^1)^2)(T-t))(1-\nu^1)^n\right]^{-\frac{\alpha^*}{\sigma^1}} >$$
$$> g b^n b^{\Pi(S_t^1, S_t^2)} \tilde{a} e^{-rT} f(S_t^1 e^{(\sigma^1 x + (r + \nu^1\lambda^* - \frac{1}{2}(\sigma^1)^2)(T-t))}(1-\nu^1)^n)\Big\}\Bigg) dx\Bigg]$$

where $\tilde{a}$, b, g, $\Pi(S_t^1, S_t^2)$ were defined in the proof of Theorem 2 and

$$\phi_{T-t}(x) = \frac{1}{\sqrt{2\pi(T-t)}} e^{-\frac{x^2}{2(T-t)}}.$$

Thus the value of the optimal strategy is a function of S_t^1, S_t^2 and t.

We are going to derive the components of the hedge and the partial differential equation (22) from the representation

$$(A.18) \qquad \frac{X_t^\pi}{B_t} = \frac{X_0^\pi}{B_0} + \int_0^t \sum_{i=1}^2 \gamma_u^i d(\frac{S_u^i}{B_u}).$$

Let us denote

$$\mathbb{C}(S_t^1, S_t^2, t) := X_t^\pi$$

and rewrite (A.18) in the form

$$(A.19) \qquad \frac{\mathbb{C}(S_t^1, S_t^2, t)}{B_t} = \frac{\mathbb{C}(S_0^1, S_0^2, 0)}{B_0} + \int_0^t \gamma_u^1 d\left(\frac{S_u^1}{B_u}\right) + \int_0^t \gamma_u^2 d\left(\frac{S_u^2}{B_u}\right).$$

The discounted price processes $Y_t^i = \frac{S_t^i}{B_t}, i = 1, 2$ satisfy the equations

$$(A.20) \qquad dY_t^i = Y_{t-}^i(\sigma^i dW_t^* - \nu^i d(\Pi_t - \lambda^* t)), i = 1, 2.$$

Appling (A.19) and (A.20) we get
(A.21)

$$\frac{\mathbb{C}(S_t^1, S_t^2, t)}{B_t} =$$
$$= \frac{\mathbb{C}(S_0^1, 0)}{B_0} + \int_0^t \frac{\gamma^1\sigma^1 S_{u-}^1 + \gamma^2\sigma^2 S_{u-}^2}{B_u} dW_u^* - \int_0^t \frac{\gamma^1\nu^1 S_{u-}^1 + \gamma^2\nu^2 S_{u-}^2}{B_u} d(\Pi_u - \lambda^* u).$$

The Ito formula yields

$$(A.22) \qquad d\frac{\mathbb{C}(S_t^1, S_t^2, t)}{B_t} = e^{-rt} d\mathbb{C}(S_t^1, S_t^2, t) - re^{-rt}\mathbb{C}(S_{t-}^1, S_{t-}^2, t)dt,$$

$$(A.23) \qquad d\frac{S_t^i}{B_t} = e^{-rt} dS_t^i - re^{-rt} S_{t-}^i dt.$$

Applying the Ito formula to the value process yields
(A.24)

$$\mathbb{C}(S_t^1, S_t^2, t) = \mathbb{C}(S_0^1, S_0^2, 0) + \int_0^t \frac{\partial}{\partial x}\mathbb{C}(S_{u-}^1, S_{u-}^2, u)dS_u^1 + \int_0^t \frac{\partial}{\partial y}\mathbb{C}(S_{u-}^1, S_{u-}^2, u)dS_u^2$$
$$+ \int_0^t \frac{\partial}{\partial t}\mathbb{C}(S_{u-}^1, S_{u-}^2, u)du + \frac{1}{2}\int_0^t \frac{\partial^2}{\partial x^2}\mathbb{C}(S_{u-}^1, S_{u-}^2, u)d\langle S^{1^c}, S^{1^c}\rangle_u$$
$$+ \frac{1}{2}\int_0^t \frac{\partial^2}{\partial y^2}\mathbb{C}(S_{u-}^1, S_{u-}^2, u)d\langle S^{2^c}, S^{2^c}\rangle_u + \int_0^t \frac{\partial^2}{\partial x \partial y}\mathbb{C}(S_{u-}^1, S_{u-}^2, u)d\langle S^{1^c}, S^{2^c}\rangle_u$$
$$+ \sum_{0<u\leqslant t}\Big[\mathbb{C}(S_u^1, S_u^2, u) - \mathbb{C}(S_{u-}^1, S_{u-}^2, u)$$
$$- \frac{\partial}{\partial x}\mathbb{C}(S_{u-}^1, S_{u-}^2, u)\Delta S_u^1 - \frac{\partial}{\partial y}\mathbb{C}(S_{u-}^1, S_{u-}^2, u)\Delta S_u^2\Big].$$

We know from (9) that

$$(A.25) \qquad \Delta S_u^i = -\nu^i S_{u-}^i \Delta\Pi_u$$

and

$$(A.26) \qquad \begin{array}{l} \mathbb{C}(S_u^1, S_u^2, u) - \mathbb{C}(S_{u-}^1, S_{u-}^2, u) = \\ \left[\mathbb{C}(S_{u-}^1(1-\nu^1), S_{u-}^2(1-\nu^2), u) - \mathbb{C}(S_{u-}^1, S_{u-}^2, u)\right] \cdot \Delta\Pi_u. \end{array}$$

Furthermore, the properties of W^* and Π imply

$$(A.27) \qquad d\left\langle S^{i^c}, S^{j^c}\right\rangle_u = (\sigma^i\sigma^j)(S_{u-}^i S_{u-}^j)du, i, j = 1, 2$$

Using (A.22)-(A.27) we obtain

$$d\mathbb{C}(S_t^1, S_t^2, t) = \tfrac{\partial}{\partial x}\mathbb{C}(S_{t-}^1, S_{t-}^2, t)dS_t^1 + \tfrac{\partial}{\partial y}\mathbb{C}(S_{t-}^1, S_{t-}^2, t)dS_t^2 + \tfrac{\partial}{\partial t}\mathbb{C}(S_{t-}^1, t)dt$$
$$+ \tfrac{1}{2}(\sigma^1 S_{t-}^1)^2 \tfrac{\partial^2}{\partial x^2}\mathbb{C}(S_{t-}^1, S_{t-}^2, t)dt + \tfrac{1}{2}(\sigma^2 S_{t-}^2)^2 \tfrac{\partial^2}{\partial y^2}\mathbb{C}(S_{t-}^1, S_{t-}^2, t)dt$$
$$+ (\sigma^1\sigma^2 S_{t-}^1 S_{t-}^2)\tfrac{\partial^2}{\partial x\partial y}\mathbb{C}(S_{t-}^1, S_{t-}^2, t)dt$$
$$+ \left[\mathbb{C}(S_{t-}^1(1-\nu^1), S_{t-}^2(1-\nu^2), t) - \mathbb{C}(S_{t-}^1, S_{t-}^2, t)\right] d\Pi_t$$
$$- \tfrac{\partial}{\partial x}\mathbb{C}(S_{t-}^1, S_{t-}^2, t)(-\nu^1 S_{t-}^1)d\Pi_t - \tfrac{\partial}{\partial y}\mathbb{C}(S_{t-}^1, S_{t-}^2, t)(-\nu^2 S_{t-}^2)d\Pi_t.$$

Finally we arrive at
(A.28)

$$d\frac{\mathbb{C}(S_t^1, S_t^2, t)}{B_t} = \left(\tfrac{\partial}{\partial x}\mathbb{C}(S_{t-}^1, S_{t-}^2, t)\tfrac{S_{t-}^1}{B_t}\sigma^1 + \tfrac{\partial}{\partial y}\mathbb{C}(S_{t-}^1, S_{t-}^2, t)\tfrac{S_{t-}^2}{B_t}\sigma^2\right)dW_t^*$$
$$+ \left[\mathbb{C}(S_{t-}^1(1-\nu^1), S_{t-}^2(1-\nu^2), t) - \mathbb{C}(S_{t-}^1, S_{t-}^2, t)\right] e^{-rt}d(\Pi_t - \lambda^* t)$$
$$+ \Big(\left[\mathbb{C}(S_{t-}^1(1-\nu^1), S_{t-}^2(1-\nu^2), t) - \mathbb{C}(S_{t-}^1, S_{t-}^2, t)\right]\lambda^* + rS_{t-}^1\tfrac{\partial}{\partial x}\mathbb{C}(S_{t-}^1, S_{t-}^2, t)$$
$$+ rS_{t-}^2\tfrac{\partial}{\partial y}\mathbb{C}(S_{t-}^1, S_{t-}^2, t) + \tfrac{\partial}{\partial t}\mathbb{C}(S_{t-}^1, S_{t-}^2, t) + \tfrac{1}{2}(\sigma^1 S_{t-}^1)^2\tfrac{\partial^2}{\partial x^2}\mathbb{C}(S_{t-}^1, S_{t-}^2, t)$$
$$+ \tfrac{1}{2}(\sigma^2 S_{t-}^2)^2\tfrac{\partial^2}{\partial y^2}\mathbb{C}(S_{t-}^1, S_{t-}^2, t) + (\sigma^1\sigma^2 S_{t-}^1 S_{t-}^2)\tfrac{\partial^2}{\partial x\partial y}\mathbb{C}(S_{t-}^1, S_{t-}^2, t) -$$
$$r\mathbb{C}(S_{t-}^1, S_{t-}^2, t) + \tfrac{\partial}{\partial x}\mathbb{C}(S_{t-}^1, S_{t-}^2, t)\nu^1\lambda^* S_{t-}^1 + \tfrac{\partial}{\partial y}\mathbb{C}(S_{t-}^1, S_{t-}^2, t)\nu^2\lambda^* S_{t-}^2\Big) e^{-rt}dt$$

The proof of Theorem 1 now follows from the comparison (A.28) with the representation (A.21).

References

[1] Föllmer H., Leukert P., *Efficient Hedges: Cost versus Shortfall Risk,* Preprint, Humboldt Universität Berlin (1998). To appear in: Finance and Stochastics 4, 2000.

[2] Föllmer H., Leukert P., *Quantile Hedging,* Finance and Stochastics, **3** (1999), 251–273.

[3] Shiryaev A.N., *Essentials of Stochastic Finance,* World Scientificy (1999).

[4] Aase K., *Contingent Claim Valuation when the security price is a combination of an Ito process and a Random point process,* Stoch. Process. Appl., **28** (1988), 185–220.

[5] Bardhan J., Chao X. , *Pricing options on securities with discontinous returns,* Stoch. Process. Appl., **48** (1993), 123–137.

[6] Colwell D.B., Elliott R., *Discontinous asset prices and non-attainable contingent claims and corporate policy,* Math. Finance, **3** (1993), 295–318.

[7] Runggaldier W., Mercurio F, *Option Pricing for jump diffusions: Approximations and their interpretation,* Math. Finance, **3** (1993), 191–200.

[8] Melnikov A.V., Shiryaev A.N., *Criteria for absence of arbitrage in Financial market,* Frontiers in Pure and Appl. Probability II, TVP, Moscow, (1996), 121–134.

[9] Volkov S.N., Kramkov D.O., *On methodology of hedging of options,* Survey of Applied and Industrial Math., **4** (1997), 18–65.

Steklov Mathematical Institute,
Gubkina str. 8,
Moscow 117966, Russia
E-mail address: melnikov@mi.ras.ru

Trends in Mathematics, © 2001 Birkhäuser Verlag Basel/Switzerland

Exponential formula and Girsanov theorem for mixed semilinear stochastic differential equations.

Yuriy Krvavych and Yuliya Mishura

Abstract. The existence and uniqueness conditions for solution of semilinear stochastic differential equations that contains differentials with respect to Wiener process and fractional Brownian motion are considered in this paper. Also, for such mixed Brownian – fractional Brownian semilinear stochastic differential equations the conditions of measure transformation are established.

1. Introduction

We consider the notion of stochastic integral $\int_0^t f_s\, dB_s^H$, where B_s^H is fractional Brownian motion (fBm) with $H > \frac{1}{2}$ ([1]), f_s is random measurable function, Hölder continuous of order H (see, for example, [2], where it is proved that such integral can be considered as a limit of Riemann-Stielties sums). The main subject of our consideration are semilinear stochastic differential equations (SDE) containing Wiener and fBm differentials. We establish the conditions of existence and uniqueness of solution of such equations. Similar questions were considered in [2], [3], [4] but in [2] only the existence and uniqueness of local solution were proved; in [3] the existence and uniqueness of global solution of equations containing only one differential of Hölder continuous process were considered; the equations with the same Wiener and fBm differentials were considered in [4] but semilinear case can't be included to their conditions.

In chapter 3 we establish the conditions of measure transformation for such semilinear SDE.

2. Existence and uniqueness of solutions of semilinear stochastic differential equations that contains differentials with respect to Wiener process and fBm.

Let us consider the semilinear differential equation in complete probability space $(\Omega, \mathcal{F}, P)$

$$\begin{cases} dX_t = \sigma_1 X_t \, dW_t + \sigma_2 X_t \, dB_t^H + b(t, X_t) \, dt, \quad t \geq t_0 \\ X_{t=t_0} = X_0 \text{ is } \mathcal{F}_{t_0} \text{ - measurable, random variable,} \end{cases} \tag{1}$$

where $\sigma_i \in \mathbb{R}, i = 1, 2,$ B_t^H is fractional Brownian Motion (fBm) with Hurst index $H \in \left(\frac{1}{2}, 1\right)$, $W_t = B_t^{\frac{1}{2}}$ is Wiener process, and function

$$b \in C\,(\mathbb{R} \times \mathbb{R}_+, \mathbb{R}) \tag{2}$$

satisfies Lipschitz condition on $x \in \mathbb{R}$ uniformly on $t \in [0, T]$, that is $\forall T > 0 \; \exists L_T > 0:$

$$\sup_{t \in [0,T]} |b(t, x_1) - b(t, x_2)| \leq L_T \, |x_1 - x_2| \tag{3}$$

Definition 2.1. *Solution of the equation (1) is a Hölder continuous with index H random process that has a form*

$$X_t = X_0 + \sigma_1 \int_{t_0}^{t} X_s \, dW_s + \sigma_2 \int_{t_0}^{t} X_s \, dB_s^H + \int_{t_0}^{t} b(X_s, s) \, ds, \quad t \geq t_0 \tag{4}$$

It should be noted that integrals $\int_{t_0}^{t} X_s \, dW_s$, $\int_{t_0}^{t} X_s \, dB_s^H$ exist if X_s is Hölder continuous with index H ([2]).

Theorem 2.2. *Let the function b satisfies (2), (3) and growth condition*

$$\forall T > 0 \; \exists C_T > 0: \; |b(t, x)| \leq C_T \, (1 + |x|). \tag{5}$$

Then the equation (1) has a unique solution on $[0, +\infty)$.

Proof. Firstly let us consider an auxiliary system of partial differential equations

$$\begin{cases} \frac{\partial h}{\partial Z_j}(Y, (Z_1, Z_2)) = \sigma_j \, h(Y, (Z_1, Z_2)), \; j = \overline{1, 2} \\ h(Y_0, \overline{Z_0}) = X_0, \quad \overline{Z_0} = \left(W_{t_0}, B_{t_0}^H\right) \end{cases}$$

This system has a solution $h(Y, \overline{Z}) = (Y - Y_0 + X_0)\mathrm{e}^{\sigma_1(Z_1 - W_{t_0}) + \sigma_2(Z_2 - B_{t_0}^H)}$, where $\overline{Z} = (Z_1, Z_2),$ $Z_1(t) = W_t,$ $Z_2(t) = B_t^H$

We will try to find the solution X_t of equation (1) in the following form $X_t = h(Y(t), \overline{Z}(t))$, where $Y \in C'(\mathbb{R})$ with $Y(t_0) = Y_0$ that is $\mathcal{F}_{t_0}$-measurable random value.

Applying the formula Itô to the function $h(Y(t), \overline{Z}(t))$ we obtain

$$dX_t = \sum_{i=1}^{2} \frac{\partial h}{\partial Z_i}(Y(t), \overline{Z}(t)) \, dZ_i(t) + \frac{\partial h}{\partial Y}(Y(t), \overline{Z}(t)) Y' \, dt +$$

$$+ \frac{1}{2} \sum_{i,j=1}^{2} \frac{\partial^2 h}{\partial Z_i \partial Z_j}(Y(t), \overline{Z}(t)) \, d\,[Z_i, Z_j]_t =$$

$$= \sum_{i=1}^{2} \sigma_i \, h(Y(t), \overline{Z}(t)) \, dZ_i(t) + Y' \frac{\partial h}{\partial Y}(Y(t), \overline{Z}(t)) \, dt +$$

$$+\frac{1}{2}\sigma_1^2\, h(Y(t), \overline{Z}(t))dt.$$

Equation (1) implies the following ordinary differential equation for Y :

$$\begin{cases} Y'(t) = (c_1(t))^{-1} b(t, (Y - Y_0 + X_0)\, c_1(t)) - \frac{1}{2}\sigma_1^2\, (Y - Y_0 + X_0) =: f(y,t) \\ Y(t_0) = Y_0, \end{cases} \tag{6}$$

where $c_1(t) = e^{\sigma_1(Z_1 - W_{t_0}) + \sigma_2(Z_2 - B^H_{t_0})}$. Further we will assume that $\omega \in \Omega$ is fixed. Put $L_1(T) := \max\limits_{t\in[0,T]}(c_1(t))^{-1} > 0$, $L_2(T) := \max\limits_{t\in[0,T]} c_1(t) > 0$, $D_1 := C_T\, L_1(T)$, $D_2 = C_T + \frac{1}{2}\sigma_1^2$.

Then under $|Y - Y_0| \le \beta$; $|t - t_0| \le \alpha$:

$$M := \max_{t\in[0,T]} |f(y,t)| \le \frac{1}{2}\sigma_1^2(\beta + |X_0|)\, (L_1(T) + (\beta + |X_0|))\, C_T \le D_1 + D_2(\beta + |X_0|)$$

and according to the Picard theorem there exists solution of (6) on $[t_0, t_0 + l^{(0)}]$, where

$$l^{(0)} = \min\left(\alpha, \frac{\beta}{M}\right) \ge \min\left(\alpha, \frac{\beta}{D_1 + D_2(\beta + |X_0|)}\right) =: l_0,$$

and therefore solution of (6) exists on the smaller segment $[t_0, t_0 + l_0]$. From this on segment $[t_0, t_0 + l_0]$ there exists a solution of (1).

Now we can estimate the solution of (1) in the point $t_0 + l_0$ in such a way:

$$h_{t_0+l_0} \le |Y - Y_0 + X_0|\, L_2(T) \le (\beta + |X_0|)\, L_2(T).$$

Further we will extend the solution h for $t \geqslant t_0 + l_0$. The value of solution h in the point $t_0 + l_0$ will be new initial value $X_0^{(1)} \le (\beta + |X_0|)\, L_2(T)$.

Now, under $|Y(t) - Y(t_0 + l_0)| \le \beta_1$, $|t - (t_0 + l_0)| \le \alpha_1$, the solution of (6) exists on segment $[t_0 + l_0, t_0 + l_0 + l_1]$, where

$$l_1 = \min\left(\alpha_1, \frac{\beta_1}{D_1 + D_2(\beta_1 + (\beta + |X_0|)L_2(T))}\right)$$

Further on the n-th step of such solution extension procedure we have:

$$l_n = \min\left(\alpha_n, \frac{\beta_n}{D_1 + D_2(\beta_n + \sum\limits_{k=0}^{n-1} \beta_{n-1-k}\, L_2^{k+1}(T) + |X_0|L_2(T))}\right)$$

$\alpha_0 := \alpha$, $\beta_0 := \beta$ and on $[t_0 + \sum\limits_{i=0}^{n-1} l_i,\ t_0 + \sum\limits_{i=0}^{n} l_i]$ there exists a solution of (6).

Now let us investigate the properties of

$z_n =: \dfrac{\beta_n}{D_1 + D_2(\beta_n + \sum\limits_{k=0}^{n-1} b_{n-1-k}\, L_2^{k+1}(T) + |X_0|L_2(T))}$, $n \ge 0$:

a) under $|X_0| \le 1$ we can chose $\beta_k \equiv 1$, $k = \overline{0,n}$ and then

$$\frac{1}{D_1 + 2D_2\frac{L_2^{n+1}(T)-1}{L_2(T)-1}} \le z_n \le \frac{1}{D_1 + D_2\frac{L_2^{n+1}(T)-1}{L_2(T)-1}}.$$

b)under $|X_0| > 1$ we can chose $\beta_k \equiv |X_0|$, $k = \overline{0,n}$ and then

$$\frac{1}{D_1 + 2D_2 \frac{L_2^{n+1}(T)-1}{L_2(T)-1}} \le z_n = \frac{1}{\frac{D_1}{|x_0|} + D_2(1 + 2\sum_{k=1}^{n} L_2^k(T))} \le \frac{1}{D_2 \frac{L_2^{n+1}(T)-1}{L_2(T)-1}}.$$

For both cases a) and b) we put $a_n = \frac{1}{D_1+2D_2\frac{L_2^{n+1}(T)-1}{L_2(T)-1}}$, $n \ge 0$, that is $l_n = \min(a_n, z_n) = a_n \Rightarrow \sum_{n\ge 0} l_n = S$, and then

1) under $L_2(T) \le 1$ the series $\sum_{n\ge 0} z_n$ is divergent and therefore the series $\sum_{i\ge 0} l_i$ is also divergent, and thus there exists a finite quantity N of steps in our solution extension procedure, such that $[t_0, T] \subset [t_0, t_0 + \sum_{i=0}^{N} l_i]$, that is on the $[t_0, T]$ there exists a solution of (6).

2) under $L_2(T) > 1$ the series $\sum_{n\ge 0} z_n$ is convergent and moreover

$$\sum_{n\ge 0} z_n \ge \sum_{n\ge 0} \frac{1}{D_1 + 2D_2 \frac{L_2(T)^{n+1}-1}{L_2(T)-1}} =: S$$

In fact we proved that on segment $[t_0, t_0 + \frac{1}{2}S]$ there exists a bounded solution h of (1). Now we can consider a finite value of h in the point $t_0 + \frac{1}{2}S$ as a new initial value of h and for it we can prove that solution h exists on $[t_0 + \frac{1}{2}S, t_0 + S]$, because the step $\frac{1}{2}S$ of solution extension doesn't depend on initial value. Therefore we can extend solution h to whole segment $[t_0, T]$. Since T is arbitrary we obtain the solution (1) on $[0, \infty)$. Uniqueness follows from theorem 7.1.1.([2]). In this theorem it is established that any two of solutions coincide on the common interval of definition. □

3. Girsanov theorem and measure transformation for mixed semilinear stochastic differential equations.

Consider the equation (1) and suppose that $b(t,x)$ satisfies the conditions (2), (3), (5) and can be presented as $b(t,x) = e(t,x)x$, where $e \in C(\mathbb{R}_+ \times \mathbb{R})$ Further we will use the following notations: $B = B(H + 1/2, 3/2 - H)$, $B(\cdot,\cdot)$ is beta-function, $B_1 = B(H - 1/2, 3/2 - H)$, $C_1 = (2HB)^{-1}$, $C_H = \left(\frac{2H\,\Gamma(3/2-H)}{\Gamma(H+1/2)\,\Gamma(2-2H)}\right)^{1/2}$, $C_2 = \frac{C_H}{2H(2-2H)^{1/2}}$, $C_0 = \frac{C_1}{C_2}$. Also, let the kernel $K(t,s) = C_1 s^{1/2-H}(t-s)^{1/2-H} I\{s \in (0,t)\}$, $\widehat{e}(t,x) := e(t,x)t^{1/2-H}$.

Let the function ψ satisfies four conditions:

1) $\int_0^T K(t,s)|\psi_s|\,ds < \infty$ a.s., $t \in [0,T]$,

2) there exists a representation

$$\int_0^T K(t,s)\psi_s\,ds = \int_0^T \delta(s)\,ds,\ \ t > 0, \tag{7}$$

with $\int_0^T |\delta_s|\,ds < \infty$ a.s., $t \in [0,T]$.

3) $\mathbb{E}\int_0^T s^{2H-1}\delta_s\,ds < \infty$,

4) $\mathbb{E}\exp\left\{Z_T - \frac{1}{2}\langle Z\rangle_T\right\} = 1,\ \ t \in [0,T]$,

where $Z_t = \frac{1}{C_2}\int_0^t s^{H-1/2}\delta_s\,dW_s$ and $\langle Z\rangle_t = \frac{1}{C_2^2}\int_0^t s^{2H-1}\delta_s^2\,ds$.

Now, we try to change the measure P for another probability measure Q such that $Q_T \ll P_T$, where $Q_T = Q|_{[0,T]}$, and that the drift $e(t,X_t)X_t\,dt$ will be annihilated under Q_T. At first, let some probability measure $\widetilde{Q}$ satisfies the assumptions

$$\frac{d\widetilde{Q}_T}{dP_T} = \exp\left\{\int_0^T \varphi_s\,dW_s - \frac{1}{2}\int_0^T \varphi_s^2\,ds\right\}$$

and

$$\mathbb{E}\left[\exp\left\{\int_0^T \varphi_t\,dW_t - \frac{1}{2}\int_0^T \varphi_t^2\,dt\right\}\right] = 1, \tag{8}$$

where

$$\mathbb{E}\int_0^T |\varphi_t|^2\,dt < \infty. \tag{9}$$

Then from Girsanov theorem $W_t - \int_0^t \varphi_s\,ds = \widehat{W}_t$ is a Wiener process under the measure $\widetilde{Q}_T$. Also, let the measure $\overline{Q}$ be such that $\frac{d\overline{Q}_T}{dP_T} = \exp\left\{Z_T - \frac{1}{2}\langle Z\rangle_T\right\}$, $\overline{Q} \ll P$, then the process $\widehat{B}_t^H = B_t^H - \int_0^t \psi_s\,ds$ is fBm relatively to $\overline{Q}$ (see theorem 4 from [5], also see [6], [7]). Of course, we are need in the inequality $\widetilde{Q}_T = \overline{Q}_T = Q_T$, whence, in particular,

$$Z_t = \int_0^t \varphi_s\,dW_s. \tag{10}$$

Therefore we want to find φ_s and ψ_s in such a way that the common drift equals

$$\sigma_1 \int_0^t \varphi_s \, ds + \sigma_2 \int_0^t \psi_s \, ds = - \int_0^t e(s, X_s) \, ds. \tag{11}$$

In this case

$$\int_0^t e(s, X_s) \, ds + \sigma_1 \int_0^t X_s \, dW_s + \sigma_2 \int_0^t X_s \, dB_s^H = \sigma_1 \int_0^t X_s \, d\widehat{W}_s + \sigma_2 \int_0^t X_s \, d\widehat{B}_s^H.$$

From (11) we obtain that

$$\sigma_1 \varphi_t + \sigma_2 \psi_t = -e(t, X_t), \quad t \in [0, T].$$

Further, we use the Abel rearrangement to the relation

$$C_1 \int_0^t (t-s)^{1/2-H} s^{1/2-H} \psi_s \, ds = \int_0^t \delta_s \, ds = C_2 \int_0^t s^{1/2-H} \varphi_s \, ds :$$

$$C_1 \int_0^t (t-u)^{H-3/2} \int_0^u (u-s)^{1/2-H} s^{1/2-H} \psi_s \, ds \, du$$

$$= C_2 \int_0^t (t-u)^{H-3/2} \int_0^u s^{1/2-H} \varphi_s \, ds \, du,$$

or

$$C_0 B_1 \int_0^t s^{1/2-H} \psi_s \, ds = \int_0^t \frac{(t-u)^{H-1/2}}{H-1/2} u^{1/2-H} \varphi_u \, du,$$

if we differentiate, then

$$C_0 B_1 t^{1/2-H} \psi_t = \int_0^t (t-u)^{H-3/2} u^{1/2-H} \varphi_u \, du,$$

and

$$\sigma_1 \varphi_t + \frac{t^{H-1/2}}{C_0 B_1} \sigma_2 \int_0^t (t-u)^{H-3/2} u^{1/2-H} \varphi_u \, du = -e(t, X_t),$$

let further $t^{1/2-H} \varphi_t = \Theta_t$, then

$$\sigma_1 \Theta_t + \frac{\sigma_2}{C_0 B_1} \int_0^t (t-u)^{H-3/2} \Theta_u \, du = -\widehat{e}(t, X_t).$$

It is Volterra equation with weak singularity, and its unique solution has a form

$$\Theta_t = -\frac{\widehat{e}(t, X_t)}{\sigma_1} - \frac{1}{\sigma_1}\int_0^t \sum_{n=1}^{\infty} \rho^n \frac{(t-s)^{n(H-1/2)-1}}{\Gamma(n(H-1/2))}\widehat{e}(s, X_s)\, ds,$$

where $\rho = -\frac{\sigma_2}{C_0 B_1}\Gamma(H-1/2)$.

So, we proved the following result.

Theorem 3.1. *Under the conditions (7), (8), (9) and (10) the equation (1) has a form*

$$dX_t = \sigma_1 X_t\, d\widehat{W}_t + \sigma_2 X_t\, d\widehat{B}_t^H$$

under measure Q.

Definition 3.2. *Financial market described by the equation (1) is in equilibrium on $[0,T]$ if both the kernel $\Theta_t t^{H-1/2}$ and likelihood ratio $\left.\frac{dQ}{dP}\right|_{\mathcal{F}_t}$ are the functions of t and W_t, and don't depend on the path of W_s, $s \le t$.*

This definition generalizes the usual definition of equilibrium ([8], [10]) in which path's independence of $\left.\frac{dQ}{dP}\right|_{\mathcal{F}_t}$ is declared, and the kernel equals simply $e(t, W_t)$. Note that financial market described by the equation (1) admits no arbitrage as it follows from [9].

Definition 3.3. *We say that Θ_t, satisfying definition 3.2, satisfies fractal Burger's equation with index H if $g(t, W_t) := \Theta_t t^{H-1/2}$ satisfies usual Burger's equation*

$$-g(s,x)g_2'(s,x) = g_1'(s,x) + \frac{1}{2}g_{22}''(s,x)$$

Evidently, fractal Burger's equation has a form

$$-s^{H-1/2}p(s,x)p_2'(s,x) = (H-1/2)s^{-1}p(s,x) + p_1'(s,x) + \frac{1}{2}p_{22}''(s,x),\ s>0,\ x\in\mathbb{R}$$

Theorem 3.4. *If financial market is in equilibrium then Θ_t satisfies fractal Burger's equation.*

Proof. Let

$$\int_0^t \Theta_s s^{H-1/2}\, dW_s - \frac{1}{2}\int_0^t \Theta_s^2 s^{2H-1}\, ds = G(t, W_t),$$

where $g, G \in C^2(\mathbb{R}_+ \times \mathbb{R})$. Then

$$\int_0^t \Theta_s g(s, W_s)\, dW_s - \frac{1}{2}\int_0^t g^2(s, W_s)\, ds = G(t, W_t),\ t\in[0,T]$$

By Itô formula,

$$G(t, W_t) = \int_0^t \left[G_1'(s, W_s) + \frac{1}{2}G_{22}''(s, W_s)\right] ds + \int_0^t G_2'(s, W_s)\, dW_s.$$

From here $g(s, W_s) = G_2'(s, W_s)$, $-\frac{1}{2}g^2(s, W_s) = G_1'(s, W_s) + \frac{1}{2}G_{22}''(s, W_s)$, or, simply,

$$g(s,x) = G_2'(s,x), \tag{12}$$
$$-\frac{1}{2}g^2(s,x) = G_1'(s,x) + \frac{1}{2}G_{22}''(s,x)$$

Further, $g_2'(s,x) = G_{22}''(s,x)$,

$$-\frac{1}{2}g^2(s,x) = G_1'(s,x) + \frac{1}{2}g_2'(s,x). \tag{13}$$

Differentiate (12) in s and (13) in x:

$$g_1'(s,x) = G_{12}''(s,x)$$
$$-g(s,x)g_2'(s,x) = G_{12}''(s,x) + \frac{1}{2}g_{22}''(s,x)$$

or

$$-g(s,x)g_2'(s,x) = g_1'(s,x) + \frac{1}{2}g_{22}''(s,x).$$

□

Similar problem for usual SDE with Wiener process was considered in [10].

References

[1] I. Norros, E. Valkeila and J. Virtamo, *An elementary approach to a Girsanov formula and other analytical results on fractional Brownian motions*, Bernoulli, **6** (1999).

[2] M. Zaehle, *On the link between fractional and stochastic calculus*, Preprint

[3] A. Rusmaikina, *Stieltjes integrals of Holder continuous functions with applications to fractional Brownian motion*, Mathematics ArXiv electronic edition: http://arxiv.org/PS_cache/math/pdf/0005/0005147.pdf

[4] John G.M. Schoenmakers and Peter E. Kloeden, *Robust option replication for a Black- Scholes model extended with nondeterministic trends*, Preprint

[5] Yu. S. Mishura and E. Valkeila, *Martingale transforms and Girsanov theorem for long-memory Gaussian processes*, submitted to Statistics and Probability Letter.

[6] L. Decreosfond and A. S. Üstunel, *Stochastic analysis on the fractional Brownian motions*, Potential Analysis **10** (1999), 177 – 214.

[7] M.L. Kleptsyna, A. Le Breton and M. C. Rouland, *On elementary approach to filtering in system with fractional Brownian observation noise*, INRIA, Preprint 349, (1998), 38.

[8] J. Karatzas and S. E. Shreve, *Methods of Mathematical Finance*, Applications for Stochastic Modelling and Applied Probability **Vol. 39** Springer Verlag (1998), 408.

[9] Yu. S. Mishura and E. Valkeila, *On arbitrage in the mixed Brownian – fractional Brownian market model*, University of Helsinki, Preprint 261,(2000), 14.

[10] S. Hodges and A. Carverhill, *Quasi mean reversion in an efficient stock market: the characterisation of economic equilibria which support Black-Scholes option pricing*, The Economic Journal, **103** (1993), 395–405.

Department of Mathematics,
Kyiv National University of Taras Shevchenko,
64, Volodymyrska str.,
Kyiv 01033, Ukraine
E-mail address: myus@mechmat.univ.kiev.ua , krvavych@yahoo.com

Trends in Mathematics, © 2001 Birkhäuser Verlag Basel/Switzerland

An introduction to optimal consumption with partial observation

David Lefèvre[1], Bernt Øksendal[2,3], and Agnès Sulem[1]

Abstract. We give a short introduction to some of the theory and methods involved in stochastic control with partial observation. As an illustration we use the stochastic maximum principle and the Kalman-Bucy filter to solve explicitly a problem about optimal consumption in an economy where the mean relative growth rate is only observed indirectly (partially).

1. Introduction

Many classic mathematical models in finance assume complete knowledge of all the parameters involved. For example, if an economic quantity $X(t)$ grows like a mean reverting Ornstein-Uhlenbeck process of the form

$$dX(t) = (\mu - \rho X(t))dt + \alpha\, dW(t) \tag{1.1}$$

where μ, $\rho > 0$ and α are constants, it is often assumed that these constants are completely observable. However, in practical situations one cannot observe these quantities directly, only indirectly (partially) through the observations of $X(s)$; $s \leq t$. Then the question is: How do we find optimal consumption and/or optimal portfolio in such a situation? This is an example of a *stochastic control problem with partial observation.*

The purpose of this paper is to give an introduction to the theory and methods of such problems. This is done by studying in detail the specific problem of optimal consumption from the economy (1.1) under a terminal constraint in the case when the mean relative growth rate μ is only partially observed. We solve this problem explicitly by using the stochastic maximum principle and the Kalman-Bucy filter. For more information about stochastic control with partial observation we refer to e.g. [B2], [EGK] and the references therein.

2. Statement of the problem

a) Complete observation case

Suppose the wealth $X(t) = X^{(c)}(t)$ of a person with consumption rate $c(t) \geq 0$ satisfies the stochastic differential equation

$$\begin{cases} dX(t) = (\mu - \rho X(t) - c(t))dt + \alpha(t)dW(t)\ ; \quad t \in [0,T] \\ X(0) = x > 0 \end{cases} \tag{2.2}$$

Here $c(t) = c(t,\omega)$ is an $\mathcal{F}_t$-adapted process (our control), $\alpha(t)$ is a given deterministic function and μ, ρ, T, x are (known) positive constants.

Suppose the expected total discounted utility $J^{(c)}$ corresponding to the consumption rate $c(t)$ is given by

$$J = J^{(c)} = E\Big[\int_0^T e^{-\delta t}\,\frac{c^\gamma(t)}{\gamma}dt\Big], \tag{2.3}$$

where $\delta > 0$, $\gamma \in (0,1)$ are (known) constants ($1-\gamma$ is the risk aversion of this person) and $E = E^x$ denotes the expectation given that $X(0) = x$.

We say that $c = c(t,\omega)$ is *admissible* and write $c \in \mathcal{A}$ if $c(t,\omega)$ is $\mathcal{F}_t$-adapted, nonnegative and satisfies the *terminal condition*

$$E[X^{(c)}(T)] = x_T\ , \tag{2.4}$$

where $x_T \in \mathbf{R}$ is a given constant. This constraint expresses that on the average a certain wealth x_T should be left for the next generation. Consider the problem

Problem 2.1. *Find* $\bar{J}$ *and* $\bar{c} \in \mathcal{A}$ *such that*

$$\bar{J} := \sup\{J^{(c)}; c \in \mathcal{A}\} = J^{(\bar{c})}. \tag{2.5}$$

This is a stochastic control problem with complete information, *because we assume that all the parameters involved* $(\mu, \rho, \alpha(\cdot), \ldots)$ *are known and the consumption rates considered are allowed to be adapted to the filtration* $\mathcal{F}_t$.

b) Partial observation case

In the partial observation case the setup and the problem is the same as in (2.2)-(2.5) above, but with one important difference:

We no longer assume that μ is a known constant. Instead we assume that $\mu = \mu(\omega)$ is a random variable with a known distribution and that μ is independent of $\{W(s)\}_{s\geq 0}$. Thus the specific value of μ in (2.2)–(2.5) is not known, and we assume that we cannot observe it directly. We only assume that we can observe it indirectly through the observations of the process $\xi(t)$ defined by

$$d\xi(t) = \mu\,dt + \alpha(t)dW(t)\ ; \qquad t \geq 0 \qquad \xi(0) = 0 \tag{2.6}$$

We claim that this is equivalent to saying that we observe the no consumption wealth $X^{(0)}(t)$ given by

$$dX^{(0)}(t) = (\mu - \rho X^{(0)}(t))dt + \alpha(t)dW(t)\ . \tag{2.7}$$

To verify this claim, first note that by (2.7), we have

$$dX^{(0)}(t) + \rho X^{(0)}(t)dt = d\xi(t) \tag{2.8}$$

or

$$d(e^{\rho t}X^{(0)}(t)) = e^{\rho t}d\xi(t) \tag{2.9}$$

which gives

$$X^{(0)}(t) = X^{(0)}(0)e^{-\rho t} + \int_0^t e^{-\rho(t-s)}d\xi(s)\,. \tag{2.10}$$

Hence, if we observe $\xi(s); s \leq t$ then we know $X^{(0)}(s); s \leq t$.

On the other hand, by (2.7) we also have

$$d\xi(t) = dX^{(0)}(t) + \rho X^{(0)}(t)dt \tag{2.11}$$

which gives

$$\xi(t) = X^{(0)}(t) - X^{(0)}(0) + \rho \int_0^t X^{(0)}(s)ds\,. \tag{2.12}$$

So, if we know $X^{(0)}(s); s \leq t$, we also know $\xi(s); s \leq t$. This proves the claim.

Let $\mathcal{G}_t$ denote the σ-algebra generated by $\{\xi(s)\}_{s\leq t}$. We say that $c(t,\omega)$ is *admissible* for the partial observation problem and write $c \in \mathcal{A}_p$ if $c(t,\omega)$ is $\mathcal{G}_t$-adapted, nonnegative and satisfies the terminal condition

$$E[X^{(c)}(T)] = x_T\,. \tag{2.13}$$

The *partial observation problem* is

Problem 2.2. *Find J^* and $c^* \in \mathcal{A}_p$ such that*

$$J^* = \sup\{J^{(c)}; c \in \mathcal{A}_p\} = J^{(c^*)} \tag{2.14}$$

where, as before,

$$J^{(c)} = E\Big[\int_0^T e^{-\delta t}\,\frac{c^\gamma(t)}{\gamma}dt\Big]. \tag{2.15}$$

Here E denotes the expectation with respect to the joint probability law of μ and $\{W(t)\}_{t\in[0,T]}$.

3. Optimal consumption with complete observation

We first solve the complete observation problem (Problem 2.1). To this end we recall the *stochastic maximum principle* (see e.g. [B1], [P] or [YZ] for proofs and more details):

Suppose $X(t) = X^{(u)}(t)$ is a controlled stochastic system of the form

$$\begin{cases} dX(t) = b(t, X(t), u(t)dt + \sigma(t, X(t))dW(t)\,; & t \in [0,T] \\ X(0) = x \end{cases} \tag{3.16}$$

where $b : [0,T]\times\mathbf{R}^n\times U \to \mathbf{R}^n$ and $\sigma : [0,T]\times\mathbf{R}^n \to \mathbf{R}^{n\times m}$ are given C^2 functions with respect to x whose partial derivatives are uniformly Lipschitz continuous w.r.t. x and uniformly continuous w.r.t. u, where $(U, d(\cdot,\cdot))$ is a metric space. $W(t) = (W_1(t), \ldots, W_m(\cdot))^T$ is m-dimensional Brownian motion and $u : [0,T] \times \Omega \to U$ is an $\mathcal{F}_t$-adapted control process.

Let $f : [0,T] \times \mathbf{R}^n \times U \to \mathbf{R}$, $g : \mathbf{R}^n \to \mathbf{R}$ and $G : \mathbf{R}^n \to \mathbf{R}^N$ be given lower bounded functions satisfying the same conditions as b and σ above and define the *performance functional*

$$J^{(u)} = E\Big[\int_0^T f(t, X(t), u(t))dt + g(X(T))\Big] \tag{3.17}$$

and the *terminal condition*

$$E[G(X^{(u)}(T))] = 0\,. \tag{3.18}$$

Let $\mathcal{A}$ denote the set of all $\mathcal{F}_t$-adapted processes $u(t,\omega) : [0,T] \times \Omega \to U$ such that $X^{(u)}(t)$ does not explode in $[0,T]$ and such that (3.18) holds. Consider the problem to find $\bar{J} \in \mathbf{R}$ and $\bar{u} \in \mathcal{A}$ such that

$$\bar{J} = \sup\{J^{(u)}; u \in \mathcal{A}\} = J^{(\bar{u})} \tag{3.19}$$

If such $\bar{u} \in \mathcal{A}$ exists, then $\bar{u}$ is called an *optimal control* and $(\bar{u}, \bar{X})$ (where $\bar{X} = X^{(\bar{u})}$) is called an *optimal pair.*

Define the *Hamiltonian* $H : [0,T] \times \mathbf{R}^n \times U \times \mathbf{R}^n \times \mathbf{R}^{n\times m} \to \mathbf{R}$ by

$$H(t,x,u,p,q) = b(t,x,u)^T p + \mathrm{tr}[\sigma(t,x)^T q] + f(t,x,u) \tag{3.20}$$

where $(\cdot)^T$ denotes matrix transposed and $\mathrm{tr}[\cdot]$ denotes matrix trace. Consider the following *stochastic backward* differential equation in the pair of unknown *adjoint* $\mathcal{F}_t$-adapted processes $p(t) \in \mathbf{R}^n$, $q(t) \in \mathbf{R}^{n\times m}$:

$$\begin{cases} dp(t) = -H_x(t, X(t), u(t), p(t), q(t))dt + q(t)dW(t)\,; & t \in [0,T] \\ p(T) = g_x(X(T)) + \lambda^T G_x(X(T)) \end{cases} \tag{3.21}$$

Here $H_x = \nabla_x H = (\frac{\partial H}{\partial x_1}, \ldots, \frac{\partial H}{\partial x_n})^T$ is the gradient of H with respect to x and similarly with g_x, G_x. $X(t) = X^{(u)}(t)$ is the process obtained by using the control $u \in \mathcal{A}$ and $\lambda \in \mathbf{R}^N$ is a constant. The equation (3.21) is the *adjoint* equation and $p(t)$ is sometimes interpreted as the *shadow price* (of a resource).

Theorem 3.1. *[The stochastic maximum principle]*

a) *(Necessary conditions for optimality)*

Let $(\bar{u}(t), \bar{X}(t))$ be an optimal pair for problem (3.19). Then there exists a constant $\lambda \in \mathbf{R}^N$ and a solution $(p(t), q(t))$ of (3.21) with $u(t) = \bar{u}(t)$, $X = \bar{X}(t)$ such that

$$H(t, \bar{X}(t), \bar{u}(t), p(t), q(t)) = \max_{u\in U} H(t, \bar{X}(t), u, p(t), q(t)) \quad \textit{for a.a. } (t,\omega) \in [0,T] \times \Omega\,. \tag{3.22}$$

b) *(Sufficient conditions for optimality)*

Suppose $\lambda \in \mathbf{R}_+^N$, $u \in \mathcal{A}$, put $X(t) = X^{(u)}(t)$ and let $p(t), q(t)$ be a solution of the corresponding adjoint equation (3.21). Assume that the functions

$$H(t, \cdot, \cdot, p(t), q(t)), g(\cdot) \text{ and } G(\cdot) \text{ are concave for all } t \in [0, T] \tag{3.23}$$

and that

$$H(t, X(t), u(t), p(t), q(t)) = \max_{u \in U} H(t, X(t), u, p(t), q(t)) \tag{3.24}$$
$$\text{for a.a. } (t, \omega) \in [0, T] \times \Omega \, .$$

Then $(u(t), X(t))$ is an optimal pair for problem (3.19).

For proofs we refer to [B1], [YZ, Chapter 3] and [P].

We now apply the stochastic maximum principle to solve Problem 2.1 (complete observation case).

Here the Hamiltonian function (3.20) is

$$H(t, x, c, p, q) = (\mu - \rho x - c)p + \alpha(t)q + e^{-\delta t}\,\frac{c^\gamma}{\gamma} \tag{3.25}$$

Hence $\frac{\partial H}{\partial x}(t, x, c, p, q) = -\rho p$ and the adjoint equation (3.21) gets the form

$$\begin{cases} dp(t) = \rho\, p(t)dt + q(t)dW(t) \, ; & t \in [0, T] \\ p(T) = \lambda \end{cases} \tag{3.26}$$

We immediately see that the solution of this equation is

$$p(t) = \lambda e^{-\rho(T-t)} \, , \qquad q(t) = 0 \, . \tag{3.27}$$

Now $c \to H(t, x, c, p, q)$ is maximal when

$$\frac{\partial H}{\partial c}(t, x, c, p, q) = 0 \qquad \text{i.e.} \qquad -p + e^{-\delta t}c^{\gamma - 1} = 0$$

This gives

$$\bar{c}(t) = \left(p(t)e^{\delta t}\right)^{\frac{1}{\gamma-1}} = \lambda^{\frac{1}{\gamma-1}} \exp\left\{\frac{1}{\gamma - 1}(-\rho(T-t) + \delta t)\right\} \tag{3.28}$$

It remains to find λ. From (2.2) we have

$$\begin{cases} d\bar{X}(t) = (\mu - \rho\bar{X}(t))dt + \alpha(t)dW(t) - \bar{c}(t)dt \, ; & t \in [0, T] \\ \bar{X}(0) = x \end{cases} \tag{3.29}$$

Integrating and taking the expectation we get (using that μ is constant)

$$E[\bar{X}(t)] = x + \mu t - \rho \int_0^t E[\bar{X}(s)]ds - \int_0^t \bar{c}(s)ds \, .$$

Or, with $y(t) = E[\bar{X}(t)]$,

$$y'(t) + \rho\, y(t) = \mu - \bar{c}(t)$$

which integrates to

$$e^{\rho T}y(T) = y(0) + \int_0^T e^{\rho s}(\mu - \bar{c}(s))ds\ .$$

Using the terminal condition (2.4) and (3.28) this gives

$$x_T = xe^{-\rho T} + e^{-\rho T}\int_0^T e^{\rho s}\Big[\mu - \lambda^{\frac{1}{\gamma-1}} \exp\{\frac{1}{\gamma-1}(-\rho(T-s)+\delta s)\}\Big]ds$$

$$= xe^{-\rho T} + \frac{\mu}{\rho}(1-e^{-\rho T}) - \frac{1-\gamma}{\delta+\gamma\rho}\lambda^{\frac{1}{\gamma-1}}\exp\Big(\frac{\rho\gamma T}{1-\gamma}\Big)\Big[1-\exp\Big(-\frac{\rho+\delta\gamma}{1-\gamma}T\Big)\Big]$$

Hence,

$$\lambda = \Big(xe^{-\rho T} + \frac{\mu}{\rho}(1-e^{-\rho T}) - x_T\Big)^{\gamma-1}\Big[\frac{1-\gamma}{\delta+\gamma\rho}\{\exp\Big(\frac{\gamma\rho T}{1-\gamma}\Big) - \exp\Big(-\frac{\delta T}{1-\gamma}\Big)\}\Big]^{1-\gamma} \quad \text{if } \delta+\gamma\rho \neq 0 \tag{3.30}$$

$$\lambda = \Big(xe^{-\rho T} + \frac{\mu}{\rho}(1-e^{-\rho T}) - x_T\Big)^{\gamma-1} T^{1-\gamma}\exp(-\delta T) \quad \text{if } \delta+\gamma\rho = 0 \tag{3.31}$$

With this value of λ the value of $\bar{c}(t)$ given by (3.28) is optimal, provided that

$$xe^{-\rho T} + \frac{\mu}{\rho}(1-e^{-\rho T}) \geq x_T\ . \tag{3.32}$$

This condition is necessary to make it possible at all to attain (2.4) for some $u(t) \geq 0$.

We have proved:

Theorem 3.2. *Assume that (3.32) holds. Then the optimal consumption rate* $\bar{c}(t)$ *for the complete observation problem (Problem 2.1) is*

$$\bar{c}(t) = \lambda^{\frac{1}{\gamma-1}}\exp\Big\{\frac{1}{\gamma-1}(-\rho(T-t)+\delta t)\Big\}; \qquad t\in[0,T] \tag{3.33}$$

where λ *is given by (3.30),(3.31).*

4. Optimal consumption with partial observation

From now on we assume that

$$\int_0^T \alpha^{-2}(t)dt < \infty \tag{4.34}$$

Then we have the following fundamental result from linear filtering theory (see e.g. [D], [Ka] or [Ø]).

Theorem 4.1. *Let $\mathcal{G}_t$ be the σ-algebra generated by $\xi(s)$, $s \le t$ (see (2.5)). Define the* innovation process *$B(t) = B(t, \omega)$ by*

$$\begin{cases} dB(t) = \alpha^{-1}(t)(\mu - E[\mu|\mathcal{G}_t])dt + dW(t) \; ; & t \ge 0 \\ B(0) = 0 \end{cases} \tag{4.35}$$

Then $B(t)$ is a Brownian motion. Moreover, $B(s)$; $s \le t$ generates the same filtration $\mathcal{F}_t$ as $W(s)$; $s \le t$.

Corollary 4.2. *With $B(t)$ as in Theorem 4.1 we have*

$$\mu dt + \alpha(t)dW(t) = E[\mu|\mathcal{G}_t]dt + \alpha(t)dB(t) \tag{4.36}$$

We also need the following result, which is a special case of the Kalman filter formula (see e.g. [D], [Ka] or [Ø, Example 6.2.11]):

Theorem 4.3. *As before let*

$$d\xi(t) = \mu dt + \alpha(t)dW(t) = E[\mu|\mathcal{G}_t]dt + \alpha(t)dB(t) \; ; \qquad t \ge 0$$

and let $\mathcal{G}_t$ be the σ-algebra generated by $\xi(s)$; $s \le t$. Then, if we put

$$\bar{\mu} = E[\mu] \qquad \text{and} \qquad A = E[(\mu - \bar{\mu})^2]^{-1} \tag{4.37}$$

and

$$dY(t) = \alpha^{-2}(t)d\xi(t) \qquad \text{and} \qquad dZ(t) = \alpha^{-2}(t)dt \; , \tag{4.38}$$

we have

$$\begin{aligned} E[\mu|\mathcal{G}_t] &= \Big(A + \int_0^t \alpha^{-2}(s)ds\Big)^{-1}\Big(\bar{\mu}A + \int_0^t \alpha^{-2}(s)d\xi(s)\Big) \\ &= (A + Z(t))^{-1}(\bar{\mu}A + Y(t)) \; . \end{aligned} \tag{4.39}$$

Using (4.36)–(4.39) we see that we can write our system $X(t)$ in (2.2) on the form

$$\begin{aligned} dX(t) &= (-\rho X(t) - c(t))dt + d\xi(t) \\ &= (E[\mu|\mathcal{G}_t] - \rho X(t) - c(t))dt + \alpha(t)dB(t) \; . \end{aligned} \tag{4.40}$$

Note that in this equation all the coefficients are adapted to the observed filtration $\mathcal{G}_t$. Thus we have transformed the original partial *observation problem (2.2), (2.14) into the* complete *observation problem (4.40), (2.14). However, the price we pay for this is that the constant μ in (2.2) is replaced by the more complicated process $E[\mu \mid \mathcal{G}_t]$ in (4.40).*

We now proceed to solve the (complete observation) stochastic control problem (4.40), (2.14). To this end, we first write the system in Markovian form, as follows:

$$dX(t) = \Big[\frac{\bar{\mu}A + Y(t)}{A + Z(t)} - \rho X(t) - c(t)\Big]dt + \alpha(t)dB(t) \; ; \qquad X(0) = x \tag{4.41}$$

$$dY(t) = \alpha^{-2}(t)(A + Z(t))^{-1}(\bar{\mu}A + Y(t))dt + \alpha^{-1}(t)dB(t) \; ; \; Y(0) = y \tag{4.42}$$

$$dZ(t) = \alpha^{-2}(t)dt \; ; \qquad Z(0) = z \tag{4.43}$$

To emphasize the initial values x, y, z we sometimes write $X^x(t)$, $Y^y(t)$ and $Z^z(t)$ and we let $E^{x,y,z}$ denote the expectation when these initial values are assumed.

In this setting it is natural to extend Problem 2.2 slightly, as follows:

Problem 4.4. *Define*

$$J^{(c)}(x) = E^{x,0,0}\Big[\int_0^T e^{-\delta t}\,\frac{c^\gamma(t)}{\gamma}dt\Big]; \qquad s \le T \tag{4.44}$$

Find $J^*(x)$ *and* $c^* \in \mathcal{A}_p = \mathcal{A}_p(x)$ *such that*

$$J^*(x) = \sup\{J^{(c)}(x);\ c \in \mathcal{A}_p\} = J^{(c^*)}(x)\ . \tag{4.45}$$

To solve this problem we again apply the stochastic maximum principle (Theorem 3.1). In this case the Hamiltonian is

$$H(s,x,y,z,c,p,q) = \Big(\frac{(\bar\mu A + y)}{A+z} - \rho x - c\Big)p_1 + \alpha^{-2}(s)(A+z)^{-1}(\bar\mu A + y)p_2$$
$$+\alpha^{-2}(s)p_3 + \alpha(s)q_1 + \alpha^{-1}(s)q_2 + e^{-\delta s}\,\frac{c^\gamma}{\gamma}\ , \tag{4.46}$$

where $p = (p_1,p_2,p_3)^T$, $q = (q_1,q_2,q_3)^T$.

The function $c \to H(s,x,y,z,c,p,q)$ is maximal when

$$c = c^*(s) = (p_1(s)e^{\delta s})^{\frac{1}{\gamma-1}}. \tag{4.47}$$

The adjoint equation system gets the form

$$dp_1(t) = \rho\, p_1(t)dt + q_1(t)dB(t) \tag{4.48}$$
$$dp_2(t) = -(A + Z^0(t))^{-1}[p_1(t) + \alpha^{-2}(t)p_2(t)]dt + q_2(t)dB(t) \tag{4.49}$$
$$dp_3(t) = (A+Z^0(t))^{-2}[(\bar\mu A+Y^0(t))p_1(t) \tag{4.50}$$
$$+\alpha^{-2}(t)(\bar\mu A + Y^0(t))p_2(t)]dt + q_3(t)dB(t)$$

with terminal values

$$p_1(T) = \lambda_1\ , \quad p_2(T) = p_3(T) = 0 \tag{4.51}$$

where $\lambda_1 \in \mathbf{R}$ is a constant to be determined.

We see that (4.48) has the solution

$$p_1(t) = \lambda_1 e^{-\rho(T-t)}\ , \qquad q_1(t) = 0\ . \tag{4.52}$$

Hence, by (4.47),

$$c^*(t) = \lambda_1^{\frac{1}{\gamma-1}} \exp\Big(\frac{1}{1-\gamma}(\rho(T-t) - \delta t)\Big)\ . \tag{4.53}$$

To determine λ_1, we now proceed as in the case of complete observations : By integrating the equation

$$dX^*(t) = (E[\mu|\mathcal{G}_t] - \rho X^*(t) - c^*(t))dt + \alpha(t)dW(t) \tag{4.54}$$

and taking the expectation we get, with

$$\bar{\mu} = E[\mu] = E[E[\mu|\mathcal{G}_t]] \,, \tag{4.55}$$

$$E[X^*(t)] = x + \bar{\mu}t - \rho \int_0^t E[X^*(s)]ds - \int_0^t c^*(s)ds \,.$$

Hence the terminal condition (2.13) gives, by the same calculation as in Section 3,

$$\lambda_1 = \Big(xe^{-\rho T} + \frac{\bar{\mu}}{\rho}(1-e^{-\rho T}) - x_T\Big)^{\gamma-1} \Big[\frac{1-\gamma}{\delta+\gamma\rho}\{\exp\Big(\frac{\gamma\rho T}{1-\gamma}\Big) - \exp\Big(-\frac{\delta T}{1-\gamma}\Big)\}\Big]^{1-\gamma} \text{if } \delta+\gamma\rho \neq 0 \tag{4.56}$$

$$\lambda_1 = \Big(xe^{-\rho T} + \frac{\bar{\mu}}{\rho}(1-e^{-\rho T}) - x_T\Big)^{\gamma-1} T^{1-\gamma}\exp(-\delta T) \quad \text{if } \delta+\gamma\rho = 0, \tag{4.57}$$

provided that

$$xe^{-\rho T} + \frac{\bar{\mu}}{\rho}(1-e^{-\rho T}) \geq x_T \,. \tag{4.58}$$

We have proved :

Theorem 4.5. *Assume that (4.58) holds. Then the optimal consumption rate $c^*(t)$ for the partial observation problem (Problem 2.2) is*

$$c^*(t) = \lambda_1^{\frac{1}{\gamma-1}} \exp\Big(\frac{1}{1-\gamma}(\rho(T-t)-\delta t)\Big) \tag{4.59}$$

where λ_1 is given by (4.56),(4.57).

Remark 4.6. *Note that the solutions $\bar{c}(t)$ and $c^*(t)$ of the complete and partial observation problem, respectively, are identical except that the constant μ in the expression for $\bar{c}(t)$ has been replaced by the constant $\bar{\mu} = E[\mu]$ in the expression for $c^*(t)$.*

Since $\bar{\mu}$ is assumed known a priori, this means -perhaps surprisingly- that observing $\xi(s); s \leq t$ has no effect at all on our consumption rate.

Remark 4.7. *In general, we say that a partial observation control problem satisfies the* certainty equivalence principle *if the following holds : Let $S(\theta)$ denote the optimal control in the complete observation case with known parameter θ. Then $S(\hat{\theta})$ is the optimal control in the partial observation case, where $\hat{\theta} = E[\theta|\mathcal{G}_t]$ is the estimate of θ based on the observation $\mathcal{G}_t$.*

It is known that the certainty equivalence principle holds in the quadratic-linear partial observation control problems (see e.g. [B2]*). On the other hand, it has been proved that in the Merton type of optimal consumption/portfolio problem in a Black-Scholes market with unobservable drift vector, the certainty equivalence principle holds if and only if the utilities are of logarithmic type (see* [Ku]*). Our problem gives another example where the certainty equivalence principle does not hold.*

5. The value function

Finally we find the value functions $\bar{J} = \bar{J}(x)$ and $J^* = J^*(x)$ of the complete observation problem (Problem 2.1) and the partial observation problem (Problem 2.2), respectively. We could compute $\bar{J}(x)$ and $J^*(x)$ by evaluating directly

$$\bar{J}(x) = E\Big[\int_0^T e^{-\delta t}\,\frac{\bar{c}^{\gamma}(t)}{\gamma}dt\Big] \qquad \text{and} \qquad J^*(x) = E\Big[\int_0^T e^{-\delta t}\,\frac{(c^*)^{\gamma}(t)}{\gamma}dt\Big]\,,$$

respectively. However, it is easier to use the connection

$$\frac{d}{dx}\bar{J}(x) = p(0)\,, \qquad \frac{d}{dx}J^*(x) = p_1(0) \tag{5.60}$$

between the maximum principle and dynamic programming (see e.g. [YZ, Theorem 5.4.1]):

In the complete observation case this gives, by (3.27)

$$\frac{d}{dx}\bar{J}(x) = \lambda\, e^{-\rho T}\,, \tag{5.61}$$

where $\lambda = \lambda(x)$ is given by (3.30),(3.31). Therefore

$$\bar{J}(x) = \frac{1}{\gamma}\Big(xe^{-\rho T} + \frac{\mu}{\rho}(1-e^{-\rho T}) - x_T\Big)^{\gamma}\psi + C\,,$$

where C is a constant and

$$\psi = \begin{cases} \Big[\frac{1-\gamma}{\delta+\gamma\rho}\big\{\exp\big(\frac{\gamma\rho T}{1-\gamma}\big) - \exp\big(-\frac{\delta T}{1-\gamma}\big)\big\}\Big]^{1-\gamma} e^{-\rho T} & \text{if} \quad \delta+\gamma\rho \neq 0 \\ T^{1-\gamma}\exp(-(\delta+\rho)T) & \text{if} \quad \delta+\gamma\rho = 0 \end{cases} \tag{5.62}$$

Since $\bar{J}(x) = 0$ when $xe^{-\rho T} + \frac{\mu}{\rho}(1-e^{-\rho T}) = x_T$ we have $C = 0$. Hence

$$\bar{J}(x) = \frac{1}{\gamma}\Big(xe^{-\rho T} + \frac{\mu}{\rho}(1-e^{-\rho T}) - x_T\Big)^{\gamma}\psi\,. \tag{5.63}$$

Similarly, for the partial observation problem we get, by (4.52),

$$\frac{d}{dx}J^*(x) = \lambda_1(x)\, e^{-\rho T} \tag{5.64}$$

where $\lambda_1 = \lambda_1(x)$ is given by (4.56),(4.57).

This gives

$$J^*(x) = \frac{1}{\gamma}\Big(xe^{-\rho T} + \frac{\bar{\mu}}{\rho}(1-e^{-\rho T}) - x_T\Big)^{\gamma}\psi_1\,, \tag{5.65}$$

where $\psi_1 = \psi$ is given by (5.62).

We have proved:

Theorem 5.1. *The value functions $\bar{J}(x)$ and $J^*(x)$ of the complete observation problem (Problem 2.1) and the partial observation problem (Problem 2.2), respectively, are given by*

$$\bar{J}(x) = \frac{1}{\gamma}\Big(xe^{-\rho T} + \frac{\mu}{\rho}(1 - e^{-\rho T}) - x_T\Big)^{\gamma}\psi \tag{5.66}$$

$$J^*(x) = \frac{1}{\gamma}\Big(xe^{-\rho T} + \frac{\bar{\mu}}{\rho}(1 - e^{-\rho T}) - x_T\Big)^{\gamma}\psi_1 \tag{5.67}$$

where $\psi = \psi_1$ are both given by (5.62).

ACKNOWLEDGEMENTS This work was partially supported by the French-Norwegian cooperation project Aur 99–050.

References

[B1] A. Bensoussan: Perturbation Methods in Optimal Control. J. Wiley & Sons 1988.

[B2] A. Bensoussan: Stochastic Control of Partially Observable Systems. Cambridge Univ. Press 1992.

[D] M.H.A. Davis: Linear Estimation and Stochastic Control. Chapman and Hall 1977.

[EGK] R.J. Elliott, H. Geman and B.M. Korkie: Portfolio optimization and contingent claim pricing with differential information. Stochastics and Stochastics Reports 60 (1997), 185–203.

[Ka] G. Kallianpur: Stochastic Filtering Theory. Springer-Verlag 1980.

[Ku] Y. Kuwana: Certainty equivalence and logarithmic utilities in consumption/investment problems. Math. Finance 5 (1995), 297–309

[Ø] B. Øksendal: Stochastic Differential Equations. Fifth edition. Springer-Verlag 2000.

[P] S. Peng: A general stochastic maximum principle for optimal control problems. SIAM J. Control & Opt. (1990), 966–979.

[YZ] J. Yong and X.Y. Zhou: Stochastic Controls. Springer-Verlag 1999.

[1]INRIA, Domaine de Voluceau-Rocquencourt, B.P. 105, F–78153 Le Chesnay Cedex, France
E-mail address: `david.lefevre@inria.fr` and `Agnes.Sulem@inria.fr`

[2]Dept. of Mathematics, University of Oslo, P. O. Box 1053 Blindern, N–0316 Oslo, Norway

[3]Norwegian School of Economics and Business Administration, Helleveien 30, N–5045 Bergen, Norway
E-mail address: `oksendal@math.uio.no`

Trends in Mathematics, © 2001 Birkhäuser Verlag Basel/Switzerland

Continuous Time CAPM, Price for Risk and Utility Maximization

Johannes Leitner

Abstract. In an arbitrage free incomplete market we consider the problem of maximizing terminal isoelastic utility. The relationship between the optimal portfolio, the optimal martingale measure in the dual problem and the optimal value function of the problem is described by an BSDE. For a *totally unhedgeable price for instantaneous risk*, isoelastic utility of terminal wealth can be maximized using a portfolio consisting of the locally risk-free bond and a *locally efficient* fund only. In a markovian market model we find a non-linear PDE for the logarithm of the value function. From the solution we can construct the optimal portfolio and the solution of the dual problem. In a market with a zero bond, we solve the mean-variance efficiency problem and find the *intertemporal price for risk* to equal the standard deviation of the discounted variance-optimal martingale measure divided by the zero bond price.

1. Preliminaries

Let a filtered probability space $\Omega_\infty := (\Omega, \mathcal{F}, (\mathcal{F}_s)_{s\geq 0}, P)$, satisfying the usual conditions be given. For simplicity we assume $\mathcal{F}_0$ to be trivial up to sets of measure 0 with respect to P and $\mathcal{F}_{\infty-} = \mathcal{F}_\infty := \mathcal{F}$. Let $S = (S_t)_{0\leq t<\infty}$ be a $\mathbf{R}^d$-valued continuous semimartingale. $\mathcal{M} := (\Omega_\infty, S)$ is a common model for a market, where S describes the price processes of d assets. We will always work with a fixed finite time horizon $0 \leq T < \infty$. This is equivalent to working with the market model $\mathcal{M}_T := (\Omega_T, S^T)$, where $\Omega_T := (\Omega, \mathcal{F}_T, (\mathcal{F}_{s\wedge T})_{0\leq s}, P_{|\mathcal{F}_T})$. We often write S instead of S^T if we are working in $\mathcal{M}_T$.

Denote by $L^1(S)$ the space of predictable $\mathbf{R}^d$-valued processes which are integrable with respect to S, see [17], [18] and [28] for a summary on vector stochastic integration. We want to model the economic activity of investing money into a portfolio of assets and changing the number of assets held over time according to a certain hedging strategy. This is achieved with the following definition:

Definition 1.1. *A* hedging strategy *in the market $\mathcal{M}$ is a $H \in L^1(S)$. The corresponding* value process *V^H of H is defined as $V^H := HS$. The* gains process *of H is defined as the semimartingale $G^H := H \cdot S$. H is called* self-financing *if*

$V^H = V_0^H + G^H$, i.e. $H_tS_t = H_0S_0 + \int_0^t H_s dS_s$, $\forall\, t \geq 0$. Denote the space of all self-financing hedging strategies in $\mathcal{M}$ by $\mathcal{SF}(\mathcal{M})$.

Definition 1.2. *A continuous semimartingale $B > 0$ is called a* numéraire *for the market $\mathcal{M}$. The market discounted with respect to B is then defined as $\mathcal{M}^B := (\Omega_\infty, S^B)$, where $S^B := \frac{S}{B}$. The discounted market restricted to the interval $[0,T]$ is defined as $\mathcal{M}_T^B := (\mathcal{M}^B)_T = \left(\Omega_T, (S^B)^T\right)$.*

See [11] and [12] for the following result:

Proposition 1.3. *Let B be a numéraire for the market $\mathcal{M}$. Then $\mathcal{SF}(\mathcal{M}^B) = \mathcal{SF}(\mathcal{M})$ holds.*

We will assume the existence of a traded numéraire B and consider the market $\bar{\mathcal{M}}_T := (\Omega_T, \bar{S}^T)$, where $\bar{S} := (S, B)$. In addition we assume that the numéraire B is a continuous semimartingale of finite variation, uniformly bounded and uniformly bounded away from 0 on finite intervals and $B_0 = 1$. Since we are working with a continuous semimartingale model, the price process admits a representation

$$S = S_0 + \mu \cdot \alpha + M, \tag{1}$$

where $\mu = (\mu^i)_{1 \leq i \leq d}$ is predictable, α is a predictable, increasing, continuous, locally integrable process such that μ is locally integrable with respect to α. Furthermore, there exists a symmetric non-negative $d \times d$-matrix-valued predictable process $C = (C^{ij})_{1 \leq i,j \leq d}$, locally integrable with respect to α, such that $[S^i, S^j] = [M^i, M^j] = < M^i, M^j > = C^{ij} \cdot \alpha$. α can be chosen such that $B = \mathcal{E}(r \cdot \alpha)$ for a predictable process r. Denote the set of uniformly integrable, resp. local martingales, living on Ω_T by $\mathcal{L}^u(\Omega_T)$, resp. by $\mathcal{L}(\Omega_T)$.

So far we did not worry about arbitrage. We define the following sets of local martingale measures:

$$\mathcal{D}(\bar{\mathcal{M}}_T) := \left\{ Z \in \mathcal{L}(\Omega_T) | Z_0 = 1, Z > 0, (S^B)^T Z \in \mathcal{L}(\Omega_T) \right\}, \tag{2}$$

$$\mathcal{D}^e(\bar{\mathcal{M}}_T) := \left\{ Z \in \mathcal{D}(\bar{\mathcal{M}}_T) | Z \in \mathcal{L}^u(\Omega_T) \right\}, \tag{3}$$

and for $q > 1$

$$\mathcal{D}^q(\bar{\mathcal{M}}_T) := \left\{ Z \in \mathcal{D}^e(\bar{\mathcal{M}}_T) | Z_T \in L^q(\Omega_T) \right\}, \tag{4}$$

resp. for $q < 1$ set $\mathcal{D}^q(\bar{\mathcal{M}}_T) := \mathcal{D}(\bar{\mathcal{M}}_T)$. We will work with the following No-Arbitrage condition:

$$\mathcal{D}^e(\bar{\mathcal{M}}_T) \neq \emptyset, \;\; \forall\, T < \infty, \tag{5}$$

resp. for $q > 1$

$$\mathcal{D}^q(\bar{\mathcal{M}}_T) \neq \emptyset, \;\; \forall\, T < \infty. \tag{6}$$

The first condition is known to be equivalent to the NFLVR-condition, see [15] and [7].

p will always denote a real number different from 1. We define $q := \frac{p}{p-1}$, such that for $p \neq 0, 1$, $p^{-1} + q^{-1} = 1$ holds, but for $p = 0$ we have $q = 0$.

In all probabilistic theories of financial markets allowing to trade at an infinitely large number of instances of time one has to exclude certain self-financing hedging strategies, e.g. doubling strategies, in order to avoid arbitrage opportunities. We will work with following arbitrage-free subsets of $\mathcal{SF}(\bar{\mathcal{M}}_T)$:

1. For $p > 1$ and $\mathcal{D}^q(\bar{\mathcal{M}}_T) \neq \emptyset$, (see [8]):

$$\mathcal{SF}^p(\bar{\mathcal{M}}_T) := \Big\{ H \in \mathcal{SF}(\bar{\mathcal{M}}_T) | V_T^H \in L^p(\Omega_T), \frac{V^H}{B^T} Z \in \mathcal{L}^u(\Omega_T), \forall\, Z \in \mathcal{D}^q(\bar{\mathcal{M}}_T) \Big\}, \tag{7}$$

2. For $p < 1$

$$\mathcal{SF}^p(\bar{\mathcal{M}}_T) := \{ H \in \mathcal{SF}(\bar{\mathcal{M}}_T) | V^H \geq 0 \}. \tag{8}$$

Define for a constant v

$$\mathcal{A}_v^{*p}(\bar{\mathcal{M}}_T) := \left\{ \frac{V_T^H}{B_T} | H \in \mathcal{SF}^p(\bar{\mathcal{M}}_T), V_0^H = v \right\}. \tag{9}$$

For $p > 1$ and $\mathcal{D}^q(\bar{\mathcal{M}}_T) \neq \emptyset$, $\mathcal{SF}^p(\bar{\mathcal{M}}_T)$ has an important property: $\mathcal{A}_1^{*p}(\mathcal{M}_T)$ is known to be closed, if $\left(\frac{S}{B}\right)^T$ is locally in $L^p(\Omega_T)$ in the sense, that there exists a sequence $U_n, n \in \mathbf{N}$ of localizing stopping times increasing to infinity such that for each n, the family $\{S_\tau^T | \tau \text{ stopping time}, \tau \leq U_n\}$ is bounded in $L^p(\Omega_T)$, see [8]. This condition certainly holds if $\bar{S}$ is continuous.

In the continuous case, $\mathcal{D}^e(\bar{\mathcal{M}}_T) \neq \emptyset$ implies $\mu = rS - C\lambda$, $P \otimes d\alpha$-almost surely for a predictable process $\lambda \in L^2_{loc}(M)$ and every $Z \in \mathcal{D}^e(\bar{\mathcal{M}}_T)$ is of the form $Z = \mathcal{E}(\lambda \cdot M + N)^T$, where $N \in \mathcal{L}(\Omega_T)$ is a not necessarily continuous local martingale orthogonal to M with $[M, N] = 0$, see [1].

2. The Price for Instantaneous Risk

We start with a lemma, that will allow us to define self-financing hedging strategies in a very convenient way:

Lemma 2.1. *Let $H \in \mathcal{SF}(\mathcal{M})$ be such that $V^H > 0$ almost surely. Set $\tilde{H} := \frac{H}{V^H}$. Then $\tilde{H} \in L^1(S)$, $\tilde{H}S = 1$ and*

$$V^H = V_0^H + V^H \cdot (\tilde{H} \cdot S) = V_0^H \mathcal{E}(\tilde{H} \cdot S), \tag{10}$$

holds. Conversely, let $\tilde{H} \in L^1(S)$ with $\tilde{H}S = 1$ be given and set $H := v_0 \mathcal{E}(\tilde{H} \cdot S)\tilde{H}$ for a $\mathcal{F}_0$-measurable random variable v_0. Then $H \in \mathcal{SF}(\mathcal{M})$ and $V^H = v_0 \mathcal{E}(\tilde{H} \cdot S)$. We call $\tilde{H}$ a generator for the self-financing strategy H and define $V^{(\tilde{H})} := V^H$.

Proof. This follows immediately from the uniqueness of the solution to the Doléan-Dade SDE defining the stochastic exponential, see [18]. □

λ is often called the risk premium vector process. We want to draw attention to the following interpretation of the quantity $\beta := \sqrt{\lambda C \lambda}$: From Lemma 2.1 and $\mu = rS - C\lambda$, $P \otimes d\alpha$-a.s., we immediately find $V^{(\hat{H})} := \mathcal{E}\left(\hat{H} \cdot \bar{S}\right)^T =$

$\mathcal{E}\left((r - HC\lambda)\cdot\alpha + H\cdot M\right)^T$ for a process $\hat{H} = \left(H, \frac{1-HS}{B}\right) \in L^1(\bar{S})$. From Cauchy-Schwarz inequality it follows that $|HC\lambda| \le \sqrt{\lambda C\lambda}\sqrt{HCH} = \beta\sqrt{HCH}$. We have

$$[V^{(\hat{H})}, V^{(\hat{H})}] = \left(V^{(\hat{H})}\right)^2 HCH\cdot\alpha^T.$$

We interpret $\sqrt{HCH}$ as a measure for the relative instantaneous risk of the portfolio generated by $\hat{H}$ and $\hat{H}(\mu, rB) = r - HC\lambda$ as a measure for the instantaneously expected relative return rate. For $\beta \neq 0$ and $\sqrt{HCH} \neq 0$, we find for the instantaneous Sharpe-ratio $\frac{\hat{H}(\mu,rB)-r}{\sqrt{HCH}} = \frac{-HC\lambda}{\sqrt{HCH}}$ of instantaneously expected relative excess return over the instantaneously risk-free return rate and the relative instantaneous risk, $-\beta \le \frac{-HC\lambda}{\sqrt{HCH}} \le \beta$ and $\frac{-HC\lambda}{\sqrt{HCH}} = \beta$ iff $H \in k\lambda + \mathrm{Ker}(C)$ for a predictable, strictly negative, process $k \in L^2_{loc}(\beta)$, resp. $\frac{-HC\lambda}{\sqrt{HCH}} = -\beta$ iff $H \in k\lambda + \mathrm{Ker}(C)$ for a predictable, strictly positive, process $k \in L^2_{loc}(\beta)$. We call these hedging strategies *locally efficient* and β can be interpreted as an instantaneous price for risk or instantaneous Sharpe-ratio. See [24] for the notion of *myopic strategies*, see [22] and [27] for the origin of this idea in a single period market model.

The obvious question is now:

When are locally efficient hedging strategies globally optimal ?

3. Utility Maximization

In this section we define what we mean by globally optimal. Assume $\mathcal{D}^q(\bar{\mathcal{M}}_T) \neq 0$ for $p > 1$, resp. $\mathcal{D}^e(\bar{\mathcal{M}}_T) \neq 0$ for $p < 1$. Consider the *isoelastic utility functions*: (see [25] and [2]),

$$U^{(p)}(x) := \begin{cases} -\frac{|x|^p}{p} & : \; x \in \mathbf{R}, \quad p > 1 \\ \frac{x^p}{p} & : \; x \ge 0\,, \quad p < 1, p \neq 0 \\ \ln(x) & : \; x > 0\,, \quad p = 0 \end{cases} \tag{11}$$

and the optimization problem, (see [23], [16], [19] and [3]),

$$\mathcal{V}(p, T) := \sup_{\substack{H \in \mathcal{SF}^p(\bar{\mathcal{M}}_T) \\ V_0^H = 1}} E\left[U^{(p)}(V_T^H)\right], \tag{12}$$

together with the dual problem, (see [14], [13])

$$\mathcal{W}(q, T) := \inf_{Z \in \mathcal{D}^q(\bar{\mathcal{M}}_T)} E\left[-U^{(q)}\left(\frac{Z_T}{B_T}\right)\right]. \tag{13}$$

Denote the unique solutions of the two problems by $V^{opt,p,T}$, resp. $Z^{opt*,q,T}$. See [26] for general duality theory and [4],[5] for stochastic duality theory. We have the following result about the relationship of the two solutions:

Theorem 3.1. *Assume $\mathcal{V}(p,T) < \infty$ for $p < 1$, and $\mathcal{D}^q(\bar{\mathcal{M}}_T) \neq \emptyset$ for $p > 1$. Then*

$$cE\left[B_T \frac{dU^{(p)}(V_T^{opt,p,T})}{dx} | \mathcal{F}_\cdot\right] = Z^{opt*,q,T} \in \mathcal{D}^q(\bar{\mathcal{M}}_T), \tag{14}$$

for a constant c. Conversely, if for a constant c

$$Z := cE\left[B_T \frac{dU^{(p)}(V_T^H)}{dx} | \mathcal{F}_\cdot\right] \in \mathcal{D}^q(\bar{\mathcal{M}}_T), \tag{15}$$

for a $H \in \mathcal{SF}^p(\bar{\mathcal{M}}_T)$ with $V_0^H = 1$ such that $\frac{V^H}{B^T} Z$ is a uniformly integrable martingale, then

$$(V^{opt,p,T}, Z^{opt*,q,T}) = (V^H, Z). \tag{16}$$

Proof. For $p < 1$, existence follows from [21] and an additional argument using Lemma 2.1, showing that $Z^{opt*,q,T} \in \mathcal{D}(\bar{\mathcal{M}}_T)$ in the continuous semimartingale model. For $p > 1$ existence of $V^{opt,p,T}$ follows from the uniform convexity of L^p spaces. Conversely, by the concavity of $U^{(p)}$ we have

$$\begin{aligned} U^{(p)}(V_T^{\tilde{H}}) &\leq U^{(p)}(V_T^H) + B_T \frac{dU^{(p)}(V_T^H)}{dx} \frac{V_T^{\tilde{H}} - V_T^H}{B_T} \\ &= U^{(p)}(V_T^H) + c^{-1} Z_T \frac{V_T^{\tilde{H}} - V_T^H}{B_T}, \end{aligned}$$

for all $V^{\tilde{H}}, \tilde{H} \in \mathcal{SF}^p(\bar{\mathcal{M}}_T)$, with $V_0^{\tilde{H}} = 1$. Hence

$$E\left[U^{(p)}\left(V_T^{\tilde{H}}\right)\right] \leq E\left[U^{(p)}\left(V_T^H\right)\right]. \tag{17}$$

Similarly we have for $\tilde{Z} \in \mathcal{D}^q(\bar{\mathcal{M}}_T)$

$$\begin{aligned} U^{(q)}\left(\frac{\tilde{Z}_T}{B_T}\right) &\leq U^{(q)}\left(\frac{Z_T}{B_T}\right) + \frac{dU^{(q)}}{dx}\left(\frac{Z_T}{B_T}\right) \frac{\tilde{Z}_T - Z_T}{B_T} \\ &= U^{(q)}\left(\frac{Z_T}{B_T}\right) + c' \frac{V_T^H}{B_T}\left(\tilde{Z}_T - Z_T\right), \end{aligned}$$

for a constant c' of appropriate sign, hence

$$E\left[-U^{(q)}\left(\frac{Z_T}{B_T}\right)\right] \leq E\left[-U^{(q)}\left(\frac{\tilde{Z}_T}{B_T}\right)\right]. \tag{18}$$

□

4. Totally Unhedgeable Price for Risk

We are now going to answer the question posed at the end of Section 2.

Definition 4.1. *For $0 \leq T < \infty$, let a constant $c > 0$ and a not necessarily continuous local martingale N orthogonal to M, (or equivalently with $[N, M] = 0$), be given such that*

$$E\left[\mathcal{E}\left(\left(pr - q\frac{\beta^2}{2}\right) \cdot \alpha\right)_T |\mathcal{F}_\cdot\right] = c\mathcal{E}(N). \tag{19}$$

We then call the instantaneous price for risk β^T in the market $\bar{\mathcal{M}}_T$ totally *p*-unhedgeable.

For the case $p = 0$ we can choose $c = 1, N = 0$. The optimization problem (12) for $p = 0$ is also known as maximizing the Kelly-criterion, see [20], [6] and [12].

Proposition 4.2. *Assume β^T to be totally p-unhedgeable in $\bar{\mathcal{M}}_T$ with a representation (19). For $p > 1$ assume $\mathcal{E}(\lambda \cdot M)^T \in \mathcal{D}^q(\bar{\mathcal{M}}_T)$, resp. for $p < 1$ assume $\mathcal{E}(q\lambda \cdot M + N)^T$ to be a uniformly integrable martingale. Then the optimal pair for the market $\bar{\mathcal{M}}_T$ is given by*

$$\left(V^{opt,p,T}, Z^{opt*,q,T}\right) = \left(V^{(H^p)}, \mathcal{E}(\lambda \cdot M + N)^T\right), \tag{20}$$

where $H^p := \left(\frac{\lambda}{p-1}, \frac{1 - \frac{\lambda S}{p-1}}{B}\right)^T$ generates the value process

$$V^{(H^p)} := \mathcal{E}\left(\left(r - \frac{\beta^2}{p-1}\right) \cdot \alpha + \frac{\lambda}{p-1} \cdot M\right)^T. \tag{21}$$

Proof. Observe that $cB_T \frac{dU^{(p)}}{dx}\left(V_T^{(H^p)}\right) = \mathcal{E}\left(\lambda \cdot M + N\right)_T$, for a constant c. For $p > 1$, $\mathcal{E}(\lambda \cdot M)^T \in \mathcal{D}^q(\bar{\mathcal{M}}_T)$ implies $V^{(H^p)} H^p \in \mathcal{SF}^p(\bar{\mathcal{M}}_T)$ and $\mathcal{E}(\lambda \cdot M + N)^T \in \mathcal{D}^q(\bar{\mathcal{M}}_T)$. Now apply Theorem 3.1. □

The case of a totally unhedgeable price for risk is a very special one. In the next section we will deal with a quite general situation.

5. The BSDE Approach

Assume the existence of a continuous local martingale N orthogonal to M such that (M, N) has the local martingale representation property and $[N, N] = C' \cdot \alpha$. We assume in this section $p \neq 0, 1$. Let $0 \leq t \leq T$ be fixed.

Consider the following ansatz:

$$cB_T U'(V_T^{opt}) = \mathcal{E}(\lambda \cdot M + \nu \cdot N)_T, \tag{22}$$

respectively

$$\ln\left(\left(cB_T U'\left(V_T^{opt}\right)\right)^{-1} \mathcal{E}(\lambda \cdot M + \nu \cdot N)_T\right) = 0, \tag{23}$$

where $V^{opt} = \left(V^{(\hat{H})}\right)^T$ for $\hat{H} = \left(H, \frac{1-HS}{B}\right)^T \in L^2(\bar{S})$ and $\nu \in L^2_{loc}(N)$. Ansatz (22) leads to a FBSDE. For the isoelastic utility functions ansatz (23) will lead to a BSDE, where (H, ν) form part of the solution, see [29] for an introduction to BSDE theory. For $0 \leq s \leq T$ define the adapted process

$$Y_s := \ln\left(\left(cB_s \frac{dU^{(p)}}{dx}\left(V_s^{opt,p,T}\right)\right)^{-1} \mathcal{E}(\lambda \cdot M + \nu \cdot N)_s\right). \tag{24}$$

Applying Itô's formula leads to the following BSDE for $0 \leq t' \leq T$:

$$\begin{aligned} Y_{t'}^{(p,T)} &= -\int_{t'}^{T} \frac{(p-1)H_s C_s H_s - \nu_s C'_s \nu_s - \lambda_s C_s \lambda_s}{2} d\alpha_s \\ &\quad - \int_{t'}^{T} ((p-1)H_s C_s \lambda_s - pr_s)\, d\alpha_s \\ &\quad - \int_{t'}^{T} (\lambda_s - (p-1)H_s)\, dM_s - \int_{t'}^{T} \nu_s dN_s. \end{aligned} \tag{25}$$

Conversely, given an adapted solution $(Y^{(p,T)}, H, \nu)$ to the BSDE (25) on $[0, T]$, we can define a self-financing hedging strategy in $\bar{\mathcal{M}}_T$ by using $\hat{H} := \left(H, \frac{1-HS}{B}\right)^T$ as a generator for $V^{(\hat{H})} := \mathcal{E}\left((r - HC\lambda) \cdot \alpha + H \cdot M\right)^T \in \mathcal{SF}^0(\bar{\mathcal{M}}_T)$. We also have $Z^\nu := \mathcal{E}(\lambda \cdot M + \nu \cdot N)^T \in \mathcal{D}(\bar{\mathcal{M}}_T)$.

Proposition 5.1. *Assume $(Y^{(p,T)}, (H, \nu))$ to be a solution to the BSDE (25) on $[0, T]$. If for $p < 1$, $\mathcal{E}\left((H + \lambda) \cdot M + \nu \cdot N\right)^T$ is a uniformly integrable martingale, respectively if for $p > 1$, $V^{(\hat{H})}\hat{H} \in \mathcal{SF}^p(\bar{\mathcal{M}}_T)$ and $Z^\nu \in \mathcal{D}^q(\bar{\mathcal{M}}_T)$, then $\left(V^{(\hat{H})}, Z^\nu\right)$ is the optimal pair for the market $\bar{\mathcal{M}}_T$. Furthermore*

$$\mathcal{V}(p, T) = -\mathrm{sgn}(q) \exp(Y_0^{(p,T)}). \tag{26}$$

Proof. The assertion follows from Theorem 3.1. □

Conversely, the existence of an optimal pair for the market $\bar{\mathcal{M}}_T$ together with the local martingale representation property of (M, N), implies the unique existence of a solution $(Y^{(p,T)}, (H, \nu))$ for the BSDE (25) on $[0, T]$ satisfying the assumptions of Proposition 5.1.

6. The Markovian Case

Consider the following Markovian Model: Let $W = (W_1, W_2) \in \mathbf{R}^d \times \mathbf{R}^{d'}$ be a standard Brownian Motion on Ω_∞ and $\hat{\mu} = (\mu, \mu') : \mathbf{R}^{d+d'} \to \mathbf{R}^d \times \mathbf{R}^{d'}$, $\sigma : \mathbf{R}^{d+d'} \to \mathbf{R}^{(d+d')\times(d+d')}$ be functions such that the (for simplicity time-homogeneous) SDE

$$dX_t = \hat{\mu}(X_t)dt + \sigma(X_t)dW_t, \quad X_0 = x_0, \tag{27}$$

has a strong solution $X = (S, S') \in \mathbf{R}^d \times \mathbf{R}^{d'}$ which defines a market model $\bar{\mathcal{M}}_T^{x_0} := \left(\Omega_T, (S^T, B^T)\right)$, where $B_t := \exp\left(\int_0^t r(X_u)du\right)$ for a bounded function $r : \mathbf{R}^{d+d'} \to \mathbf{R}$ and S' is to be interpreted as a non-traded state variable. Assume

$\sigma^*\sigma = C \times C'$ for $C : \mathbf{R}^{d+d'} \to \mathbf{R}^{d\times d}$, resp. $C' : \mathbf{R}^{d+d'} \to \mathbf{R}^{d'\times d'}$, (hence in the general setting above $M = \sqrt{C} \cdot W_1, N = \sqrt{C'} \cdot W_2, \alpha_t = t$), and $\mu = rS - C\lambda$ for a $\lambda : \mathbf{R}^{d+d'} \to \mathbf{R}^d$ such that $\beta^2 := \lambda C \lambda$ is bounded. We expect the solution $Y^{(p,T)}$ of the BSDE (25) to be of a markovian form:

$$Y_t^{(p,T)} = Y^p(S_t, S'_t, T - t). \tag{28}$$

From the BSDE (25) we derive the following non-linear PDE for Y^p:

$$-\frac{\partial Y^p}{\partial t} + \mathcal{L}Y^p = \mathcal{L}^{(p)}Y^p + \mathcal{L}'Y^p + q\frac{\beta^2}{2} - pr, \tag{29}$$

with boundary condition $Y^p(s, s', 0) = 0$, $\forall\, s, s'$, where $\mathcal{L}$ is the Itô operator and for f differentiable in s and s':

$$\begin{aligned}
\mathcal{L}'f &:= -\frac{1}{2}\sum_{i,j=1}^{d'} \frac{\partial f}{\partial s'_i} C'_{i,j} \frac{\partial f}{\partial s'_j} \\
\mathcal{L}^{(p)}f &:= \frac{1}{2(p-1)}\sum_{i,j=1}^{d} \frac{\partial f}{\partial s_i} C_{i,j} \frac{\partial f}{\partial s_j} - q\sum_{i,j=1}^{d} \frac{\partial f}{\partial s_i} C_{i,j}\lambda_j.
\end{aligned}$$

Assume Y^p to be a $C^{2,2,1}$-solution of the PDE (29), with $Y^p(\cdot,\cdot,0) = 0$. Applying Itô's formula to $Y_t^{(p,T)} := Y^p(S_t, S'_t, T - t)$ we find

$$\begin{aligned}
&\left(Y^{(p,T)}, (H, \nu)\right) := \\
&\left(Y^{(p,T)}, \left(\frac{\lambda(S_\cdot, S'_\cdot) - \frac{\partial Y^p}{\partial s}(S_\cdot, S'_\cdot, T - \cdot)}{p-1}, \frac{\partial Y^p}{\partial s'}(S_\cdot, S'_\cdot, T - \cdot)\right)\right),
\end{aligned}$$

to be a solution for the BSDE (25) on $[0, T]$. Set

$$V^{p,T} := \mathcal{E}\left((r - HC\lambda) \cdot \alpha + H \cdot M\right)^T, \tag{30}$$

and

$$Z^{p,T} := \mathcal{E}(\lambda \cdot M + \nu \cdot N)^T. \tag{31}$$

Under appropriate integrability assumptions these two process solve the optimization problems (12) resp. (13):

Theorem 6.1. *If for $p > 1$, $Z^{p,T} \in \mathcal{D}^q(\bar{\mathcal{M}}_T^{x_0})$ and $V^{p,T}H \in \mathcal{SF}^p(\bar{\mathcal{M}}_T^{x_0})$, resp. if for $0 \neq p < 1$, $\mathcal{E}((\lambda + H) \cdot M + \nu \cdot N)^T$ is a uniformly integrable martingale, then $(V^{p,T}, Z^{p,T})$ is the optimal pair for the market $\bar{\mathcal{M}}_T^{x_0}$. Furthermore*

$$\mathcal{V}(p,T)(x_0) := \sup_{\substack{H \in \mathcal{SF}^p(\bar{\mathcal{M}}_T^{x_0}) \\ V_0^H = 1}} E^{x_0}\left[U^{(p)}(V_T^H)\right] = -\operatorname{sgn}(q)\exp(Y^p(S_0, S'_0, T)), \tag{32}$$

where $x_0 = (S_0, S'_0)$ and E^{x_0} denotes the expectation with respect to the measure induced by the Markov process X solving the SDE (27).

Proof. The assertion follows directly from Proposition 5.1. □

For $p = 2$, this allows to calculate the Hedging Numéraire explicitly, see [13].

Example 6.2. *For $q\frac{\beta^2}{2} - pr$ constant, set $Y^p(s, s', T-t) := -(T-t)\left(q\frac{\beta^2}{2} - pr\right)$, which solves (29). Hence the minimal martingale measure, see* [10], *is the L^q-optimal martingale measure $Z^{opt*,q,T} = \mathcal{E}(\lambda \cdot M)^T$ and $V^{opt,p,T}$ is given by the locally efficient strategy $V^{(H^p)}$.*

7. Mean-Variance Efficiency and Intertemporal Price for Risk

Consider the optimization problem

$$\mathcal{V}(T,e) := \inf_{\substack{H \in \mathcal{SF}^2(\bar{\mathcal{M}}_T) \\ V_0^H = 1}} E\left[\left(V_T^H\right)^2\right], \tag{33}$$

under the constraint

$$E\left[V_T^H\right] = e,$$

for $e \in \mathbf{R}$. Denote the solution if it exists by $V^{T,e}$. This problem is equivalent to the optimization problem

$$\tilde{\mathcal{V}}(T,e) := \inf_{\substack{H \in \mathcal{SF}^2(\bar{\mathcal{M}}_T) \\ V_0^H = 1}} E\left[\operatorname{Var}\left(V_T^H\right)\right], \tag{34}$$

under the constraint $E\left[V_T^H\right] = e$, i.e. we want to determine the mean-variance efficient frontier. Assume the existence of a zero bond B^T with maturity in T in the market $\bar{\mathcal{M}}_T$.

Definition 7.1. *The* intertemporal price for risk *for maturity time T in the market $\bar{\mathcal{M}}_T$ is defined as*

$$\beta_{[0,T]} := \frac{\sqrt{\operatorname{Var}\left(\frac{Z^{opt*,2,T}}{B_T}\right)}}{B_0^T}. \tag{35}$$

Solving the above constrained optimization problems using convex combinations $sV^{opt,2,T} + (1-s)B^T$, $s \in \mathbf{R}$, we find the following result for the intertemporal Sharpe-ratio:

Theorem 7.2. *Assume the existence of the zero bond B^T in $\bar{\mathcal{M}}_T$. Then the following inequality holds for all $H \in \mathcal{SF}^2(\bar{\mathcal{M}}_T)$ with $V_0^H = 1$:*

$$\frac{1}{B_0^T} - \beta_{[0,T]}\sqrt{\operatorname{Var}\left(V_T^H\right)} \le E\left[V_T^H\right] \le \frac{1}{B_0^T} + \beta_{[0,T]}\sqrt{\operatorname{Var}\left(V_T^H\right)}. \tag{36}$$

In particular, $\beta_{[0,T]} = 0$ implies $E[V_T^H] = \frac{1}{B_0^T}$ for all $H \in \mathcal{SF}^2(\bar{\mathcal{M}}_T)$ with $V_0^H = 1$, $V^{opt,2,T} = \frac{B^T}{B_0^T}$ and the so-called Return-to-Maturity Expectation Hypothesis *for the zero bond price in $t = 0$ holds:*

$$B_0^T = \frac{1}{E[B_T]}. \tag{37}$$

Furthermore, if $\beta_{[0,T]} \neq 0$, *then*

$$V^{T,e} = s(e)V^{opt,2,T} + \big(1 - s(e)\big)\frac{B^T}{B_0^T}, \tag{38}$$

where $s(e) := -\frac{eB_0^T - 1}{1-(B_0^T)^2\mathcal{V}(2,T)}$, *and if*

$$\left| E\left[V_T^H\right] - \frac{1}{B_0^T} \right| = \beta_{[0,T]}\sqrt{\operatorname{Var}\left(V_T^H\right)}, \tag{39}$$

for a $H \in \mathcal{SF}^2(\bar{\mathcal{M}}_T)$ *with* $V_0^H = 1$, *then* $V^H = V^{T,E[V_T^H]}$.

Remark 7.3. *The result can be proved for a not necessarily continuous but locally* L^2*-semimartingale (in the sense described in Section 1) price process and in particular for the original one-step CAPM and its multi-period generalizations.*

References

[1] J. P. Ansel and C. Stricker, *Lois de martingale, densités et décomposition de Föllmer Schweizer,* Ann. Inst. H. Poincaré **28** (1992), 375–392.

[2] K. J. Arrow, *Essays in the Theory of Risk-Bearing,* North-Holland Publishing Company, Amsterdam, **1976**.

[3] T. R. Bielecki and S. R. Pliska, *Risk-Sensitive Dynamic Asset Management,* Appl. Math. Optim. **39** (1999), 337–360.

[4] J. M. Bismut, *Conjugate convex functions in optimal stochastic control,* J. Math. Anal. Appl. **44** (1973), 384–404.

[5] J. M. Bismut, *Growth and optimal intertemporal allocations of risk,* J. Economic Theory **10** (1975), 239–287.

[6] L. Breiman, *The Capital Growth Critirion and Continuous-Time Models,* Naval Research Logistics Quarterly **7** (1960), 647–651. (Reprinted in [30]).

[7] F. Delbaen and W. Schachermayer, *A general version of the fundamental theorem of asset pricing,* Math. Ann. **300** (1994), 463–520.

[8] F. Delbaen and W. Schachermayer, *Attainable claims with p'th moments,* Ann. Inst. Henri Poincaré **32(6)** (1996), 743–763.

[9] F. Delbaen, P. Monat, W. Schachermayer, M. Schweizer and C. Stricker, *Weighted norm inequalities and hedging in incomplete markets,* Finance and Stochastics **1** (1997), 181–227.

[10] H. Föllmer and M. Schweizer, *Hedging of continuous claims under incomplete information,* in: M. Davis and R. Elliot, Eds., Applied Stochastic Analysis (Gordon and Breach, London, New York) (1990), 389–414.

[11] H. Geman, N. El Karoui and J. Rochet, *Changes of Numéraire, Changes of Probability Measure and Option Pricing,* J. Appl. Prob. **32** (1995), 443–458.

[12] T. Goll and J. Kallsen, *Optimal portfolios for logarithmic utility,* Stoch. Proc. Appl. **89** (2000), 31–48.

[13] C. Gourieroux, J. P. Laurent and H. Pham, *Mean-Variance Hedging and Numéraire,* Mathematical Finance **8(3)** (1998), 179–200.

[14] P. Grandits and L. Krawczyk, *Closedness of some spaces of stochastic integrals,* in: J. Azéma, M. Émery, M. Ledoux and M. Yor, Eds., Séminaire de Probabilités XXXII (Springer, Berlin, New York) (1998), 73–85.

[15] J. Harrison and R. Pliska, *Martingales and Stochastic Integrals in the Theory of Continuous Trading,* Stoch. Proc. Appl. **11** (1981), 215–260.

[16] H. He and N. D. Pearson, *Consumption and portfolio policies with incomplete markets and short-sale constraints: the infinite dimensional case,* Mathematical Finance **1** (1991), 1–10.

[17] J. Jacod, *Calcul Stochastique et Problèmes de Martingales,* Springer, Berlin, New York, **1979**.

[18] J. Jacod and A. N. Shiryaev, *Limit Theorems for Stochastic Processes,* Springer, New York, **1987**.

[19] I. Karatzas, J. P. Lehoczky, S. E. Shreve and G.-L. Xu, *Martingale and Duality Methods for Utility Maximization in an Incomplete Market,* SIAM Journal of Control and Optimization **29(3)** (1991), 702–730.

[20] J. L. Kelly, *A new interpretation of the information rate,* Bell System Technical Journal **35** (1956), 917–926.

[21] D. Kramkov and W. Schachermayer, *The Asymptotic Elasticity of Utility Functions and Optimal Investment in Incomplete Markets,* Annals of Applied Probability **9(3)** (1999), 904–950.

[22] H. M. Markowitz, *Portfolio Selection,* Journal of Finance **7** (1952), 77–91.

[23] R. C. Merton, *Continuous-Time Finance,* Basil Blackwell, Oxford, **1990**.

[24] J. Mossin, *Optimal Multiperiod Portfolio Policies,* The Journal of Business **1968**, 215–229.

[25] J. W. Pratt, *Risk aversion in the small and in the large,* Econometrica **32** (1964), 122–136.

[26] R. T. Rockafellar, *Convex Analysis,* Princeton University Press, Princeton, **1970**.

[27] W. F. Sharpe, *Capital Asset Prices: A Theory of Market Equilibrium under Conditions of Risk,* Journal of Finance **19** (1964), 425–442.

[28] A. N. Shiryaev, *Essentials of stochastic Finance: Facts, Models, Theory,* World Scientific Publishing, Singapore, **1999**.

[29] J. Yong and X. Y. Zhou, *Stochastic controls: Hamiltonian systems and HJB equations,* Springer, Berlin, New York, **1999**.

[30] W. T. Ziemba and R. G. Vickson, *Stochastic Optimization Models in Finance,* Academic Press, New York, **1975**.

Center of Finance and Econometrics,
University of Konstanz,
D-78457 Konstanz, Germany
E-mail address: `johannes.leitner@uni-konstanz.de`

Trends in Mathematics, © 2001 Birkhäuser Verlag Basel/Switzerland

LQ control and mean–variance portfolio selections: The stochastic parameter case

Andrew E.B. Lim[1] and Xun Yu Zhou[2]

Abstract. This paper concerns the continuous-time, mean–variance portfolio selection problem with random interest rate, appreciation rates and volatility coefficients. The problem is tackled using the results of indefinite stochastic linear–quadratic (LQ) optimal control and backward stochastic differential equations (BSDEs), two theories that have been extensively studied and developed in recent years. Specifically, the mean–variance problem is formulated as a linearly constrained stochastic LQ control problem. Solvability of this LQ problem is reduced, in turn, to proving global solvability of a stochastic Riccati equation. The proof of existence and uniqueness of this Riccati equation, which is a fully nonlinear and singular BSDE with random coefficients, is interesting in its own right and relies heavily on the structural properties of the equation. The optimal investment strategy as well as the mean–variance efficient frontier are then analytically derived in terms of the solution of this equation. In particular, it is demonstrated that the efficient frontier in the mean–standard deviation diagram is still a straight line or, equivalently, risk-free investment is still possible, even when the interest rate is random.

1. Introduction

This paper is concerned with continuous–time dynamic mean–variance portfolio selection problem in a complete market. An important feature of our problem is that the interest rate of the bond and the appreciation and volatility rates of the stocks may be random processes. This contrasts with the model that is commonly adopted in which these quantities are taken as deterministic (and in many cases, time-invariant).

The single-period mean–variance problem was formulated as a quadratic programming problem in [5, 6]; see also [7] for a derivation of the efficient frontier. On the other hand, work on multi-period portfolio selection has focused predominantly on maximizing the expected utility of terminal wealth $EU(x(T))$. Note however that mean–variance analysis and expected utility maximization are two different tools for dealing with portfolio selection, accounting for different degrees of risk-aversion. For example, the former enjoys a better performance when the market is less volatile, while the latter performs better when the outcome of the stock prices occurs at the tails of the distribution; see, for example, [10].

In this paper, we generalize the results of [11] to the case of random coefficients. (See [3] for discrete-time results). In making this generalization, the following difficulties are encountered, and need to be addressed. Firstly, dynamic programming is difficult to apply since the usual HJB equation is not valid when the coefficients are random. Secondly, if one is to employ the LQ approach as done in [11], then global solvability of a so-called *stochastic Riccati equation* (SRE) is non-trivial. In the deterministic case, this SRE reduces to a *linear* deterministic *ordinary differential equation* (ODE); existence and uniqueness follows immediately from standard results. When the coefficients are random, however, the SRE is a fully *nonlinear*, singular *backward stochastic differential equation* (BSDE) for which the usual assumptions (such as the Lipschitz and linear growth conditions; see [2, 8, 9]) are *not* satisfied. For these reasons, global solvability of this equation can not be established using standard techniques and relies on exploiting the special structure of the SRE associated with the mean–variance problem. In this paper, an outline of these results is presented. Afterwards, the efficient frontier and the optimal investment policy of the original mean–variance problem are given in terms of the solution to the above SRE. Due to constraints on length, the interested reader is directed to the paper [4] for detailed proofs, which have been omitted from this discussion. Finally, this paper demonstrates again that the stochastic LQ control and BSDEs can serve as a powerful tools for dealing with certain financial application problems. (See also [2]).

2. Formulation of mean–variance portfolio selection

Let $(\Omega,\ \mathcal{F},\ \{\mathcal{F}_t\}_{t\geq 0},\ P)$ be a complete filtered probability space such that $\mathcal{F}_0$ is augmented by all the P-null sets of $\mathcal{F}$ (and hence $\{\mathcal{F}_t\}_{t\geq 0}$ is right continuous on $[0,T)$). Let $W(t) = (W^1(t), \cdots, W^m(t))$ be an $\mathbb{R}^m$-valued standard Brownian motion (with $W(0) = 0$) on this space, and assume that $\{\mathcal{F}_t\}_{t\geq 0}$ is generated by $W(\cdot)$. Throughout this paper, for any $1 \leq q < \infty$ we denote the set of $\{\mathcal{F}_t\}_{t\geq 0}$-adapted processes f on $[0,\ T]$ such that $E\int_0^T |f(t)|^q\,dt < \infty$ by $L^q_{\mathcal{F}}(0,\ T;\ \mathbb{R}^m)$, the set of $\{\mathcal{F}_t\}_{t\geq 0}$-adapted processes which are uniformly bounded by $L^\infty_{\mathcal{F}}(0,\ T;\ \mathbb{R}^m)$, the set of continuous $\{\mathcal{F}_t\}_{t\geq 0}$ adapted processes such that $E\{\sup_{t\in[0,\ T]} |f(t)|^2\} < \infty$ by $L^2_{\mathcal{F}}(\Omega;\ C(0,T;\mathbb{R}^m))$, and the set of all $\{\mathcal{F}_t\}_{t\geq 0}$-adapted, uniformly bounded continuous processes by $L^\infty_{\mathcal{F}}(\Omega;\ C(0,T;\mathbb{R}^m))$. Finally, the set of bounded, $\mathbb{R}^n$-valued, $\mathcal{F}_T$-adapted random variables will be denoted by $L^\infty_{\mathcal{F}_T}(\Omega,\ \mathbb{R}^n)$.

Consider a market with $m+1$ securities, consisting of a bond and m stocks. The bond price $P_0(t)$ satisfies the (stochastic) ordinary differential equation:

$$\begin{cases} dP_0(t) = r(t)\,P_0(t)\,dt, & t \in [0,\ T], \\ P_0(0) = p_0 > 0, \end{cases} \tag{1}$$

where the interest rate $r(t) > 0$ is a bounded, $\{\mathcal{F}_t\}_{t\geq 0}$-adapted, scalar-valued stochastic process. The price of each of the stocks, $P_1(t), \cdots, P_m(t)$, satisfies the stochastic differential equation (SDE):

$$\begin{cases} dP_i(t) = P_i(t)\Big\{b_i(t)\,dt + \sum_{j=1}^m \sigma_{ij}(t)\,dW^j(t)\Big\}, & t \in [0, T], \\ P_i(0) = p_i > 0, \end{cases} \tag{2}$$

where $b_i(t) > 0$ and $\sigma_i(t) = [\sigma_{i1}, \cdots, \sigma_{im}(t)]$ are the appreciation rate and dispersion (or volatility) rate of the ith stock, respectively. Once again, we assume that $b_i(t)$ and $\sigma_{ij}(t)$ are scalar-valued, $\{\mathcal{F}_t\}_{t\geq 0}$-adapted, bounded stochastic processes. Denoting

$$\sigma(t) = \begin{bmatrix} \sigma_1(t) \\ \vdots \\ \sigma_m(t) \end{bmatrix} \in \mathbb{R}^{m\times m}, \tag{3}$$

we assume throughout that there exists $\delta > 0$ such that

$$\sigma(t)\,\sigma(t)' \geq \delta\, I, \quad \forall\, t \in [0, T],\ P - a.s.. \tag{4}$$

In particular, $\sigma(t)$ must be non-singular a.e. $t \in [0, T]$, P-a.s.. We assume that the total wealth of an agent $x(t)$ follows the SDE :

$$\begin{cases} dx(t) &= \Big\{r(t)\,x(t) + \sum_{i=1}^m [b_i(t) - r(t)]\,u_i(t)\Big\}\,dt \\ & \quad + \sum_{j=1}^m \sum_{i=1}^m \sigma_{ij}(t)\,u_i(t)\,dW^j(t), \quad t \in [0, T], \\ x(0) &= x_0 > 0, \end{cases} \tag{5}$$

where $u_i(t)$ is the total market value of the agent's wealth in the ith asset. A portfolio $u(\cdot)$ is *admissible* if it is $\mathbb{R}^m$-valued, integrable, $\{\mathcal{F}_t\}_{t\geq 0}$-adapted and (5) has a unique solution $x(\cdot)$ corresponding to $u(\cdot)$. Such a pair $(x(\cdot), u(\cdot))$ will be referred to as an *admissible pair.* The *mean–variance portfolio selection problem* can be formulated as follows:

$$\begin{cases} \min J(u(\cdot)) := E\,\frac{1}{2}\,(x(T) - d)^2, \\ \text{subject to:} \\ \quad Ex(T) = d, \\ \quad u(\cdot) \text{ is admissible.} \end{cases} \tag{6}$$

3. A stochastic LQ framework

In the first subsection, we formulate a general linear–quadratic (LQ) control problem. A special case of this LQ problem comes about with the introduction of Lagrange multipliers in the mean–variance problem (6). For this reason, solvability of LQ problems is an important part of our study of (6). In subsequent sections, we focus more closely on the mean–variance problem. In much of our discussion, the fundamental role of the Stochastic Riccati Equation is emphasized. Due to constraints on length, formulas and proofs have in most cases been omitted. These can be found in the paper [4].

3.1. A general unconstrained LQ problem

Let $(\Omega, \mathcal{F}, \{\mathcal{F}_t\}_{t\geq 0} P)$ be a filtered probability space, and $W(\cdot)$ be a k-dimensional Brownian motion on this space. We assume once again that $\{\mathcal{F}_t\}_{t\geq 0}$ is generated by $W(\cdot)$ and augmented by the null sets of $\mathcal{F}$. Consider the following linear SDE:

$$\begin{cases} dx(t) &= [A(t)\,x(t) + B(t)\,u(t)\,]\,dt \\ & \quad + \sum_{j=1}^{k}[C_j(t)\,x(t) + D_j(t)\,u(t)]\,dW^j(t), \quad t \in [0, T], \\ x(0) &= x_0, \end{cases} \tag{7}$$

where A, B, C_j and D_j are $\{\mathcal{F}_t\}_{t\geq 0}$-adapted processes, and $x_0 \in \mathbb{R}^n$ is non-random. Precise assumptions on this data will be specified below. The class of *admissible controls* is the set

$$\mathcal{U} = \Big\{u(\cdot) \in L^1_{\mathcal{F}}(0, T; \mathbb{R}^m)\,\Big|\,(7) \text{ has a unique solution under } u(\cdot)\Big\}.$$

If $u(\cdot) \in \mathcal{U}$ and $x(\cdot)$ is the associated solution of (7), then we refer to $(x(\cdot), u(\cdot))$ as an *admissible pair*. The cost functional is given by

$$J(u(\cdot)) = \\ E\frac{1}{2}\Big\{\int_0^T \Big(x(t)'Q(t)x(t) + u(t)'R(t)u(t)\Big)\,dt + x(T)'Hx(T) + 2c'x(T)\Big\}. \tag{8}$$

Throughout this section, we shall assume the following:

Assumption:

(A1)

$$\begin{cases} A,\, C_j,\, Q & \in & L^\infty_{\mathcal{F}}(0, T; \mathbb{R}^{n\times n}), \\ B,\, D_j & \in & L^\infty_{\mathcal{F}}(0, T; \mathbb{R}^{n\times m}), \\ R & \in & L^\infty_{\mathcal{F}}(0, T; \mathbb{R}^{m\times m}), \\ H & \in & L^\infty_{\mathcal{F}_T}(\Omega; \mathbb{R}^{n\times n}), \\ c & \in & L^\infty_{\mathcal{F}_T}(\Omega; \mathbb{R}^{n}), \\ x_0 & \in & \mathbb{R}^n. \end{cases}$$

In particular, we do not assume that $Q \geq 0$, $R > 0$ or $H \geq 0$. In addition, A, B, C_j, D_j, Q, R, H and c may be random. The (unconstrained) *stochastic LQ problem* associated with (7)-(8) is as follows:

$$\begin{cases} \min J(u(\cdot)), \\ \text{subject to:} \\ \quad (x(\cdot), u(\cdot)) \text{ admissible for (7).} \end{cases} \tag{9}$$

We mention again that the HJB partial differential equation does not apply since the coefficients of (9) are random. In [4], an approach based on the 'completion of squares' technique is used to study (9). It is shown that the crucial step in solving (9) is proving existence and uniqueness of the so called *Stochastic Riccati*

Equation (SRE). The SRE is a nonlinear, singular *Backward Stochastic Differential Equation* (BSDE) which does not satisfy the usual assumption of linear growth and Lipschitz continuity. Global solvability of the general SRE is still an open question. (For further details, the reader should consult [1]).

3.2. An unconstrained mean–variance problem

In this subsection we consider an unconstrained version of the mean–variance portfolio selection problem (6) (i.e. the problem obtained after deleting the constraint $Ex(T) = d$). Unconstrained problems of this type come about after the introduction of Lagrange multipliers in the constrained problem (6). We see that this is a special case of the general stochastic LQ problem formulated in the previous subsection with the following coefficients

$$\left\{ \begin{array}{l} n = 1,\ m = k,\ A(t) = r(t),\ B(t) = [b_1(t) - r(t), \cdots, b_m(t) - r(t)], \\ C_j(t) = 0,\ D_j(t) = [\sigma_{1j}(t), \cdots, \sigma_{mj}(t)],\ Q = R = 0,\ H = 1,\ c = -d. \end{array} \right. \tag{10}$$

To ease the notation we will continue to use $A(t), B(t)$, etc. to denote the coefficients (though one should keep in mind that the specific values of them are specified in (10)). The only exception (again for the sake of notational simplicity) is that we shall use $\sigma(t) \equiv \left[\ D_1(t)'\ \middle|\ \cdots \middle|\ D_m(t)'\ \right] \in \mathbb{R}^{m \times m}$ instead of the D_j's.

The following equation is a special case of the SRE associated with the stochastic LQR problem (9). It plays a fundamental role in the solution of (10):

$$\left\{ \begin{array}{rcl} dp & = & -\Big\{\Big[2A - B(\sigma\,\sigma')^{-1}B'\Big]p - 2B(\sigma^{-1})'\,\Lambda - \frac{1}{p}\Lambda'\Lambda\Big\}\,dt \\ & & +\sum_{j=1}^{m} \Lambda_j\, dW^j, \quad t \in [0, T], \\ p(T) & = & H, \\ p(t) & > & 0, \quad \forall\, t \in [0, T]. \end{array} \right. \tag{11}$$

Existence and uniqueness of a solution $(p, \Lambda) \in L^\infty_{\mathcal{F}}(\Omega; C(0, T; \mathbb{R})) \times L^2_{\mathcal{F}}(0, T; \mathbb{R}^m)$ of (11) is a sufficient condition for solvability of (10); see [4]. When the coefficients A, B, etc., are all deterministic, we have $\Lambda = 0$ and (11) becomes linear ODE with bounded coefficients. In this case, existence and uniqueness follows from the standard theory. In the random coefficient case, however, things are not so straight forward. In this case, the SRE (11) is a complicated nonlinear BSDE. In particular, the right hand side of this equation is not globally Lipschitz continuous nor linearly growing (due to the term $\Lambda'\Lambda/p$) and hence, the standard results on existence and uniqueness of solutions can not be applied. For details on the role of the SRE in the mean–variance problem, and expressions for the optimal control and cost associated with (10) (which are valid, of course, under the assumption that (11) is solvable) the reader should consult [4]).

3.3. Constrained mean–variance problem

We turn our attention now to the original mean–variance problem (6). We keep the same setup, notation and assumptions of the previous subsection, with an additional constraint represented by the following functional

$$J_1(u(\cdot)) = E\Big\{c_1\, x(T)\Big\}, \quad u(\cdot) \in \mathcal{U}. \tag{12}$$

Let $d \in \mathbb{R}$ be given and fixed. Consider the following linearly constrained LQ problem:

$$\begin{cases} J^* = \inf J(u(\cdot)), \\ \text{subject to:} \\ \quad J_1(u(\cdot)) = d, \\ \quad (x(\cdot),\, u(\cdot)) \text{ admissible for (7).} \end{cases} \tag{13}$$

Recall that for the original mean–variance problem (6) the values of the coefficients A, B, etc. are specified in (10). Moreover, $c_1 = 1$. However, as before, we keep the general notation to ease the exposition as well as to see more general properties of the solution.

An admissible control $u(\cdot) \in \mathcal{U}$ is said to be a *feasible control* for (13) if it satisfies the constraint in (13). If there exists a feasible control, then the problem (13) is said to be *feasible.* Finally, if (13) is finite and the optimal cost is achieved by a feasible control $u^*(\cdot)$, then (13) is said to be *solvable* and $u^*(\cdot)$ is an *optimal control.*

Once again, existence and uniqueness of solutions for the SRE (11) plays an important role in the solution of (13). (In particular, solvability of the SRE implies that the unconstrained problem (which is of the form (10)) that comes about with the introduction of Lagrange multipliers in (13) is solvable). On the other hand, the introduction of Lagrange multipliers only makes sense if the original constrained problem (13) is feasible. Necessary and sufficient conditions for feasibility of (13) are as follows.

Proposition 3.1. *Suppose that* $(\psi,\, \xi) \in L^\infty_{\mathcal{F}}(\Omega;\, C(0,\, T;\, \mathbb{R})) \times L^2_{\mathcal{F}}(0,\, T;\, \mathbb{R}^m)$ *is the unique solution of the BSDE*

$$\begin{cases} d\psi(t) &=& -A(t)\, \psi(t)\, dt + \sum_{j=1}^m \xi_j(t)\, dW^j(t), \quad t \in [0,\, T], \\ \psi(T) &=& c_1. \end{cases} \tag{14}$$

Then the LQ problem (13) is feasible for every $d \in \mathbb{R}$ *if and only if*

$$E \int_0^T \Big|\psi(t)\, B(t) + \xi(t)'\, \sigma(t)'\Big|^2 dt > 0. \tag{15}$$

From Proposition 3.1 and the condition (15), we obtain certain minimum requirements for the LQ problem (13) to be feasible for every $d \in \mathbb{R}$. In particular, it is necessary that $c_1 \neq 0$ (since $(\psi,\, \xi) = (0,\, 0)$ is the unique solution of (14) when $c_1 = 0$) and $B \neq 0$ or $\sigma \neq 0$. In view of (13), these requirements for feasibility are as expected. On the other hand, if $(\psi,\, \xi) \neq (0,\, 0)$ (which is the case when $c_1 \neq 0$), then (15) is easily satisfied since $(\psi,\, \xi)$ is independent of B and σ. In particular, the assumption of feasibility of (13) for every $d \in \mathbb{R}$ (or equivalently, that (15) holds) is very mild. For a proof of this result and expressions for the optimal control and the optimal cost (under the assumption of solvability of (11)) for (13), the reader should consult [4].

4. Solvability of stochastic Riccati equation

Global solvability of the SRE (11) is the fundamental issue in the solution of (13) (and hence the mean–variance problem (6)). In particular, solvability of (11) is sufficient for solvability of the unconstrained stochastic LQ problem (of form (10)) that arises when Lagrange multipliers are introduced (13). The following result for the SRE (11) is obtained in [4]. The proof relies on the fact that the state variable $p(\cdot)$ is scalar–valued, and the control cost R is identically zero.

Theorem 4.1. *Suppose all the parameters involved are specified by (10). Then there exists a unique solution* $(p, \Lambda) \in L^\infty_{\mathcal{F}}(\Omega, C(0, T; \mathbb{R})) \times L^2_{\mathcal{F}}(0, T; \mathbb{R}^m)$ *of the stochastic Riccati equation (11).*

Finally, we also mention that solvability results for the multi-dimensional SRE are obtained in [1]. In particular, global solvability is proven in the case $R > 0$. On the other hand, in the absence of such a constraint on R, local solvability (in general) can only be guaranteed.

5. Optimal portfolio and efficient frontier

Finally we are at the stage of solving the original mean–variance problem (6). The problem is a special case of the constrained LQ problem (13) with system parameters:

$$\begin{cases} A(t) = r(t), & B(t) = [b_1(t) - r(t), \cdots, b_m(t) - r(t)], \\ C_j(t) = 0, & D_j(t) = [\sigma_{1j}(t), \cdots, \sigma_{mj}(t)], \end{cases} \tag{16}$$

and cost parameters

$$H = 1, \quad c = -d, \quad c_1 = 1. \tag{17}$$

Denoting

$$\rho(t) = B(t)(\sigma(t)\,\sigma(t)')^{-1}B(t)',$$

the SRE (11) can be rewritten as

$$\begin{cases} dp &= -\Big\{(2r - \rho)\,p - 2B(\sigma^{-1})'\Lambda - \frac{1}{p}\Lambda'\Lambda\Big\}\,dt + \sum_{j=1}^m \Lambda_j\,dW^j, \\ p(T) &= 1. \end{cases} \tag{18}$$

Consider the BSDE:

$$\begin{cases} dg &= \Big\{rg + B(\sigma^{-1})'\zeta\Big\}\,dt + \sum_{j=1}^m \zeta_j\,dW^j, \\ g(T) &= 1. \end{cases} \tag{19}$$

and the SDEs:

$$\begin{cases} dx &= \Big\{\Big[r - B(\sigma\sigma')^{-1}\Big(B' + \sigma\frac{\Lambda}{p}\Big)\Big]x \\ &\quad +d\,B(\sigma\,\sigma')^{-1}\Big[\Big(B' + \sigma\frac{\Lambda}{p}\Big)g + \sigma\zeta\Big]\Big\}\,dt \\ &\quad -\Big\{\Big[\sigma^{-1}B' + \frac{\Lambda}{p}\Big]\,(x - d\,g) - d\zeta\Big\}\,dW, \\ x(0) &= x_0 \end{cases} \tag{20}$$

$$\begin{cases} dx^{(1)} &= \Big\{\Big[r - B(\sigma\sigma')^{-1}\Big(B' + \sigma\frac{\Lambda}{p}\Big)\Big]x^{(1)} \\ &\quad -B(\sigma\,\sigma')^{-1}\Big[\Big(B' + \sigma\frac{\Lambda}{p}\Big)g + \sigma\zeta\Big]\Big\}\,dt \\ &\quad -\Big\{\Big[\sigma^{-1}B' + \frac{\Lambda}{p}\Big]\,(x^{(1)} + g) + \zeta\Big\}\,dW, \\ x^{(1)}(0) &= 0. \end{cases} \tag{21}$$

We have the following result.

Theorem 5.1. *Assume that (15) holds (with the parameters specified in (16)-(17)). Then*

$$p(0)g(0)^2 < 1 \;\; \text{and} \;\; g(0) > 0. \tag{22}$$

Moreover, the mean–variance problem (6) is finite for every $d \in \mathbb{R}$ and the optimal value of Var $x(T)$, *amongst all the wealth processes $x(\cdot)$ satisfying $Ex(T) = d$, is*

$$\operatorname{Var} x(T) = \frac{p(0)\,g(0)^2}{1 - p(0)\,g(0)^2}\Big[d - \frac{x_0}{g(0)}\Big]^2. \tag{23}$$

Furthermore, if there exists a policy $u^*(\cdot)$ that results in a wealth process $x^*(\cdot)$ that achieves $Ex^*(T) = d$ and (23), then

$$x^*(t) = x(t) + \lambda^* x^{(1)}(t) \tag{24}$$

and the optimal portfolio is

$$\begin{aligned} u^*(t) \;&=\; -(\sigma(t)\sigma(t)')^{-1}\Big[\Big(B(t)' + \sigma(t)\frac{\Lambda(t)}{p(t)}\Big)(x^*(t) - dg(t)) - d\sigma(t)\zeta(t) \\ &\quad +\lambda^*\Big(\Big(B(t)' + \sigma(t)\frac{\Lambda(t)}{p(t)}\Big)g(t) + \sigma(t)\zeta(t)\Big)\Big] \end{aligned} \tag{25}$$

where

$$\lambda^* = -\frac{p(0)\,g(0)^2}{1 - p(0)\,g(0)^2}\Big[d - \frac{x_0}{g(0)}\Big]. \tag{26}$$

In the case when $r(t)$ is deterministic, we recover the equations presented in [11].

We have seen that the process $g(t)$ plays a central role in the solution to the mean–variance problem; see Theorem 5.1. In particular, $g(t)$ is the price process of a contingent claim with a unit terminal payoff, while $(\sigma(t)^{-1})'\zeta(t)$ is the replicating portfolio of the claim; see Proposition 6.1, p. 394 of [9].

The optimal variance (23) involves a perfect square. In particular, choosing $d = x_0/g(0)$ gives $Var\ x(T) = 0$. The associated optimal policy can be shown to be

$$u^*(t) = \frac{x_0}{g(0)}\,(\sigma(t)^{-1})'\zeta(t). \tag{27}$$

If $r(t)$ is deterministic, then $\zeta(t) = 0$ and hence, $u^*(t) = 0$, meaning that the risk-free investment corresponds to putting all the wealth in the bond and nothing into shares, which is precisely the result obtained in [11]. On the other hand, if $r(t)$ is random, then $\zeta(t)$ is generally non-zero and hence $u^*(t) \neq 0$. Note that as discussed earlier, the part $(\sigma(t)^{-1})'\zeta(t)$ in (27) is nothing but the replicating portfolio for the contingent claim as specified in (19). In other words, if the interest rate is random, then a risk-free investment is still possible, which uses the above replicating portfolio to perfectly hedge the risk arising from the random interest rate.

If we denote by $\sigma_{x(T)}$ the standard deviation of the terminal wealth, then (23) gives

$$Ex(T) = \frac{x_0}{g(0)} + \sqrt{\frac{1-p(0)\,g(0)^2}{p(0)}}\,\frac{1}{g(0)}\sigma_{x(T)}. \tag{28}$$

Hence the efficient frontier in the mean–standard deviation diagram is still a straight line, which is also termed the *capital market line*. The slope of this line, $k = \sqrt{\frac{1-p(0)\,g(0)^2}{p(0)}}\frac{1}{g(0)}$, is called the *price of risk*.

6. Conclusion

This paper summarizes some recent results for the continuous-time mean–variance portfolio selection problem with random interest rate, appreciation rates and volatility coefficients. A key element of these results is a proof existence and uniqueness of solutions of a certain nonlinear BSDE. The mean–variance efficient frontier for this problem is a perfect square, suggesting that risk-free investment is still possible when interest rates are random. This paper again demonstrates that the stochastic LQ control can serve as a powerful framework for dealing with certain financial application problems. Also, it represents another application of BSDEs in finance.

Acknowledgment. This research was partially supported by the RGC Earmarked Grants CUHK 4125/97E, CUHK 4054/98E, and CUHK 4435/99E. While this research was being carried out, Andrew Lim received financial support from a Chinese University of Hong Kong Postdoctoral Research Fellowship.

References

[1] S. Chen and J. Yong. Stochastic linear quadratic optimal control problems. *Appl. Math. Optim.* (to appear).

[2] N. El Karoui, S. Peng, and M.C. Quenez. Backward stochastic differential equations in finance. *Math. Fin.* **7**, 1997, 1 – 71.

[3] D. Li and W.L. Ng. Optimal dynamic portfolio selection: Multi-period mean–variance formulation. *Math. Fin.* (to appear).

[4] A.E.B. Lim and X.Y. Zhou. Mean variance portfolio selection with random parameters. Working paper, Department of Systems Engineering & Engineering Management, The Chinese University of Hong Kong.

[5] H. Markowitz. Portfolio selection. *J. of Fin.*, **7**, 1952, 77 – 91.

[6] H. Markowitz. *Portfolio Selection: Efficient Diversification of Investment*, John Wiley & Sons, New York, 1959.

[7] R.C. Merton. An analytic derivation of the efficient frontier. *J. Fin. Quant. Anal.* **7**, 1972, 1851 – 1872.

[8] E. Pardoux and S. Peng. Adapted solution of backward stochastic equation. *Syst. & Contr. Lett.* **14**, 1990, 55 – 61.

[9] J. Yong and X.Y. Zhou. *Stochastic Controls: Hamiltonian Systems and HJB Equations*, Springer, New York, 1999.

[10] Y. Zhao, Y. and W.T. Ziemba. Mean–variance versus expected utility in dynamic investment analysis. Working paper, Faculty of Commerce and Business Administration, University of British Columbia.

[11] X.Y. Zhou and D. Li. Continuous-time mean-variance portfolio selection: A stochastic LQ framework. *Appl. Math. Optim.* **42**, 2000, 19 – 33.

[1]Center for Applied Probability,
Columbia University,
New York, NY 10027
U.S.A.
E-mail address: `lim@ieor.columbia.edu`

[2]Department of Systems Engineering & Engineering Management
The Chinese University of Hong Kong
Shatin, N.T.
Hong Kong.
E-mail address: `xyzhou@se.cuhk.edu.hk`

Trends in Mathematics, © 2001 Birkhäuser Verlag Basel/Switzerland

Liquidity Risk in Energy Markets

S.Nagornii and G.Dozeman

Abstract. We describe how to incorporate the dynamic effects connected with the Liquidity Risk (LR) in the standard "static" Value at Risk (VaR) methodology. An integral-like equation was derived to define a *liquidation time* for a particular instrument (any financial derivative). Furthermore, we show that dynamic LR for the total portfolio may be calculated through the special *renormalization* of the weights/positions of each of the portfolio constituents.

1. Introduction

Many financial institutions now use Value-at-Risk (VaR) to evaluate the market risk inherent in a portfolio of investments. Introduction of VaR is an attempt to provide for senior management a single number summarizing the "total" risk, associated with price fluctuation, in a complex portfolio of financial instruments. The VaR number is supposed to reflect the maximum possible reduction in the portfolio market value that may occur within some (*fixed*!) time horizon, given a certain confidence level. Time-intervals (horizons) are generally short, often only one day, and the same for all portfolio constituents/instruments, while confidence levels are typically 95% or higher. The VaR calculations are aimed at making a statement that we are $p\%$ certain that we will not lose more than V (dollars/euros) in the next N days. In this case V is the VaR of the portfolio, which is a function of two parameters (in a simplified case): N is a time-horizon and p is a confidence level. It should also be mentioned that VaR, as defined above, certainly depends on the assumptions made about the (portfolio) Probability Density Function (PDF), and the *estimators* chosen to calculate its moments (mean, variance) and covariances/correlations, since the real PDF is unknown.

In the framework of the general VaR- methodology there are several widely used methods for calculating VaR. Each of them is subject to various customizations and optimizations. A lot of commercial VaR applications are based on the Variance-Covariance methods or $RiskMetrics^{TM}$ [1], originated by J.P.Morgan. The $RiskMetrics$ dataset is now being published daily by Reuters. This contains recent volatilities and correlations for more then 400 values derived from highly liquid markets. In general, such a *static* VaR is designed to capture the prices' fluctuations within a (given) fixed time interval. We call it "static" to stress that time-horizon i) is the same for all portfolio instruments, ii) does not depend upon the time, i.e. the time-horizon is fixed, iii) cannot be derived within the VaR methodology itself and iv) should be chosen from some *uncertain* principles. However, this is not completely true, since very often the time-horizon is chosen from the risk control perspective. This implicitly suggests that the current positions may be liquidated (if necessary) within a chosen time-interval. Such an assumption, however, has no proved background often and may only be valid in a very liquid market. In the reality such a *static* VaR looks like a "theoretical" value only. Indeed, it is not evident that current positions may be liquidated within the chosen (without background!) time-horizon. As a result, the financial institution is exposed to an *additional risk* which is not captured by standard (*static*) VaR. Recently, many risk managers have begun to focus on risks associated with the lack of liquidity in financial markets. Unfortunately, most VaR models tend to underestimate liquidity risk, while many models ignore it altogether (see, for example, discussions in [2,3]). New approaches in asset pricing have demonstrated how liquidity plays an important role in security valuation and optimal portfolio choice by imposing (effectively) endogenous borrowing and short-selling constraints (see [4] and references there). The role of the liquidity in determining an optimal hedging strategy, and as a source of an additional "model risk" in dynamic hedging was recently discussed in the literature (see, for instance, [5,6]). Possible influence of market liquidity on implied volatilities (their "smile", for instance) was also demonstrated recently [6].

In general, over-simplified assumptions about *liquidity* may hide potentially big problems for those institutions using standard VaR risk measure. To incorporate additional "Liquidity Risk" in the static VaR, the dynamical effects associated with *liquidation – time*, which may be very different for different instruments, and strongly dependent upon the time to delivery, should be taken into account. This is certainly correct for any market, but it is especially important for the *energy market*, where portfolios often contain strongly correlated constituents (financial instruments) with very different times to delivery, and very different liquidity, as a result. To demonstrate this we presented in fig.1, as an example, time-dependence of the liquidity for three futures contracts traded at the "Nord Pool", the most developed European Power-Exchange at the moment. We define **liquidity** as the **total volume** (in $MegaWatt$ (MW) for power market) for each financial derivative traded at the exchange on a particular day. Several common properties may be seen from fig.1: i) liquidity increases when time approaches to the *delivery – time*; ii) while liquidity remains very small when time is far away from delivery; iii) liquidity becomes much more volatile, as it increases; iv) within last few weeks before *delivery* liquidity decreases rapidly. So, it is not evident that a particular portfolio instrument may be liquidated at any time within the same short time interval (which is very common for standard VaR practice) and financial institutions are, indeed, exposed to an additional risk. The problem becomes more significant in connection with the decentralization of the European energy market: several European Power-Exchanges where recently opened (APX in the Netherlands, LPX and EEX in Germany) and others are also under way.

A distinction between *endogenous* and *exogenous* liquidity risk was made in the literature (see, for instance, [7]). The former is associated to the size of an individual dealer position in relation to the depth of a (typically liquid) market, while the latter is a structural (or temporary) condition of an entire market. At least three distinct liquidity effects may seriously "damage" standard VaR calculations:

i) **Endogenous LiquidityRisk**: the effect which may have an impact on the particular market itself. This means the risk that the very act of liquidating a (large) position will send a "signal" to other market participants, and will change the behaviour of the total market.

ii) **Spread Volatility**: the value of most investments depends not only upon the fluctuations in prices, interest rates (i.e. yield curves) and so on, but also on the spreads to those curves / fluctuations. The effect of the widening and tightening of spreads, which is chiefly driven by changes in market liquidity, is completely ignored by most models.

iii) **Time to Liquidation**: it takes longer to liquidate illiquid investments, and the *dynamic* VaR should reflect the corrections coming from the *dynamic* effects which connect time-horizon with the *liquidation−time*, make time-horizons different for different instruments, and make the time-horizons dependent upon the time to delivery (expiration time), as well. While most VaR models allow the user to vary the horizon for an entire portfolio, it is often not possible to vary the horizons for individual positions (i.e., standard VaR is **static**!).

Although these effects are not entirely independent, we estimate their importance in accordance with the possible size of the impacts in the case of energy markets. Based on the observations from the Nord Pool Exchange we conclude that positions in general are not so large as to initiate a serious "endogenous liquidity risk". Similarly, "spread volatility" (SV) was not observed to be a significant issue (SV is small enough and more or less stable). Consequently, in this paper, contrary to [7] where SV was under examination, we concentrate on the LR effects associated with the *liquidation − time*, which we believe, may generates the most significant risk for particular energy markets. Of course, further development of the European Power Market may change the current situation. However, *finite liquidity* and corresponding non-zero *liquidation − time* is a permanent property of the "real life" energy market. The approach developed here to incorporate "liquidation-time risk" in the static VaR, is not however restricted to the energy market.

2. Correlations Within "Static" VaR Methodology

First let us restate the standard VaR definition for the isolated i- instrument:

$$Var_i(t,\Delta t) = P_i(t)[1-\exp\{-m_i\sigma_i(t,\Delta t)+\mu_i(t+\Delta t,\sigma_i,...)]. \qquad (1)$$

In eq.(1) t is a current time; Δt is a (chosen) time-horizon; $P_i(t)$ is a current position at time t; m_i is a "multiplier" which depends on the (chosen) confidence level and the type of PDF (for instance, for normally distributed returns and 5% confidential level m_N=1.67, for logistic distribution and the same confidential level $m_L = 1.62$); $\sigma_i(t, t+\Delta t)$ is a (predicted) volatility at time $t + \Delta t$, while $\mu_i(t + \Delta t, \sigma_i, ...)$ is an expectation of the return at time $t + \Delta t$, which could be a function of σ_i(and even something else). To specify the dependence of the volatility upon the time-horizon, standard Wiener's arguments have often been used:

$$\sigma_i(t, \Delta t) \quad = \quad \sigma_i(t) \sqrt{\Delta t}. \tag{2}$$

Although $\sqrt{t}$ - "scaling" was sometimes criticized in the literature (see, for instance, [8] and references), and other definitions for time-scaling of variance were proposed (see, for instance, Dorst and Nijman [9]), one should remember that in general "time-scaling" is a pure property of the (underlying) stochastic process. For instance, $\sqrt{t}$ - "scaling" is a property of a Wiener process and it is certainly correct for this particular type of processes. While the Dorst-Nijman formula, for instance, was derived for a $GARCH(1,1)$ process and it is correct in this particular framework only. So, to find a "correct" time-scaling one should find a "correct" underlying process first. In the present paper we consider (for simplicity only) a Wiener process and, as a result, use corresponding time-scaling (2). We do not say that a Wiener process is a good approximation for energy markets. It is simply very common among practitioners and convenient for our goals: to demonstrate how the dynamic effects may be incorporated into the standard (static) VaR methodology.

Instead of eq.(1), a simplified expression is commonly used [1]. Supposing that the expectation of the returns is close to zero ($\mu \to 0$), and expanding the exponential function (when the argument is small enough), we arrive at a well known (simplified) result:

$$VaR_i(t, \Delta t) \quad \Rightarrow \quad m_i \ \sigma_i(t, \Delta t) \ P_i(t), \tag{3}$$

The standard *Variance-Covariance* arguments allow us to express the total portfolio ("reduced") variance through the variances and covariances of the constituents:

$$\sigma^2 \ = \sum_i \sigma_i^2 \omega_i^2 \ + \sum_{i,j} cov(i,j) \ \omega_i \ \omega_j, \qquad cov(i,j) \ = \ \rho_{i,j} \ \sigma_i \ \sigma_j, \tag{4}$$

where σ_i^2 is a variance of the i- constituent, and $cov(i,j)$ ($\rho_{i,j}$) is a covariance (correlation) between i, j- constituents, while σ^2 is a ("reduced") variance for the portfolio. The "normalized" (reduced) weight of the i- instrument is defined in a standard way: $\omega_i = P_i/\Sigma|P_i|$, where P_i is a position for i- contract.

Since the PDF in practice is usually unknown, one has to use *estimators* to calculate σ_i^2 and $cov(i,j)$:

$$\sigma_i^2 \ = \ T^{-1} \sum_{t=1}^{T} (r_{i,t} - \overline{r_i})^2, \qquad \overline{r_i} \ = \ T^{-1} \sum_{t=1}^{T} r_{i,t}, \tag{5}$$

$$cov(i,j) = T^{-1}\sum_{t=1}^{T}(r_{i,t}-\overline{r_i})(r_{j,t}-\overline{r_j}). \tag{6}$$

An important requirement for the use of eqs.(5,6) is that *estimators* for variance and covariance should be *consistent*. This guarantees that the total portfolio variance is **always positive**. In matrix representation this means that the variance-covariance matrix is *always positively defined* (PD) for consistent estimators. So, if one gets a non-PD matrix, this may indicate an inconsistency in the definitions of the estimators for variance and covariance (see below).

Let us show that if estimators for variance and covariance of the portfolio constituents are consistent (as in eqs.(5,6)), the total portfolio variance may be only positive. Indeed, using eqs.(5,6) and eq.(4), after simple, but long algebra we arrive with another representation for the total portfolio (reduced) variance which is **always positive** explicitly (N is the total number of the portfolio constituents):

$$\sigma^2(t) = T^{-1}\sum_{t=1}^{T}(R_t-\overline{R})^2 \tag{7}$$

$$R_t = \sum_{i=1}^{N}\omega_i\, r_{i,t}\ ; \qquad \overline{R} = T^{-1}\sum_{t=1}^{T}R_t, \tag{8}$$

Here we introduced R_t - total return for portfolio at time t. Although, correlations are not presented in eqs.(7,8) explicitly, they are certainly included in an implicit way. Using eqs.(7,8), and following a simplified expression (3), it is very easy to calculate a (static) portfolio VaR for fixed time-horizon Δt:

$$VaR(t,\Delta t) = m(p\%, PDF)\ \sigma(t,\Delta t)\ \sum_i |P_i(t)|, \tag{9}$$

where dependence of $\sigma(t,\Delta t)$ on the time-horizon (Δt) is the same as in eq.(2), since in a standard VaR approach Δt is fixed and is the same for all portfolio constituents. Note, the multiplicative structure of the eq.(2) allows us *formally* to transfer factor $\sqrt{\Delta t}$ from $\sigma_i(t,\Delta t)$ to the corresponding i-weight:

$$\sigma_i(\Delta t)\ \omega_i \ \rightarrow\ \sigma_i\ \omega_i(\Delta t)\ ; \quad \omega_i(\Delta t) = \omega_i\sqrt{\Delta t}. \tag{10}$$

This formal transformation (10) will later be used to incorporate dynamic effects, associated with the *liquidation – time*, into the static VaR approach.

From eq.(7) we see that the portfolio variance is always positive (like it should be), since the consistency condition was explicitly used from the beginning. Although correlations do not appear in eq.(7) explicitly, they are certainly taken into account in an implicit way. This reflects a well known fact that correlations depend upon the representation. We shortly demonstrate this using the matrix form of eq.(4):

$$\sigma^2 \;=\; V^T\,\rho\,V,$$

where the "state-vector" is $V^T = (\sigma_1, \sigma_2, ..., \sigma_N)$, while the correlation matrix ρ was defined in eq.(4) through its matrix elements $\rho_{i,j}$.

Introducing an arbitrary matrix C, such that $CC^{-1} = 1$, we get:

$$\sigma^2 = V^T \rho V = V^T (C^{-1}C)\rho C^{-1} C V = (CV)^T (C^{-1T} \rho C^{-1})(CV) = \tilde{V}^T \tilde{\rho} \tilde{V}$$

Where $\tilde{V} = CV$ and $\tilde{\rho} = C^{-1T}\rho C^{-1}$ are a new state-vector and a new correlation matrix. Both differ from their first form, but the total result is essentially the same. In general, for any symmetric matrix, there always exists a unitary transformation which reduces it to diagonal form.

3. Liquidation Time for Financial Derivatives

To derive an equation for the *liquidation – time* ($t_L^{(i)}$), we introduce a *liquidity* (L_i) stochastic process for i- instrument:

$$dL_i \;=\; f_i(t, T, ...)\; dt \;+\; \sigma_i^L(t, T, ...)\; dz_i. \tag{11}$$

Here $L_i(t, T)$ is a liquidity (total volume for i- contract traded at the particular day). t and T are current and "delivery" time, respectively. The function $f_i(t, T, ...)$ is a regular part, while z corresponds to a stochastic (let us say, Wiener like) process. Let us stress that σ_i^L is a volatility of the liquidity process (do not mix with the volatilities of price-returns σ_i defined in the previous section). The solution of eq.(11) may be presented in a following general form:

$$L_i(t, T, ...) \;=\; \mu_i(t, T, ...) \;+\; \sigma_i^L(t, T, ...)\; D_i(0, 1). \tag{12}$$

Where $\mu_i(t, T) = E[L_i(t, T)]$ is an expectation of liquidity at time t and given delivery time T. $D_i(0, 1)$ is a distribution function.

From the standard VaR arguments, but taking into account that *liquidity* may be only **positive** or equal to **zero**, we can expect with $p\%$ confidence level that the lowest possible liquidity at time t is as follows:

$$max\{\; (E[L_i(t, T)] \;-\; \alpha^{(i)}(p)\; \sigma_i^L(t, T));\;\; 0\}. \tag{13}$$

Here $\alpha^{(i)}(p)$ is an analogy of the "multiplier" m_i in eq.(1) which is a function of the chosen confidence level (p) and the type of PDF, i.e. function $D_i(0, 1)$. The sign "-" in (13) indicates that we allow only $p\%$ of events to have smaller liquidity. This guarantees the *lowest* possible value of the liquidity with (1-p)% probability. For instance, 5% confidence level (p = 0.05) means that we can be sure for 95% that eq.(13) indeed gives us the lowest estimation of the liquidity at time t.

As usual, we suggest that within a very short time interval: from t to $t+dt$ ($dt \rightarrow 0$) the liquidity is a constant. So, the partial value of the position (dP) which may be liquidated within time-interval dt is:

$$dP = \{E[L_i(t,T)] - \alpha^{(i)}(p)\sigma_i^L(t,T)\}\ \Theta\{E[L_i(t,T)] - \alpha^{(i)}(p)\sigma_i^L(t,T)\}\ dt. \quad (14)$$

Here Θ is a "step"- function: $\Theta(x) = 1$ if $x \geq 0$, and $\Theta(x) = 0$ if $x < 0$.

As usually eq.(14) may be generalized for a finite time-interval, and we arrive with an "integral-like" equation which defines the *liquidation time* (t_L) necessary to liquidate current position P_i ($\tau \equiv t$ is a current time here):

$$P_i = \int_{T-\tau}^{T-\tau+t_L^{(i)}} ds\{E[L_i(s)] - \alpha^{(i)}(p)\sigma_i^L(s)\}\ \Theta\{E[L_i(s)] - \alpha^{(i)}(p)\sigma_i^L(s)\}. \quad (15)$$

To solve (15) for $t_L^{(i)}$ we need a model for the liquidity process. For simplicity, let us consider popular *Geometric Brownian Motion* (GBM) as a model for $L_i(t)$. We do not say that in the "real life" the $L(t)$- process looks like GBM, it is only an example. In the standard GBM framework eq.(11) becomes:

$$dL \quad = \quad \mu\ L\ dt\ +\ \sigma_L\ L\ dz. \quad (16)$$

The solution of eq.(16) corresponds to the well known lognormal PDF:

$$\ln L(t)\ -\ \ln L_{t=0}\ \ =\ \{(\mu - \sigma_L^2/2)t\ +\ N(0, \sigma_L\sqrt{t})\}. \quad (17)$$

Here $N(\mu, \sigma)$ is a "normal" PDF with mean μ and variance σ^2. Substituting the expectation of the liquidity from eq.(17) into the eq.(15), we define time t_L necessary to liquidate P- position in GBM model($\Delta\tau = T - \tau$):

$$P = \int_{\Delta\tau}^{\Delta\tau+t_L} ds\{L_0 e^{\mu s} - \alpha(p)\sigma_L\sqrt{s}\}\ \Theta\{L_0 e^{\mu s} - \alpha(p)\sigma_L\sqrt{s}\} \quad (18)$$

Using parameters P, L_0, μ and σ as "input", one can easily calculate function $t_L(\tau)$ for any $\tau < T$. In fig.2 we presented an estimation of the liquidation time for a "real life" contract traded at the Nord Pool exchange (we put $P = 5$ MW).We see that even at a point not far from delivery, the time which we need to eliminate the position is not very short, and this certainly increases real risk in comparison with the risk obtained within standard "static" VaR methodology (for short time-horizon). Note, an assumption that the PDF for the L- process is symmetric was made for simplicity only. Our preliminary analysis shows that due to several reasons the "real life" PDF of the liquidity process is practically always *skewed* in a "positive direction" where $L(t) > E[L(t)]$. For a skewed PDF the lowest and highest confidence borders are different. Therefore, instead of one parameter $\alpha(p)$ in a symmetric case, two parameters $\alpha_{\pm}(p, \gamma)$ describe the highest/lowest p- confidence borders for a skewed PDF. To account for the skewness (γ) we need only to change $\alpha^{(i)}(p)$ in eq.(15) by $\alpha_-^{(i)}(p, \gamma)$ (for a symmetric PDF, like "normal", $\gamma \equiv 0$). The "positive" skewness increases the lowest p- confidence border of the liquidity (for a "positively skewed" PDF $\alpha_{-}^{(i)}(p,\gamma) < \alpha^{(i)}(p)$) and decreases the liquidation-time, as a result. By the way, possible *skewness* of the price-returns distribution may also seriously damage standard VaR results mainly based on the assumption of a *symmetric* PDF, as in *RiskMetrics* [1], for instance.

4. From "Static" To "Dynamic" VaR

In this chapter we consider how to incorporate dynamic effects associated with LR into the "static" VaR approach. There are, at least, two options for this:

1. We can simply replace the fixed time-horizon Δt in the standard VaR equations by the dynamic liquidation-time $t_L^{(i)}$:

$$\Delta t \quad \rightarrow \quad t_L^{(i)}(\tau). \tag{19}$$

This means that we introduce a new definition for the *time − horizon*, connecting it with the *liquidation time*. After that, the *time − horizon* is not a *free* parameter any more, but it is a dynamic variable which is defined in the framework of the approach, and should be calculated every day for each particular portfolio instrument. Indeed, in accordance with definition (19) each constituent of the portfolio has its own time horizon, which is a function of the time to delivery: see eq.(15). Such a *dynamic* VaR has only one *free* parameter: the confidence level, and instead of eq.(2) we have:

$$\sigma_i(t,\Delta t) \quad \rightarrow \quad \sigma_i(t,t_L^{(i)}) \quad = \quad \sigma_i(t)\,\sqrt{t_L^{(i)}}. \tag{20}$$

This new (dynamic) VaR has a slightly different meaning than the static one: it shows with a certain confidence level the maximum possible loss for the financial institution to eliminate (completely or partially) its current portfolio.

2. In certain cases, however, it may be not satisfactory to change the standard VaR definition. That is why we propose a second variant to adjust "static" VaR for dynamic effects. In this case the definition and meaning of a standard VaR will be preserved, while the corrections for LR will be included. This may be achieved combining eq.(2) and eq.(20):

$$\sigma_i(t,t_L^{(i)}) \quad = \quad \sigma_i(t)\,\sqrt{t_L^{(i)}}\,\sqrt{\frac{\Delta t}{\Delta t}} \quad = \quad \sigma_i(t,\Delta t)\,\sqrt{\frac{t_L^{(i)}}{\Delta t}}. \tag{21}$$

$$\sigma_i(t,t_L^{(i)}) \quad = \quad \sigma_i(t,\Delta t)\;\kappa_i(\Delta t,t_L^{(i)})$$

So, to incorporate dynamic effects in a static portfolio VaR we need to multiply the volatility of each constituent by factor $\kappa_i = (t_L^{(i)}/\Delta t)^{1/2}$, as it is in eq.(21). Furthermore, to preserve volatility structure (in eq.(21) $\sigma_i(t,\Delta t)$ is the same as in eq.(2)), and to make it easier to incorporate LR effects in the portfolio VaR through the eqs.(7),(8), we transfer factor κ from σ_i to ω_i, using eqs.(4),(10). As a result, the dynamic effects associated with LR may be included through only the transformation ("renormalization") of the *weights* of the portfolio constituents in eq.(8):

$$\omega_i \rightarrow \tilde{\omega}_i = \omega_i \{1 + \frac{(t_L^{(i)} - \Delta t)}{\Delta t}\}^{\frac{1}{2}}. \quad (22)$$

Here Δt is a (chosen) static time-horizon, and $t_L^{(i)}$ is a liquidation time for i-instrument. Eq.(22) together with eq.(7,8) allows us to calculate a *dynamic* VaR:

$$VaR_{dynamic} = VaR_{static} + \Delta(VaR)_{LR}. \quad (23)$$

Note, the $\Delta(VaR)_{LR}$ correction comes from the transformation (22). This accounts for the (dynamic) effect that liquidation time for various constituents may essentially deviate from the (chosen) time-horizon (for static VaR), and to a different extent every day.

From eq.(23) we see that $VaR_{dynamic}$ may not only be higher than VaR_{static}, but it may also be lower because the $\Delta(VaR)_{LR}$ correction may be positive or negative. Eq.(22) shows that accounting for the liquidation-time may increase the corresponding i- weight ($\tilde{\omega}_i > \omega_i$, if $t_L^{(i)} > \Delta t$), and also decrease it ($\tilde{\omega}_i < \omega_i$, if $t_L^{(i)} < \Delta t$). If the chosen (fixed) time horizon is too short (in comparison with some "characteristic liquidation-time" for the portfolio), then accounting for LR will, likely, increase the static VaR. However, if the chosen time horizon is larger than "characteristic liquidation-time", the dynamic VaR may be lower than static one. In general, if we see that *dynamic* VaR is systematically (from day to day) higher than the *static* VaR, it means that our chosen (fixed) time-horizon is **too short** for this particular portfolio (at this particular time). If *dynamic* VaR is systematically lower than the *static* one, the opposite is true. The above described situations may not, however, be stable in time, especially for energy markets, where liquidity is a strong function of the time to delivery. That is why *dynamic* VaR, which captures both sources of Risk: **price fluctuations** and **liquidation − time** simultaneously, is an important risk-measure for the energy Risk Management.

5. Conclusions

In the present paper we presented a simple (practical) method of how to incorporate dynamic effects associated with liquidity risk into the already popular "static" VaR methodology. Introducing a liquidity stochastic process, a practical equation for the liquidation time was derived. Two different ways to transform liquidation time into the additional risk-effect were discussed. It was shown that for the financial derivatives traded in the energy market the actual liquidation time is often not short enough to be neglected, especially if a short time-horizon was chosen for the "static" VaR. Moreover, strong correlations between derivatives with different liquidity, which is the property of (European) energy market, may essentially increase the role of dynamic effects and impact of the liquidity risk for the portfolio.

We have shown that accounting for the "liquidation" time may seriously modify "static" VaR estimation, and dynamic (liquidity) effects should certainly be taken into account in Energy Risk Control/Management of the financial institutions (however, one should be very careful when applying the VaR methodology to the *physical positions* in the energy industry: see, for instance, [10]). Although we discussed only the energy market, this method may also be used in other markets.

We are grateful to J.van der Pol who collected "Nord Pool" data, P. Saunders for the useful comments and R. Frey for an overview of the literature. We also would like to thank A. Fox and A. Borsari how helped us with the preparation of the manuscript.

References

1. J.P. Morgan. "RiskMetrics". *Technical Document.* Forth Edition, 1996, New York.

2. F.X. Diebold, A. Hickman, A. Inoue and T. Schuermann. "Scale Models". *Risk*, 1998, *N* 11 (January), p.104-107.

3. R. Jarrow and A. Subramanian. "Mopping up Liquidity". *Risk*, 1997, *N* 10 (December), p.170-173.

4. F. Longstaff. "Optimal Portfolio Choice and Valuation of Illiquid Securities". *Preprint.* 1998. UCLA, Department of Finance.

5. A.S. Mello and J.E. Parsons. "Hedging and Liquidity". *The Review of Financial Studies*, 2000, Vol. 13, *N* 1, p. 127-153.

6. R. Frey. "Market Illiquidity as a Source of Model Risk in Dynamic Hedging". *Preprint*, Swiss Banking Institute, University of Zürich. 2000 February, 14 p.; *Conference on Intertemporal Finance* - Workshop: *Mathematical Finance*, Oct. 5-7, 2000, Center of Finance and Econometrics, University of Konstanz, Germany, p.53-54.

7. A. Bangia, F.X. Diebold, T. Schuermann and J.D. Stroughair. "Modeling Liquidity Risk With Implications for Traditional Market Risk Measurement and Management". *Preprint.* The Wharton School, Financial Institutions Center, University of Pennsylvania, 1999, June, 16 p.

8. P.F.Christoffersen, F.X. Diebold and T. Schuermann. "Horizon Problems and Extreme Events in Financial Risk Management". *Economic Policy Review.* 1998, October, p.109-118.

9. F.C. Dorst and T.E. Nijman. "Temporal Aggregation of GARCH Processes". *Econometrica* 61: 909-27, 1993.

10. G. Keers. "The over value of VaR".
Energy & Power Risk Management, February 2000, p.30-31.

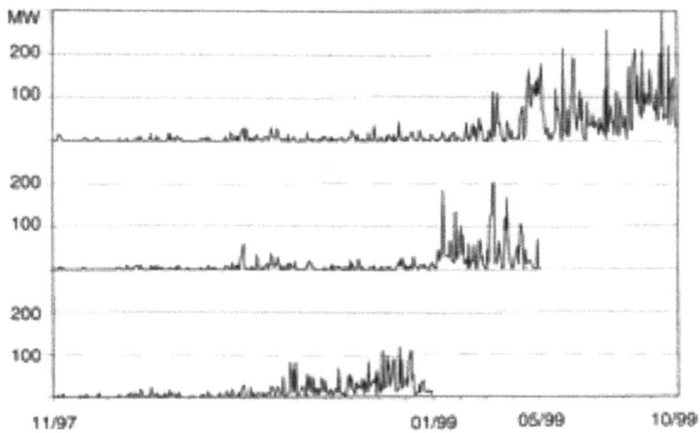

FIGURE 1. *Liquidity in MegaWatt for three futures contracts traded at Nord Pool Exchange: Winter-I/1999 (bottom); Summer-1999 (middle); Winter-II/1999 (top).*

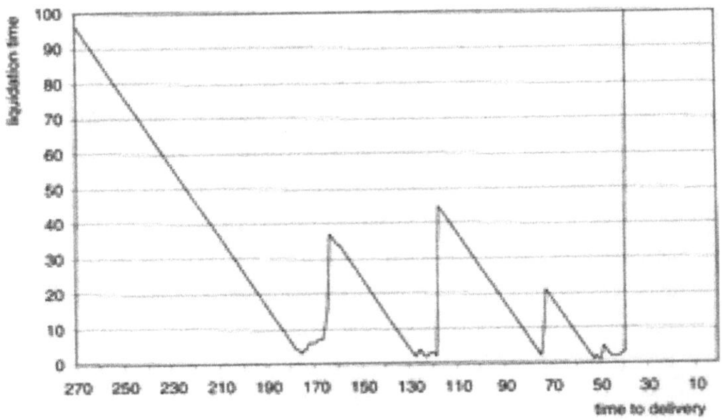

FIGURE 2. *Liquidation time (days) for Winter-I/1999 contract as a function of the time to delivery.*

Southern Energy Europe, Dr. Willem Dreesweg 2, Amstelveen, The Netherlands
E-mail address: sergey.nagornii@southernenrgy-europe.nl

Trends in Mathematics, © 2001 Birkhäuser Verlag Basel/Switzerland

Riccati Equation and Viscosity Solutions in Mean Variance Hedging

Bernhard Peisl

Abstract. It is well known in the literature that in a Markovian market the Riccati Backward Stochastic Differential Equation arising from the Mean Variance Hedging Problem can be characterized by a parabolic partial differential equation. We consider the solution of this Riccati Backward Stochastic Differential Equation at the initial time as function of the initial parameters. For a complete Markovian market we show that this function is the unique viscosity solution of the parabolic partial differential equation.

1. Introduction

One of the major problems in finance is the hedging of a future cash flow ξ, delivered at time T. The market consists of $m+1$ assets: the non-risky bond and m risky assets driven by a d-dimensional Brownian motion. In general it is not possible to construct a self-financed portfolio based on the $m+1$ assets, so as to attain as terminal value the $\mathcal{F}_T$-measurable random variable ξ.

However one can determine a "good" hedging strategy by solving the mean-variance hedging problem introduced by Föllmer and Sondermann (1986):

$$\min_{\pi \in \mathcal{L}^2_{\mathcal{F}}(t,T;\mathbf{R}^m)} E\left|X^{t,x;\pi}(T) - \xi\right|^2,$$

where $X^{t,x;\pi}(T)$ is the terminal value of the self-financed portfolio.

There are many different approaches to solve the mean-variance hedging problem (see for example Duffie and Richardson [3], Föllmer and Sondermann [5], Gourieroux, Laurent and Pham [7], Monat and Stricker [18], Pham, Rheinländer and Schweizer [20], Schweizer [21]). In a recent paper Kohlmann and Zhou [15] obtain an explicit representation of the optimal portfolio using the theory of backward stochastic differential equations (BSDE for short). This representation is closely related to a backward Riccati differential equation. The global existence and uniqueness result for such a Riccati BSDE is proven by Kohlmann and Tang [12] even for the more general case with stochastic coefficients.

The support by the Center of Finance and Econometrics, University of Konstanz, is gratefully acknowledged.

In this article we connect the Riccati BSDE to viscosity solutions. The theory of viscosity[1] solutions deals with certain partial differential equations of the form $F(x, u, Du, D^2u) = 0$ where $F : \mathbf{R}^n \times \mathbf{R} \times \mathbf{R}^n \times S(n) \to \mathbf{R}$ and $S(n)$ is the set of symmetric $n \times n$ matrices. In the expression $F(x, u, Du, D^2u)$ u is a real-valued function and Du and D^2u correspond to the gradient of u and the matrix of second derivatives of u, respectively. However, Du and D^2u have no classical meaning as in general u will be only continuous.

In the following we apply the theory of viscosity solution on the parabolic equation $u_t + F(t, x, u, Du, D^2u) = 0$ where now $u : \mathbf{R} \times \mathbf{R}^n \to \mathbf{R}$, $(t, x) \mapsto u(t, x)$, and u_t, Du, D^2u mean D_tu, D_xu, D_x^2u.

It is well known in the literature that in a Markovian market the Riccati Backward Stochastic Differential Equation arising from the Mean Variance Hedging Problem can be characterized by such a parabolic partial differential equation. We consider the solution of this Riccati BSDE at the initial time as function u of the initial parameters. Applying the inverse transformation on the parabolic equation, for complete markets we derive a linear partial differential equation. As the inverse transformation preserves the property "viscosity solution" the problem is reduced to a simpler one which was solved by Pardoux and Peng [19]. It follows that the function u is the unique viscosity solution of the parabolic partial differential equation.

The organization of this article is as follows. In section 2 we give a definition of viscosity solutions and cite a result of Pardoux and Peng [19] connecting viscosity solutions and stochastic differential equations with Lipschitz-continuous generator. In section 3 we introduce the mean-variance hedging problem and cite the explicit solution given by Kohlmann and Tang [14]. In section 4 we show the main result of this article.

2. Viscosity solutions

2.1. Definition

Let F be a continuous function satisfying the ellipticity condition

$$F(t, x, p, A + B, V) \leq F(t, x, A, V) \tag{1}$$

for all $(t, x) \in [0, T] \times \mathbf{R}^d$, $T > 0$, $p \in \mathbf{R}^d$, $V \in \mathbf{R}$ and symmetric matrices A, B with $B \geq 0$. Consider the partial differential equation

$$\left\{ \begin{array}{rcl} 0 & = & \frac{\partial}{\partial t} v(t, x) - F(t, x, D_x v(t, x), D_x^2 v(t, x), v(t, x)), \\ & & \quad (t, x) \in [0, T] \times \mathbf{R}^d, \\ v(T, x) & = & g(x), \qquad x \in \mathbf{R}^d, \end{array} \right. \tag{2}$$

where $g \in C(\mathbf{R}^d)$.

[1]Viscosity solutions were introduced by Crandall and Lions [2] in 1983. During the following years this theory was further investigated by Crandall, Evans, Ishii, Lions and others. For an extensive presentation of the theory of viscosity solutions, and also for historical comments, we refer to Crandall, Ishii and Lions [1].

Definition 2.1. *Let $u \in C([0,T] \times \mathbf{R}^d)$ satisfy $u(T,x) = g(x), x \in \mathbf{R}^d$, $g \in C(\mathbf{R}^d)$. u is said to be a viscosity sub-solution (resp. super-solution) of equation (2) in $(0,T) \times \mathbf{R}^d$ if for each $\varphi \in C^\infty((0,T) \times \mathbf{R}^d)$*

$$\frac{\partial}{\partial t}\varphi(\bar{t},\bar{x}) - F\left(\bar{t},\bar{x},D_x\varphi(\bar{t},\bar{x}),D_x^2\varphi(\bar{t},\bar{x}),\varphi(\bar{t},\bar{x})\right) \geq 0 \quad \textit{(resp. } \leq 0\textit{)}$$

at every $(\bar{t},\bar{x}) \in (0,T) \times \mathbf{R}^d$ which is a local minimum (resp. maximum) of $\varphi - u$ on $(0,T) \times \mathbf{R}^d$, with $\varphi(\bar{t},\bar{x}) = u(\bar{t},\bar{x})$.

u is called a viscosity solution of (2) in $(0,T) \times \mathbf{R}^d$ if it is both a viscosity sub-solution and viscosity super-solution.

2.2. Connection to stochastic differential equations

In this subsection we cite a result given by Pardoux and Peng [19] in 1992.

We work on a fixed time interval $[0,T]$. On a probability space $(\Omega, \mathcal{F}, P)$ we are given a d-dimensional Brownian motion $\{W_t; t \in [0,T]\}$. We assume that for $0 \leq t \leq T$, $\{\mathcal{F}_t, 0 \leq t \leq T\}$ is the natural filtration generated by the d-dimensional Brownian motion W, completed by the P-null sets of $\mathcal{F}$. Consider the Forward Backward Stochastic Differential Equation (FBSDE for short) system

$$\left\{\begin{array}{rcl} X_s^{t,x} & = & x + \int_t^s b(s, X_\tau^{t,x})d\tau + \int_t^s \sigma(s, X_\tau^{t,x})dW_\tau \\ Y_s^{t,x} & = & g(X_T^{t,x}) + \int_s^T f(X_\tau^{t,x}, Y_\tau^{t,x}, Z_\tau^{t,x})d\tau + \int_s^T Z_\tau^{t,x} dW_\tau \\ & & t \leq s \leq T \end{array}\right. \tag{3}$$

where $x \in \mathbf{R}^d$, b and σ are smooth deterministic functions, g is a deterministic Lipschitz-continuous function and f is a stochastic function which is globally Lipschitz-continuous in (x,y,z) uniformly in t:

$$|f(t,x,y,z) - f(t,x',y',z')| \leq c(|x-x'| + |y-y'| + |z-z'|) \quad \text{for all } t \in [0,T].$$

We define

$$u(t,x) := Y_t^{t,x}. \tag{4}$$

Theorem 2.2 (Pardoux/Peng). *The function u defined by (4) is the unique viscosity solution of the backward parabolic partial differential equation*

$$\left\{\begin{array}{rcl} -\frac{\partial v}{\partial t}(t,x) & = & \frac{1}{2}\sum_{i,j=1}^d (\sigma\sigma^*)_{i,j}(t,x)\frac{\partial^2 v}{\partial x_i \partial x_j}(t,x) + \sum_{i=1}^d b_i(t,x)\frac{\partial v}{\partial x_i}(t,x) \\ & & +f(t,x,v(t,x),(\frac{\partial v}{\partial x_i}\sigma)(t,x)), \qquad (t,x) \in [0,T] \times \mathbf{R}^d \\ v(T,x) & = & g(x), \qquad x \in \mathbf{R}^d. \end{array}\right.$$

Remark 2.3. *As Pardoux and Peng needed that f is Lipschitz-continuous for technical reasons this theorem cannot be easily extended to the general non-linear case.*

3. Mean Variance Hedging in a Markovian market

3.1. The general market model

We consider the financial market in which there are $m+1$ primitive assets: one bond (the non-risky asset) of price process

$$S_0(s) = \exp \int_t^s r(\tau)d\tau, \quad t \le s \le T, \tag{5}$$

and m stocks (risky assets)

$$dS(s) = \operatorname{diag}(S(s))(\mu(s)ds + \sigma(s)dW(s)), \quad t \le s \le T. \tag{6}$$

Where $W = (W_1, \dots, W_d)^*$ is a d-dimensional Brownian motion defined on a complete probability space $(\Omega, \mathcal{F}, P)$, and $\{\mathcal{F}_t, 0 \le t \le T\}$ is the P-augmentation of the natural filtration generated by the d-dimensional Brownian motion W. We assume that the instantaneous interest rate r, the m-dimensional appreciate vector process μ and the volatility $m \times d$ matrix process σ are progressively measurable with respect to $\{\mathcal{F}_t, 0 \le t \le T\}$. We further assume they are uniformly bounded and there exists a positive constant ε such that

$$\sigma\sigma^*(s) \ge \varepsilon I_{m\times m}, \quad 0 \le s \le T, a.s. \tag{7}$$

With $\tilde{\mu} := \mu - re_m$, where $e_m = (1, \dots, 1)^* \in \mathbf{R}^m$, the risk premium process is given by

$$\lambda(s) = \sigma^*(\sigma\sigma^*)^{-1}\tilde{\mu}(s), \quad 0 \le s \le T. \tag{8}$$

3.2. Formulation of the Mean Variance Hedging Problem

For any $x \in \mathbf{R}$ and $\pi \in \mathcal{L}^2_{\mathcal{F}}(t, T; \mathbf{R}^m)$, we define the self-financed wealth process X by

$$\left\{ \begin{array}{rcl} dX & = & (rX + (\tilde{\mu}, \pi))\, ds + \pi^*\sigma dW, \quad t \le s \le T, \\ X(t) & = & x, \quad \pi \in \mathcal{L}^2_{\mathcal{F}}(t, T; \mathbf{R}^m), \end{array} \right. \tag{9}$$

where x is the initial capital and π the amount invested in the risky assets S.

Problem 3.1. *Given a random variable $\xi \in L^2(\Omega, \mathcal{F}_T, P)$, consider the control problem*

$$\min_{\pi \in \mathcal{L}^2_{\mathcal{F}}(t,T;\mathbf{R}^m)} E\left|X^{t,x;\pi}(T) - \xi\right|^2 \tag{10}$$

where $X^{t,x;\pi}$ is the solution to the wealth equation (9).

This problem is called the mean variance hedging problem in financial economics. In the next subsection we will see the complete solution given by Kohlmann and Tang [12].

3.3. Complete solution

In September 2000 Kohlmann and Tang [12] obtained the global existence and uniqueness result for a one-dimensional backward stochastic Riccaty equation whose generator contains a quadratic term of L (the second unknown component). They applied this result to the mean variance hedging problem with stochastic market conditions.

In this setting the associated Riccati equation is a non-singular BSDE (see [12])

$$\left\{\begin{array}{rcl} dK & = & -\left((2r-|\lambda|^2)K - 2(\lambda, L) - K^{-1}L^*\sigma^*(\sigma\sigma^*)^{-1}\sigma L\right) ds \\ & & +(L, dW), \quad t \le s \le T, \\ K(T) & = & 1. \end{array}\right. \tag{11}$$

Let (ψ, ϕ) the $\mathcal{F}_t$-adapted solution of the BSDE

$$\left\{\begin{array}{rcl} d\psi & = & -\left((r-|\lambda|^2-(\lambda, K^{-1}L))\psi - (\lambda + K^{-1}\sigma^*(\sigma\sigma^*)^{-1}\sigma L, \phi)\right) ds \\ & & +(\phi_i, dW), \quad t \le s \le T, \\ \psi(T) & = & \xi. \end{array}\right. \tag{12}$$

Then the optimal hedging portfolio is

$$\pi = -(\sigma\sigma^*)^{-1}\left((\tilde{\mu} + \sigma K^{-1}L)X - \tilde{\mu}K^{-1}\psi - \sigma K^{-1}\phi\right). \tag{13}$$

3.4. Markovian market

In a Markovian market we assume the following structure for the randomness of the market parameters:

$$r(s,\omega) := r(s, Y_s), \qquad \mu(s,\omega) := \mu(s, Y_s), \qquad \sigma(s,\omega) := \sigma(s, Y_s)$$

where $\{Y_s : t \le s \le T\}$ is defined by the stochastic differential equation

$$\left\{\begin{array}{rcl} dY & = & \eta(s,Y)ds + \gamma(t,Y)dW, \quad t \le s \le T, \\ Y_t & = & y \in \mathbf{R}^m. \end{array}\right. \tag{14}$$

Hence equation (14) for Y and the Riccati equation (11) constitute a Forward-Backward Stochastic Differential Equation.

3.5. Characterization of the solution of the Riccati equation

We define the generator of the Riccati equation by

$$h(s,y,z,v) \;:=\; z(2r-|\lambda|^2)(s,y) - 2v^*\lambda(s,y) - z^{-1}v^*\sigma^*(\sigma\sigma^*)^{-1}\sigma(s,y)v \tag{15}$$

Then it is well known in the literature that the solution of the Riccati equation (11) can be characterized by the parabolic differential equation (where indices on Z indicate partial derivatives of Z)

$$\left\{\begin{array}{rcl} 0 & = & Z_s + (\eta(s,y), Z_y) + \frac{1}{2}tr\,(\gamma\gamma^*(s,y)Z_{yy}) \\ & & +h(s,y,Z,Z_y\gamma(s,y)), \qquad y \in \mathbf{R}^d, t \le s \le T, \\ Z(T,y) & = & 1, \quad y \in \mathbf{R}^d, \end{array}\right. \tag{16}$$

through the relation

$$K(s) = Z(s, Y_s), \qquad L(s) = (Z_y\gamma(s, Y_s))^*. \tag{17}$$

To indicate the dependency of K on the initial parameters in the following we write $K^{t,y}$.

In the next section we want to show that there exists a unique viscosity solution of equation (16).

4. Viscosity property of the solution of the Riccati equation

4.1. Proof

In this subsection we assume that the market is complete. This means that there are as many stocks as Brownian motions: $m = d$. Furthermore the matrix σ is invertible. In this setting we have

Theorem 4.1. *The function u defined by $u(t,y) := K^{t,y}(t)$, $(t,y) \in [0,T] \times \mathbf{R}^d$, is the unique viscosity solution of the parabolic PDE*

$$\begin{cases} 0 &= v_t(t,y) + (\eta(t,y), v_y(t,y)) + \frac{1}{2} tr\,(\gamma\gamma^*(t,y)v_{yy}(t,y)) \\ & \quad +h(t,y,v(t,y),(v_y\gamma)(t,y)), \qquad y \in \mathbf{R}^d, 0 \le t \le T, \\ v(T,y) &= 1, \quad y \in \mathbf{R}^d, \end{cases} \tag{18}$$

where

$$\begin{aligned} h(t,y,z,\zeta) &= z(2r - |\lambda|^2)(t,y) - 2\zeta^*\lambda(t,y) - z^{-1}\zeta^*\sigma^*(\sigma\sigma^*)^{-1}\sigma(t,y)\zeta, \\ & \qquad (t,y,\zeta) \in [0,T] \times \mathbf{R} \times \mathbf{R}^d, z \neq 0. \end{aligned}$$

Proof. As Kohlmann and Tang [12] showed that $K^{t,y}$ is positive we may apply the inverse transformation

$$w(t,y) := \frac{1}{v(t,y)} \quad , \qquad (t,y) \in [0,T] \times \mathbf{R}^d.$$

Assume on some suitable set $M \subset \mathbf{R} \times \mathbf{R}^d$ we are given a parabolic differential equation by

$$\begin{cases} 0 &= v_t(t,y) - F(t,y,v_y(t,y),v_{yy}(t,y),v(t,y)) \\ v(T,y) &= g(y) \end{cases} \tag{19}$$

where F is an operator satisfying (1). Then applying the inverse transformation we formally derive the transformed PDE

$$\begin{cases} w_t(t,y) &= -w^2(t,y)F(t,y,-\frac{w_y(t,y)}{w^2(t,y)}, \frac{2(w_y(t,y))^*w_y(t,y)}{w^3(t,y)} - \frac{w_{yy}(t,y)}{w^2(t,y)}, \frac{1}{w(t,y)}) \\ w(T,y) &= \frac{1}{g(y)}. \end{cases} \tag{20}$$

Thus we define the operator G by

$$G(t,x,p,A,V) := -V^2F(t,y,-\frac{p}{V^2}, \frac{2p^*p}{V^3} - \frac{A}{V^2}, \frac{1}{V}).$$

The first step is to show that G is elliptic:
By the ellipticity of F, with $F(\cdot,\cdot,A-B,\cdot) \geq F(\cdot,\cdot,A-B+B,\cdot) = F(\cdot,\cdot,A,\cdot) \Leftrightarrow -F(\cdot,\cdot,A-B,\cdot) \leq F(\cdot,\cdot,A,\cdot)$ for $B \geq 0$ we derive

$$\begin{aligned} G(t,x,p,A+B,V) &= -V^2F(t,y,-\tfrac{p}{V^2},\tfrac{2p^*p}{V^3}-\tfrac{A}{V^2}-\tfrac{B}{V^2},\tfrac{1}{V}) \\ &\leq -V^2F(t,y,-\tfrac{p}{V^2},\tfrac{2p^*p}{V^3}-\tfrac{A}{V^2},\tfrac{1}{V}) \end{aligned}$$

which means that G is elliptic.

In the second step we show that if w is a viscosity solution of (20) then $u = \frac{1}{w}$ is a viscosity solution of (19):
We assume that w is a viscosity solution of (20). Let $(\widetilde{t},\widetilde{y})$ be a local maximum of $\varphi - u$ ($\varphi \in C^\infty([0,T]\times\mathbf{R}^d)$) with $\varphi(\widetilde{t},\widetilde{y}) = u(\widetilde{t},\widetilde{y})$. Then there exists a $\varepsilon > 0$ such that $\varphi(t,y) - u(t,y) \leq \varphi(\widetilde{t},\widetilde{y}) - u(\widetilde{t},\widetilde{y}) = 0$ for $(t,x) \in U_\varepsilon(\widetilde{t},\widetilde{y})$. As u is positive and u and φ are continuous and $\varphi(\widetilde{t},\widetilde{y}) = u(\widetilde{t},\widetilde{y})$ we can choose an ε such that $\varphi(t,y) \neq 0$ in $U_\varepsilon(\widetilde{t},\widetilde{y})$. Hence we have $\frac{1}{\varphi(t,y)} \geq \frac{1}{u(t,y)} = w(t,y)$ in $U_\varepsilon(\widetilde{t},\widetilde{y})$. And finally with $\frac{1}{\varphi(t,y)} - w(t,y) \geq 0 = \frac{1}{\varphi(\widetilde{t},\widetilde{y})} - w(\widetilde{t},\widetilde{y})$ we derive that $(\widetilde{t},\widetilde{y})$ is a local minimum of $\frac{1}{\varphi} - w$.
As $\psi := \frac{1}{\varphi}|_{U_\varepsilon(\widetilde{t},\widetilde{y})} \in C^\infty(U_\varepsilon(\widetilde{t},\widetilde{y}))$ and as by assumption w is a viscosity solution of (20) we have $\psi_t(\widetilde{t},\widetilde{y}) - G(\widetilde{t},\widetilde{y},\psi_y(\widetilde{t},\widetilde{y}),\psi_{yy}(\widetilde{t},\widetilde{y}),\psi(\widetilde{t},\widetilde{y})) \geq 0$. Hence

$$\begin{aligned} &\varphi_t(\widetilde{t},\widetilde{y}) - F(\widetilde{t},\widetilde{y},\varphi_y(\widetilde{t},\widetilde{y}),\varphi_{yy}(\widetilde{t},\widetilde{y}),\varphi(\widetilde{t},\widetilde{y})) \\ =\ & \frac{-\psi_t(\widetilde{t},\widetilde{y})}{\psi^2(\widetilde{t},\widetilde{y})} - F(\widetilde{t},\widetilde{y},\frac{-\psi_y(\widetilde{t},\widetilde{y})}{\psi^2(\widetilde{t},\widetilde{y})},-\frac{\psi^2\psi_{yy}(\widetilde{t},\widetilde{y}) - 2(\psi_y(\widetilde{t},\widetilde{y}))^*\psi_y(\widetilde{t},\widetilde{y})}{\psi^4(\widetilde{t},\widetilde{y})},\frac{1}{\psi(\widetilde{t},\widetilde{y})}) \\ =\ & -\frac{1}{\psi^2(\widetilde{t},\widetilde{y})}\left(\psi_t(\widetilde{t},\widetilde{y}) + \psi^2(\widetilde{t},\widetilde{y})F(\cdots)\right) \\ =\ & -\frac{1}{\psi^2(\widetilde{t},\widetilde{y})}\left(\psi_t(\widetilde{t},\widetilde{y}) - G(\widetilde{t},\widetilde{y},\psi_y(\widetilde{t},\widetilde{y}),\psi_{yy}(\widetilde{t},\widetilde{y}),\psi(\widetilde{t},\widetilde{y}))\right) \\ \leq\ & 0 \end{aligned}$$

Hence u is viscosity super-solution of (19). In the same way one shows that u is viscosity sub-solution of (19).

It is clear that if the solution of equation (20) is unique then the solution of equation (19) is unique and vice versa. Otherwise assume there would be two solutions v_1 and v_2 of (19). Then $w_1 := \frac{1}{v_1}$ and $w_2 := \frac{1}{v_2}$ satisfy (20). But as the solution of (20) is unique we have $w_1 = w_2$ and hence $v_1 = v_2$.

In the last step we apply the inverse transformation on equation (18): In this case the transformed PDE is

$$\begin{aligned} 0 &= w_t(t,y) + (\eta(t,y), w_y(t,y)) \\ &\quad -tr\left(\gamma\gamma^*(t,y)\tfrac{1}{w(t,y)}(w_x(t,y))^* w_x(t,y)\right) + \tfrac{1}{2}tr\left(\gamma\gamma^*(t,y)w_{yy}(t,y)\right) \\ &\quad -w(t,y)(2r-|\lambda|^2)(t,y) - 2\gamma^*(w_x(t,y))^*\lambda(t,y) \\ &\quad +\tfrac{1}{w(t,y)}\gamma^*(t,y)(w_x(t,y))^*\sigma^*(\sigma\sigma^*)^{-1}\sigma(t,y)w_x(t,y)\gamma(t,y), \\ &\qquad y\in \mathbf{R}^d, 0\le t\le T, \\ w(T,y) &= 1, \quad y\in\mathbf{R}^d. \end{aligned}$$

As the matrix σ is invertible we have $\sigma^*(\sigma\sigma^*)^{-1}\sigma = I_{d\times d}$. And an easy calculation shows that

$$tr\left(\gamma\gamma^*\frac{1}{w}(w_x)^*w_x\right) = \frac{1}{w}\gamma^*(w_x)^*\sigma^*(\sigma\sigma^*)^{-1}\sigma w_x\gamma.$$

Thus the quadratic terms vanish and the PDE becomes linear. Applying now Theorem 2.2 we derive that there exists a unique viscosity solution of (20). And hence by the considerations above there exists a unique viscosity solution of (19). □

Remark 4.2. *In the case of an incomplete market the method of the inverse transformation cannot be applied successfully. One crucial point in the proof above was the fact that $\sigma^*(\sigma\sigma^*)^{-1}\sigma = I_{d\times d}$. If the market is incomplete, of course, this equality is not satisfied. Hence after applying the inverse transformation the quadratic terms won't vanish.*

Now one could have the idea to complete the market (see for example Karatzas et al. [8]) in such a way that one does not want to invest in the additional stocks at all. Or to introduce constraints in a complete market that one is not allowed to trade some stocks (see El Karoui et al. [4]). In both cases the market is complete and the matrix σ is split into $\sigma = \binom{\sigma_1}{\sigma_2}$. Now one can solve the mean variance hedging problem in a complete market and the optimal portfolio for the incomplete market is that part of the portfolio which corresponds to the stocks one is allowed or willing to invest in.

But the Riccati BSDE of the incomplete market situation and the Riccati BSDE of the situation of a special complete market described as above are different. For that reason we may not conclude that there exists a viscosity solution of the Riccati BSDE of the incomplete market case if there is one for that special complete market.

However, working with this special complete market we can find another representation of the Riccati BSDE of the incomplete market case which avoids the product $\sigma_1^*(\sigma_1\sigma_1^*)^{-1}\sigma_1$. But now a term with the product $\sigma_2^*(\sigma_2\sigma_2^*)^{-1}\sigma_2$ appears. By rank reasons this product is not equal to $I_{d\times d}$. Hence we have a similar problem as in the original incomplete case.

References

[1] M.G. Crandall, H. Ishii and P.-L. Lions, *User's guide to viscosity solutions of second order partial differential equations,* Bull. Am. Math. Soc., New Ser. 27, No.1, 1-67 (1992)

[2] M.G. Crandall and P.-L. Lions, *Viscosity solution of Hamilton-Jacobiequations,* Trans. Amer. Math. Soc 277 (1983), 1-42

[3] D. Duffie and H.R. Richardson, *Mean-variance hedging in continuous time,* The Annals of Applied Probability 1991, Vol. 1, No. 1, 1-15

[4] N. El Karoui, S. Peng and M.C. Quenez, *Backward Stochastic Differential Equations in Finance,* Mathematical Finance, Vol.7, No.1 (January 1997), 1-71

[5] W.H. Fleming and H.M. Soner, *Controlled Markov Processes and Viscosity Solutions,* Springer-Verlag (1993)

[6] H. Föllmer and D. Sondermann, *Hedging of Non-redundant Contingent Claims,* Contributions to Mathematical Economics (1986). Ed. Werner Hildenbrand, Anreu Mas-Colell, 205-223

[7] C. Gourieroux, J.P. Laurent and H. Pham, *Mean-variance hedging and numéraire,* Mathematical Finance, Vol.8, No.3 (July 1998), 179-200

[8] I. Karatzas, J.P. Lehoczky, S.E. Shreve and G.-L. Xu, *Martingale and duality methods for utility maximization in an incomplete market,* SIAM J. Control and Optimization, Vol. 29, No. 3, pp. 702-730, May 1991

[9] I. Karatzas and S.E. Shreve, *Brownian Motion and Stochastic Calculus,* Springer-Verlag (1991), Second Edition

[10] M. Kohlmann and B. Peisl, *A Note on Mean-Variance Hedging of Non-Attainable Claims,* CoFE Working Paper No.00/06 (2000), University of Konstanz

[11] M. Kohlmann and S. Tang, *Optimal Control of Linear Stochastic Systems with Singular Costs, and the Mean-Variance Hedging Problem with Stochastic Market Conditions,* CoFE Working Paper No.00/13 (2000), University of Konstanz

[12] M. Kohlmann and S. Tang, *Global Adapted Solution of One-Dimensional Backward Stochastic Riccati Equation, with Application to the Mean-Variance Hedging,* CoFE Working Paper No.00/26 (2000), University of Konstanz

[13] M. Kohlmann and S. Tang, *Multi-Dimensional Backward Stochastic Riccati Equations, and Applications,* CoFE Working Paper No.00/29 (2000), University of Konstanz

[14] M. Kohlmann and X.Y. Zhou, *Backward Stochastic Differential Equations and Stochastic Controls: A New Perspective,* CoFE Working Paper No.99/09 (1999), University of Konstanz

[15] M. Kohlmann and X.Y. Zhou, *Relationship between Backward Stochastic Differential Equations and Stochastic Control: An LQ Approach,* SIAM Journal on Control and Optimization, Vol. 38, Number 5, pp. 1392-1407, 2000

[16] J.P. Laurent and H. Pham, *Dynamic programming and mean-variance hedging,* Finance Stochast. 3, 83-110 (1999)

[17] J.-P. Lepeltier and J. San Martin, *Existence for BSDE with superlinear-quadratic coefficient,* Stochastics and Stichastics Reports, Vol. 63, pp. 227-240 (1998)

[18] P. Monat and C. Stricker, *Föllmer-Schweizer decomposition and mean-variance hedging for general claims,* The Annals of Probability, 1995, Vol. 23, No. 2, 605-628

[19] E. Pardoux and S. Peng, *Backward Stochastic Differential Equations and Quasilinear Parabolic Partial Differential Equations,* Rozovskii, B. L. and Sowers, R. S. (eds), Stochastic partial differential equations and their applications (Lect. Notes Control inf. Sci. 176, pp. 200-217), Springer 1992

[20] H. Pham, T. Rheinländer and M. Schweizer, *Mean-variance hedging for continous processes: New proofs and examples,* Finance Stochast. 2, 173-198 (1998)

[21] M. Schweizer, *Mean-variance hedging for general claims,* The Annals of Applied Probability, 1992, Vol. 2, No. 1, 171-179

[22] D.W. Stroock, *Topics in stochastic differential equations,* Tata Institute of Fundamental Research, Bombay, published by Springer-Verlag (1982)

[23] J. Yong and X.Y. Zhou, *Stochastic Controls: Hamiltonian Systems and HJB Equations,* Springer-Verlag (1999)

Center of Finance and Econometrics
Fachbereich Mathematik
Universität Konstanz
78457 Konstanz
Germany
E-mail address: Bernhard.Peisl@uni-konstanz.de

Trends in Mathematics, © 2001 Birkhäuser Verlag Basel/Switzerland

A Minimal Financial Market Model

Eckhard Platen

1. Introduction

An advanced financial market model should be analytically tractable and must reflect with a minimal number of factors essential stylised empirical facts. It must work equally well for derivative pricing and hedging as well as for risk measurement and portfolio management. All factors in such a market model should represent directly interpretable or observable quantities.

The well-known Black-Scholes model (BSM) assumes that geometric Brownian motions generate the asset price dynamics. Although the theoretical and practical importance of the BSM cannot be underestimated, it is far from being satisfactory. The BSM has the drawback that it assumes deterministic volatility. This does not match historically observed stochastic volatility. For instance, in the context of option pricing, practitioners have to correct for implied volatility skews and smiles due to stochastic volatility. The at the money short term implied volatility of an index has typically a strong negative correlation with the index itself, which is also known as the leverage effect. Furthermore, it has been observed that log-returns of indices over long time periods tend to be Gaussian distributed. On the other hand, short term log-returns of asset prices have been shown to be leptokurtic with specific distributional properties. These stylised empirical facts should be well reflected in a sophisticated financial market model. Such a model should also acknowledge natural symmetries that exist in a benchmarked market as we will see below.

There is a rich literature on asset price modelling. For surveys on the modelling of stochastic volatility and other empirical stylised facts in the context of asset price dynamics, see [11] and [9]. A well studied group of asset price models covers the subordinated models, see, for instance, [2], [16], [10] and [14] to mention just a few. These models use non-decreasing stochastic processes, usually with independent increments, to generate an independent stochastic operational time as directing process. A disadvantage of subordinated asset price models is that they typically assume independence between an asset price and the directing process. This means they have difficulties to model the above mentioned leverage effect.

It is highly desirable that at least an idealised version of a financial market model should form a complete market. In [7] and [5] volatility function approaches are developed that lead to complete market models. These are in some sense generalisations of the constant elasticity of variance model, see, e.g., [3]. Unfortunately, as shown in [4], such models may, in general, not guarantee strict positivity of asset prices.

This paper proposes a diffusion model, the minimal market model (MMM), that uses a minimal number of directly interpretable factors. These factors are modelled as square root and Ornstein-Uhlenbeck processes under the real world probability measure. The basic building blocks of the proposed model are the different denominations of the benchmark portfolio measured in units of primary assets. They determine all financial quantities including short rates, dividend rates, volatilities and risk premia. The MMM reflects major stylised empirical facts as will be discussed at the end of the paper. Most importantly, it generates stochastic volatility endogenously without using any additional stochastic volatility process. This is in contrast to most stochastic volatility models, see, for instance, [13] or [8], which use typically unobservable exogeneous processes to generate stochastic volatility. The MMM is computationally highly tractable and allows a good fit to historical and derivative data, as shown in [12].

2. Minimal Market Model

2.1. Savings Accounts and Benchmark Portfolio

Let us define a *primary asset* as an income or loss producing asset, for instance, a stock or currency. We consider in a market the evolution of the prices of $d+1$ primary assets, $d \in \{1, 2, \ldots\}$, that are modelled on a filtered probability space $(\Omega, \mathcal{A}_T, \underline{\mathcal{A}}, P)$. Here the filtration $\underline{\mathcal{A}} = (\mathcal{A}_t)_{t\in[0,T]}$ fulfills the usual conditions with $\mathcal{A}_0$ being trivial, see [17].

We assume that each primary asset has its own time value. The time value of the domestic currency is expressed via the corresponding *savings account process* $B^0 = \{B^0(t),\, t \in [0,T]\}$, where

$$B^0(t) = \exp\left\{\int_0^t f^0(s)\, ds\right\}$$

for $t \in [0,T]$, $T \in (0,\infty)$. This savings account accumulates continuously interest according to the *domestic interest rate process* $f^0 = \{f^0(t),\, t \in [0,T]\}$ which describes the income rate from holding the domestic currency. Thus the 0th primary asset can be interpreted as the domestic currency. The time value of the jth primary asset is similarly modelled by the jth *savings account process* $B^j = \{B^j(t),\, t \in [0,T]\}$, where

$$dB^j(t) = B^j(t)\, f^j(t)\, dt \tag{2.1}$$

for $t \in [0,T]$ with $B^j(0) = 1$, $j \in \{1,\ldots,d\}$. The jth *income rate process* $f^j = \{f^j(t),\, t \in [0,T]\}$ can be, for instance, a dividend rate or domestic or foreign interest rate. In summary, the jth savings account measures accumulated income or loss generated by the jth asset in units of the jth asset, $j \in \{0,\ldots,d\}$.

The *exchange price* $X^{i,j}(t)$ is the price of one unit of the jth asset at time t measured in units of the ith asset. The jth *savings account value* $S^{i,j}(t)$ at time t, when measured in units of the ith primary asset, is given by the formula

$$S^{i,j}(t) = X^{i,j}(t)\, B^j(t). \tag{2.2}$$

We assume that there exists a *best benchmark portfolio* (BBP) which is a strictly positive $\underline{A}$-adapted, self-financing portfolio that when used as numeraire makes any benchmarked price process to an $(\underline{A}, P)$-martingale. Let us denote by $D^j = \{D^j(t),\, t \in [0,T]\}$ the jth denomination of the BBP, that is when it is measured in units of the jth primary asset, $j \in \{0,\ldots,d\}$. In the MMM the jth denomination $D^j(t)$ of the BBP at time t is specified as a functional of the form

$$D^j(t) = (Y^j(t))^{q_j}\, \xi^j(t), \tag{2.3}$$

with jth *average growth*

$$\xi^j(t) = \xi_j \exp\left\{\int_0^t \eta^j(s)\, ds\right\} \tag{2.4}$$

for $t \in [0,T]$ and $j \in \{0,\ldots,d\}$. The jth *initial parameter* $\xi_j > 0$ and the jth *exponent* $q_j \in (0,\infty]$ are given constants. The jth *square root process* $Y^j = \{Y^j(t),\, t \in [0,T]\}$, that appears in (2.3), is characterised by the stochastic differential equation (SDE)

$$dY^j(t) = \frac{\nu^j}{4}\, \varphi^j(t)\, (1 - Y^j(t))\, dt - \sum_{k=1}^{d} \gamma^{j,k}(t)\, \sqrt{Y^j(t)}\, dW^k(t) \tag{2.5}$$

with jth *diffusion parameter* $\varphi^j(t) = \sum_{k=1}^d (\gamma^{j,k}(t))^2$ for $t \in [0,T]$ and initial value $Y^j(0) > 0,\ j \in \{0,\ldots,d\}$.

Here $W^1,\ldots,W^d$ are independent standard Wiener processes. The jth *dimension* $\nu^j \in (2 + 4q_j, \infty)$ is a constant and the j,kth *volatility parameter* $\gamma^{j,k} : [0,T] \to (-\infty,\infty)$ is a deterministic function of time for $j \in \{0,\ldots,d\}$, $k \in \{1,\ldots,d\}$. Obviously, the square root process Y^j fluctuates around its long term average value of one. The diffusion parameter φ^j controls the speed of its evolution and the dimension ν^j the magnitude of extreme fluctuations. For larger dimension ν^j, extreme fluctuations are less likely.

The jth *growth rate process* $\eta^j = \{\eta^j(t),\, t \in [0,T]\}$ governs, according to (2.3), the average growth of the jth denomination of the BBP. We assume that η^j is an Ornstein-Uhlenbeck process with

$$d\eta^j(t) = \alpha^j(t)\, (\bar{\eta}^j(t) - \eta^j(t))\, dt + \sum_{k=1}^{d} \beta^{j,k}(t)\, dW^k(t) \tag{2.6}$$

for $t \in [0,T]$ and initial value $\eta^j(0) \in (-\infty,\infty)$, $j \in \{0,\ldots,d\}$. Here the jth *speed of adjustment parameter* $\alpha^j : [0,T] \to (0,\infty)$, j,kth *diffusion parameter* $\beta^{j,k} : [0,T] \to (-\infty,\infty)$ and jth *average growth rate* $\bar{\eta}^j : [0,T] \to (-\infty,\infty)$ are assumed to be given deterministic functions of time, $j \in \{0,\ldots,d\}$, $k \in \{1,\ldots,d\}$.

Note that the SDEs for the square root and growth rate processes have unique, strong solutions with explicitly known transition densities, see [17].

2.2. Asset Price Dynamics

The jth *benchmarked savings account process* $\hat{S}^j = \{\hat{S}^j(t),\, t \in [0,T]\}$ is formed by the ratio

$$\hat{S}^j(t) = \frac{B^j(t)}{D^j(t)} \tag{2.7}$$

for $t \in [0,T]$ and $j \in \{0,\dots,d\}$. We remark that there is a structural similarity between all benchmarked savings accounts in the model.

By application of the Itô formula, see [22], we obtain from (2.7), (2.1), (2.3) and (2.5) for the jth benchmarked savings account the SDE

$$\begin{aligned} d\hat{S}^j(t) &= \hat{S}^j(t)\left[f^j(t) - \eta^j(t) + q_j\,\frac{\nu^j}{4}\,\varphi^j(t) + \frac{\varphi^j(t)\,q_j}{Y^j(t)}\left(\frac{q_j+1}{2} - \frac{\nu^j}{4}\right)\right] dt \\ &\quad + \hat{S}^j(t)\sum_{k=1}^{d}\sigma^{j,k}(t)\,dW^k(t) \end{aligned} \tag{2.8}$$

for $t \in [0,T]$ with initial value

$$\hat{S}^j(0) = \frac{1}{D^j(0)} \tag{2.9}$$

and j,kth *volatility* of the jth benchmarked savings account

$$\sigma^{j,k}(t) = \frac{q_j\,\gamma^{j,k}(t)}{\sqrt{Y^j(t)}} \tag{2.10}$$

for $j \in \{0,\dots,d\}$, $k \in \{1,\dots,d\}$ and $t \in [0,T]$.

By assumption, the BBP makes all benchmarked savings account processes martingales. To ensure that $\hat{S}^j$ is a martingale we require from (2.8) that the jth income rate has to fulfill the relation

$$f^j(t) = \eta^j(t) + q_j\,\varphi^j(t)\left\{\frac{1}{Y^j(t)}\left[\frac{\nu^j}{4} - \frac{q_j+1}{2}\right] - \frac{\nu^j}{4}\right\} \tag{2.11}$$

for $t \in [0,T]$ and $j \in \{0,\dots,d\}$.

From (2.7), (2.8) and (2.1) we obtain by application of the Itô formula the dynamics of the jth denomination of the BBP in the form

$$\begin{aligned} dD^j(t) &= d\left(\frac{B^j(t)}{\hat{S}^j(t)}\right) \\ &= D^j(t)\left[\left(f^j(t) + \sum_{k=1}^{d}(\sigma^{j,k}(t))^2\right) dt - \sum_{k=1}^{d}\sigma^{j,k}(t)\,dW^k(t)\right] \end{aligned} \tag{2.12}$$

for $t \in [0,T]$ and $j \in \{0,\dots,d\}$.

Using the ith and jth denominations of the BBP, the i,jth exchange price can be expressed by the ratio

$$X^{i,j}(t) = \frac{D^i(t)}{D^j(t)}. \tag{2.13}$$

It follows by application of the Itô formula and (2.12) that

$$\begin{aligned} dX^{i,j}(t) &= X^{i,j}(t) \Bigg[(f^i(t) - f^j(t))\, dt \\ &\quad + \sum_{k=1}^{d} (\sigma^{j,k}(t) - \sigma^{i,k}(t)) \left\{ -\sigma^{i,k}(t)\, dt + dW^k(t) \right\} \Bigg] \end{aligned} \tag{2.14}$$

for $t \in [0,T]$ with initial value $X^{i,j}(0) = \frac{D^i(0)}{D^j(0)}$ and $i,j \in \{0,\ldots,d\}$.

Thus we derive by the Itô formula, (2.14) and (2.1) for the jth savings account, when measured in units of the ith asset, the SDE

$$dS^{i,j}(t) = S^{i,j}(t) \left[f^i(t)\, dt + \sum_{k=1}^{d} (\sigma^{j,k}(t) - \sigma^{i,k}(t)) \left\{ -\sigma^{i,k}(t)\, dt + dW^k(t) \right\} \right] \tag{2.15}$$

for $t \in [0,T]$ with $S^{i,j}(0) = X^{i,j}(0)$ and $i,j \in \{0,\ldots,d\}$.

Note that in the appreciation rate of (2.15) a premium appears in excess of the income rate of the denominating ith asset. More precisely, the i,jth *risk premium* is obtained in the form

$$p^{i,j}(t) = \sum_{k=1}^{d} (\sigma^{i,k}(t) - \sigma^{j,k}(t))\, \sigma^{i,k}(t) \tag{2.16}$$

for $t \in [0,T]$ and $i,j \in \{0,\ldots,d\}$. For $i = 0$ the equation (2.15) describes the dynamics of the jth savings account expressed in units of the domestic currency.

2.3. Pricing and Hedging

We introduce the $(n+1) = 4(d+1)$ factors $Z^0,\ldots,Z^n$ which are assumed to be given by the SDE

$$dZ^\ell(t) = \alpha^\ell(t, Z^0(t),\ldots,Z^n(t))\, dt + \sum_{k=1}^{d} \beta^{\ell,k}(t, Z^0(t),\ldots,Z^n(t))\, dW^k(t) \tag{2.17}$$

for $t \in [0,T]$ with appropriate initial value $Z^\ell(0)$, $\ell \in \{0,\ldots,n\}$. The first $d+1$ factors are the square root processes, that is $Z^\ell(t) = Y^\ell(t)$, see (2.5), for $t \in [0,T]$ and $\ell \in \{0,\ldots,d\}$. The following $d+1$ factors play the role of the growth rate processes, that means $Z^{\ell+d+1}(t) = \eta^\ell(t)$, see (2.6), for $t \in [0,T]$ and $\ell \in \{0,\ldots,d\}$. The third set of factors is chosen to generate the average growth of the denominations of the BBP, that is $Z^{\ell+2d+2}(t) = \xi^\ell(t)$, see (2.4), for $t \in [0,T]$ and $\ell \in \{0,\ldots,d\}$. The fourth group of factors represents the savings accounts, that is $Z^{\ell+3d+3}(t) = B^\ell(t)$, see (2.1), for $t \in [0,T]$ and $\ell \in \{0,\ldots,d\}$. In a stylised version of the MMM, see [21], one only uses the first $d+1$ factors which are already sufficient to produce the key stylised features of the model, that is its stochastic volatility, associated leverage effect and the appropriate leptokurtic log-return distribution.

We call a nonnegative, integrable benchmarked payoff

$$H = H(\bar{T}, Z^0(\bar{T}), \ldots, Z^n(\bar{T})) \in [0, \infty)$$

that matures at $\bar{T} \in (0, T]$, a *European contingent claim*. One can show, see [21], that the benchmarked price

$$u(t, Z^0(t), \ldots, Z^n(t))$$

at time $t \in [0, \bar{T}]$ of the contingent claim H is given by the conditional expectation

$$u(t, Z^0(t), \ldots, Z^n(t)) = E\left(H \mid \mathcal{A}_t\right). \tag{2.18}$$

Let us denote by $b(t) = [b^{k,j}(t)]_{k,j=1}^d$ a matrix with $b^{k,j}(t) = \hat{S}^j(t)\,(\sigma^{j,k}(t) - \sigma^{0,k}(t))$. For given H we introduce the vector $c(t) = (c^1(t), \ldots, c^d(t))^\top$ with

$$\begin{aligned} c^k(t) \quad = \quad & \sum_{\ell=0}^{n} \beta^{\ell,k}(t, Z^0(t), \ldots, Z^n(t)) \, \frac{\partial u(t, Z^0(t), \ldots, Z^n(t))}{\partial Z^\ell} \\ & - u(t, Z^0(t), \ldots, Z^n(t)) \, \sigma^{0,k}(t). \end{aligned} \tag{2.19}$$

If the matrix $b(t)$ has an inverse $b^{-1}(t)$ for all $t \in [0, \bar{T}]$, then the numbers $\delta^0(t), \ldots, \delta^d(t)$ of units of the corresponding savings account, which must be held at time t in a self-financing hedge portfolio that replicates H, are determined by the relations

$$(\delta^1(t), \ldots, \delta^d(t))^\top = b^{-1}(t)\, c(t) \tag{2.20}$$

and

$$\delta^0(t) = \frac{1}{\hat{S}^0(t)} \left(u(t, Z^0(t), \ldots, Z^n(t)) - \sum_{j=1}^{d} \delta^j(t)\, \hat{S}^j(t) \right). \tag{2.21}$$

To give an example for the price of a contingent claim, it follows from [21] that a European call option with strike price K and maturity $\bar{T}$ written on the BBP D^0 has, in units of domestic currency, the price

$$C^0(t, Z^0(t), K, \bar{T}) = E\left(\frac{D^0(t)}{D^0(\bar{T})} \, (D^0(\bar{T}) - K)^+ \,\middle|\, \mathcal{A}_t \right).$$

for $t \in [0, \bar{T}]$. Since the benchmarked BBP has constant value one, the equations (2.19) - (2.21) determine with the specification

$$u(t, Z^0(t), \ldots, Z^n(t)) = 1$$

the numbers of units of savings accounts in the BBP. It can be shown that the BBP is the inverse of the state-price deflator, see [6], and that it is the optimal growth portfolio, see [18].

3. Properties of the MMM

Let us now discuss some properties of the MMM that are responsible for its good fit to empirical data. Most importantly, the MMM generates stochastic volatility without the use of an external volatility process. Note from (2.10), that volatilities are proportional to the square root of inverted square root processes. This feature causes negative correlation between the jth denomination of the BBP and its volatility. Since the BBP is like an index, the MMM generates a leverage effect as is typically observed for indices but also for equities. The resulting volatility process exhibits natural persistence and generates threshold exceedances of log-returns that are clustered and have leptokurtic distributions that are similar to those observed in reality, as we will see below.

From (2.12) and (2.10) it follows that the log-return of the jth BBP over a small time interval $[t, t+\varepsilon]$ with $\varepsilon > 0$, has the form

$$\log(D^j(t+\varepsilon)) - \log(D^j(t)) \approx \frac{1}{\sqrt{Y^j(t)}} \sum_{k=1}^{d} \gamma^{j,k}(t)\, q_j\, (W^k(t+\varepsilon) - W^k(t)),$$

which is approximately a Student t distributed random variable since the invariant distribution of Y^j is the gamma distribution. Thus under the MMM, log-returns over small time intervals of the jth denomination of the BBP are Student t distributed. This feature has been shown to be consistent with observations obtained from most stock market indices, see [15], and index benchmarked stock prices, see [19].

Under the MMM log-returns of the BBP over long time intervals are, in principle, characterised by the corresponding Gaussian growth rates η^j and thus are essentially Gaussian distributed. This matches the widely observed Gaussianity of long term log-returns.

Since the leverage effect is modelled naturally by the MMM, implied volatility skews for European puts and calls on stock indices, stocks and currencies match those observed in practice, see [12].

Under the MMM the short rate (2.11) involves the inverse of a square root process, and is thus stochastic, with properties that are similar to those actually observed, see [20]. Note that the inverse of a square root process generates a diffusion coefficient function with exponent 1.5 for the short rate as suggested by different empirical studies, see, e.g., [1].

The MMM in its version with constant parameters provides a good fit to real data. For example, movements in the at-the-money implied volatility are naturally captured by movements in the underlying factor, see [12]. In addition all of the parameters in the model can be estimated almost directly from observed quantities. Another important advantage of the model is its computational tractability due to the fact that the transition densities of the factors have an explicit form.

References

[1] Chan, K. C., G. A. Karolyi, F. A. Longstaff, & A. B. Sanders (1992). An empirical comparison of alternative models of the short-term interest rate. *J. Finance* **47**, 1209–1227.

[2] Clark, P. K. (1973). A subordinated stochastic process model with finite variance for speculative prices. *Econometrica* **41**, 135–159.

[3] Cox, J. C. & S. A. Ross (1976). The valuation of options for alternative stochastic processes. *J. Financial Economics* **3**, 145–166.

[4] Delbaen, F. & H. Shirakawa (1997). Squared Bessel processes and their applications to the square root interest rate model. Preprint. Department of Industrial Engineering and Management, Tokio Institute of Technology.

[5] Derman, E. & I. Kani (1994). Riding on a smile. *Risk* **7**, 32–39.

[6] Duffie, D. (1996). *Dynamic Asset Pricing Theory* (2nd ed.). Princeton, University Press.

[7] Dupire, B. (1994). Pricing with a smile. *Risk* **7**, 18–20.

[8] Fouque, J. P., G. Papanicolau, & K. R. Sircar (1999). Financial modelling in a fast mean-reverting stochastic volatility environment. *Asian-Pacific Financial Markets* **6**, 37–48.

[9] Frey, R. (1997). Derivative asset analysis in models with level-dependent and stochastic volatility. Mathematics of Finance, Part II. *CWI Quarterly* **10**(1), 1–34.

[10] Geman, H., D. Madan, & M. Yor (1998). Asset prices are Brownian motion: only in business time. University Paris IX Dauphine, (working paper).

[11] Ghysels, E., A. Harvey, & E. Renault (1996). Stochastic volatility. In *Statistical Methods in Finance*, Volume 14 of *Handbook of Statist.*, pp. 119–191. North-Holland.

[12] Heath, D. & E. Platen (2000). Pricing and hedging of derivatives under an alternative asset price model with stochastic volatility. University of Technology Sydney, (working paper).

[13] Heston, S. L. (1993). A closed-form solution for options with stochastic volatility with applications to bond and currency options. *Rev. Financial Studies* **6**(2), 327–343.

[14] Heyde, C. C. (1999). A risky asset model with strong dependence through fractal activity time. Columbia University, New York, (working paper).

[15] Hurst, S. R. & E. Platen (1997). The marginal distributions of returns and volatility. In Y. Dodge (Ed.), *L_1-Statistical Procedures and Related Topics*, Volume 31 of *IMS Lecture Notes - Monograph Series*, pp. 301–314. Institute of Mathematical Statistics Hayward, California.

[16] Hurst, S. R., E. Platen, & S. T. Rachev (1997). Subordinated market index models: A comparison. *Financial Engineering and the Japanese Markets* **4**, 97–124.

[17] Karatzas, I. & S. E. Shreve (1988). *Brownian Motion and Stochastic Calculus.* Springer.

[18] Karatzas, I. & S. E. Shreve (1998). *Methods of Mathematical Finance*, Volume 39 of *Appl. Math.* Springer.

[19] Platen, E. (1999a). On the log-return distribution of index benchmarked share prices. Technical report, University of Technology, Sydney. QFRG Research Paper 22.

[20] Platen, E. (1999b). A short term interest rate model. *Finance and Stochastics* **3**(2), 215–225.

[21] Platen, E. (2000). Financial modelling with benchmark portfolio. University of Technology Sydney, (working paper).

[22] Protter, P. (1990). *Stochastic Integration and Differential Equations.* Springer.

University of Technology Sydney, School of Finance, Economics and Department of Mathematical Sciences, PO Box 123, Broadway, NSW, 2007, Australia
E-mail address: eplaten@uts.edu.au

Trends in Mathematics, © 2001 Birkhäuser Verlag Basel/Switzerland

A note on equivalent martingale measures with bounded density

Miklós Rásonyi

Abstract. Let S be a disrete-time martingale with finite time horizon. We show that the set of equivalent martingale measures with bounded density is dense in the set of equivalent martingale measures with respect to the total variation norm.

1. Introduction

Let $(\Omega, \mathbf{F} = (\mathcal{F}_t)_{0 \le t \le T}, \mathcal{F}, P)$ be a discrete-time stochastic basis. For the sake of simplicity we suppose that each $\mathcal{F}_t$ is complete, $\mathcal{F}_0$ is trivial and $\mathcal{F} = \mathcal{F}_T$.

Let $S_t, 0 \le t \le T$ be an adapted process with values in $\mathbb{R}^d$. We may interpret S as the discounted price process of d traded securities in a financial market. In this case each self-financing portfolio policy corresponds to a predictable d-dimensional process ϕ and to a value process

$$V_t^\phi := V_0^\phi + \sum_{i=1}^{T} \langle \phi_i, S_i - S_{i-1} \rangle.$$

Here $\langle \cdot, \cdot \rangle$ denotes scalar product in $\mathbb{R}^d$.

An efficient market shouldn't contain arbitrage opportunities so we expect our model to satisfy

$$\mathbf{NA}: \; \forall \phi \; \left(V_0^\phi = 0, V_T^\phi \ge 0 \; a.s. \Rightarrow V_T^\phi = 0 \; a.s. \right)$$

The following landmark result was first obtained by R. C. Dalang, A. Morton and W. Willinger in [2]:

Theorem 1.1. **NA** *is equivalent to the existence of a measure $Q \sim P$ such that S is a Q-martingale (with respect to the filtration* $\mathbf{F}$*).* □

From now on we suppose that P itself is a martingale measure, i.e. S is a P-martingale with respect to the filtration $\mathbf{F}$. Let us define

$$\mathcal{Q} := \{Q \sim P : S \text{ is a } Q - \text{martingale}\},$$

$$\mathcal{Q}_b := \{Q \in \mathcal{Q} : ess.sup \frac{dQ}{dP} < \infty\}.$$

$$\mathcal{Q}_{bb} := \{Q \in \mathcal{Q} : ess.sup \frac{dQ}{dP} \vee \frac{dP}{dQ} < \infty\}.$$

The elements of $\mathcal{Q}$ correspond to linear pricing rules on the set of (bounded) contingent claims (i.e. $\mathcal{F}_T$-measurable random variables). The supper-replication cost (or the seller price, see e.g. [6]) of a nonnegative claim X is given by

$$\sup_{Q \in \mathcal{Q}} E_Q X. \tag{1}$$

The above listed facts demonstrate that in mathematical finance the set $\mathcal{Q}$ is of the utmost importance. It is therefore desirable that we explore the structure of $\mathcal{Q}$. The main theorem of the present paper asserts that $\mathcal{Q}_b$ is dense in $\mathcal{Q}$ in the topology induced by the total variation norm of singned measures.

This result was announced in [3] for the first time, then reappeared in [4]. A complete proof has recently been obtained by Yu. M. Kabanov and Ch. Stricker in a continuous-time framework using advanced stochastic calculus (see [5]). Here we give a rather transparent proof in a discrete time environment. As in [5] and [2], our argument is based on the measurable selection theorem (see III. 44-45. in [1]). We think that this proof may serve as a good illustration of that technology.

Throughout the paper conditional disconditional expectations are taken in the generalized sense, see e.g. [1].

2. The main theorem

$\overline{\mathbb{R}}^d$ will denote the one-point compactification of $\mathbb{R}^d$. $C(\overline{\mathbb{R}}^d)$ is the set of continuous functions on $\overline{\mathbb{R}}^d$. If W is a topological space, $\mathcal{B}(W)$ will stand for its Borel-σ-field.

Theorem 2.1. *$\mathcal{Q}_b$ is dense in $\mathcal{Q}$ in the topology induced by the total variation norm.*

The proof of this theorem can be reduced to the one-step case, more precisely to the following lemma:

Proposition 2.2. *Let $\mathcal{G} \subset \mathcal{H} \subset \mathcal{F}$ σ-algebras which are complete with respect to P. Let $Y : \Omega \to \mathbb{R}^d$, $D : \Omega \to \mathbb{R}_+ \setminus \{0\}$ $\mathcal{H}$-measurable random variables with*

$$E|Y| < \infty, E(Y|\mathcal{G}) = 0,$$
$$E(D|\mathcal{G}) = 1, E(YD|\mathcal{G}) = 0. \tag{2}$$

Then there is a sequence of strictly positive $\mathcal{H}$-measurable bounded *random variables G_n such that*

$$E(G_n|\mathcal{G}) = 1, E(YG_n|\mathcal{G}) = 0$$

and we have also

$$G_n \to D \text{ a.s.}$$

Idea of proof. Firstly, let us consider the case when $\mathcal{G}$ is trivial. Denoting by $\mu(dx, dy), dx \in \mathcal{B}(\mathbb{R}_+), dy \in \mathcal{B}(\mathbb{R}^d)$ the distribution of (D, Y) we can rewrite (2) as

$$\int_{\mathbb{R}^{d+1}} \mu(dx, dy)x = 1, \int_{\mathbb{R}^{d+1}} \mu(dx, dy)xy = 0.$$

We are trying to find a bounded function

$$l_n : \mathbb{R}^{d+1} \to \mathbb{R}_+ \setminus \{0\}$$

satisfying

$$\int_{\mathbb{R}^{d+1}} \mu(dx, dy) l_n(x, y) = \int_{\mathbb{R}^{d+1}} \mu(dx, dy)x,$$

$$\int_{\mathbb{R}^{d+1}} \mu(dx, dy) y l_n(x, y) = \int_{\mathbb{R}^{d+1}} \mu(dx, dy)yx.$$

$$\int_{\mathbb{R}^{d+1}} \mu(dx, dy)|x - l_n(x, y)| \leq \frac{1}{n}.$$

Such a function will be provided by Lemma 3.1 if we take $h := p := d + 1$,

$$f(x, y) := x, \phi(x, y) := (1, y^1, \dots, y^d) \in \mathbb{R}^p, (x, y) \in \mathbb{R}^{d+1}.$$

Taking $G_n(\omega) := l_n(D(\omega), Y(\omega))$ we will get a sequence satisfying the requirements:

$$E|D - G_n| = \int_{\Omega \times \mathbb{R}^{d+1}} \mu(dx, dy)|x - l_n(x, y)| \leq \frac{1}{n},$$

hence it suffoces to subtract a subsequence so as to get convergence almost surely.

For general $\mathcal{G}$, we apply the same argument for fixed ω, then take a measurable selector. □

3. An approximation result and corollaries

Let μ be a probability measure on the Borel sets of $\mathbb{R}^h$. Let $\phi : \mathbb{R}^h \to \mathbb{R}^p$ be a Borel-measurable function satisfying

$$\int_{\mathbb{R}^h} \mu(dz)|\phi(z)| < \infty.$$

Let U be the set of μ-almost surely strictly positive Borel-measurable functions r on $\mathbb{R}^h$ such that

$$\int_{\mathbb{R}^h} \mu(dz)|\phi(z)|r(z) < \infty.$$

Let $U_0 := U \cap C(\overline{\mathbb{R}}^h)$. The following lemma is akin to the key lemma of [5].

Lemma 3.1. *Let $f \in U, \varepsilon > 0$ be fixed. Then there exists $l \in U_0$ such that*

$$\int_{\mathbb{R}^h} \mu(dz)|f(z) - l(z)| < \varepsilon,$$

$$\int_{\mathbb{R}^h} \mu(dz)\phi(z)f(z) = \int_{\mathbb{R}^h} \mu(dz)\phi(z)l(z).$$

□

Finally we list two corollaries of the main theorem:

Corollary 3.2. *$\mathcal{Q}_{bb}$ is dense in $\mathcal{Q}$, too.*

Proof. If $Q \in \mathcal{Q}_b$ then $(1-\frac{1}{n})Q + \frac{1}{n} \to Q, n \to \infty$ in the total variation norm topology. □

Another consequence of the above is that we may express the super-replication cost (1) as follows (cf. Corollary 1.4 in [5]):

Corollary 3.3. *For any nonnegative $\mathcal{F}_T$-measurable random variable X:*

$$\sup_{Q\in\mathcal{Q}} E_Q X = \sup_{Q\in\mathcal{Q}_b} E_Q X = \sup_{Q\in\mathcal{Q}_{bb}} E_Q X.$$

Proof. It suffices to show that for any $R \in \mathcal{Q}$

$$\sup_{Q\in\mathcal{Q}_{bb}} E_Q X \geq E_R X. \tag{3}$$

Let us fix $n \in \mathbb{N}$ and a sequence $Q_k \in \mathcal{Q}_{bb}, k \in \mathbb{N}$ such that $Q_k \to R, k \to \infty$ in the total variation norm. We notice that

$$E_{Q_k}(X \wedge n) \to E_R(X \wedge n), k \to \infty.$$

In particular

$$\sup_{Q\in\mathcal{Q}_{bb}} E_Q(X \wedge n) \geq E_R(X \wedge n).$$

An application of the theorem on monotone convergence of nonnegative functions yields (3). □

Acknowledgements. I'd like to thank Yu. M. Kabanov and Ch. Stricker for providing me with [5].

References

[1] Dellacherie C., Meyer P. A. Probabilities and Potential. North-Holland, Amsterdam, 1978

[2] Dalang R. C., Morton A., Willinger W. Equivalent martingale measures and no-arbitrage in stochastic securities market models. *Stoch. and Stoch. Rep.*, **29**, (1990), 185–201.

[3] Jacka S. A., A martingale representation result and an appplication to incomplete financial markets *Mathematical Finance*, **2**, (1992), 23–34.

[4] Kramkov D. O., On closure of a family of martingale measures and an optional decomposition of supermartingales *Theory Probab. Appl.* **41**, (1996), 788–791.

[5] Kabanov Yu. M., Stricker Ch. On equivalent martingale measures with bounded densities. *preprint*, Université de Franche-Comté, Besançon, (2000)

[6] Stettner L. Option pricing in discrete-time incomplete market models. *Mathematical Finance*, **10**, (2000), 305–321.

Computer and Automation Institute of the Hungarian Academy of Sciences,
13-17 Kende utca,
1111 Budapest,
Hungary
E-mail address: rasonyi@sztaki.hu

Trends in Mathematics, © 2001 Birkhäuser Verlag Basel/Switzerland

Local optimality in the multi-dimensional multi-period mean-variance portfolio problem

Manfred Schäl

Abstract. A variant of the mean-variance problem is studied where the conditional variances of the gain in the last periods are minimized by a backward iteration.

1. Introduction

In discrete time $t = 0, 1, ..., T$, a frictionless financial market is studied which is free of arbitrage. The investor can invest in a bond and d stocks with unrestricted short selling. The bond is described by a price process $\{B_t,\ 0 \le t \le T\}$ and serves as a numeraire. The stocks are described by a d-dimensional price process $\{S_t,\ 0 \le t \le T\}$. An initial wealth x and a (self-financing dynamic) portfolio plan ξ define the value process $\{V_t^\xi(x),\ 0 \le t \le T\}$. The following mean-variance problem is classical (see [9] for the single-period case and [1], [4], [10], [13] for the multi-period case):

$$\min_\xi Var[V_T^\xi(x)/B_T] \quad \text{subject to} \quad E[V_T^\xi(x)/B_T] = v.$$

If the underlying probability measure P is not just a so-called martingale measure, then the constraint can be satisfied for all given numbers v (by some ξ). Of course, if B_T is deterministic then the problem is the same for the undiscounted and the discounted value process. But we don't make such an assumption and consider the discounted values. Then the problem is the same if we consider the gain process $\{G_t(\xi)\}$ in the discounted model defined by $x + G_t(\xi) := V_t^\xi(x)/B_t$ and study the problem:

$$\min_\xi Var[G_T(\xi)] \quad \text{subject to} \quad E[G_T(\xi)] = g. \tag{1.1}$$

It is known that the solution can be obtained by a solution of a utility maximization problem with a quadratic utility function $U(x) = -\{1 - x\}^2$. For the utility maximization problem one can use the classical criterion of maximizing the expected utility of the gain $G_T(\xi)$ which is considered here as a global optimization problem (see [16] – [21]). If ξ^* is a solution of

$$E[U(G_T(\xi^*))] = \max_\xi E[U(G_T(\xi))] \ , \tag{1.2}$$

then $c \cdot \xi^*$ is a solution of (1.1) with $c = g/(1 - \gamma_0)$ where $\gamma_0 := E[1 - G_T(\xi^*)]$. Obviously, $1 - G_T(\xi^*)$ is the orthogonal projection in $L^2(P)$ of 1 on the orthogonal complement of the linear space $\mathcal{G} = \{G_T(\xi),\ \xi \text{ any plan}\}$. Given the quadratic utility U, one can also define a locally optimal portfolio plan as a modification of Bellman's backward iteration which is different from the classical optimality when the utility is not linear. Let $G_{t,T}(\xi) := G_T(\xi) - G_t(\xi)$ describe the gain between t and T and $E_t[\cdot]$ the conditional expectation given the history of the market up to time t. Then we consider an investor who will not care about the past at any time t but will only consider the future gains. Thus at time $T-1$ the investor considers $\max_\xi E_{T-1}[U(G_{T-1,T}(\xi))] =: E_{T-1}[U(G_{T-1,T}(\xi^0))]$. At time t the investor knows how he will decide at the times $t+1, ..., T-1$ and thus he knows $G_{t+1,T}(\xi^0)$. Then he wants to solve the following problems by backward iteration :

$$E_t[U(G_{t,T}(\xi^0))] = \max_\xi E_t[U(G_{t,T}(\xi))] \tag{1.3}$$

$$\text{subject to } G_{t+1,T}(\xi) = G_{t+1,T}(\xi^0)\ ,\ 0 \leq t < T.$$

The portfolio plan ξ^0 is obtained by a sequence of orthogonal projections for the one-period models concerning the periods $(t-1, t]$. We will use here the condition 3.2 that the so-called (extended) mean-variance tradeoff process (in the sense of [19]) is deterministic and we will present some sufficient conditions. Condition 3.2 is satisfied if the (vector valued) returns in different periods are independent. But 3.2 is also fulfilled in situations with a deterministic relative risk return process. Under this condition 3.2, it is known that the variance-optimal (signed) martingale measure and the minimal (signed) martingale measure coincide (see [20], [21]). Here, the following will be shown: The expected gains $E[G_T(\xi^*)]$ and $E[G_T(\xi^0)]$ coincide. Thus, given the constraint $E[G_T(\xi)] = 1 - \gamma_0$, the investor can choose between ξ^* and ξ^0. Moreover, $E_t[G_{t,T}(\xi^0)] =: 1 - \gamma_t$ is deterministic for all t and $c \cdot \xi^0$ solves the following iterative mean-variance problems for the conditional variances Var_t :

$$Var_t[G_{t,T}(c \cdot \xi^0)] = \min_\xi Var_t[G_{t,T}(\xi)] \tag{1.4}$$

$$\text{subject to } E_m[G_{m,T}(\xi)] = c \cdot (1 - \gamma_m),\ t \leq m < T,\ 0 \leq t < T.$$

Both ξ^* and ξ^0 can be defined in terms of the adjustment process $\{\lambda_t\}$ (in the sense of [19] – [21]), i.e., the mutual fund principle holds with $\{\lambda_t\}$ as dynamic mutual fund. Here, under condition 3.2, the adjustment process is obtained by simple local (one-step) optimization problems. In (1.4) one can choose c such that $c \cdot \xi^0$ satisfies the desired constraint for $m = 0$. Then the other constraints are given through the numbers γ_m. The situation is particularly simple in the important (stationary) case where the (vector-valued) returns in different periods are independent and identically distributed. Then one has

$$\lambda_t^k = \lambda_0^k \cdot S_0^k \cdot B_t / S_t^k, \quad \gamma_t = (1 - \kappa)^{T-t} \quad \text{for some} \quad 0 \leq \kappa < 1.$$

Under ξ^0 the situation is more stable in the sense that $E_t[G_{t,T}(\xi^0)] = 1 - \gamma_t$ is already known from the beginning. Moreover, an investor using ξ^0 is more risk averse at the beginning than under ξ^* in the sense that he uses a portfolio which is smaller by the factor $0 < \gamma_1 \leq 1$. Local optimization in the sense of (1.3) is similar to the concept of local risk minimization (see [7]). In the latter procedure one solves the following problems again by backward iteration for an arbitrary square-integrable contingent claim H (formulated here for self-financing plans):

$$Var_t[\{H - G_{t,T}(\xi^H)\}^2] = \min_{\xi} Var_t[\{H - G_{t,T}(\xi)\}^2]$$

$$\text{subject to } G_{t+1,T}(\xi) = G_{t+1,T}(\xi^H),\ 0 \leq t < T.$$

There one has $Var_t[\{H - G_{t,T}(\xi)\}^2] = \min_x E_t[\{H - x - G_{t,T}(\xi)\}^2]$. Of course, in the case $H = 1$ this would lead to the trivial solution $\xi^H = 0$.

2. The financial market model

The evolution of the price of $1 + d$ assets will be modeled by a 1-dimensional and a d-dimensional stochastic process $\{B_t,\ t = 0, 1, ..., T\}$ and $\{S_t,\ t = 0, 1, ..., T\}$ on some probability space (Ω, F, P) where T is the time horizon. Then B_t and the components S_t^k, $1 \leq k \leq d$, of S_t are positive. The asset with price process $\{B_t\}$ is called the **bond** (or bank account). We assume (w.l.o.g.) $B_0 = 1$. The **interest rates** of the bond are then defined through $B_t =: (1 + r_1) \cdot ... \cdot (1 + r_t)$. The other d assets are called **stocks**. In many cases B_t will be deterministic or predictable. But we don't need such an assumption. The **discounted stock price process** $\breve{S}_t = (\breve{S}_t^1, ..., \breve{S}_t^d)$ is defined by $\breve{S}_t^k := S_t^k / B_t$. It is assumed that $\breve{S}_t^k$ is square-integrable for all k and t. For any vector-valued process $\{Z_t\}$ we define the backward increment by $\Delta Z_t := Z_t - Z_{t-1}$. Further, we write $x^\top$ for the transposed vector and $x^\top \cdot y$ for the inner product of $x, y \in \mathbf{R}^d$. There is some filtration $\{F_t,\ t = 0, 1, ..., T\}$ on (Ω, F) such that B_t and S_t are F_t-measurable, *i.e.*, $\{B_t\}$ and $\{S_t\}$ are adapted to $\{F_t\}$. Here, F_t describes the events of the **history** up to time t, where (w.l.o.g.) $F_0 = \{\emptyset, \Omega\}$ and $F_T = F$. Then S_0 is deterministic. A **portfolio plan** is given by an $\mathbf{R}^d$-valued adapted stochastic process $\xi = \{\xi_t,\ 0 \leq t < T\}$ such that $\xi_{t-1}^\top \cdot \Delta\breve{S}_t$ is square-integrable for $1 \leq t \leq T$. During $(t-1, t]$ the investor holds a portfolio $\xi_{t-1} = (\xi_{t-1}^1, ..., \xi_{t-1}^d)$ where ξ_{t-1}^k denotes the number of shares of the $k - th$ stock. We will consider self-financing portfolio plans. Given the initial wealth x, the number η_{t-1} of shares of the bond in $(t-1, t]$ is then specified for $1 \leq t \leq T$ by ξ according to the **budget equations** $\eta_0 + \xi_0^\top \cdot S_0 = x$, $\eta_t B_t + \xi_t^\top \cdot S_t = \eta_{t-1} B_t + \xi_{t-1}^\top \cdot S_t$. Then the **value process** $\{V_t^\xi(x)\}$ is given through $V_t^\xi(x) := \eta_t B_t + \xi_t^\top \cdot S_t$. As is well known, one can use a reduction to the discounted model where the interest rates of the bond are zero. This leads to

$$V_t^\xi(x)/B_t = x + G_t(\xi)\ ,\ \text{where}\ \ G_t(\xi) := \sum_{1 \leq m \leq t} \xi_{m-1}^\top \cdot \Delta\breve{S}_m\ ,\ 0 \leq t \leq T, \quad (2.1)$$

is the accumulated gain in the discounted model. As usual, we assume the well-known **no-arbitrage condition** (see [6], [11]) :

for $1 \leq t \leq T$ and any ξ : $\xi_{t-1}^{\top} \cdot \Delta\breve{S}_t \geq 0$ a.s. $\Longrightarrow \xi_{t-1}^{\top} \cdot \Delta\breve{S}_t = 0$ a.s.. (NA)

3. The one-period models

Throughout the paper we will work with fixed regular conditional probabilities

$$P_t[B](\omega) := P[B|F_t](\omega)\ ,\ B \in F,\ \omega \in \Omega,\ 0 \leq t < T. \tag{3.1}$$

There $P_0 = P$. We may assume (w.l.o.g.)

$$E_t[\|\Delta\breve{S}_{t+1}\|^2](\omega) < \infty \text{ for all } \omega \in \Omega,\ 0 \leq t < T, \text{ where } E_t[Z] := E[Z|F_t] \tag{3.2}$$

denotes the expectation of a random variable Z w.r.t. the fixed conditional probability. Thus (3.2) means that

$$\Delta\breve{S}_{t+1}^k \in L_t^2(\omega) := \{Z \in L^2(P);\ E_t[Z^2](\omega) < \infty\}\ ,\ 1 \leq k \leq d, \tag{3.3}$$

where $L_t^2(\omega)$ is endowed with the inner product $< Z, Z' >:= E_t[Z \cdot Z'](\omega)$. We may also assume (w.l.o.g.) that the no-arbitrage condition holds locally (see [2] Lemma 2.3, [11](3.22), [16]§2), i.e.,

$$P_t[\vartheta^{\top} \cdot \Delta\breve{S}_{t+1} \geq 0](\omega) = 1 \Longrightarrow P_t[\vartheta^{\top} \cdot \Delta\breve{S}_{t+1} = 0](\omega) = 1 \qquad (NA)^*$$
$$\forall\ \ \omega \in \Omega,\ \ 0 \leq t < T,\ \ \vartheta \in \mathbf{R^d}.$$

Let $\mathcal{L}_t(\omega)$ be the smallest linear space L in $\mathbf{R}^d$ such that $P_t[\Delta\breve{S}_{t+1} \in L](\omega) = 1$. Now we define a finite-dimensional linear subspace of $L_t^2(\omega)$ by

$$\mathcal{G}_t := \{\vartheta^{\top} \cdot \Delta\breve{S}_{t+1};\ \vartheta \in \mathbf{R}^d\}\ . \tag{3.4}$$

Further define $\Pi_t(\omega)Z$ as the orthogonal projection of $Z \in L_t^2(\omega)$ onto $\mathcal{G}_t$ and $\Pi_t^{\perp}(\omega)Z$ as the orthogonal projection onto the orthogonal complement $\mathcal{G}_t^{\perp}(\omega)$ of $\mathcal{G}_t$ where
$\mathcal{G}_t^{\perp}(\omega) = \{Z \in L_t^2(\omega);\ E_t[Z{\cdot}G](\omega) = 0 \forall G \in \mathcal{G}_t\} = \{Z \in L_t^2(\omega);\ E_t[Z{\cdot}\Delta\breve{S}_{t+1}^k](\omega) = 0 \forall k\}$.
There $\mathcal{G}_t^{\perp}(\omega)$ is a closed subset of $L_t^2(\omega)$. Since $\mathcal{G}_t$ is a finite-dimensional linear space, it is also closed and we have $\mathcal{G}_t^{\perp\perp}(\omega) = \mathcal{G}_t$. We define $\lambda_t(\omega) \in \mathcal{L}_t(\omega) \subset \mathbf{R}^d$ such that

$$\Pi_t 1 = \lambda_t^{\top} \cdot \Delta\breve{S}_{t+1},\ i.e.\ \min_{\vartheta \in \mathbf{R}^d} E_t[\{1 - \vartheta^{\top} \cdot \Delta\breve{S}_{t+1}\}^2] \text{ is attained at } \vartheta = \lambda_t, \tag{3.5}$$

where we know:

$$\min_{\vartheta \in \mathbf{R}^d} E_t[\{1 - \vartheta^{\top} \cdot \Delta\breve{S}_{t+1}\}^2] = \min_{\vartheta \in \mathcal{L}_t} E_t[\{1 - \vartheta^{\top} \cdot \Delta\breve{S}_{t+1}\}^2]. \tag{3.6}$$

Proposition 3.1. (a) λ_t *is unique.*

(b) $1 - \lambda_t^{\top} \cdot \Delta\breve{S}_{t+1} = \Pi_t 1$.
(c) $0 < E_t[\{1 - \lambda_t^{\top} \cdot \Delta\breve{S}_{t+1}\}^2] = E_t[1 - \lambda_t^{\top} \cdot \Delta\breve{S}_{t+1}] \leq 1$. *Proof.* a) Assume that the minimum in (3.6) is attained at $\vartheta_t \in \mathcal{L}_t$. Then $P_t[\lambda_t^{\top} \cdot \Delta\breve{S}_{t+1} = \vartheta_t^{\top} \cdot \Delta\breve{S}_{t+1}] = 1$

and hence $P_t[(\lambda_t - \vartheta_t)^\top \cdot \Delta\breve{S}_{t+1} = 0] = 1$. This implies that $\lambda_t - \vartheta_t \perp \mathcal{L}_t$ where $\lambda_t - \vartheta_t \in \mathcal{L}_t$ and hence $\lambda_t - \vartheta_t = 0$. b) We have $1 = \Pi_t^\perp 1 + \Pi_t 1$ *a.s.* .
c) We have $E_t[\Pi_t^\perp 1] = E_t[\Pi_t^\perp 1 \cdot (\Pi_t 1 + \Pi_t^\perp 1)] = E_t[\{\Pi_t^\perp 1\}^2]$. Further we know that $1 = E_t[\{\Pi_t^\perp 1\}^2] + E_t[\{\Pi_t 1\}^2]$. Now assume that $E_t[\{\Pi_t^\perp 1\}^2] = 0$; then $P_t[\lambda_t^\top \cdot \Delta\breve{S}_{t+1} = 1] = 1$ which is a contradiction to (NA)*. $\square$
The arguments for 3.1c are the same as in [21](1.5). We set

$$\kappa_t := E_t[\lambda_t^\top \cdot \Delta\breve{S}_{t+1}]. \tag{3.7}$$

Then by (3.5) and 3.1c we have

$$0 < 1 - \kappa_t = \min_{\vartheta \in \mathbf{R}^d} E_t[\{1 - \vartheta^\top \cdot \Delta\breve{S}_{t+1}\}^2] \leq 1. \tag{3.8}$$

There $1 - \kappa_t$ can be looked upon as the measure of a local risk. For the main results of the paper we will use the following condition:
Condition 3.2 . *The quantities κ_t are deterministic for $0 \leq t < T$.* Under this condition, the process $\{\lambda_t\}$ coincides with the *adjustment process* in the sense of [19] – [21]. Moreover, Schweizer calls the process $\tilde{K}_t := \sum_{m=0}^{t-1} \kappa_m$ the *extended mean-variance tradeoff.* Now we provide sufficient conditions for 3.2. Of course κ_0 is always deterministic. The **relative risk process** $\{R_t = (R_t^1, ..., R_t^d),\ 1 \leq t \leq T\}$ is defined by

$$1 + R_t^k := 1 + \Delta\breve{S}_t^k / \breve{S}_{t-1}^k = (1 + \Delta S_t^k /\ S_{t-1}^k)/(1 + r_t).$$

The condition of the next proposition is fulfilled in many market models.
Proposition 3.3. *For $1 \leq t < T$, κ_t is deterministic if R_{t+1} and F_t are stochastically independent.* The proof will be given for the more general situation of 3.4 . It shows that in the situation of 3.3:

$$\lambda_t^k = \delta_t^k / \breve{S}_t^k, 1 \leq k \leq d, \text{for some deterministic numbers } \delta_t^k. \tag{3.9}$$

Condition 3.4. *There exist some $n \geq 1$, a sequence $\{\theta_t, 0 \leq t < T\}$ of n-dimensional vectors, an adapted $\mathbf{R}^n$-valued process $\{W_t,\ 0 \leq t \leq T\}$ and an adapted $\mathbf{R}^{d \times n}$-valued process $\{\sigma_t, 0 \leq t < T\}$, such that:*
(1) $R_t^k = \sum_{j=1}^n \sigma_{t-1}^{k,j} \cdot (\theta_{t-1}^j + \Delta W_t^j)$, $1 \leq k \leq d$;
(2) *ΔW_t and F_{t-1} are stochastically independent for $2 \leq t \leq T$;*
(3) *the range $\{\sigma_{t-1}^\top(\omega)\vartheta; \vartheta \in \mathbf{R}^d\}$ of $\sigma_{t-1}^\top(\omega)$ does not depend on $\omega \in \Omega$, $1 \leq t < T$.*
Obviously, condition 3.4 (3) is satisfied if $n = d$ and $\sigma_{t-1}(\omega)$ has full rank for all ω and t. The situation in 3.3 is included; we just have to choose $\Delta W_t = R_t$. Another interesting example is a discrete-time stochastic difference equation where $\{W_t\}$ is a discrete-time white noise, i.e., ΔW_t is independent of F_{t-1} and for each t the random variables $\Delta W_t^1, ..., \Delta W_t^d$ are uncorrelated and standardized (with mean zero and variance one). Then θ_t is called **relative risk vector** or **unit price for**

risk which is deterministic under condition 3.4. Then we have for the conditional expectation and the conditional covariance matrix of R_t :

$$E_{t-1}[R_t] = \sigma_{t-1}\theta_{t-1} \ , \ Cov_{t-1}[R_t] = \sigma_{t-1}\sigma_{t-1}^\top \ . \tag{3.10}$$

For $n = d = 1$, the latter situation can also be found in GARCH-M-models (see [5]).

Theorem 3.5. *Under condition* 3.4, κ_t *is deterministic for all* t, *i.e., Condition* 3.2 *is fulfilled. Proof.* We set $I_{t-1} := \{\sigma_{t-1}^\top \vartheta;\ \vartheta \in \mathbf{R}^d\} \subset \mathbf{R}^n$ and $\theta_{t-1}^j + \Delta W_t^j =: U_t^j$. Then $U_t = (U_t^1, ..., U_t^d)^\top$ is independent of F_{t-1} and I_{t-1} is deterministic by assumption. According to (3.8) we have $1 - \kappa_t = \min_{\vartheta \in \mathbf{R}^d} E_t[\{1 - \vartheta^\top \cdot \Delta\breve{S}_{t+1}\}^2] = \min_{\delta \in \mathbf{R}^d} E_t[\{1 - \sum_{k=1}^d (\delta^k / \breve{S}_t^k) \cdot \Delta\breve{S}_{t+1}^k\}^2] = \min_\delta E_t[\{1 - \sum_{k=1}^d \delta^k \cdot \left(\sum_{j=1}^n \sigma_t^{k,j} \cdot U_{t+1}^j\right)\}^2] = \min_\delta E_t[\{1 - \delta^\top \sigma_t U_{t+1}\}^2] = \min_{\zeta \in I_t} E_t[\{1 - \zeta^\top \cdot U_{t+1}\}^2] = \min_{\zeta \in I_t} E[\{1 - \zeta^\top \cdot U_{t+1}\}^2]$.□

Lemma 3.6. *For any* $Z \in L^2(P)$ *we have:* $\Pi_t(Z) = \Pi_t(E_{t+1}[Z])$, $0 \le t < T$. The proof follows from

$$E_t[\{Z - \vartheta^\top \cdot \Delta\breve{S}_{t+1}\}^2] = E_t[\{Z - E_{t+1}[Z]\}^2] + E_t[\{E_{t+1}[Z] - \vartheta^\top \cdot \Delta\breve{S}_{t+1}\}^2] \ . \tag{3.11}$$

4. Global optimality Definition 4.1. *A portfolio plan* ξ^* *is called globally optimal for* $c \in \mathbf{R}$ *if*:

$$E[\{c - G_T(\xi^*)\}^2] \le E[\{c - G_T(\xi)\}^2] \text{ for all } \ \xi. \tag{4.1}$$

By the following relation we can restrict attention to the case $c = 1$. There we set $c \cdot \xi := (c \cdot \xi_t)$.

Lemma 4.2. *If* ξ^* *is globally optimal for* 1,*then* $c \cdot \xi^*$ *is globally optimal for* $c \in \mathbf{R}$.

For a proof of the following result see [20] or the proof of Theorem 5.4 below.

Proposition 4.3. *If* ξ^* *globally optimal for* 1 , *then*

$$Var[G_T(c \cdot \xi^*)] \le Var[G_T(\xi)] \quad \text{for all } \ \xi \text{ with } \ E[G_T(\xi)] = c \cdot E[G_T(\xi^*)].$$

We will compute $E[G_T(\xi^*)]$ in 4.4 below. In order to solve (4.1) for $c = 1$, one can write $1 - G_T(\xi) = 1 + G_T(-\xi)$. There $1 + G_T(-\xi)$ is the discounted value process *of* $-\xi$ starting with an initial value 1 and we want to study

$$\min_\xi E[\{1 - G_T(\xi)\}^2] = \min_\xi E[\{1 + G_T(\xi)\}^2]. \tag{4.2}$$

It is convenient to describe a portfolio plan $\pm\xi$ by a *proportional portfolio plan* $\pm\varphi = (\pm\varphi_t,\ 0 \le t < T)$. As a portfolio plan, a proportional portfolio plan φ is an $\mathbf{R}^d$-valued adapted process where $\pm\varphi_t^k$ represents the proportion $\pm\xi_t^k/(1 \pm G_t(\xi))$ for $1 \pm G_t(\xi) \neq 0$ and $\varphi_t^k = 0$ for $1 \pm G_t(\xi) = 0$. Then

$$1 \pm G_t(\xi) = (1 \pm \varphi_0^\top \cdot \Delta\breve{S}_1) \cdot ... \cdot (1 \pm \varphi_{t-1}^\top \cdot \Delta\breve{S}_t) \ . \tag{4.3}$$

Note that the property "$1 \pm G_{t-1}(\xi) = 0 \implies 1 \pm G_t(\xi) = 0$", $1 \le t < T$, is both necessary and sufficient to describe a portfolio plan by a proportional portfolio plan. But obviously, when studying (4.2) we can restrict attention to such plans.

Theorem 4.4. *Under condition* 3.2 *we have for* $\xi_t^* := (1 - G_t(\xi^*))\lambda_t$, $0 \le t < T$:

ξ^* *is a portfolio plan which is globally optimal for* 1 *and*

$$1 - G_t(\xi^*) = (1 - \lambda_0^\top \cdot \Delta\breve{S}_1) \cdot \ldots \cdot (1 - \lambda_{t-1}^\top \cdot \Delta\breve{S}_t),$$

$$E[1 - G_t(\xi^*)] = E[\{1 - G_t(\xi^*)\}^2] = \Pi_{m=0}^{t-1}(1 - \kappa_m) \ , \ 1 \le t \le T.$$

Proof. In view of (3.5) and (3.8) we know that $E_t[\{1+\varphi_{t-1}^\top \cdot \Delta\breve{S}_t\}^2] \ge E_t[\{1-\lambda_{t-1}^\top \cdot \Delta\breve{S}_t\}^2] = 1-\kappa_t$. Thus $E[\{\Pi_{t=1}^T(1+\varphi_{t-1}^\top \cdot \Delta\breve{S}_t)\}^2] \ge E[\{(1-\lambda_{T-1}^\top \cdot \Delta\breve{S}_T) \cdot \Pi_{t=1}^{T-1}(1+\varphi_{t-1}^\top \cdot \Delta\breve{S}_t)\}^2] = (1-\kappa_{T-1}) \cdot E[\{\Pi_{t=1}^{T-1}(1+\varphi_{t-1}^\top \cdot \Delta\breve{S}_t)\}^2]$. By induction we obtain: $E[\{\Pi_{t=1}^T(1+\varphi_{t-1}^\top \cdot \Delta\breve{S}_t)\}^2] \ge E[\{\Pi_{t=1}^T(1-\lambda_{t-1}^\top \cdot \Delta\breve{S}_t)\}^2]$ and $E[\{\Pi_{m=1}^t(1-\lambda_{m-1}^\top \cdot \Delta\breve{S}_m)\}^i] = \Pi_{m=0}^{t-1}(1-\kappa_m)$ for $i = 1, 2$. Applying well-known selection theorems, it is easy to see that λ_t can be chosen as F_t-measurable random variable (see [12] Proposition 2.2, [16] Proposition 4.7). Moreover, we have $E[\{\xi_t^{*\top} \cdot \Delta\breve{S}_{t+1}\}^2] = E[\{(\Pi_{m=1}^t(1-\lambda_{m-1}^\top \cdot \Delta\breve{S}_t) \cdot (\lambda_t^\top \cdot \Delta\breve{S}_{t+1})\}^2] = \Pi_{m=0}^{t-1}(1-\kappa_m) \cdot \kappa_t \le 1$. []

5. Local optimality In addition to (2.1) let us define :

$$G_{t,T}(\xi) := \sum_{t<m\le T} \xi_{m-1}^\top \cdot \Delta\breve{S}_m \ , \ W_{t,T}(\xi) := E_t[G_{t,T}(\xi)] \ . \tag{5.1}$$

Definition 5.1. *A portfolio plan* ξ^0 *is called locally optimal for* $c \in \mathbf{R}$ *if* $\forall 0 \le t < T$:

$$E_t[\{c - G_{t,T}(\xi^0)\}^2] \le E_t[\{c - G_{t,T}(\xi)\}^2] \quad \text{for all } \xi \text{ with } G_{t+1,T}(\xi) = G_{t+1,T}(\xi^0).$$

For a plan ξ as in 5.1 we have $G_{t,T}(\xi) = \xi_t^\top \cdot \Delta\breve{S}_{t+1} + G_{t+1,T}(\xi^0)$. The following lemma shows that local optimality can be based on the conditional future gains $W_{t,T}(\xi)$. Similarly, the concept of recursive utility is based on the conditional utilitiy of the future gains (see [8]).

Lemma 5.2. *A portfolio plan* ξ^0 *is locally optimal for* c *if and only if* $\forall 0 \le t < T$, $\forall \vartheta \in \mathbf{R}^d$:

$$E_t[\{c - \xi_t^{0\top} \cdot \Delta\breve{S}_{t+1} - W_{t+1,T}(\xi^0)\}^2] \le E_t[\{c - \vartheta^\top \cdot \Delta\breve{S}_{t+1} - W_{t+1,T}(\xi^0)\}^2] \ .$$

Proof. In view of 3.6, ξ^0 is locally optimal if and only if $\xi_t^{0\top} \cdot \Delta\breve{S}_{t+1} = \Pi_t(c - G_{t+1,T}(\xi^o)) \forall 0 \le t < T \iff \xi_t^{0\top} \cdot \Delta\breve{S}_{t+1} = \Pi_t(E_{t+1}[c - G_{t+1,T}(\xi^0)]) = \Pi_t(c - W_{t+1,T}(\xi^0)) \quad \forall 0 \le t < T$ which is equivalent to the condition of the lemma. []

For many considerations it is again sufficient to consider the case $c = 1$. Indeed we have:

Lemma 5.3. *If* ξ^0 *is locally optimal for* 1, *then* $c \cdot \xi^0$ *is locally optimal für* c. Now we can give a characterization of local optimality in a larger set of plans and the relation to the minimization of the conditional variances given the conditional means where we set

$$Var_t[Z] := E_t[Z^2] - E_t[Z]^2.$$

Theorem 5.4. *If ξ^0 is locally optimal for* 1, *then* for $0 \leq t < T$:

(a) $E_t[\{1 - G_{t,T}(\xi^0)\}^2] \leq E_t[\{1 - G_{t,T}(\xi)\}^2]$ $\forall \xi$ *with* $W_{m,T}(\xi) = W_{m,T}(\xi^0)$, $t < m < T$;
(b)$Var_t[G_{t,T}(\xi^0)] \leq Var_t[G_{t,T}(\xi)]$ $\forall \xi$ *with* $W_{m,T}(\xi) = W_{m,T}(\xi^0)$, $t \leq m < T$.
The quantities $W_{m,T}(\xi^0)$ will be computed in 5.6 below.

Proof. First consider the case $t = T - 1$. Then (a) is clear and (*b*) follows as below. Now assume that (*b*) holds for $t + 1$ and $W_{m,T}(\xi) = W_{m,T}(\xi^0)$, $t < m < T$. Then from (3.11) and Lemma 5.2 we obtain $E_t[\{1 - G_{t,T}(\xi)\}^2] = E_t[\{1 - \xi_t^\top \cdot \Delta\breve{S}_{t+1} - W_{t+1,T}(\xi)\}^2] + E_t[\{G_{t+1,T}(\xi) - W_{t+1,T}(\xi)\}^2] = E_t[\{1 - \xi_t^\top \cdot \Delta\breve{S}_{t+1} - W_{t+1,T}(\xi^0)\}^2] + E_t[Var_{t+1}[G_{t+1,T}(\xi)]] \geq E_t[\{1 - \xi_t^{0\top} \cdot \Delta\breve{S}_{t+1} - W_{t+1,T}(\xi^0)\}^2] + E_t[Var_{t+1}[G_{t+1,T}(\xi^0)]] = E_t[\{1 - G_{t,T}(\xi^0)\}^2]$. If in addition $W_{t,T}(\xi) = W_{t,T}(\xi^0)$, then $Var_t[G_{t,T}(\xi)] = Var_t[1 - G_{t,T}(\xi)] = E_t[\{1 - G_{t,T}(\xi)\}^2] - \{1 - W_{t,T}(\xi)\}^2 \geq E_t[\{1 - G_{t,T}(\xi^0)\}^2] - \{1 - W_{t,T}(\xi^0)\}^2 = Var_t[G_{t,T}(\xi^0)]$. []

Proposition 5.5. *A plan ξ^0 is locally optimal for* 1 *if and only if*

$\xi^0_{t-1} \cdot \Delta\breve{S}_t = \Pi_{t-1}(1 - G_{t,T}(\xi^0)) = \Pi_{t-1}(E_t[1 - G_{t,T}(\xi^0)])$. *Then*
(a) $1 - G_{t,T}(\xi^0) = \Pi_t^\perp(1 - G_{t+1,T}(\xi^0)) = \Pi_t^\perp(\Pi_{t+1}^\perp(...\Pi_{T-1}^\perp(1)))$;
(b) $E_t[1 - G_{t,T}(\xi^0)] = E_t[\Pi_t^\perp(E_{t+1}[1 - G_{t+1,T}(\xi^0)])]$
$= E_t[\Pi_t^\perp(E_{t+1}[\Pi_{t+1}^\perp(...E_{T-1}[\Pi_{T-1}^\perp(1)])])])]$. *Proof.* First we consider the case $t = T - 1$. Then we have: $\xi^{0\top}_{T-1} \cdot \Delta\breve{S}_T = \Pi_{T-1}(1)$ and $1 - G_{T-1,T}(\xi^0) = 1 - \Pi_{T-1}(1) = \Pi_{T-1}^\perp(1)$, hence $E_{T-1}[1 - G_{T-1,T}(\xi^0)] = E_{T-1}[\Pi_{T-1}^\perp(1)]$. According to 5.2 and 3.6 we know: $\xi^{0\top}_{t-1} \cdot \Delta\breve{S}_t = \Pi_{t-1}(1 - G_{t,T}(\xi^0)) = \Pi_{t-1}(E_t[1 - G_{t,T}(\xi^0)])$. Thus we obtain $1 - G_{t-1,T}(\xi^0) = 1 - G_{t,T}(\xi^0) - \xi^{0\top}_{t-1} \cdot \Delta\breve{S}_t = 1 - G_{t,T}(\xi^0) - \Pi_{t-1}(1 - G_{t,T}(\xi^o))$ hence $1 - G_{t-1,T}(\xi^0) = \Pi_{t-1}^\perp(1 - G_{t,T}(\xi^0))$. Similarly we get $E_t[1 - G_{t-1,T}(\xi^0)] = E_t[1 - G_{t,T}(\xi^0)] - \xi^{0\top}_{t-1} \cdot \Delta\breve{S}_t = E_t[1 - G_{t,T}(\xi^0)] - \Pi_{t-1}(E_t[1 - G_{t,T}(\xi^0)])$, hence $E_t[1 - G_{t-1,T}(\xi^0)] = \Pi_t^\perp(E_t[1 - G_{t,T}(\xi^0)])$. Now we can use backward induction. []

We remark that in 5.1, 5.2, 5.3, and 5.5 one can replace the function 1 by an arbitrary random variable $X \in L^2$. The result in 5.4 can be compared with the following relation for the plan ξ^* being globally optimal for 1: $1 - G_T(\xi^*) = \Pi^\perp(1)$ where $\Pi^\perp$ is the projection in $L^2(P)$ on $\mathcal{G}^\perp$, and $\mathcal{G}$ is the linear space of gains $G_T(\xi)$ under all plans ξ (see [21]). Now we will use our Condition 3.2 and we set

$$0 < \gamma_t := \Pi_{m=t}^{T-1}(1 - \kappa_m) \leq 1, 0 \leq t < T, \gamma_T := 1. \tag{5.5}$$

Theorem 5.6. *Under Condition* 3.2 *we have for* $\xi^0_t = \gamma_{t+1}\lambda_t$, $0 \leq t < T$: ξ^0 *is locally optimal for* 1. *Then:* $1 - W_{t,T}(\xi^0) = \gamma_t$, $0 \leq t < T$, *in particular* $E[1 - G_T(\xi^0)] = \gamma_0$. *Proof.* According to 5.5 we have $\xi^{0\top}_{t-1} \cdot \Delta\breve{S}_t = \Pi_{t-1}(\gamma_t) = \gamma_t \cdot \Pi_{t-1}(1) = \gamma_t\lambda^\top_{t-1} \cdot \Delta\breve{S}_t$. From 5.8 below we conclude that ξ^0 is a plan. Then we know that $E_{T-1}[1 - G_{T-1,T}(\xi^0)] = E_{T-1}[\Pi_{T-1}^{\perp}(1)] = 1 - \kappa_{T-1} = \gamma_{T-1}$. Now assume that $1 - W_{t+1,T}(\xi^0) = \gamma_{t+1}$, then we obtain by use of 5.5b: $E_t[1-G_{t,T}(\xi^0)] = E_t[\Pi_t^{\perp}(E_{t+1}[1-G_{t+1,T}(\xi^0)])] = E_t[\Pi_t^{\perp}(\gamma_{t+1})] = \gamma_{t+1} \cdot E_t[\Pi_t^{\perp}(1)] = \gamma_{t+1} \cdot (1-\kappa_t) = \gamma_t$. ▯

From 4.4 and 5.6 we obtain

$$\xi_t^0 = \gamma_{t+1} \cdot \lambda_t,$$
$$\xi_t^* = (1 - G_t(\xi^*)) \cdot \lambda_t, \quad 0 \le t < T.$$

Corollary 5.7. *Assume Condition* 3.2. *If* ξ^0 *and* ξ^* *are the locally and the globally optimal plan for* 1, *respectively, then:* $E[G_T(\xi^*)] = E[G_T(\xi^0)] = 1 - \gamma_0$. *Moreover,* $E_t[\xi_t^{0\top} \cdot \Delta\breve{S}_{t+1}] = \kappa_t \cdot \gamma_{t+1}$ *is deterministic whereas* $E_t[\xi_t^{*\top} \cdot \Delta\breve{S}_{t+1}] = (1 - G_t(\xi^*) \cdot \kappa_t = (1 - \lambda_0^\top \cdot \Delta\breve{S}_1) \cdot \ldots \cdot (1 - \lambda_{t-1}^\top \cdot \Delta\breve{S}_t) \cdot \kappa_t$. Now we want to compare the conditional variances of ξ^0 and ξ^*. From 3.1c we conclude:

Lemma 5.8. $E_t[\{\lambda_t^\top \cdot \Delta\breve{S}_{t+1}\}^2] = \kappa_t$, $Var_t[\lambda_t^\top \cdot \Delta\breve{S}_{t+1}] = (1 - \kappa_t) \cdot \kappa_t$. Now we get:

Corollary 5.9. *Assume Condition* 3.2, *then we obtain for* ξ^* *and* ξ^0 *as in* 5.7 : $Var_t[\xi_t^{0\top} \cdot \Delta\breve{S}_{t+1}] = (1-\kappa_t) \cdot \kappa_t \cdot \gamma_{t+1}^2$; $Var_t[\xi_t^{*\top} \cdot \Delta\breve{S}_{t+1}] = (1-G_t(\xi^*))^2 \cdot (1-\kappa_t) \cdot \kappa_t$. Comparing ξ^0 and ξ^* for $t = 0$ and $t = T - 1$ we see that:

$$\xi_0^0 = \gamma_1 \lambda_0, \quad \xi_0^* = \lambda_0, \quad \xi_{T-1}^0 = \lambda_{T-1}, \quad \xi_{T-1}^* = (1 - G_{T-1}(\xi^*))\lambda_{T-1},$$

where $0 < \gamma_1 \le 1$, $0 < E[1 - G_{T-1}(\xi^*)] \le 1$.

Thus, at the beginning the investor is less risk averse under ξ^0 than under ξ^*. At the end, the opposite property holds but "only in the mean".

References.

[1] E.J. Elton and M.J. Gruber, *Finance as a Dynamic Process*, Englewood cliffs: Prentice Hall, 1975.

[2] R.C. Dalang, A. Morton and W. Willinger, *Equivalent martingale measures and no-arbitrage in stochastic securities market models*, Stochastics and Stochastic Reports **29** (1990), 185 — 201, .

[3] P. Grandits, *The p-optimal martingale measure and its asymptotic relation with the minimal-entropy martingale measure*", Bernoulli **5** (1999), 225-248.

[4] R.R. Grauer and N.H. Hakansson, *On the use of mean-variance and quadratic approximations in implementing dynamic investment policies*, Manag. sci., **39** (1993), 856-871.

[5] W. Härdle and C.M. Hafner, *Discrete time option pricing with flexible volatility estimation*, Finance Stochast. **4** (2000), 189-208.

[6] J. Jacod and A. N. Shiryaev, *Local martingales and the fundamental asset pricing theorems in the discrete-time case*, Finance Stochast. **3** (1998), 259-273.

[7] D. Lamberton, H. Pham and M. Schweizer, *Local risk-minimization under transaction costs, Math. Oper. Res.*, **23** (1998), 585-612.

[8] C. Ma, *A discrete-time intertemporal asset pricing model*, Mathematical Finance **8** (1998), 249-275.

[9] H.M. Markowitz , *Portfolio selection*, J. of Fin., **7** (1952), 77-91.
[10] J. Mossin, *Optimal multiperiod portfolio policies*, J. of Bus., **41**, (1068), 215-229.
[11] S.R. Pliska, *Introduction to Mathematical Finance*, Blackwell publisher, Malden, USA, Oxford, UK, 1997.
[12] L.C.G. Rogers, *Equivalent martingale measures and no-arbitrage*, Stochastics and Stochastic Reports **51** (1994), 41-49.
[13] P.A. Samuelson, *Lifetime portfolio selection by dynamic stochastic programming*, Rev. Econ. Stat., **51** (1969), 239-246.
[14] M. Schäl, *On quadratic cost criteria for option hedging*, Math. Oper. Res. **19** (1994), 121-131.
[15] M. Schäl, *Martingale measures and hedging for discrete-time financial markets*, Math. Oper. Res. **24** (1999), 509-528.
[16] M. Schäl, *Portfolio optimization and martingale measures*, Mathematical Finance **10** (2000), 289-304.
[17] M. Schäl, *Price systems constructed by optimal dynamic portfolios*, Math. Meth. Oper. Res. **51** (2000), 375-397.
[18] M. Schäl, *Markov decision theory in finance and dynamic options.* To be published in Shwartz A, Feinberg E (eds.) Markov decision processes, (2000) Kluwer.
[19] M. Schweizer, *Approximating random variables by stochastic integrals*, Ann. Probab. **22** (1994), 1536-1575.
[20] M. Schweizer, *Variance-optimal hedging in discrete time*, Math. Oper. Res. **20** (1995), 1-32.
[21] M. Schweizer, *Approximation pricing and the variance-optimal martingale measure,* Ann. Probability **24** (1996), 206-236.

Institut für Angewandte Mathematik, Universität Bonn, Wegelerstr. 6, $D-53115$ Bonn, Germany
E-mail address: `schael@uni-bonn.de`

Trends in Mathematics, © 2001 Birkhäuser Verlag Basel/Switzerland

Transaction Processes among Autonomous Traders

Julia Schmelz

Abstract. In this paper, we propose and analyze a model that is oriented towards simulation of simplified, but sufficient realistic transaction processes. We model a trader who is interested as well in buying as in selling of an article sort. He advertises his quantifiable interest on a market place. Market place and other traders are the environment of any trader. The trader decides for a trade initiative using the information on the actual situation at the market place and his actual means. Such simulation models for the analysis of the interaction of traders and market place will build a useful experimental scope. Especially, we can quantify and influence the traders behavior. A comparision with another trade modell, (see [1]), is given in this paper, too.

1. Introduction

Price policy or price formation is an often discussed issue in the scope of trading processes. We often buy a product or an article with the object of using up, but sometimes we want to sell the article we purchased. We can recognize the success of a trader by his gain in cash or material asset if there was a successful transaction.

Traders who want to sell simultaneously the same goods are in competition. Likewise we can declare competitive buyers.

Price competition between rival information (goods) sellers in the web is investigated in some interesting papers from 1998,(see [2]). There we can learn something about the situation where the seller gives the price autonomously. And we can enjoy a successful transaction if the buyers interest enables the seller a positive gain given the seller's price. In this paper we lay the foundations for an alternative model where we don't need the invariable arrangement of consumers and brokers in two groups.

We have traders as market members with the same rights and duties. We extend the scenario, that will be described in chapter 2, by a mathematical model in chapter 3 which is suitable to manage the actual offering situation.

Here, we will see the utility function of Kephart and Hanson et.al as a special case of the utility function in our model. A possible interpretation of our model comes now:

Instead of a group of priori pure sellers we have a dynamic constellation of a seller group and a buyer group while the trading process is running.
Chapter 4 deals with the first simulation experiments with respect to such trading processes.
With our model view, the modeling of several market place models is possible, but it is disregarded in this paper. We confine to a model with only one article category and only a few traders for pointing out the main working principles of our trading model.

2. Model of Autonomous Transactions on a Single Market Place

2.1. Market Place and Transaction Process

Traders and market place are arranged in a system with variables and parameters. The analysis of such a system is oriented towards modeling a process of purchasing and selling.

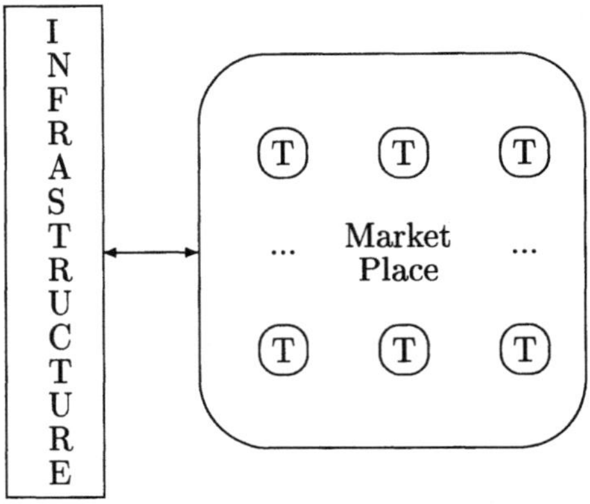

Trading can be affected by the market place in several grades. Either way the market place takes in a positive amount as a trade dependent charge and has to pay as well trade dependent costs as trade independent costs to the infrastructure of the environment. For demonstration we can remember the trading of goods that can be put down to the world-wide internet shopwindows. If the market place helps in bringing traders together in dependence of traders interest, we have a higher grade of influence.
In chapter 3 a model is established, so control and navigation of traders profit in one article category is possible by turning (playing with) parameters. Additionally, we can choose between an open or complete market with respect to the number of articles in each article category. Now we can observe a trading process after introducing suitable (time) rhythm.
The driving force in the system's process is the individual actual interest of each

market member. These for generality different interests are used for determination of utility in each time point. The formulea for evaluation of individual utility are connected with overall values V and P. (The values of V and P are fixed in simulation experiments.)

2.2. Rules of Transaction

The model describes the trading behavior of traders on a certain market place. A trader purchases some units of an article category of another trader, or he sells some units of his interest to another trader.

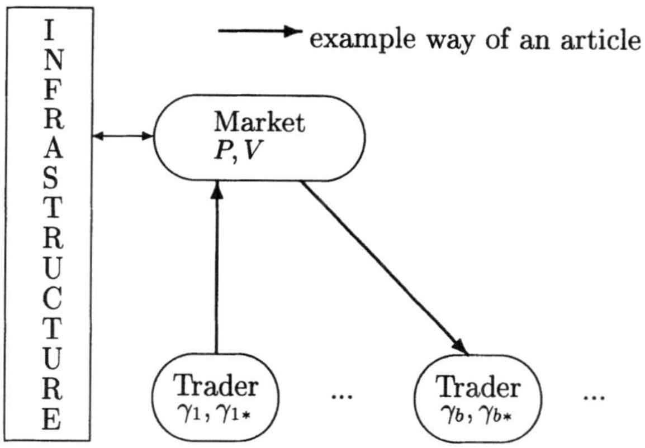

Here, successful transactions depend essentially on Bid and Ask in conjunction with the comparision of possessing and cashing in. Each trader b, $b\,\{1, ..., B\}$, has an autonomous impression of the value and the price of a certain article category on the market place. More in detail, the individual price offer with regard to purchasing be a multiplication of an individual coefficient of interest to buy γ_b, $\gamma_b \in [0, 1]$, and an overall price P. The product $P_b = \gamma_b \cdot P$ we denote by *Bid* of trader b. Here, a *Bid* is a price offer to purchase the article unit as typically used.

The offered selling price consists in a product of individual interest coefficient to sell γ_{b*}, $\gamma_{b*} \in [0, 1]$, with P. Then $P_{b*} = \gamma_{b*} \cdot P$ is the so-called *Ask* of trader b.

The overall price P represent a price orientation of one article unit on the market place. We use an real overall value V of a certain article category to explain individual loss of value caused by article selling by $V_{b*} = \gamma_{b*} \cdot V$. Similar we describe an individual value rise caused by article purchasing by $V_b = \gamma_b \cdot V$.

In this place, we explain the star notation used in this text. We write * in connection with a trader name b, for example b*, if we deal with a seller feature, for instance the interest coefficient to sell γ_{b*} of trader b. A trader name without * refers to a buyer feature, such as γ_b or V_b.

If the value of trader's γ_b is high, we say that the interest to buy the particular

article category is high. If the value of γ_{b*} is small, then the interest to possess the certain article category is small. In this case the interest to resell is high. Each trader decides by optimization of his utility in connection with the offers of the other traders for a transaction or no transaction. That means, he decides to be active or passive between the other market members. The utility in buyer view in regard to the certain article category can be determined by comparision of individual anticipated value V_b and the sum of realizable transaction price and costs per unit.
In chapter 3 it will be shown that the values of γ_{b1} and γ_{b2*} of two traders b_1 and b_2 in $\{1, ..., B\}$ determine in dependence of fixed V, P and C an intervall, so transaction prices in this intervall are advantageous for both of the traders. Such transcation prices we call *realizable*. The utility in seller view can be computed in analogous way by comparision of the realizable price with the sum of loss of value and costs.

3. Successful Transaction

3.1. Transaction Price

A successful transaction between trader 1 as buyer and trader 2 as seller needs that the inequality $\gamma_{2*} < \gamma_1$ holds.

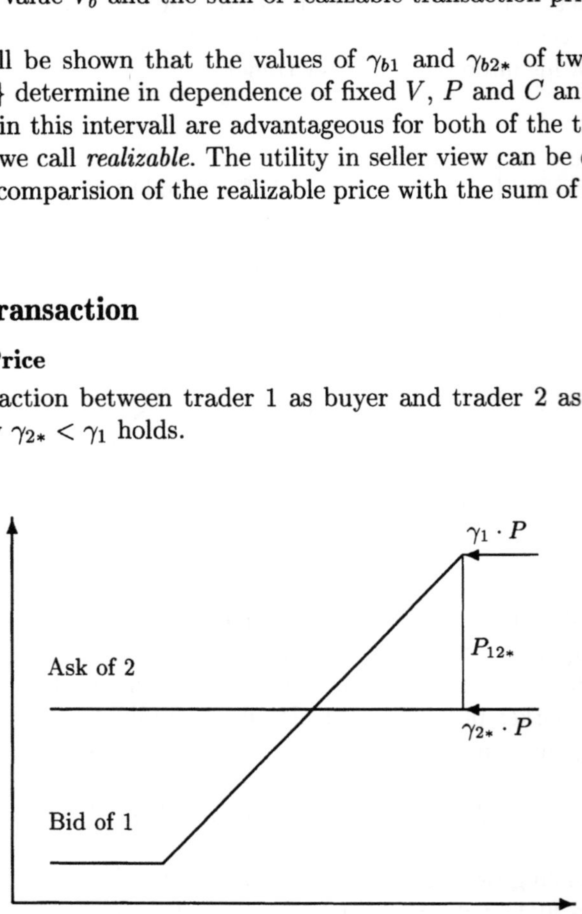

A possible transaction price for both of the traders lies in the interval $[P_{2*}, P_1]$. The transaction costs and the change in value determine wether there is a non-empty set of realizable prices in $[P_{2*}, P_1]$. A certain transaction price for trader 1 as buyer and trader 2 as seller can be described by means of a convex function of the price offers P_1 and P_{2*}:

$$P_{12*}(a) = ((1-a)\gamma_1 + a\gamma_{2*})P, \ a \in [0, 1].$$

That's a weighted mean of the buyer's price offer $P_1 = \gamma_1 \cdot P$ and and the seller's price offer $P_{2*} = \gamma_{2*} \cdot P$ (see picture above). The interval of realizable transaction prices is non-empty if there values $a,\ a \in [0,1]$ exist, for that

$$U_{12*} = max\,\{0, V_1 - P_{12*} - C\} > 0 \text{ and}$$
$$U_{1*2} = max\,\{0, P_{1*2} - V_{1*} - C*\} > 0 \text{ hold.}$$

A transaction in the opposit direction needs the assumption $\gamma_{1*} < \gamma_2$, and we use $P_{1*2}(a) = ((1-a)\gamma_{1*} + a\gamma_2)P$ analogous.

3.2. Utility Functions of Traders

Using the notation of subsections 2.1 and 3.1 we can see the values that have to be compared with the costs C or $C*$ respectively in the following scheme:

	buy	sell
Trader 1	$V_1 - P_{12*}(a)$	$P_{1*2} - V_{1*}$
Trader 1	$V_2 - P_{1*2}(a)$	$P_{12*} - V_{2*}$

The two diagonals contain the value-price-comparision of the alternativ roles of buyer and seller. Two cells in a row of the scheme can not be positive simultaneously. If a diagonal contains only positive differences and if these are greater than the costs C or $C*$ respectively, so we have the enjoy of a successful transaction. For the representation of the utility function as a whole, we use indicator functions $S_{..}$ for both of the traders. These functions indicate activity or passivity of a trader associated with potential transactions. For example the indicator S_{12*} points out wether trader 1 wants to buy from trader 2 or not.

Now we use indicator functions to write the utility functions of trader 1:
$U_1 = S_{1\,2*}(a)[V_1 - P_{12*}(a) - C] + S_{1*\,2}(a)[-V_{1*} + P_{1*2}(a) - C*]$.
Trader 2 computes its utility function with:
$U_2 = S_{2\,1*}(a)[V_2 - P_{1*2}(a) - C] + S_{2*\,1}(a)[-V_{2*} + P_{12*}(a) - C*]$.
We define $\theta(x) = \{ \begin{array}{ll} 1 & x > 0 \\ 0 & x \leq 0 \end{array}$.
Then the indicator functions $S_{..}$ are declared in the following manner:
$S_{1\,2*} = \theta(V_1 - P_{12*}(a) - C)$,
$S_{1*\,2} = \theta(-V_{1*} + P_{1*2}(a) - C*)$,
$S_{2\,1*} = \theta(V_2 - P_{1*2}(a) - C)$,
$S_{2*\,1} = \theta(-V_{2*} + P_{12*}(a) - C*)$.

3.3. Spread Splitting

The difference of prices $(\gamma_1 - \gamma_{2*})P$ can be shared out complete or partial and with different or the same shares for the two traders. If we assume a complete sharing and a fixed splitting parameter value $a, a \in [0,1]$, we can investigate several cases.

3.3.1. THE CASES $a = 0$ AND $a = 1$ Here, we discuss two extreme special cases with respect to the spread splitting. The fourfold table in 3.2 can be specialized as

	buy	sell
trader 1	$V_1 - P_1$	$P_{1*} - V_{1*}$
trader 1	$V_2 - P_{1*}$	$P_1 - V_{2*}$

where $V_b = \gamma_b V$ and $P_b = \gamma_b P$ or $V_{b*} = \gamma_{b*} V$ and $P_{b*} = \gamma_{b*} P$.
If we have $V > P$, then the trader 1 will be the buyer and plays the part of the consumer of Kephart (see [1]) . If the interest of trader 1 is high enough, such that $V_1 - P_1 = \gamma_1(V - P) > C$, then the trader 1 buys the article and pays the price P_1. If we have $P_1 - V_{2*} > C*$, then trader 2 is agreed to sell one article unit. Trader 2 is named broker in the model description in [1]. Here, the seller gets the price P_1. There is a permanent dichotomousness in brokers who only sell and consumers who only buy. Kephart's broker can be characterized that way he don't want to possess the article unit, but only want to put into liquidation. The effect in our model is that we have a selling coefficient of interest$\gamma_{2*} = 0$. So we also have $\gamma_2 = 0$ because of $\gamma_2 \leq \gamma_{2*}$. So the broker gets the price P_1. The value of the parameter γ_{1*} is now of secondary importance, for the secondary diagonal of the scheme is elementwise negative for $V > P$. The remaining model has then the parameters γ_1, V, P and Kephart's interpretation of these parameters is:

- $\gamma_1 \hat{=}$ consumer's coefficient of interest
- $V \hat{=}$ overall value
- $P \hat{=}$ price the broker asked for

The case $a = 1$ turns out to be a trade model of Kephart in [1], too. Here, trader 1 and trader 2 will have changed their roles.

3.4. The Case $a = \frac{1}{2}$

In the following, the values of C and $C*$ shall be equal and all the interest coefficients shall stem from the interval $(0, 1)$.
Obviously, in the middle of the handled two extremes cases we observe the case $a = \frac{1}{2}$. We can interpret this case as fair spread splitting, where each of the two traders give his offer without information about the other offers. After this the offers will be opened. If there exists a non-empty spread, it will be shared in two equal parts for the two traders. If we have the empirical prerequisits $\gamma_1 < \gamma_{1*}$ and $\gamma_2 < \gamma_{2*}$, we can make the following statements.

1. For any successful transaction where trader 2 sells to trader 1 the following conditions are necessary:

   (a) $\gamma_{2*} < \gamma_1$,
   (b) $\frac{\gamma_{2*}}{\gamma_1} < \frac{2V - P}{P} < \frac{\gamma_1}{\gamma_{2*}}$,

   and the following is sufficient:

(c) $\gamma_1 > \dfrac{\frac{1}{2}\gamma_{2*}P + C}{V - \frac{1}{2}P}$ und $\gamma_1 > \dfrac{\gamma_{2*}(V - \frac{1}{2}P) + C}{\frac{1}{2}P}$.

2. For any successful transaction where trader 1 sells to trader 2 the following conditions are necessary:

(a) $\gamma_{1*} < \gamma_2$,

(b) $\frac{\gamma_{1*}}{\gamma_2} < \frac{2V - P}{P} < \frac{\gamma_2}{\gamma_{1*}}$,

and the following is sufficient:

(c) $\gamma_2 > \dfrac{\frac{1}{2}\gamma_{1*}P + C}{V - \frac{1}{2}P}$ und $\gamma_2 > \dfrac{\gamma_{1*}(V - \frac{1}{2}P) + C}{\frac{1}{2}P}$.

3. An event where trader 1 sells to trader 2 and buys from trader 2 simultaneously is not possible.

Proof. With 1(a) it follows $\frac{2V - P}{P} < \frac{\gamma_1}{\gamma_{2*}}$. Further we see with $\gamma_1 \leq \gamma_{1*}$: $\frac{2V - P}{P} < \frac{\gamma_{1*}}{\gamma_{2*}}$. Finally, we can conclude $\frac{2V - P}{P} < \frac{\gamma_{1*}}{\gamma_2}$, because of $\gamma_2 \leq \gamma_{2*}$ holds.
That is obviously in contradiction to 2(a).

□

3.5. Trader Types

If the coefficients of interest are model as random variables, so the joint distribution of interest coefficients has to fulfil the empirical assumption $\gamma_b \leq \gamma_{b*}$. This assumption is well-founded by the convenient postulat that trader b wants to get at least so much as he would be agreed to pay. Salient characteristics of trader b can be caused by the choice of a concrete distribution for the tupel (γ_b, γ_{b*}) of the trader. Some traders with the same characteristics form a group. The distribution of tupel (γ_b, γ_{b*}) of trader b allocates a *type* to the trader in a sense. For example, a type is conceivable which extreme and exlusive interest is selling of goods. Hence his ask will be low relatively to the actual bid distribution. This type trades with high probability and low transaction prices.
The extreme buyer type as contrary of the previous extreme seller type is easy imaginable. A third type wants to get an ambitious minimum price in modification to the first mentioned. It's probability of a successful transaction decreases and the higher price entices.
A survey of six possible types, which can be distinguished by means of the interest situation, is given in the table below:

	high Bid	medium Bid	low Bid
high Ask	1	2	3
medium Ask	-	4	5
low Ask	-	-	6

With regard to potential transaction prices we have in a sense of stochastic order: $P^1_{12*}, P^2_{12*}, P^3_{12*} > P^4_{12*}, P^5_{12*} > P^6_{12*}$ and $P^1_{1*2} > P^2_{1*2}, P^4_{1*2} > P^3_{1*2}, P^5_{1*2}, P^6_{1*2}$.
With regard to probability of a successful transaction we have the following ranking: $sell^6 > sell^4, sell^5 > sell^1, sell^2, sell^3$ and
$buy^1 > buy^2, buy^4 > buy^3, buy^5, buy^6$. If the process is going on, a particular trader could change his type in dependence of his financial an material situation.

4. Computer Experiments and Outlook

There is an algorithm given for managing the actual offering situation of any two traders in chapter 3. In fact, there are more than only two traders at a market place. Then we have to decide how we bring the traders together in couples at each time point of our running process.
At first we couple the traders with the greatest spread from all the traders. The next traders, whom we couple, are the traders with the next greatest spread. We do this recursively until we have no more positive spread. Until now, experimental target of our simulations was the investigation of the number of successful transactions and the development of the cummulative utility.
These experimiients are carried out with several market parameter values V, P and C and with $a = \frac{1}{2}$. It is done for a set of traders with equal rights and for the comparision of two trader groups with different interest distributions.
Using uniform distributed interest, the number of successful transactions goes down if the value of costs C increases. Likewise less surprisingly, the number of successful transactions goes up if the time horizon becomes larger or the number of market members increases.
We should have a look to the following experiment examples that were carried out with my own C programs especially for this paper. The pictures were made by means of S+.
The pictures 1 and 2 represent numbers of transactions in 100 discrete time points. The comparison of both graphs indicates a reduction of this number if the costs increases. The system state on which I centered was determined by 5 market members who were interested in only one sort of articles. A black dot in the graph at level 1.0 corresponds to exactly one successful transaction event.

figure 1

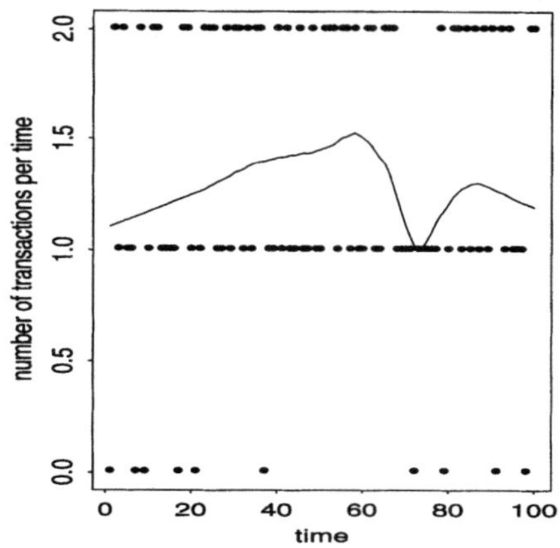

figure 2

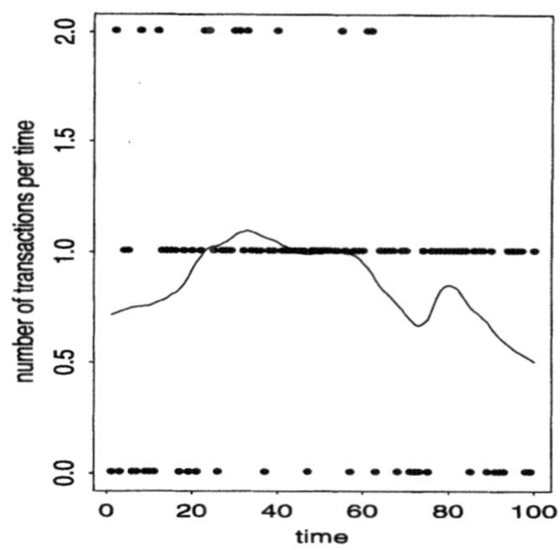

A black dot at level 2.0 corresponds to 2 transactions per time point. A dot at level 0.0 cooresponds to no transaction at this time point. The continous curve in the picture is a smoothed representation of the process of points.
The pictures 3 and 4 supply us with the representation of mean transaction price per time point in a given horizon of 100 time points.

figure 3

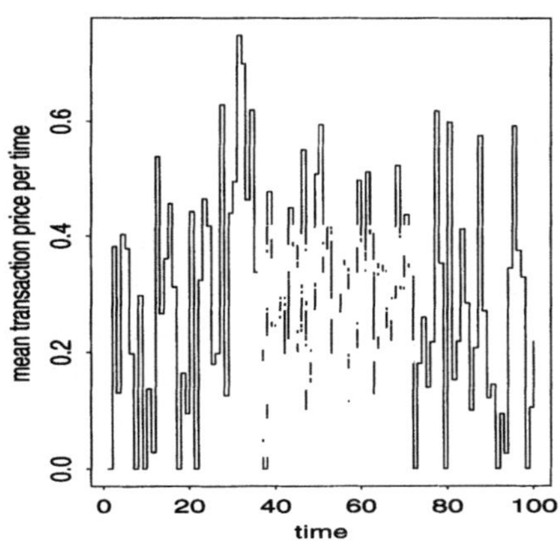

figure 4

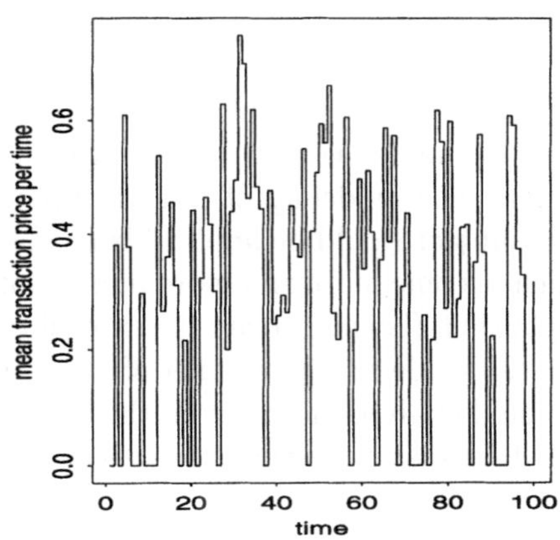

This mean will have the value 0.0 if there is no transaction in the time point. The mean transactions price per time point will be near the expected value of price $(\gamma_{.} + \gamma_{.*})P$ if the number of successful transactions per time point is larger. The mean transaction price of time points will become less if the costs increases.

The mean transaction price given all the time points with at least one transaction will go up if the costs grow up. The picture 5 shows curves of cummulative utility in 100 time points. We see the development of utility of two groups wich had different costs to pay. The dashed curve group had to pay a less value of costs C.

figure 5

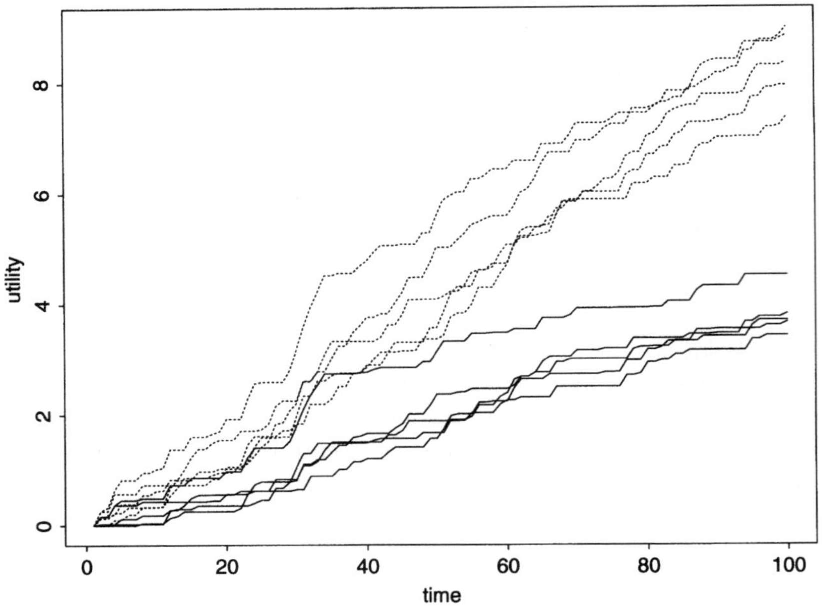

The topic with which we are dealing in this paper motivates interesting aspects for modeling of commercial transactions on the base of the utility formalization. The modeling of price competitions of sellers and buyers is conceivable. This subject would be applicable to the development of intelligent software agents which become more and more important in the electronic commerce, (see [3] , [4]). Kephart et.al demonstrate the extension of such investigations to numerous traders who can trade a large number of article categorys. In the context of my model type we could analyse the effect on the individual success of traders or the system's success if we play with new state variables such as budget of articles or money.

References

[1] Jeffrey O. Kephart, James E. Hanson and Jakka Sairamesh, *Price-War Dynamics in a Free-Market Economy of Software Agents,* Proceedings of ALIVE VI, Los Angeles,26.-29. Juni, (1998), 1073–1077.

[2] James E. Hanson and Jeffrey O. Kephart, *Spontaneous Specialization in a Free-Market Economy of Agents* , Presentet at Artificial Societies and Computational

Markets Workshop at the Second International Conference on Autonomous Agents (AA '98), Minneapolis/St. Paul, 9. Mai,(1998)

[3] Dirk Schmelz, Margitta Schmelz und Julia Schmelz, *M Utility Metrics for Economical Software Agents*, to appear at the 10th International Workshop on Software Measurement in Berlin, Germany
Oct. 4-6, (2000)

[4] L. M. Rocha,*From Artificial Life to Semiotic Agent Models*
www:http://www.c3.lanl.gov/~rocha, (1999)

Zentrum Mathematik, Technische Universität, 23 , Barerstr. 80299 München, D
E-mail address: `schmelz@mathematik.tu-muenchen.de`

Trends in Mathematics, © 2001 Birkhäuser Verlag Basel/Switzerland

The Laplace transform approach to valuing exotic options: the case of the Asian option

Michael Schröder

Abstract. This talk discusses integral representations for the Black–Scholes price of arithmetic–average Asian options. They are obtained by Laplace inversion using complex analytic methods and are based on the results of [**GY**]. It is discussed how these results are to be adapted to now valuing Asian options, and how restrictions to their validitity can be lifted.

1. Introduction

This talk pursues that interplay between finance, stochastics, and analysis initiated by the work of Yor on the valuation of exotic options. Using the Laplace transform technique the focus is on the so-called Asian option. These are options on the arithmetic average of the price of the underlying. The main result is the closed form solution for its Black-Scholes price of [**SA**].

Indeed, we identify this Black-Scholes price as a single integral over the product of two well-known functions built up using so–called Hermite functions. These functions generalize the complementary error function, and they come from boundary value problems in potential theory for domains whose surface is an infinite parabolic cylinder. It seems to be the first time that such parabolic cylinder functions are explicitly identified to characterize solutions to problems of Finance origin.

Our result is obtained using results of Geman–Yor in [**GY**]. Unfortunately, the valuation problem they consider is from a Finance perspective unrelated to valuing Asian options. Using a crucial insight of Peter Carr (Banc of America Securities, New York) into the working of Asian options and with the kind support of Yor we have been able to adapt this approach to now valuing Asian options. The modified results, however, are valid only under restrictions. These have their origin in the still limited knowledge of Bessel processes. At the current state of affairs, they are unfortunately such that the pertinent Laplace transforms do not seem to be available in those typical situations where Asian options are desirable to use.

For our valuation formula we have been able to lift these restrictions using analytic techniques. This illustrates a typical application of integral representations. They are good for establishing results of a structural nature about the special function they represent. In another typical application we have used our integral representation in [**SE**] to derive explicit series and asymptotic expansions for computing the value of the Asian option. Since these are seemingly the first and only such results ever derived, an example is briefly discussed at the end of the talk.

2. Basic notions

First we discuss the notions basic for the analysis of Asian options. We work in the Black–Scholes framework using the risk–neutral approach to the valuation of contingent claims. In this set–up there are two securities. There is a riskless security, a bond, that has the continuously compounding positive interest rate r. Then there is a risky security whose price process S is modelled as follows. Start with a complete probability space equipped with the standard filtration of a standard Brownian motion on it that has the time set $[0,\infty)$. On this filtered space one has the risk neutral measure Q, a probability measure equivalent to the given one, and a standard Q–Brownian motion B such that S is the strong solution of the following stochastic differential equation:

$$dS_t = \varpi \cdot S_t \cdot dt + \sigma \cdot S_t \cdot dB_t, \qquad t \in [0,\infty).$$

Herein the positive constant σ is the volatility of S. The specific form of the otherwise arbitrary constant ϖ depends on the security modelled. For instance it is the interest rate if S is a non–dividend–paying stock.

Fix any time t_0 and consider the accumulation process J given for any time t by:

$$J(t) = \int_{t_0}^{t} S_u \, du \, .$$

The European–style *arithmetic–average Asian option* written at time t_0, with maturity the time T, and fixed–strike price K is then the contingent claim on the closed time interval from t_0 to T paying

$$\left(\frac{J(T)}{T-t_0} - K\right)^+ = \max\left\{0, \frac{J(T)}{T-t_0} - K\right\}$$

at time T. Recall that points in time are taken to be non–negative real numbers.

3. Normalized valuation of the Asian option

In the setting of the previous section, the price C_t of the Asian option at any time t between t_0 and T is given as the following risk neutral expectation

$$C_t = e^{-r(T-t)} E^Q \left[\left(\frac{J(T)}{T-t_0} - K\right)^+ \middle| \mathscr{F}_t \right]$$

which is conditional on the information $\mathscr{F}_t$ available at time t. However, following [**GY**, §3.2], do not focus on this price. It is not natural in a mathematical sense!

This can be achieved essentially by normalizing the volatility to a dimensionless 200 percent. The precise result is the factorization:

$$C_t = \frac{e^{-r(T-t)}}{T-t_0} \cdot \frac{4S_t}{\sigma^2} \cdot C^{(\nu)}(h,q)$$

which reduces the general valuation problem to computing

$$C^{(\nu)}(h,q) = E^Q\left[(A_h^{(\nu)} - q)\right],$$

the *normalized time–t price* of the Asian option. To explain the notation, $A^{(\nu)}$ is Yor's 200 percent volatility accumulation process

$$A_h^{(\nu)} = \int_0^h e^{2(B_w+\nu w)}\,dw\,,$$

and the normalized parameters are as follows:

$$\nu = \frac{2\varpi}{\sigma^2} - 1, \qquad h = \frac{\sigma^2}{4}(T-t), \qquad q = kh+q^*,$$

where

$$k = \frac{K}{S_t}, \qquad q^* = q^*(t) = \frac{\sigma^2}{4S_t}\left(K\cdot(t-t_0) - \int_{t_0}^t S_u\,du\right).$$

To interpret these quantities, ν is the *normalized adjusted interest rate*, h is the *normalized time to maturity*, which is non–negative, and q is the *normalized strike price*. Notice how q becomes affine linear in the time variable h with coefficients k and q^* depending only on time–t magnitudes.

4. Statement of results

There is now a dichotomy in computing the normalized time–t price

$$C^{(\nu)} := E^Q\left[(A_h^{(\nu)} - q)^+\right]$$

of the Asian option according to the normalized strike price q being positive or not.

4.1. The case $q \le 0$

If q is non–positive, Asian options loose their option feature. Computing their normalized time–t prices $C^{(\nu)}$ is straightforward:

$$C^{(\nu)} = E^Q\left[A_h^{(\nu)}\right] - q$$

with

$$E^Q\left[A_h^{(\nu)}\right] = \frac{e^{2h(\nu+1)} - 1}{2(\nu+1)},$$

for ν any real number. For the proof apply Fubini's theorem, then do an explicit computation using the time–w densities $x \mapsto (2\pi w)^{-1/2}\exp(-(2w)^{-1}x^2)$ of B.

4.2. Yor's triple integral for $C^{(\nu)}$ in the case $q > 0$

We proceed chronologically and with a bias towards integral representations. If q is positive, it was Yor who in [**Y**, (6.e), p.528] first came up with the following triple integral for the normalized price $C^{(\nu)}$ of the Asian option

$$C^{(\nu)} = c \cdot \int_0^\infty x^\nu \int_0^{1/q} f(x,y) \int_0^\infty g(x,y,w)\, dw\, dy\, dx\,.$$

Here f and g are the functions in three positive variables x, y, w given by

$$f(x,y) = e^{-\frac{1}{2}(1+x^2)y} \cdot \left(y^{-1} - q\right)^+,$$

$$g(x,y,w) = e^{-xy\cosh(w)} e^{-\frac{w^2}{2h}} \cdot \sinh(w) \cdot \sin\left(\frac{\pi}{h} w\right),$$

and c is a constant in ν and h. Yor's formula is valid for all ν.

4.3. Schröder's integral for $C^{(\nu)}$ in the case $q > 0$

The main result of this talk is my 1997 closed form solution for $C^{(\nu)}$. It is valid for any real number ν, and expresses this function as the sum of five single integrals of products of two well–studied functions:

Theorem: *If q is positive, the normalized price $C^{(\nu)}$ of the Asian option is given by the following five–term sum:*

$$C^{(\nu)} = C_{\text{trig}} + \sum_{b \in \{\pm(\nu+2)\}} C_{\text{hyp},b} - \sum_{b \in \{\pm\nu\}} C_{\text{hyp},b}.$$

Postponing the explanation of the Hermite functions H_μ of degree any complex number μ, the *trigonometric term* C_{trig} is:

$$C_{\text{trig}} = \gamma \cdot \int_0^\pi H_{-(\nu+4)}\left(-\frac{\cos(\theta)}{\sqrt{2q}}\right) \cos(\nu \cdot \theta)\, d\theta\,.$$

The *hyperbolic terms* $C_{\text{hyp},b}$ with b equal to $\pm\nu$ or $\pm(\nu+2)$ are are given as single integrals of products of Hermite functions $H_{-\mu}$ times the imaginary part of weighted complementary error functions Erfc:

$$C_{\text{hyp},b} = \gamma_b \cdot \int_0^\infty H_{-(\nu+4)}\left(\frac{\cosh(y)}{\sqrt{2q}}\right) \cdot \operatorname{Im}\left(e^{b(y+\pi i)} \operatorname{Erfc}\left(\frac{y+i\pi}{\sqrt{2h}} + b\sqrt{2h}\right)\right) dy.$$

Herein γ and γ_b are constants in ν and h. The *complementary error function* Erfc is given by

$$\operatorname{Erfc}(z) = \frac{2}{\sqrt{\pi}} \int_z^\infty e^{-\xi^2}\, d\xi$$

for any complex number z, where the integral is over the parallel to the real axis starting at z with real parts to ∞. The *Hermite function* $H_{-\mu}$ of degree $-\mu$ with

μ any complex number of positive real part generalizes the complementary error function and is given by

$$H_{-\mu}(z) = \frac{1}{\Gamma(\mu)} \int_0^\infty e^{-x^2-2zx} x^{\mu-1} dx$$

for any complex number z. For $\mu = 1$ in particular we have,

$$H_{-1}(z) = \frac{\sqrt{\pi}}{2} \cdot e^{z^2} \operatorname{Erfc}(z).$$

In the general case with μ any complex number, $H_{-\mu}$ is defined using the confluent hypergeometric function Φ as

$$H_{-\mu}(z) = \frac{\Gamma(1/2)}{2^\mu \Gamma\left(\frac{1+\mu}{2}\right)} \cdot \Phi\left(\frac{\mu}{2}, \frac{1}{2}; z^2\right) + \frac{\Gamma(-1/2)}{2^\mu \Gamma\left(\frac{\mu}{2}\right)} \cdot \Phi\left(\frac{1+\mu}{2}, \frac{3}{2}; z^2\right),$$

for any complex number z. Thus Hermite functions of degree any non–negative integer n coincide with the corresponding n–th Hermite polynomials, whence in particular $H_0 = 1$, $H_1(z) = 2z$, $H_2(z) = 4z^2 - 2$, $H_3(z) = 8z^3 - 12z$.

In the sequel we sketch the key steps of the proof of the Theorem following [**SA**].

5. Proof of the Theorem, I: intuition about the Laplace transform

The first key idea for computing the normalized time–t price

$$C^{(\nu)} = E^Q\left[\left(A_h^{(\nu)} - (kh+q^*)\right)^+\right]$$

of the Asian option is, roughly speaking, to make exercise time a variable, and compute the Laplace transform of this function. I want to make precise this idea of [**GY**] first and, with the mathematical background summarized in [**SA**, §0] or [**SF**, Part 0], then try to give a Finance intuition for the Laplace transform.

Making exercise time a variable translates into considering $C^{(\nu)}$ via $h(y) = (\sigma^2/4) \cdot (y–t)$ as a function in the "exercise time" y with y from t to ∞. It seems reasonable to give the higher a weight to an option the more imminent its exercise. Translate this intuition by continuously discounting $y \mapsto C^{(\nu)}(h(y))$ with a constant discount rate z. There is of course no natural choice for z. The trick is not to stick with a single such discount rate but to consider it as a variable. In this way associate to the function in exercise time $y \mapsto C^{(\nu)}(h(y))$ a function in this discount rate z. Written in full length this *Laplace transform* of $C^{(\nu)}$ is thus formally given by

$$\mathscr{L}(C^{(\nu)})(z) = \int_0^\infty e^{-zx} E^Q\left[\left(A_x^{(\nu)} - (kx+q^*)\right)^+\right] dx$$

on changing variables $x = h(y) = (\sigma^2/4) \cdot (y-t)$.

It is a characteristic feature of the "normalized strike price" of the Asian option that it depends on the transformation variable x via $q(x) = kx + q^*$. In [**GY**], however, options are considered whose "normalized strike price" is independent of x, and this is crucial for the validity of their Laplace transform computations indeed. So these results are unfortunately unrelated to valuing the Asian option on *a priori* reasons. Luckily, in November 1999 Peter Carr came up with the idea how they can nevertheless be used for exactly this purpose.

6. Proof of the Theorem, II: Peter Carr's key reduction

Enter Peter Carr: "Why not try to reconstruct the normalized time–t price

$$C^{(\nu)} = E^Q\left[\left(A_h^{(\nu)} - (kh+q^*)\right)^+\right]$$

of the Asian option from a family of functions whose single members may be unrelated to the problem of valuing the Asian option!" For any real number a, thus let $f_{GY,a}$ be the function on the positive real line that sends any $x > 0$ to

$$f_{GY,a}(x) = E^Q\left[(A_x^{(\nu)} - a)^+\right].$$

Then we have the following

Key Reduction: *If $q = kh+q^*$ is positive, computing the normalized time–t price $C^{(\nu)}$ of the Asian option reduces to computing all $f_{GY,a}$ with $a > 0$.*

Indeed, $C^{(\nu)}$ is obtained evaluating the function $g(x,w) = E^Q[(A_x^{(\nu)} - w)^+]$ on the graph of the function $\phi(x) = kx+q^*$. So the reduction to be proved is a consequence of the following special case of the general result Peter Carr communicated to me in November 1999

Fibration principle: *Let the function on the positive real line $f : \mathbb{R}_{>0} \to \mathbb{R}$ be such that*

$$f(x) = g(x, \phi(x))$$

for all $x > 0$ with functions $\phi : \mathbb{R}_{>0} \to \mathbb{R}$ and $g : \mathbb{R}_{>0} \times \mathbb{R} \to \mathbb{R}$. For any real number a, define the function on the positive real line $f_a : \mathbb{R}_{>0} \to \mathbb{R}$ for any $x > 0$ by

$$f_a(x) = g(x, a).$$

The value of f at any $x_0 > 0$ is then obtained as the value of $f_{\phi(x_0)}$ at x_0.

7. Proof of the Theorem, III: the Geman–Yor Laplace transform

Peter Carr's key reduction of §6 shows the way to apply the Laplace transform to valuing Asian options: *change functions!* Consider the family of all functions $f_{GY,a}$ with $a > 0$ that send any $x > 0$ to

$$f_{GY,a}(x) = E^Q\left[(A_x^{(\nu)} - a)^+\right].$$

Its single members are unrelated to valuing the Asian option, but as a whole they allow to reconstruct the normalized time–t price $C^{(\nu)}$ of the Asian option. So try to compute the *Laplace transform* of any $f_{GY,a}$ which is formally defined by

$$F_{GY,a}(z) = \int_0^\infty e^{-zx} f_{GY,a}(x)\,dx = \mathscr{L}(f_{GY,a})(z)$$

for any complex number z with $\mathrm{Re}(z) \gg 0$. With the "strike price" a of $f_{GY,a}$ being a constant, the arguments of [**GY**] now apply *mutatis mutandis* to give the

Theorem (Geman–Yor 1993, modified): *If* $\nu = 2\sigma^{-2}\varpi - 1 \geq 0$, *the integrals defining any* $F_{GY,a}$ *exist for any complex number* z *with* $\mathrm{Re}(z) > 2(\nu+1)$, *and*

$$F_{GY,a}(z) = \frac{D_\nu(a,z)}{z(z-2(\nu+1))}$$

where on choosing the principal branch of the logarithm

$$D_\nu(a,z) = \frac{e^{-\frac{1}{2a}}}{a} \int_0^\infty e^{-\frac{x^2}{2a}} x^{\nu+3} I_{\sqrt{2z+\nu^2}}\left(\frac{x}{a}\right) dx$$

with I_μ *the modified Bessel function of order any complex number* μ.

For this version I am very indebted to Yor for correspondence and discussions, and I am very grateful for his support. An exposition of the astute proof of this result taking care of his tutorials in particular can be found in [**SA**]. And hopefully no pitfalls have remained undetected.

Unfortunately, the condition $\nu \geq 0$, which is equivalent to $2\varpi \geq \sigma^2$, places restrictions on the Finance applicability of the result. It is the more likely to be violated the higher is the volatility, as, for instance, if $\sigma \geq 30\%$ and $\varpi < 4.5\%$. Volatile markets, however, furnish the textbook example for justifying the use of Asian options. Nevertheless the result is so strong mathematically that it opens up a natural way for Laplace inversion.

8. Proof of the Theorem, IV: Laplace inverting $F_{GY,a}$ if $\nu \geq 0$

To summarize the current state of affairs, if $\nu = 2\sigma^{-2}\varpi - 1$ is non–negative, we have from §7 explicit functions $F = F_{GY,a}$ that come as Laplace transforms of the functions $f = f_{GY,a}$ on the time axis $\mathbb{R}_{>0}$

$$F(z) = \int_0^\infty e^{-zx} f(x)\,dx = \mathscr{L}(f)(z).$$

Minding Peter Carr's key reduction of §6, the task is to get $f(h)$ as explicitly as possible. This we want to do by *Laplace inverting* F: $f = \mathscr{L}^{-1}(F)$. The classical *Riemann inversion formula* gives this inverse $\mathscr{L}^{-1}$ of $\mathscr{L}$ by integration too:

$$\mathscr{L}^{-1}(F)(h) = \frac{1}{2\pi i} \int_{x_0 - i\infty}^{x_0 + i\infty} e^{hz} F(z)\,dz\,.$$

The integration is over the parallel to the imaginary axis through any positive real number x_0 such that all singularities of F are to its left.

The key idea now is to express in the numerator

$$D_\nu(a,z) = \frac{e^{-\frac{1}{2a}}}{a}\int_0^\infty e^{-\frac{x^2}{2a}} x^{\nu+3} I_{\sqrt{2z+\nu^2}}\left(\frac{x}{a}\right) dx\,,$$

of $F(z)$ the modified Bessel functions I_μ as integrals too. Indeed, for any complex number μ, we have the classical integral

$$I_\mu(z) = \frac{1}{2\pi i}\int_{\log C_R} e^{-\mu\cdot\xi + z\cdot\cosh(\xi)}\, d\xi\,,$$

for any complex number z with positive real part, and where for any $R > 0$ the contour of integration $\log C_R$ is

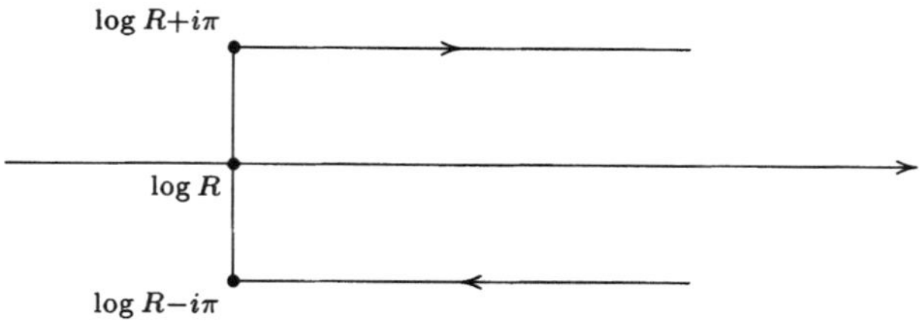

Figure 1 The contour $\log C_R$

Handwaving, I think, is no longer adequate for the remaining main part of the inversion argument. The punchline of the argument, however, is as follows. Consider $\mathscr{L}^{-1}(F)(h)$ as triple integral on substitution for $\mathscr{L}^{-1}$ and I_μ. Prove that under certain technical conditions the order of theses three integrals can be interchanged. Laplace inversion of F then is seen to reduce to that of

$$z \mapsto \frac{e^{-\xi\sqrt{z}}}{z-b^2}.$$

Using Peter Carr's reduction of §6, the integrals for the normalized time–t price of the Asian option follow after some work, and the proof is complete.

9. Proof of the Theorem, V: hints on lifting the restriction $\nu \geq 0$

Lifting of the restriction $\nu \geq 0$ under which Laplace inversion was effected in §8 is by analytic continuation. This is based on the following two results

- The normalized time–t price of the Asian option

  $$L(\nu) = E^Q\left[\left(A_h^{(\nu)} - q\right)^+\right]$$

  is an entire function in ν on $\mathbb{C}$.
- The right hand side R of the identity to be proved

  $$R(\nu) = C_{\text{trig}} + \sum_{b\in\{\pm(\nu+2)\}} C_{\text{hyp},b} - \sum_{b\in\{\pm\nu\}} C_{\text{hyp},b}$$

  is a meromorphic function in ν on $\mathbb{C}$.

To complete the proof of §4.3 Theorem given these two results, recall that the Laplace inversion of §8 gives $L = R$ on $[0, \infty)$. Now literally apply the *identity theorem* to get $L = R$ as meromorphic functions on the complex plane. Since L is continuous and bounded in each point, R cannot have any poles. Hence $L = R$ as entire functions, as was to be shown.

The proof that L is an entire function in ν is instructive, furnishes an upper bound for the normalized price of the Asian option, and will be presented in the sequel. Indeed, set $f(x) = (x-q)^+$. Applying Girsanov such that $W_h = \nu h + B_h$ becomes a standard Brownian motion, and, dropping reference to this new measure, we get

$$L(\nu) = E\Big[f(A_h^{(0)})e^{\nu W_h}\Big] \cdot e^{+\frac{h}{2}\nu^2}.$$

It is sufficient to show that the expectation factor is entire in ν. This is true by definition if we have the convergent series

$$E\Big[f(A_h^{(0)})e^{\nu W_h}\Big] = \sum_{m=0}^{\infty} \frac{\nu^m}{m!} E\Big[f(A_h^{(0)})W_h^m\Big]$$

for all ν. For this it is sufficient to show that the series is absolutely convergent for all ν. Using the Cauchy–Schwarz inequality this is implied by

$$\sum_{m=0}^{\infty} \frac{|\nu|^m}{m!} \sqrt{E\Big[f^2(A_h^{(0)})\Big]} \sqrt{E\Big[W_h^{2m}\Big]}$$

being convergent for all ν. This finally follows using the ratio test since

$$E\Big[W_h^{2m}\Big] \leq \frac{1}{\sqrt{\pi}} \cdot (2h)^m \cdot m!$$

for all $m \geq 0$, and $E[f^2(A_h^{(0)})]$ is finite. Indeed, all higher moments of all $A^{(\nu)}$ have been explicitly computed and can be found in [**Y**, §4].

10. Remarks on computing if $q > 0$

In [**SE**] we found asymptotic expansions instrumental for computing the normalized time–t price of the Asian option. Roughly speaking, such asymptotic expansions are highly divergent series whose first few terms, however, approximate a given function value up to a certain maximal accuracy. This is why explicit estimates for this approximation error are a prerequisite for computation.

I want to illustrate this with the trigonometric term C_{trig} of §4.3 Theorem. As the counterpart of §4.1 it filters out the non–risky part of the value of the Asian option. If $\nu > -2$, we have the expansion in powers of $q > 0$

$$C_{\text{trig}} = E\Big[A_h^{(\nu)}\Big] + \sum_{n=0}^{N-1} a_n \cdot e^{-\frac{1}{2q}} \cdot (2q)^{\nu+n+3} + R_N$$

for any non–negative integer N. Any coefficient a_n satisfies

$$a_n \approx \varepsilon_\nu \cdot \frac{(2n+4+[\nu])!}{n! \cdot (n+\nu+2)},$$

and the error R_N of this N–th order approximation satisfies

$$\left|R_N\right| \leq |\varepsilon_\nu| \cdot \frac{(2N+4+[\nu])!}{N! \cdot (N+\nu+2)} \cdot e^{-\frac{1}{2q}} \cdot (2q)^{\nu+N+3}.$$

Thus the absolute value of R_N is majorized by the absolute value of the N–th summand of the asymptotic expansion, and the smaller q the better the approximation behaviour of the asymptotic expansion.

I want to close my talk with some numerical illustrations of this asymptotic expansion. Assuming an interest rate of 5% p.a., consider an Asian option issued today at time zero at par on a non–dividend paying stock. The normalized time–t strike price q then is equal to the normalized time to maturity $h = \sigma^2 T/4$. In dependency on time to maturity T and the volatility σ we have

$q = h$	$T = 1$ year	$T = 6$ months	$T = 3$ months
$\sigma = 20\%$	0.01	0.005	0.0025
$\sigma = 30\%$	0.0225	0.0113	0.0056

Table 1 $q = h$ as function of σ and T

Thus the smaller T or σ the smaller h and q. If $T = 1$ year, we have the following estimates for the approximation errors R_N of the asymptotic expansion in C_{trig}

$\lvert R_N\rvert \leq \bullet$	$\sigma = 20\%$	$\sigma = 30\%$
$N = 0$	$1.151 \cdot 10^{-32}$	$1.815 \cdot 10^{-17}$
$N = 4$	$7.18 \cdot 10^{-36}$	$5.250 \cdot 10^{-20}$
$N = 8$	$1.448 \cdot 10^{-38}$	$1.252 \cdot 10^{-21}$
$N = 12$	$9.897 \cdot 10^{-41}$	$1.328 \cdot 10^{-22}$
$N = 16$	$1.77 \cdot 10^{-42}$	$4.193 \cdot 10^{-23}$
$N = 20$	$6.915 \cdot 10^{-44}$	$3.127 \cdot 10^{-23}$
$N = 24$	$5.211 \cdot 10^{-45}$	$4.727 \cdot 10^{-23}$
$N = 28$	$6.910 \cdot 10^{-46}$	$1.304 \cdot 10^{-22}$
$N = 32$	$1.505 \cdot 10^{-46}$	$6.064 \cdot 10^{-22}$
$N = 36$	$5.094 \cdot 10^{-47}$	$4.478 \cdot 10^{-21}$
$N = 40$	$2.568 \cdot 10^{-47}$	$5.004 \cdot 10^{-20}$
$N = 44$	$1.859 \cdot 10^{-47}$	$8.139 \cdot 10^{-19}$

Table 2 Accuracy of N–th order approximation in C_{trig} as function of σ for $T = 1$

The resulting error curves have the typical U–shape. And with q decreasing their minimum wanders south–east, i.e., the maximally achievable accuracies increase.

Acknowledgements

I wish to thank very much my friends Joachim Ballmann, Peter Carr, Dan Fulea, Gerald Höhn, and Uwe Weselmann for their moral support over the last months. Invitations to present our results at l'Université Louis Pasteur, Strasbourg, at the Carnegie Mellon University Risk Management Conference, Pittsburgh, at the DMV Tagung in Dresden, and at this Workshop in Konstanz are gratefully acknowledged.

References

[**GY**] H. Geman, M. Yor: Bessel processes, Asian options, and perpetuities, *Math. Finance* **3** (1993), 349–375.

[**RY**] D. Revuz, M. Yor: *Continuous martingales and Brownian motion*, 2nd ed., Springer 1994.

[**SA**] M. Schröder: On the valuation of arithmetic–average Asian options: integral representations, Universität Mannheim, Oktober 1997, revised November 1999.

[**SE**] M. Schröder: On the valuation of arithmetic–average Asian options: explicit formulas, Universität Mannheim, März 1999.

[**SF**] M. Schröder: Fact and fantasy in recent applications of transform methods to the valuation of exotic options, Universität Mannheim, September 1999.

[**Y80**] M. Yor: Loi d'indice du lacet Brownien, et distribution de Hartman–Watson, *Z. Wahrscheinlichkeitstheorie* **53** (1980), 71–95.

[**Y92**] M. Yor: Sur certaines fonctionnelles exponentielles du mouvement Brownien réel, *J. Appl. Prob.* **29** (1992), 202–208.

[**Y**] M. Yor: On some exponential functionals of Brownian motion, *Adv. Appl. Prob.* **24** (1992), 509–531.

Michael Schröder
Lehrstuhl Mathematik III
Universität Mannheim
D–68131 Mannheim
Germany
E-mail address: schroeder@math.uni-mannheim.de

Trends in Mathematics, © 2001 Birkhäuser Verlag Basel/Switzerland

Reversible Real Options

Mark Shackleton and Rafał Wojakowski

Abstract. In this paper we price perpetual reversible real options and examine the critical exercise threshold. Because of the nature of this critical threshold, this boundary is also applicable for finite horizon real options for which a formula is derived.

1. Introduction

For an underlying price process S with expected capital gain $\mu - \delta$ and dividend yield δ

$$dS = (\mu - \delta) S dt + \sigma S dW$$

no arbitrage implies that the price of any time homogeneous option C on S must satisfy a partial differential equation

$$\frac{1}{2}\sigma^2 S^2 \frac{\partial^2 C}{\partial S^2} + (r - \delta) S \frac{\partial C}{\partial S} - rC = 0 .$$

The general solution to this consists of calls and puts and has the form $AS^a + BS^b$. Constants A and B are arbitrary, to be determined by the boundary conditions whereas $a > 1$ and $b < 0$ solve the following fundamental equation

$$\frac{1}{2}\sigma^2 \beta (\beta - 1) + (r - \delta) \beta - r = 0 , \tag{1}$$

which yields

$$a, b = \frac{1}{2} - \frac{r - \delta}{\sigma^2} \pm \sqrt{\left(\frac{r - \delta}{\sigma^2} - \frac{1}{2}\right)^2 + \frac{2r}{\sigma^2}} .$$

Note that for the call and put to be defined we require $\delta > 0$ and $r > 0$ respectively.

January 15, 2001. The authors are grateful to Konstanz Workshop on Mathematical Finance and to Math-Finance seminar participants in Lancaster for useful discussions. Usual disclaimers apply.
Contact: Department of Accounting and Finance, Lancaster University, LA1 4YX, UK. `mailto:m.shackleton@lancaster.ac.uk` or `mailto:r.wojakowski@lancaster.ac.uk` URL: `http://www.lancs.ac.uk/staff/shacklem/` or `http://www.lancs.ac.uk/staff/wojakows/` .

2. Costly Reversion and Reversible Real Options

We can formulate a reversible problem like Brennan and Schwartz (1985).[1] Suppose the stock can be called by paying a sum $\overline{X}$ with further rights to put the stock back and retrieve a different sum $\underline{X}$ i.e. a ***bond plus call*** $\overline{X}+C$ can be switched for the ***stock plus put*** $S+P$ and vice versa

$$\begin{aligned} \overline{X}+C &\rightarrow S+P\,, \\ \underline{X}+C &\leftarrow S+P\,. \end{aligned}$$

Forward and reverse exercise amounts may not be equal $\underline{X} \neq \overline{X}$ so forward and reverse exercise may be separated in S. Value matching and smooth pasting at upper $\overline{K}$ *and* lower $\underline{K}$ critical thresholds yield a system of four equations in four variables $\{A, B, \overline{K}, \underline{K}\}$. The exercise amounts $\{\overline{X}, \underline{X}\}$ are normally the inputs although it is actually easy to evaluate $\{A, B, \overline{X}, \underline{X}\}$ as a function of $\{\overline{K}, \underline{K}\}$. We have

$$\begin{array}{r} A\overline{K}^a + \overline{X} = B\overline{K}^b + \overline{K} \\ A\underline{K}^a + \underline{X} = B\underline{K}^b + \underline{K} \\ Aa\overline{K}^a = Bb\overline{K}^b + \overline{K} \\ Aa\underline{K}^a = Bb\underline{K}^b + \underline{K} \end{array} \Leftrightarrow \begin{bmatrix} 1 & 0 & \overline{K}^a & -\overline{K}^b \\ 0 & 1 & \underline{K}^a & -\underline{K}^b \\ 0 & 0 & a\overline{K}^a & -b\overline{K}^b \\ 0 & 0 & a\underline{K}^a & -b\underline{K}^b \end{bmatrix} \begin{bmatrix} \overline{X} \\ \underline{X} \\ A \\ B \end{bmatrix} = \begin{bmatrix} \overline{K} \\ \underline{K} \\ \overline{K} \\ \underline{K} \end{bmatrix},$$

thus

$$\begin{bmatrix} \overline{X} \\ \underline{X} \\ A \\ B \end{bmatrix} = \begin{bmatrix} 1 & 0 & \overline{K}^a & -\overline{K}^b \\ 0 & 1 & \underline{K}^a & -\underline{K}^b \\ 0 & 0 & a\overline{K}^a & -b\overline{K}^b \\ 0 & 0 & a\underline{K}^a & -b\underline{K}^b \end{bmatrix}^{-1} \begin{bmatrix} \overline{K} \\ \underline{K} \\ \overline{K} \\ \underline{K} \end{bmatrix}$$

$$= \begin{bmatrix} 1 & 0 & \frac{-\overline{K}^a b\underline{K}^b + \overline{K}^b a\underline{K}^a}{ab\left(\overline{K}^a\underline{K}^b - \overline{K}^b\underline{K}^a\right)} & -\overline{K}^a\overline{K}^b \frac{-b+a}{ab\left(\overline{K}^a\underline{K}^b - \overline{K}^b\underline{K}^a\right)} \\ 0 & 1 & \underline{K}^a\underline{K}^b \frac{-b+a}{ab\left(\overline{K}^a\underline{K}^b - \overline{K}^b\underline{K}^a\right)} & -\frac{-\underline{K}^a b\overline{K}^b + \underline{K}^b a\overline{K}^a}{ab\left(\overline{K}^a\underline{K}^b - \overline{K}^b\underline{K}^a\right)} \\ 0 & 0 & \frac{\underline{K}^b}{a\left(\overline{K}^a\underline{K}^b - \overline{K}^b\underline{K}^a\right)} & -\frac{\overline{K}^b}{a\left(\overline{K}^a\underline{K}^b - \overline{K}^b\underline{K}^a\right)} \\ 0 & 0 & \frac{\underline{K}^a}{b\left(\overline{K}^a\underline{K}^b - \overline{K}^b\underline{K}^a\right)} & -\frac{\overline{K}^a}{b\left(\overline{K}^a\underline{K}^b - \overline{K}^b\underline{K}^a\right)} \end{bmatrix} \begin{bmatrix} \overline{K} \\ \underline{K} \\ \overline{K} \\ \underline{K} \end{bmatrix}.$$

This says that if $\{\overline{K}, \underline{K}\}$ were known it would be an easy task to evaluate the vector $\{A, B, \overline{X}, \underline{X}\}$ as a function of $\{\overline{K}, \underline{K}\}$ present on the right hand side. Multiplying this matrix out and defining the ratio of the critical thresholds

$$\gamma \triangleq \frac{\underline{K}}{\overline{K}}$$

[1]See also a similar problem in Dixit and Pindyck (1994).

yields

$$\begin{bmatrix} \overline{X} \\ \underline{X} \\ A \\ B \end{bmatrix} = \frac{\overline{K}}{ab\left(\gamma^b - \gamma^a\right)} \begin{bmatrix} ab\left(\gamma^b - \gamma^a\right) - b\gamma^b + a\gamma^a - (a-b)\,\gamma \\ ab\left(\gamma^{b+1} - \gamma^{a+1}\right) + b\gamma^{a+1} - a\gamma^{b+1} + (a-b)\,\gamma^{a+b} \\ \frac{b\left(\gamma^b - \gamma\right)}{\overline{K}^a} \\ \frac{a\left(\gamma^a - \gamma\right)}{\overline{K}^b} \end{bmatrix}. \tag{2}$$

Observe also that

$$ab = -\frac{2r}{\sigma^2} \qquad a + b = \frac{r - \delta - \frac{1}{2}\sigma^2}{\frac{1}{2}\sigma^2}.$$

Now if $\overline{K} \neq 0$, $\gamma^b \neq \gamma^a$ and $ab \neq 0$, a new variable, the ratio of the exercise amounts, defined as

$$\alpha \triangleq \frac{\underline{X}}{\overline{X}},$$

can be expressed as a function of γ (ratio of exercise thresholds) by dividing the first two lines in (2) yielding

$$\alpha(\gamma) = \frac{ab\left(\gamma^{b+1} - \gamma^{a+1}\right) + b\gamma^{a+1} - a\gamma^{b+1} + (a-b)\,\gamma^{a+b}}{ab\left(\gamma^b - \gamma^a\right) + a\gamma^a - b\gamma^b - (a-b)\,\gamma}.$$

Obviously we would prefer $\gamma(\alpha)$ in order to determine the thresholds as a function of the investment ratio α but a numerical solution for the inverse is easy to obtain: for a given $\alpha^\star$ one can find the value $\gamma^\star$ that satisfies $\alpha(\gamma^\star) = \alpha^\star$.

2.1. Asymptotes of $\alpha(\gamma)$

This function $\alpha(\gamma)$ (ratio of exercise thresholds) has three asymptotes at $\gamma = (\infty, 0, 1)$. For large and small γ the limiting behaviour is

$$\alpha(\gamma)\big|_{\gamma\uparrow\infty} \propto \gamma\frac{b(1-a)}{a(1-b)} \qquad \alpha(\gamma)\big|_{\gamma\downarrow 0} \propto \gamma\frac{a(b-1)}{b(a-1)}.$$

These correspond to the standard Merton (1973) irreversible puts and calls where exercise is one way only and no return is permitted. If $\underline{K}$ is finite then $\overline{K}$ is infinite or if $\overline{K}$ is finite then $\underline{K}$ is zero. This can be seen by taking the highest or lowest power of γ in denominator or numerator, in which case the standard put and call thresholds are retrieved

$$\begin{bmatrix} \overline{K} \\ \underline{K} \end{bmatrix} = \begin{bmatrix} \frac{ab\left(\gamma^b - \gamma^a\right)}{ab(\gamma^b - \gamma^a) - b\gamma^b + a\gamma^a - (a-b)\gamma}\overline{X} \\ \frac{ab\left(\gamma^b - \gamma^a\right)}{ab(\gamma^b - \gamma^a) + b\gamma^a - a\gamma^b + (a-b)\gamma^{a+b-1}}\underline{X} \end{bmatrix}$$

$$\lim_{\gamma\to\infty} \begin{bmatrix} \overline{K} \\ \underline{K} \end{bmatrix} = \begin{bmatrix} \frac{b}{b-1}\overline{X} \\ \frac{a}{a-1}\underline{X} \end{bmatrix} \qquad \lim_{\gamma\to 0} \begin{bmatrix} \overline{K} \\ \underline{K} \end{bmatrix} = \begin{bmatrix} \frac{a}{a-1}\overline{X} \\ \frac{b}{b-1}\underline{X} \end{bmatrix}.$$

2.2. Special case $\gamma \to 1$

There is, however, the third special case when γ is close to one, implying

$$\lim_{\gamma\to 1} \alpha(\gamma) = 1 .$$

This corresponds to a case of full reversibility where upper and lower thresholds converge to each other ($\underline{K} \to \overline{K} \to K$) as the exercise amounts converge ($\underline{X} \to \overline{X} \to X$)

$$\lim_{\gamma\to 1} \underline{K} = K = \lim_{\gamma\to 1} \overline{K} \qquad\qquad \lim_{\gamma\to 1} \underline{X} = X = \lim_{\gamma\to 1} \overline{X} .$$

Using de l'Hospital's rule to evaluate the limiting case $\gamma \to 1$ we get

$$\begin{bmatrix} \overline{K} \\ \underline{K} \end{bmatrix} = \begin{bmatrix} \frac{ab\left(b\gamma^{b-1}-a\gamma^{a-1}\right)}{ab(b\gamma^{b-1}-a\gamma^{a-1})-b^2\gamma^{b-1}+a^2\gamma^{a-1}-(a-b)}\overline{X} \\ \frac{ab\left(b\gamma^{b-1}-a\gamma^{a-1}\right)}{ab(b\gamma^{b-1}-a\gamma^{a-1})+ba\gamma^{a-1}-ab\gamma^{b-1}+(a-b)(a+b-1)\gamma^{a+b-2}}\underline{X} \end{bmatrix} .$$

With $\gamma = 1$, $\alpha = 1$ ($\overline{K} = \underline{K} = K$ and $\overline{X} = \underline{X} = X$) this implies

$$K = \frac{ab}{ab-(a+b)+1} X = \frac{rX}{\delta}$$

and

$$\lim_{\gamma\to 1} \begin{bmatrix} A \\ B \end{bmatrix} = \frac{1}{ab\,(b-a)} \begin{bmatrix} b\,(b-1)\left(\frac{rX}{\delta}\right)^{1-a} \\ a\,(a-1)\left(\frac{rX}{\delta}\right)^{1-b} \end{bmatrix} .$$

3. Costless reversion

For costless reversion (for the sum X), the critical threshold is based on a current yield criteria $\delta S \lessgtr rX$, not a naive present value condition $S \lessgtr X$. This is similar to the Jorgensonian (1963) user cost of capital. This is because the reversible real option effectively generates the maximum of the flow on the stock δS or the flow on the exercise proceeds rX. Because there is no benefit to waiting, it is always optimal to switch to enjoy the maximum cash yield at any point in time. Thus switching occurs around $\delta K = rX$ and values of the fully reversible options are given by

$$\begin{aligned} C^R(S_0, X, T) &= AS^a = \frac{rX}{\delta}\frac{b-1}{a\,(b-a)}\left(\frac{\delta S_0}{rX}\right)^a \qquad (3) \\ P^R(S_0, X, T) &= BS^b = \frac{rX}{\delta}\frac{a-1}{b\,(b-a)}\left(\frac{\delta S_0}{rX}\right)^b . \end{aligned}$$

These can be shown to be greater than the standard Merton call or put values (because these options allow more flexibility). Furthermore because early exercise incurs no penalty (in the sense that it can always be undone) exercise occurs earlier in the reversible case than the non–reversible case

$$\frac{b}{b-1}X = \underline{K} < K = \frac{r}{\delta}X < \overline{K} = \frac{a}{a-1}X .$$

4. Finite Reversible Real Options

Although the infinite asymptotes for the thresholds are available as the solution to a non-linear equation, for any finite horizon problem (non time homogeneous) the true boundaries can only be found using numerical techniques. However for costless reversion, the boundaries are flat for all time because there is no penalty for early exercise and the problem is again time homogeneous. This means that the boundaries are known a priori for the reversible case and that the problem solution can be formulated. For a finite horizon T, the reversible real option has an optimal policy that is to hold the stock until time T, if the stock is priced more than rX/δ, or the bond, if the stock is priced less than rX/δ; then, at time T, to exercise a European option (once only at T) with a strike of X.

This optimal policy can be written as a risk-neutral world expectation of the discounted flows until time T and the discounted payoff at time T because optimal exercise would yield holdings consistent with these flows and payoffs

$$C^R(S_0, X, T) + X = E_0^Q\left[\int_0^T e^{-rt}\max\left(rX, \delta S_t\right)dt + e^{-rT}\max\left(X, S_T\right)\right] .$$

Note that

$$\max\left(rX, \delta S\right) = \delta \max\left(S - \frac{r}{\delta}X, 0\right) + rX ,$$

so that this expression can be further represented using the Black Scholes (1973) European option formula

$$C^R(S_0, X, T) = \delta\int_0^T C^E(S_0, K, t)dt + C^E(S_0, X, T) \tag{4}$$

where

$$\begin{aligned} C^E(S_0, X, T) &= S_0 e^{-\delta T} N\left(\frac{\ln S_0/X + \left(r - \delta + \frac{1}{2}\sigma^2\right)T}{\sigma\sqrt{T}}\right) \\ &\quad - X e^{-rT} N\left(\frac{\ln S_0/X + \left(r - \delta - \frac{1}{2}\sigma^2\right)T}{\sigma\sqrt{T}}\right) . \end{aligned}$$

Equation (4) has the interpretation of the call value as being equal to a Black Scholes option with final maturity T and strike of X plus extra flexibility value represented by the dividend yield on a continuum of new Black Scholes options of intermediate lives $(0, T)$ struck at the new flexibility threshold $K = \frac{r}{\delta}X$. Thus the reversible option does everything the T maturity one does, plus gives a continuous yield δ on the excess of S_t over the yield threshold $\frac{r}{\delta}X$. This continuous yield has an ex ante expected value given by a continuum of Black Scholes options. As $T \to \infty$ this expression converges toward the known infinite horizon solution, given by equation (3), because the final Black Scholes component goes to zero.

5. Concluding remarks

1. The standard Merton options are irreversible (reversible only at $0, \infty$).
2. Reversible calls and puts can be priced and are worth more than the irreversible.
3. Reversible real options are exercised earlier than irreversible and are not subject to hysteresis.
4. Real options are not subject to straight NPV criteria $S \lessgtr X$.
5. Reversible real options are exercised according to a yield criteria $\delta S \lessgtr rX$.
6. The reversible threshold can be less than X if $\delta > r$.

References

Black, F., and M. Scholes (1973): "The Pricing of Options and Corporate Liabilities," *Journal of Political Economy*, **81**(3), 637–654.

Brennan, M. J., and E. S. Schwartz (1985): "Evaluating Natural Resource Investments," *Journal of Business*, **58**(2), 135–158.

Dixit, A. K., and R. S. Pindyck (1994): *Investment under Uncertainty.* Princeton University Press.

Jorgenson, D. W. (1963): "Capital Theory and Investment Behavior," *American Economic Review*, **53**(2), 247–259.

Merton, R. C. (1973): "The Theory of Rational Option Pricing," *Bell Journal of Economics*, **4**(1), 141–183.

Department of Accounting and Finance, Lancaster University, LA1 4YX, UK

Trends in Mathematics, © 2001 Birkhäuser Verlag Basel/Switzerland

A Toolbox for Generalized Relative Entropies, EMM and Contingent Claim Valuation

Wolfgang Stummer

Abstract. We consider price models which are one-dimensional stochastic differential equations of the form $dX_t = b(t, X_t)\, dt + \sigma_t\, X_t\, dW_t$. We present a toolbox of handy-to-verify non-stochastic conditions which guarantee the finiteness of the $\mathfrak{I}_\alpha$−divergence between the law of the solution X_t and the corresponding drift-less measure (the special case $\alpha = 1$ is the relative entropy, a concept which is known to play an important role in mathematical finance). We also give further properties of these divergences. This will be applied to the contexts of equivalent martingale measures and contingent claim valuation. It is shown by example that also locally unbounded drifts b and exploding volatility functions σ are covered. Furthermore, Samuelson's geometric Brownian motion and thus the seminal option pricing theorems of Black-Scholes and Merton, appear as a special case.

1. Introduction

One prominent model for the time-behaviour of financial asset prices is the geometric Brownian motion X_t, which is the unique (strong) solution of the stochastic differential equation (SDE)

$$dX_t = C_1\, X_t\, dt + \sigma^{co}\, X_t\, dW_t\,, \qquad (1.1)$$

with some constants C_1 and $\sigma^{co} > 0$, and Brownian motion W (see [15], [10], [11]). This model was also used by [1] and [12] as the underlying in their seminal investigations on option pricing. Although the geometric Brownian motion model has deficiences for daily or intra-day observations, it serves relatively well for describing the long-term behaviour of asset prices or indices, see e.g. the empirical studies in [5] and [4]. Also, in the context of "real options", it is well known that the underlying quantity X can only be approximated by a geometric Brownian motion; see e.g. [7], [2].

Inspired by such facts, it is reasonable to study some mathematical finance of a financial market $\mathcal{M} = \{\mathcal{BC}, X\}$, which consists of

(i) a bond $\mathcal{BC}$, whose price evolution is given by $\mathcal{BC}_t = e^{\int_0^t r_v dv}$, where the deterministic short rate process r_t is nonnegative and continuous in t,

(ii) a financial asset X, whose price process is a weak solution of the SDE

$$dX_t = b(t, X_t)\, dt + \sigma_t X_t\, dW_t , \quad t \in [s, T], \quad (1.2)$$
$$X_s \text{ has probability distribution } \mu \text{ on }]0, \infty[,$$

where $s \in [0, T]$ is the starting time, μ *on* $]0, \infty[$ is the starting distribution, and $T > 0$ is the fixed final time horizon. The corresponding solution measure will be denoted by $Q_{(s,\mu)}$, and the solution measure for the case $b(t, X_t) \equiv 0$ (with some different Brownian motion $\overline{W}$) will be denoted by $P_{(s,\mu)}$. The underlying measurable space and filtrations will be omitted for the sake of brevity. Also, the financial asset X is assumed to continuously pay dividends of the amount $\delta_t X_t dt$ between time t and $t+dt$, where the dividend yield δ_t is a deterministic, continuous, nonnegative function of t.

Furthermore, we use the standard assumptions that the lending (interest) rate is equal to the borrowing (interest) rate, that there are no transaction costs and no taxes, and that trading takes place continuously.

The SDE (1.2) can be thought of as a "generalized Samuelson-Merton-Black-Scholes world"; its solutions $(X_t, Q_{(s,\mu)})$ are — in contrast to the geometric Brownian motion — typically non-lognormally distributed.

As usual, we investigate the discounted price process $\widetilde{X}_t := X_t\, e^{\int_0^t (\delta_v - r_v) dv}$. This amounts to studying the discounted SDE

$$d\widetilde{X}_t = \widetilde{b}(t, \widetilde{X}_t)\, dt + \sigma_t \widetilde{X}_t\, d\widetilde{W}_t , \quad t \in [s, T], \quad (1.3)$$
$$\widetilde{X}_s \text{ has probability distribution } \widetilde{\mu} \text{ on }]0, \infty[,$$

with discounted drift $\widetilde{b}(t, y) := (\delta_t - r_t) y + e^{\int_0^t (\delta_v - r_v) dv}\, b\Big(t, e^{\int_0^t (r_v - \delta_v) dv} y\Big)$.

The corresponding solution measure will be denoted by $\widetilde{Q_{(s,\widetilde{\mu})}}$; we also use the notations $\widetilde{P_{(s,\widetilde{\mu})}}$ and $\widetilde{\overline{W}}$ for the case $\widetilde{b}(t, y) \equiv 0$.

It is well known that the relative entropy is an important tool in mathematical finance. In this paper, we deal with the following kind of "generalized relative entropies", namely the (for our context adapted form of the) "$\mathfrak{I}_\alpha$–divergences"

$$\mathfrak{I}_\alpha(\widetilde{Q_{(s,\widetilde{\mu})}} || \widetilde{P_{(s,\widetilde{\mu})}}) := \int f_\alpha\left(\frac{d\widetilde{Q_{(s,\widetilde{\mu})}}}{d\widetilde{P_{(s,\widetilde{\mu})}}}\right) d\widetilde{P_{(s,\widetilde{\mu})}} ,$$

with the nonnegative functions $f_\alpha : [0, \infty[\to [0, \infty[$ defined by

$$f_\alpha(\rho) := \begin{cases} -\log \rho + \rho - 1, & if\ \alpha = 0, \\ \frac{\alpha\rho + 1 - \alpha - \rho^\alpha}{\alpha(1-\alpha)}, & if\ \alpha \in I\!R \backslash \{0, 1\}, \\ \rho\ \log \rho + 1 - \rho, & if\ \alpha = 1; \end{cases}$$

for investigations on $\mathfrak{I}_\alpha$–divergences for general measures, see e.g. [8]. As usual, one makes the conventions $-\log 0 = \infty$, and $0\ \log 0 = 0$. The case $\alpha = 1$ corresponds equivalently to the relative entropy. Clearly, the integrals in $\mathfrak{I}_\alpha$ always exist (if the involved density exists); however, it is well known that, for general measures, one can only guarantee the finiteness of the $\mathfrak{I}_\alpha$–divergences for $\alpha \in]0,1[$ (i.e. the relative entropy is not covered).

The main goal of this paper are :

- (1) to give sufficient, handy-to-verify, far-reaching non-stochastic conditions such that for all parameters $\alpha \in I\!R$ the divergences $\mathfrak{I}_\alpha(\widetilde{Q_{(s,\widetilde{\mu})}} || \widetilde{P_{(s,\widetilde{\mu})}})$ are finite. We also investigate their uniform short-time behaviour.
- (2) to show — with the help of the results of (1) — by "direct" methods that there exists a unique equivalent martingale measure for the discounted price process $(\widetilde{X}, \widetilde{Q_{(s,\widetilde{\mu})}})$. Afterwards, the corresponding no-arbitrage result and the unique arbitrage-based prices of European contingent claims on the underlying instrument X will be presented.
- (3) to demonstrate the wide range of the context by presenting an example with locally exploding drift and volatility function.

It is a pleasure to thank A.D. Barbour and E.S.W. Shiu for inspiring discussions.

2. Generalized Relative Entropies

We always suppose that the drift $b : [0,T] \times [0,\infty[\mapsto [-\infty,\infty]$ and the volatility function $\sigma : [0,T] \mapsto [0,\infty]$ are deterministic, (possibly exploding) Borel measurable functions. Furthermore, for the special case with the Dirac-measure $\vartheta_{\widetilde{x}}$ as the starting distribution (i.e., the process $\widetilde{X}$ starts in $\widetilde{x} > 0$) we will write $\widetilde{Q_{(s,\widetilde{x})}}$ instead of $\widetilde{Q^{IV}_{(s,\vartheta_{\widetilde{x}})}}$, etc. Throughout this paper, the following toolbox of non-stochastic assumptions will be used :

Assumptions 2.1. *Suppose that the drift b can be estimated by the product $|b(t,y)| \leq b_1(t)\ b_2(y)$ for all $(t,y) \in [0,T] \times]0,\infty[$, where $b_1 : [0,T] \mapsto [0,\infty]$ and $b_2 :]0,\infty[\mapsto [0,\infty]$ are Borel measurable functions. Let b_1 and b_2 satisfy the two conditions*

$$\int_0^T b_1(v)^{q_1}\ dv\ <\ \infty$$

$$\textit{and} \qquad \sup_{a \in I\!R} \int_{a-1}^{a+1} \frac{\{b_2(e^{\varsigma})\}^{q_2}}{e^{\varsigma q_{2.1}}}\ d\varsigma\ <\ \infty\ ,$$

for some powers $q_{2.1} \in [2,\infty]$ and $q_{2.1} \in [2,\infty]$ which are connected by the inequality $\frac{1}{q_{2.1}} + \frac{1}{q_{2.1}} \leq \frac{1}{2}$. Furthermore, assume that the volatility function σ_t satisfies the two conditions

$$\int_0^T \sigma_v^2\ dv\ <\ \infty \qquad (2.1)$$

$$\textit{and} \qquad \lim_{\Delta t \downarrow 0} \sup_{u \in [0,T]} \int_u^{(u+\Delta t)\wedge T} \frac{1}{\sigma_v^{q_{2.1}} \sqrt{\int_u^v \sigma_{v_1}^2\ dv_1}}\ dv\ =\ 0\ . \qquad (2.2)$$

Remark 2.2. *Here, we denote $q_{2.1} = \infty$ for the case that b_1 is bounded; in contrast, we denote $q_{2.1} = \infty$ for the case that both (i) $b_2(y)/y$ is bounded (in y), and (ii) σ_t is bounded (in t) from below by a strictly positive constant.*

Let us first investigate the existence and uniqueness of the involved SDE :

Theorem 2.3. *Suppose that the drift b and the volatility function σ_t satisfy the Assumptions 2.1. Then the following assertions hold:*
(a) For all starting times $s \in [0,T]$ and all starting distributions $\widetilde{\mu}$ on $]0,\infty[$, there exists a weak solution $(\widetilde{X}, \widetilde{Q_{(s,\widetilde{\mu})}}, \widetilde{W})$ of the SDE (1.3). Furthermore, $\widetilde{X}_t > 0$ holds $\widetilde{Q_{(s,\widetilde{\mu})}}-$almost surely for all $t \in [s,T]$.

(b) Within the class of weak solutions $(\widetilde{\widetilde{X}}, \widetilde{\widetilde{Q_{(s,\widetilde{\mu})}}}, \widetilde{\widetilde{W}})$ of (1.3) with the additional property $\widetilde{\widetilde{Q_{(s,\widetilde{\mu})}}}\left[\int_s^T \frac{\widetilde{b}(v,\widetilde{\widetilde{X}}_v)^2}{\sigma_v^2\ \widetilde{\widetilde{X}}_v^2}\, dv < \infty\right] = 1$, the solution $(\widetilde{X}, \widetilde{Q_{(s,\widetilde{\mu})}}, \widetilde{W})$ is unique.

Remark 2.4. *The according results for the undiscounted world (i.e. without $\tilde{}$) follow as a special case.*

Proof. Let us first show that the Assumptions 2.1 imply

$$\lim_{\Delta t \downarrow 0} \sup_{(u,\widetilde{x}) \in [0,T]\times]0,\infty[} \widetilde{EP_{(u,\widetilde{x})}}\left[\int_u^{(u+\Delta t)\wedge T} \frac{\widetilde{b}(v,\widetilde{X}_v)^2}{\sigma_v^2\ (\widetilde{X}_v)^2}\, dv\right] = 0\,. \qquad (2.3)$$

Since, by straightforward calculations, the condition (2.3) is equivalent to

$$\lim_{\Delta t \downarrow 0} \sup_{(u,x) \in [0,T]\times]0,\infty[} EP_{(u,x)}\left[\int_u^{(u+\Delta t)\wedge T} \frac{\widetilde{b}(v,X_v)^2}{\sigma_v^2\ (X_v)^2}\, dv\right] = 0\ ,$$

it suffices to show the two conditions

$$\lim_{\Delta t \downarrow 0} \sup_{u \in [0,T]} \int_u^{(u+t)\wedge T} \frac{(\delta_v - r_v)^2}{\sigma_v^2}\, dv = 0 \qquad (2.4)$$

$$\text{and} \quad \lim_{\Delta t \downarrow 0} \sup_{(u,x) \in [0,T]\times]0,\infty[} EP_{(u,x)}\left[\int_u^{(u+\Delta t)\wedge T} \frac{b(v,\Upsilon_1(v))^2}{\sigma_v^2\ \Upsilon_2(v)^2}\, dv\right] = 0, \quad (2.5)$$

where $\Upsilon_2(v) := \Upsilon_2(v)\ X_v$ with $\Upsilon_2(v) := e^{\int_0^v \left(r_{v_1} - \delta_{v_1}\right) dv_1}$. By using the two assumptions (2.1) and (2.2), the condition (2.4) follows from the estimate

$$\int_u^{(u+t)\wedge T} \frac{(\delta_v - r_v)^2}{\sigma_v^2}\, dv \le C_2\ t^{2/q_{2.1}} \left\{\int_u^{(u+t)\wedge T} \frac{1}{\sigma_v^{q_{2.1}}}\, dv\right\}^{2/q_{2.1}}$$

$$\le C_2\ t^{2/q_{2.1}} \left\{\int_u^{(u+t)\wedge T} \frac{\sqrt{\Upsilon_3(0,T)}}{\sigma_v^{q_{2.1}}\ \sqrt{\Upsilon_3(u,v)}}\, dv\right\}^{2/q_{2.1}}$$

for some constant $C_2 > 0$ and $\Upsilon_3(u,v) := \int_u^v \sigma_{v_1}^2 \, dv_1$. Furthermore, one gets

$$\sup_{(u,x)\in[0,T]\times]0,\infty[} EP_{(u,x)} \left[\int_u^{(u+\Delta t)\wedge T} \frac{b(v,\Upsilon_2(v))^2}{\sigma_v^2 \, \Upsilon_2(v)^2} \, dv \right]$$

$$\leq \left(\int_0^T b_1(v)^{q_{2.1}} \, dv \right)^{2/q_{2.1}}$$

$$\times \left(\sup_{(u,x)\in[0,T]\times]0,\infty[} \int_u^{(u+\Delta t)\wedge T} EP_{(u,x)} \left[\frac{b_2(\Upsilon_2(v))^{q_{2.1}}}{\sigma_v^{q_{2.1}} \, \Upsilon_2(v)^{q_{2.1}}} \right] dv \right)^{2/q_{2.1}}$$

$$\leq C_3 \left(\sup_{(u,x)\in[0,T]\times]0,\infty[} \int_{]0,\infty[} \frac{b(y)^{q_{2.1}}}{y^{q_{2.1}}} \right.$$

$$\left. \times \left(\int_u^{(u+\Delta t)\wedge T} \frac{\Upsilon_2(v) \, exp\left(- \frac{(logy - logx - log\Upsilon_2(v) + \Upsilon_3(u,v)/2)^2}{2\Upsilon_3(u,v)} \right)}{y \, \sigma_v^{q_{2.1}} \sqrt{2\pi \Upsilon_3(u,v)}} \, dv \right) dy \right)^{\frac{2}{q_{2.1}}}$$

for some constant $C_3 > 0$. From this, the condition (2.5) — and hence, (2.3) — follo ws by applying some estimation techniques of [18] for the here-used context. But, analogously to [18], the condition (2.3) implies that, for all starting times $s \in [0,T]$ and all starting distributions $\widetilde{\mu}$ *on* $]0,\infty[$, the Girsanov condition

$$\widetilde{EP_{(s,\widetilde{\mu})}} \left[\widetilde{Z_{s,T}} \right] = 1 \qquad (2.6)$$

holds, with $\widetilde{Z_{s,t}} := \exp\left(\int_s^t \frac{\widetilde{b}(v,\widetilde{X}_v)}{\sigma_v \, \widetilde{X}_v} \, d\widetilde{\widetilde{W}}_v - \frac{1}{2} \int_s^t \frac{\widetilde{b}(v,\widetilde{X}_v)^2}{\sigma_v^2 \, (\widetilde{X}_v)^2} \, dv \right)$, $t \in [s,T]$.

Hence, one can apply the Girsanov theorem; in particular, the solution measure can be constructed by $\widetilde{Q_{(s,\widetilde{\mu})}} := \widetilde{Z_{s,T}} * \widetilde{P_{(s,\widetilde{\mu})}}$ ha ving density $\widetilde{Z_{s,T}}$ with respect to $\widetilde{P_{(s,\widetilde{\mu})}}$. The uniqueness follows by applying (an adapted version of) the uniqueness result from [6], p.304 (see also [14]). □

Now we are ready to study the $\mathfrak{I}_\alpha$−div ergences for the discounted world:

Theorem 2.5. *Suppose that the drift b and the volatility function σ_t satisfy the Assumptions 2.1. Then the following assertions hold :*
(a) for all $\alpha \in I\!R$, all $s \in [0,T]$, and all prob ability measur es$\widetilde{\mu}$ on $]0,\infty[$:

$$0 \leq \mathfrak{I}_\alpha\left(\widetilde{Q_{(s,\widetilde{\mu})}} \middle\| \widetilde{P_{(s,\widetilde{\mu})}} \right) < \infty \, .$$

(b) for all $\alpha \in I\!R$:

$$\lim_{\Delta t \downarrow 0} \sup_{(u,\widetilde{x})\in[0,T]\times]0,\infty[} \mathfrak{I}_\alpha\left(\widetilde{Q_{(u,\widetilde{x})}^{(u+\Delta t)\wedge T}} \middle\| \widetilde{P_{(u,\widetilde{x})}^{(u+\Delta t)\wedge T}} \right) = 0 \, , \qquad (2.7)$$

where $\widetilde{Q_{(u,\widetilde{x})}^{(u+\Delta t)\wedge T}}$ is the r estriction of$\widetilde{Q_{(u,\widetilde{x})}}$ to the time interval $[u,(u+\Delta t)\wedge T]$.

Remark 2.6. *For an application of such kind of $\mathfrak{I}_\alpha-$divergences properties (in an undiscounted world) to decision problems upon the involved drift, see [18]; there, one can also find estimates for large Δt.*

Proof. In the proof of Theorem 2.3 above, we have verified the condition (2.3). ¿From this, one can proceed analogously to [18] in order to obtain the assertions (a) and (b). □

3. EMM and Contingent Claim Valuation

As an application of the results of the previous section, the next theorem gives handy-to-verify, far-reaching non-stochastic conditions such that one obtains the existence of a unique equivalent martingale measure by application of the Girsanov theorem in the opposite direction:

Theorem 3.1. *Suppose that the drift b and the volatility function σ_t satisfy the Assumptions 2.1. Then one has for all starting times $s \in [0,T]$ and all starting distributions $\widetilde{\mu}$ on $]0,\infty[$:*
(a) the Girsanov property

$$\widetilde{EQ_{(s,\widetilde{\mu})}} \left[\widetilde{ZO_{s,T}} \right] = 1 \qquad (3.1)$$

holds, where

$$\widetilde{ZO_{s,t}} := \exp\left(- \int_s^t \frac{\widetilde{b}(v,\widetilde{X}_v)}{\sigma_v\, \widetilde{X}_v}\, d\widetilde{W}_v - \frac{1}{2} \int_s^t \frac{\widetilde{b}(v,\widetilde{X}_v)^2}{\sigma_v^2\, (\widetilde{X}_v)^2}\, dv \right), \qquad t \in [s,T]\, .$$

(b) $\widetilde{P_{(s,\widetilde{\mu})}}$ is an equivalent martingale measure for $(\widetilde{X}, \widetilde{Q_{(s,\widetilde{\mu})}})$.
(c) Within the class described in Theorem 2.3 (b) above, the equivalent martingale measure $\widetilde{P_{(s,\widetilde{\mu})}}$ is unique.

Remark 3.2. *(1) We have called the condition (3.1) Girsanov* **property**, *for the sake of a better verbal distinguishability with its opposite counterpart, the Girsanov condition (2.6).*
(2) Notice that (3.1) is usually more difficult to verify than (2.6). Indeed, [14] mentions that "it can be an extremely difficult task" to prove a Girsanov property like (3.1). Let us also add that in the case of locally exploding drifts, things might even be worse: in fact, for the Bessel process with dimension three — which has the particular one-dimensional, locally exploding drift $b(t,X_t) = \frac{1}{X_t}$ (and unit diffusion coefficient) — there does not even exist an EMM; see [3], also for stronger arbitrage investigations.
(3) In contrast to (2), for the validity of the Girsanov property (3.1) we have given non-stochastic assumptions which — as it will be shown by example later on — cover locally exploding drifts.

Proof. It is straightforward to see that, for the case $\alpha = 1$, the condition (2.7) in Theorem 2.5 is equivalent to the condition

$$\lim_{\Delta t \downarrow 0} \sup_{(u,\widetilde{x}) \in [0,T] \times]0,\infty[} \widetilde{EQ_{(u,\widetilde{x})}} \left[\int_u^{(u+\Delta t)\wedge T} \frac{\widetilde{b}(v,\widetilde{X}_v)^2}{\sigma_v^2 \, (\widetilde{X}_v)^2} \, dv \right] = 0 .$$

¿From this, one can apply — like in [16] or [19] — the space-time version of the well-known Khas'minskii Lemma in order to obtain the "uniform Novikov property"

$$\sup_{(u,\widetilde{x}) \in [0,T] \times]0,\infty[} \widetilde{EQ_{(u,\widetilde{x})}} \left[exp\Big(\frac{1}{2} \int_u^T \frac{\widetilde{b}(v,\widetilde{X}_v)^2}{\sigma_v^2 \, (\widetilde{X}_v)^2} \, dv \Big) \right] < \infty .$$

Thus, by the Novikov Theorem one obtains the part (a). Furthermore, by virtue of the corresponding version of the Girsanov theorem, one gets from (a) that the measure $\widetilde{P_{(s,\widetilde{\mu})}}$ is absolutely continuous with respect to (and hence, equivalent to) the solution measure $\widetilde{Q_{(s,\widetilde{\mu})}}$. Since $(\widetilde{X}, \widetilde{P_{(s,\widetilde{\mu})}})$ is a martingale, we have proved the part (b). Because of (2.3) and part (b) of Theorem 2.3, the assertion (c) follows e.g. by using the uniqueness result of [14]. □

Let us apply, as usual, the above result for investigations upon the absence of arbitrage opportunities for the market $\mathcal{M} = \{\mathcal{BC}, X\}$, where the asset price process (X, Q) is modelled as a weak solution of the undiscounted SDE (1.2). For this, and the rest of this section, we fix $s = 0$ as the starting (trading) time. By considering portfolio processes of the form $\mathcal{PO}_t := \phi_t^{(0)} \; \mathcal{BC}_t + \phi_t^{(1)} X_t, \; t \geq 0$, one gets the following arbitrage theorem:

Theorem 3.3. *Suppose that the drift b and the volatility function σ_t satisfy the Assumptions 2.1. Then, for all starting prices $x \in]0,\infty[$ at time 0, there are no arbitrage opportunities in the class of all self-financing, progressively measurable trading strategies $\phi^{\mathcal{M}} := \{\phi^{(0)}, \phi^{(1)}\}$ which satisfy the three assumptions*

$$(i) \;\; Q_{(0,x)} \left[\int_0^T \left| \phi_v^{(0)} \right| \, dv < \infty \right] = 1 ,$$

$$(ii) \;\; Q_{(0,x)} \left[\int_0^T \left(\sigma_v \; \phi_v^{(1)} \right)^2 \, dv < \infty \right] = 1 ,$$

(iii) the discounted gains proces $\widetilde{\mathcal{G}}_t := \int_0^t \frac{\phi_v^{(1)} \sigma_v X_v}{\mathcal{BC}_v} d\widetilde{\widetilde{W}}_v$ is a $\widetilde{P_{(0,x)}}$-supermartingale.

Proof. It is straightforward to see that under (i) and (ii), all the involved quantities in the corresponding gains process

$$\mathcal{G}_t = \int_0^t \phi_v^{(0)} \; r_v \; \mathcal{BC}_v \; dv + \int_0^t \phi_v^{(1)} \; dX_v + \int_0^t \phi_v^{(1)} \; \delta_v \; X_v \; dv$$

do ($Q_{(0,x)}-$ almost surely) exist; recall that this is already needed in order to define self-financing trading strategies. The rest follows by standard techniques involving the characterization formula $\frac{\mathcal{PO}_t}{\mathcal{BC}_t} - \mathcal{PO}_0 = \widetilde{\mathcal{G}}_t$. □

As an application of the above results, let us now provide a European contingent claim valuation theorem with non-stochastic, handy-to-verify assumptions, which cover a large class of underlyings X whose price processes are non-lognormally distributed generalizations of the geometric Brownian motion:

Theorem 3.4. *Suppose that the drift b and the volatility function σ_t satisfy the Assumptions 2.1. Then, for all starting prices $x \in]0,\infty[$, one gets the following statements:*
(a) For every European contingent claim $\mathfrak{H} \in L^1(\widetilde{P_{(0,x)}})$ with expiration date T, the unique arbitrage-based price $\mathcal{V}_t$ at time $t \in [0,T]$ is given by the formula ($Q_{(0,x)}-$almost surely)

$$\mathcal{V}_t \;=\; e^{-\int_t^T r_v\,dv}\; E\widetilde{P^{IV}_{(0,x)}}[\,\mathfrak{H}\mid\mathcal{F}_t] \;=\; e^{-\int_t^T r_v\,dv}\; E\widetilde{P^{IV}_{(t,\widetilde{X}_t)}}[\,\mathfrak{H}\,]\,. \qquad (3.2)$$

(b) In particular, the unique arbitrage-based price $\mathcal{V}_t^{CALL}$ of a European call option (on the underlying X) with strike price $\mathcal{SP} > 0$, is given by

$$\mathcal{V}_t^{CALL} \;=\; X_t\, e^{-\int_t^T \delta_v\,dv}\; F_N(d_1) \;-\; \mathcal{SP}\, e^{-\int_t^T r_v\,dv}\; F_N(d_2)\,, \qquad (3.3)$$

$Q_{(0,x)}-$almost surely, with

$$d_1 \;:=\; \frac{\log\left(\frac{X_t}{\mathcal{SP}}\right) \;+\; \int_t^T \left(r_v - \delta_v + \frac{\sigma_v^2}{2}\right)dv}{\sqrt{\int_t^T \sigma_v^2\,dv}} \quad \textit{and} \quad d_2 \;:=\; d_1 \;-\; \sqrt{\int_t^T \sigma_v^2\,dv}\,.$$

Proof. This follows by standard techniques from the above results and the martingale representation theorem. □

Remark 3.5. *(1) The original Black-Scholes theorem* [1] *can be derived as a special case of the part (b) of Theorem 3.4, by taking the linear drift and constant volatility of the geometric Brownian motion SDE (1.1), constant short rate $r_t \equiv r^{co}$, and zero dividend yield $\delta_t \equiv 0$. (The non-stochastic interest-rate version of) Merton's theorem* [12] *deals with the same SDE set-up (1.1), but with non-constant short rates r_t and constant dividend yield $\delta_t \equiv \delta^{co}$;* [13] *uses non-constant dividend yields δ_t. Those cases are also covered by Theorem 3.4.*
(2) In the context of "real options" one is sometimes using the Black-Scholes or Merton's formula, although one knows that the underlying quantity can only be approximated by a geometric Brownian motion; see e.g. [7], [2]. *As a technical tool to support such a practical action plan, the Theorem 3.4 delivers a handy-to-verify, non-stochastic toolbox to obtain a variety of non-lognormally distributed underlyings X, such that one can still valuate the corresponding call options with the Black-Scholes formula or Merton's formula.*
(3) The valuation formulae (3.2) and (3.3) do not depend on the modelling drift b. However, the knowledge of b is important for the implementation of, say, the formula (3.3), if for the input σ_t one uses a discretely sampled historical estimate. For a general discussion on this subject, the reader is referred to [9]. *Furthermore, although the formula (3.3) is drift-less, it can not be taken for granted to be valid*

for all underlyings X with arbitrary drifts b. For very exploding drifts b there may not even exist an EMM, and thus the formula (3.3) is not valid anymore. Hence, it is important to have assumptions on b.

Let us finally demonstrate by example that in our toolbox, exploding drifts and exploding volatility functions are covered:

Example 3.6. *Let us fix the constants $\sigma^{co} > 0$ and $\kappa \in]0, 1/2[$. Suppose that $(\widetilde{C}_j)$ is a series of nonnegative numbers with $\sum_{j=0}^{\infty} \widetilde{C}_j < \infty$. For any arbitrarily small (but fixed) level $\mathcal{L} > 0$, let $\{\mathfrak{r}_0, \mathfrak{r}_1, \mathfrak{r}_2, ...\}$ be an enumeration of the rationals in $[\mathcal{L}, \infty[$. Consider the SDE*

$$dX_t = \left(\sum_{j=0}^{\infty} \frac{\widetilde{C}_j}{|X_t - \mathfrak{r}_j|^{\kappa}} \, 1_{[\mathfrak{r}_j/(\kappa+1), \infty[}(X_t) \right) dt + \frac{\sigma^{co} \, X_t}{t^{1/4}} \, dW_t \, .$$

Although the drift is even exploding on a dense set, one can still verify the Assumptions 2.1 with $q_{2.1} = \infty$ and $q_{2.1} = 2$. Consequently, all the assertions of the theorems above can be applied.

Remark 3.7. *Again, notice the essential difference to the Bessel process which has another kind of exploding drift and diffusion coefficient, and for which already the EMM does not exist.*

References

[1] F. Black and M. Scholes, *The pricing of options and corporate liabilities,* J. Political Econ., **81** (1973), 637–654.

[2] P. Carr, *The valuation of American exchange options with application to real options,* Real Options in Capital Investment, Edited by L. Trigeorgis, Praeger Pub. (Westport CT) **1995**, 109–120.

[3] F. Delbaen and W. Schachermayer, *Arbitrage possibilities in Bessel processes and their relations to local martingales,* Probab. Theory Relat. Fields, **102** (1995), 357–366.

[4] E. Eberlein, *More realistic modeling in finance,* Proceedings of the "First International Conference on Risk Management in Finance 1998, Risklab Germany (Munich) **1999**.

[5] E. Eberlein and U. Keller, *Hyperbolic distributions in finance,* Bernoulli, **1** (1995), 281–299.

[6] I. Karatzas and S.E. Shreve, *Brownian Motion and Stochastic Calculus, 2nd ed.,* 1991, Springer-Verlag, Berlin-Heidelberg-New York.

[7] A. Kemna, *Case studies on real options,* Fin. Manag., **22** (1993), 259–270.

[8] F. Liese and I. Vajda, *Convex Statistical Distances,* 1987, Teubner-Verlag, Leipzig.

[9] A.W. Lo and J. Wang, *Implementing option pricing models when asset returns are predictable,* J. Finance, **50** (1995), 87–129.

[10] R.C. Merton, *Lifetime portfolio selection under uncertainty: the continuous-time case,* Rev. Econom. Statist., **51** (1969), 247–257.

[11] R.C. Merton, *Optimum consumption and portfolio rules in a continuous-time model,* J. Econom. Theory, **3** (1971), 373–413.

[12] R.C. Merton, *Theory of rational option pricing,* Bell J. Econom. Manag. Sci., **4** (1973), 141–183.

[13] M. Rubinstein, *The valuation of uncertain income streams and the pricing of options,* Bell J. Econom., **7** (1976), 407–425.

[14] T.H. Rydberg, *A note on the existence of unique equivalent martingale measures in a Markovian setting,* Finance Stochast., **1** (1997), 251–257.

[15] P.A. Samuelson, *Rational theory of warrant pricing,* Ind. Manag. Rev., **6** (1965), 13–31.

[16] W. Stummer, *The Novikov and entropy conditions of multidimensional diffusion processes with singular drift,* Probab. Theory Relat. Fields, **97** (1993), 515–542.

[17] W. Stummer, *On Novikov and arbitrage properties of multidimensional diffusion processes with exploding drift,* Statist. Probab. Letters, **46** (2000), 43–51.

[18] W. Stummer, *On a statistical information measure for a generalized Samuelson-Black-Scholes model,* Statistics & Decisions, (2000), 26 pages, to appear.

[19] W. Stummer and K.–Th. Sturm, *On exponentials of additive functionals of Markov processes,* Stochastic Processes Appl., **85** (2000), 45–60.

Faculty of Mathematics and Economics,
Department of Finance,
University of Ulm,
Helmholtzstrasse 18,
D-89069 Ulm, Germany
E-mail address: `ws@mathematik.uni-ulm.de`

Trends in Mathematics, © 2001 Birkhäuser Verlag Basel/Switzerland

Incremental Value-at-Risk: traps and misinterpretations

Luisa Tibiletti

Abstract. In recent financial literature the Incremental Value-at-Risk (IVaR), i.e., the incremental effect on VaR of adding a new instrument to the existing portfolio, has become a standard tool for making portfolio-hedging decisions. Since, calculating the exact IVaR value could be computationally very costly, approximate formulas have been developed. According to the most commonly used formula, IVaR is approximately equal to the current VaR multiplied by the beta coefficient of the candidate asset. A spontaneous question arises: could the beta sign be a qualitative indicator of a profitable (non-profitable) trade? Fallacy of this conjecture is proved. Even an elliptically distributed asset with a positive beta and variance greater than that of the existent portfolio may produce favourable effects on the overall VaR. These results seem to cast shadows on the above approximate formula reliability even for small changes in portfolio composition.

1. Introduction

In recent financial literature Value-at-Risk (VaR) has become a standard risk tool for designing portfolio strategies. The Incremental VaR (IVaR), i.e., the incremental effect on VaR of adding a new instrument to the existing portfolio, is the gear for making hedging decisions. Since calculating exact IVaR could be cumbersome and computationally very costly, approximate formulas suitable for small portfolio composition changes have been suggested in the literature ([7]; [4]; [10]; [3]; [9]). According to the most commonly used formula, IVaR is approximately equal to the current VaR multiplied by the beta coefficient (with respect to the current portfolio) of the candidate asset. This result seems to support the intuitive idea that the beta coefficient sign could be a qualitative indicator of a profitable new entry trade. The opposite strategy should be suggested, if the beta coefficient is positive.

The aim of this paper is just to test the ground of this conjecture. Through a number of counter-examples, the fallacy of this misleading argument is demonstrated. In order to highlight the real causes of these traps, two cases are analysed. Firstly, the current portfolio and the candidate asset are assumed to have the same elliptically contoured distribution. The hedging decision crucially depends on: (1)

the beta coefficient, (2) the relative difference between the variances and (3) the relative weight of the candidate asset in the re-calibrated portfolio. For example, it can be shown that properly balancing above gears, the introduction of a zero-mean asset even with a positive beta coefficient and a variance greater than that of the current portfolio, may reduce the overall VaR. Secondly, no assumptions on the underlying statistical distributions are made. Now, the impact of a new asset may produce unforeseeable modifications on all the high order moments of the augmented portfolio. It is noteworthy, that the sum of random variables does not even preserve the skewness and the kurtosis direction of the addenda (see [13]). Some examples (even counter-intuitive) cast shadows on the possibility to set out simple approximanting formulas on IVaR.

The paper is organized as follows. Section 2 presents the notion of IVaR. In Section 3 a common approximating formula is discussed. Some common misinterpretations in evaluating IVaR sign is discussed in Section 4. Section 5 concludes the note.

2. The Incremental Value-at-Risk

Value-at-Risk (VaR) is a probability-based metric for quantifying the market risk of assets and portfolios. Its theoretical origins can be traced back long time both in Finance and in Non-life Actuarial Science. In last decade, its adoption is continuing to become widespread not only among securities houses and investment banks, but also among commercial banks, pension funds and in every financial institution. VaR refers to the maximum amount we are likely to lose at some specific confidence level $q\%$. Let X be the random current portfolio return with mean μ_x and finite varaince σ_x^2 Thus, the *relative* VaR is defined as

$$\Pr\left(X - \mu_x \leq -VaR\right) = q\% \tag{1}$$

Clearly, the smaller its VaR, the more preferable the portfolio. In order to design profitable portfolio strategies, a logical step is to calculate the VaR of the portfolio either including the candidate asset or not. The Incremental VaR (IVaR) associated with a position in asset Y is, for a fixed value $q\%$:

$$\boxed{IVaR = VaR \text{ (with the asset } Y\text{) - } VaR\text{(without the asset } Y\text{)}}$$

If IVaR is positive, the candidate asset entry upgrades the portfolio risk, vice versa if it is negative. Unfortunately, any method of calculating IVaR that involves before and after calculations of portfolio VaRs is open to serious practical obstacles: if the assets involved are large, even a fast computer would take time to carry out the necessary matrix operations. So, easy-to-use approximating formulas for a rapid evaluation of IVaR are often used.

3. Approximate Solutions for IVaRs: shortcomings and traps

A sketchy notation follows

X return of the current portfolio

Y return of the candidate new entry asset Y

a size of the new entry asset

$S = \frac{X+aY}{1+a}$ final re-balanced portfolio with marginal position in asset Y

σ_{yx} covariance between Y and the old portfolio X

$\sigma^2_{(.)}$ variance of the random variable (.)

$\beta_{yx} = \frac{\sigma_{yx}}{\sigma^2_x}$ beta coefficient of Y with respect to the current portfolio X

Garman (see [7], [8]) was one of the first to identify the relevance of calculating IVaR for discriminating profitable investments. Recently a simpler formula providing the same numerical answer has been proposed by Dowd (see [4] pg. 49 and also [5]). For a fixed security threshold $q\%$ and for a "small" portion a of the new entry:

$$\boxed{\text{IVaR}=\text{VaR}(S)\text{-VaR}(X) \cong a\beta_{yx}\text{VaR}(X)} \tag{2}$$

(see [10] pg 154 for further formulas to decompose a portfolio's VaR into IVaR). By (2), the impact of the candidate new entry, *seems* to be promptly calculated with negligible computer and human-time wasting.

In spite of the advantages stemming from the computational simplifications, two spontaneous questions arise:

1. From the quantitative point of view, is (2) a satisfactorily approximating formula?

2. May the sign of the marginal effect of the candidate asset be ascertained by the sign of the right-hand side of (2)? In other words, will the signs of betas provide an acceptable tool for discriminating profitable assets?

With reference to the former question, above mentionated Authors have already highlighted that the size a of the candidate trade should be insignificant with respect to the size of the current portfolio, due to the fact that Taylor approximations are used.

The latter question seems to be less plain. Intuition seems to suggest a positive answer. The aim of this paper is just to test the foundation of this conjecture and prove its fallacy.

4. How to sign IVaR: traps and misinterpretations

Formula (2) seems to have a supplementary operative attractiveness. A preliminary rough screening among candidate profitable/non-profitable new entry assets through VaR, can be carried out by signing the correspondent IVaR. According to (2) IVaR seems to have the same sign of the beta. Is this discriminating rule correct? Unfortunately, the answer is no in nearly all the cases.

The stressed points are:

1. The sign of betas *does not* signal the sign of the IVaR, even in the case of elliptically contoured distributions.

2. A decreasing in variance *does not* produce a decreasing in VaR: is that surprising?

3. Stable and non-stable distributions: different results.

5. Old/new portfolios: VaR relations

A common idea throughout the literature concerns the possibility of obtaining information about the $VaR(S)$ only on the basis of the $VaR(X)$ and the relative standard deviation $\frac{\sigma_S}{\sigma_X}$. In the following we will show how this conjecture may be highly misleading.

For a fixed level of confidence $q\%$ we have

$$\Pr(X \leq -VaR(X)) = q\% = \Pr(S \leq -VaR(S))$$

Normalizing the variables involved, under the hypothesis of *null-mean*, we obtain

$$\Pr\left(\frac{X}{\sigma_X} \leq \frac{-VaR(X)}{\sigma_X}\right) = q\% = \Pr\left(\frac{S}{\sigma_S} \leq \frac{-VaR(S)}{\sigma_S}\right)$$

5.1. Normal case

If both variables$X$ and S are *normally distributed,* $VaR(S) = \frac{\sigma_S}{\sigma_X} VaR(X)$. Then

$$IVaR = VaR(X)\left[\frac{\sigma_s}{\sigma_x} - 1\right] \tag{3}$$

it is worthwhile remarking that (3) is the *exact* value of IVaR. In conclusion, under *normality assumptions*, the introduction of Y is profitable

$$IVaR < 0 \ \ iff \ \ \sigma_s < \sigma_x \tag{4}$$

Then, we will answer the following questions:

1. could the sign of beta give a clear-cut tool for screening candidate new entries?

2. If the new asset is positively correlated with the current portfolio and its variance is greater than the current one, does its entry always produce an unfavorable arise in VaR?

3. If variables are no longer normally distributed, but only belong to the same family of stable distributions, is formula (4) still valid?

In contrast with our intuition would suggest, all above questions are negatively answered.

Fallacy 1 *The sign of beta enable us to discriminate between profitable and non-profitable candidates new entries.*

Above conjecture turns out to be false even for normal distributions. Firstly, suppose that the current portfolio X and the candidate new entry Y are normally, or more in general, elliptically distributed. From (4) a fall in VaR goes with a contraction in variance. By straightforward calculations, we have $\sigma_s^2 =$

$\frac{\sigma_x^2+a^2\sigma_y^2+2a\sigma_x\sigma_y r_{xy}}{(1+a)^2}$ where r_{xy} is the Pearson correlation coefficient between X and Y. So

$$\sigma_s < \sigma_x \quad iff \quad r_{xy} < \frac{a\left(\sigma_x^2-\sigma_y^2\right)+2\sigma_x^2}{2\sigma_x\sigma_y}, \qquad a \neq 0 \tag{5}$$

or equivalently, $\beta_{yx} < 1+\frac{a}{2}\left(1-\frac{\sigma_y^2}{\sigma_x^2}\right)$. It is trivial to remark that no matter what the sign of r_{xy} and the quantity a are, if $\sigma_y^2 \leq \sigma_x^2$ the introduction of any zero-mean asset is always advisable.

Remark 5.1. *If $\sigma_y^2 \geq \sigma_x^2$, the VaR decrease depends on the fulfillment of formula* (5).

So, the knowledge of the only beta sign is not sufficient for discriminating favorable investments.

Fallacy 2 *If $r_{xy} > 0$ and $\sigma_y^2 > \sigma_x^2$ the candidate new entry Y increases VaR and variance, so its entry is always to be rejected.*

Note that the right-hand side of (5) may be positive, even if $\sigma_y^2 > \sigma_x^2$. A necessary and sufficient condition for its positiveness is $1 < \frac{\sigma_Y^2}{\sigma_X^2} < 1+\frac{2}{a}$.

Example 5.2. *Let $X \sim N(0,1)$ the current portfolio and $Y \sim N(0,1.5)$ the candidate new entry. The new entry may produce a reduction in the overall variance even if $r_{xy} > 0$*

A possible contraction in the portfolio variance can be obtained when $a < 4$ and $r_{xy} < \frac{a(-0.05)+2}{2\sqrt{1.5}}$. For example, let $a = \frac{1}{10}$ for any $r_{xy} < 0.796$, corresponding to $\beta_{yx} = \sigma_{xy} < 0.975$. It is worthwhile noting that even a great portion of Y characterized by a high variance and a strong positive correlation, may have a desirable diversification effect on the portfolio.

Remark 5.3. *In order to design favorable investment plan, a through analysis is required for taking into account the favorable effect of diversification, which is not captured by the beta coefficients.*

5.2. Non-normal case

Whenever the distributions involved are no longer normal, the discussion needed to be stressed along three different lines:

5.2.1. BOTH THE VARIABLES X AND Y BELONG TO THE SAME ELLIPTICALLY CONTOURED FAMILY. Two peculiarities still hold:

- the augmented portfolio distribution still belongs to the same family
- the distribution function is fully known when the first two moments are given.

and formula (3) is still valid: VaR degrades iff the variance of the overall portfolio drives to the same direction.

5.2.2. BOTH THE VARIABLES X AND Y HAVE THE SAME STABLE DISTRIBUTION, BUT THEY ARE NO LONGER ELLIPTICALLY CONTOURED. Although the augmented portfolio distribution still belongs to the same stable family, the final distribution function is *no longer* fully described by the first two moments.Then, the normalized variables $\frac{S-\mu_x}{\sigma_X}$ and $\frac{S-\mu_S}{\sigma_S}$ have equal mean and variance, but may possess different moments of higher order. Formula (3) is no longer acceptable.

5.2.3. THE VARIABLES X AND Y HAVE DIFFERENT DISTRIBUTIONS. No information is available over the final distribution S. A complete re-calculation is required.

Fallacy 3 *For non-elliptically contoured variables, a lower variance induces a lower VaR.*

Above is an idea that is still deeply-rooted in the literature. Outside the elliptically contoured world, the strategy of minimizing the variance for reducing VaR, may be absolutely misleading. An intuitive explanation is sketched in the following.

Remark 5.4. *The strategy of minimizing the variance is addressed to reducing the "small" risks (those in the neighbourhood of the mean) and vice versa the slight importance of the "catastrophical" risks is deserved. According to the strategy of variance minimization, the more data are concentrated on the mean the more the distribution is appreciated. On the other hand, VaR minimization is devoted to shrinking the left-tail below the q-quantile of the distribution to zero, no attention is paid to the risk distribution above the q-quantile. Clearly, no advantages in VaR terms, derive from a possible data concentration around the mean. Finally, we should not be surprising at all if such differently aimed strategies lead to different solution paths. The paths may coincide in very special circumstances, such as for elliptically countered distributions, where fatness of the tails goes hand-in-hand with the variance, but this must be considered only an exception! Whenever we leave behind the "artificial" elliptical world, many desirable properties fall short.*

Example 5.5. *Symmetrical distributions*

In order not to spread out our attention in cumbersome calculations, we will deal with very elementary discrete random variables. *A fortiori*, analogous counter-examples can be worked out for continuous, non-symmetrical random variables. Let

$$X = \begin{cases} -1, & p = 0.1 \\ 0, & p = 0.8 \\ 1, & p = 0.1 \end{cases} \qquad Y = \begin{cases} -1, & p = 0.5 \\ 1, & p = 0.5 \end{cases},$$

where $\mu_x = \mu_y = 0$ and $\sigma_x^2 = 0.2$, $\sigma_y^2 = 1$. Clearly $\sigma_y^2 > \sigma_x^2$. For $q\% = 0.1\%$, it results $VaR(X) = 1$. Let consider $S = \frac{X+aY}{1+a}$ in two special cases.

1. X and Y *independently distributed*

$$S_0 = \begin{cases} -1, & p = 0.05 \\ \frac{-1+a}{1+a}, & p = 0.05 \\ \frac{-a}{1+a}, & p = 0.4 \\ \frac{+a}{1+a}, & p = 0.4 \\ \frac{+1-a}{1+a}, & p = 0.05 \\ +1, & p = 0.05 \end{cases} \qquad \text{so } r_{xy} = 0 \text{ and } \beta_{yx} = 0$$

For every portion $a > 0$ and at $q\% = 0.1\%$, VaR decreases. Note $VaR(S_0) = -\frac{-1+a}{1+a} < VaR(X)$.

2. X and Y *positively correlated*

$$S_+ = \begin{cases} -1, & p = 0.08 \\ \frac{-1+a}{1+a}, & p = 0.02 \\ \frac{-a}{1+a}, & p = 0.4 \\ \frac{+a}{1+a}, & p = 0.4 \\ \frac{+1-a}{1+a}, & p = 0.02 \\ +1, & p = 0.08 \end{cases} \qquad \text{so, } r_{xy} = 0.27 > 0 \text{ and } \beta_{yx} = 6$$

Once again, even if $r_{xy} > 0$, for every portion $a > 0$ and at $q\% = 0.1\%$, VaR is decreasing. Note $VaR(S_+) = -\frac{-1+a}{1+a} < VaR(X)$. Finally, the introduction of any portion a of the asset Y *always produces a favorable VaR contraction.* So, in contrast with the approximating formula (2)

For any level $a > 0$, $IVaR$ and β_{yx} are always *opposite* in sign!

Besides, it is worthwhile noting that the introduction of Y is always advisable *even if*

- $\sigma_y^2 > \sigma_x^2$, i.e.

  the variance of Y is greater than that of current portfolio X

- the decreasing of VaR *does not go* with a decreasing of variance, in fact correlation and amount a play a crucial role. For example, in the null-correlation case $\sigma_x^2 > \sigma_{S_0}^2$ if $a < 0.5$ and, vice versa $\sigma_x^2 \leq \sigma_{S_0}^2$ if $a \geq 0.5$. In conclusion, it may happen that

  VaR *decreases* even if the variance *increases*

In conclusion, variance and VaR minimizing strategy may lead to a completely different results: portfolios minimizing the variance, *i.e.*, the relatively "small" risks, often increase larger risks as measured by VaR (a seminal idea of this undesirable effect has been already guessed by [1]).

An explanation of previous examples comes from the fact that VaR approach is aimed at yielding a "safety position": the left-hand side of the risk distribution should be shrunk as flat as possible. A more suitable tool for measuring the tail "fatness" are no longer the central moments of "low" order, such as the variance, but, vice versa, those of "high" order, which emphasize the weight of data lying "far" from the mean. Among the central moments, we will cast a glance over those of third and fourth order. The former determines the skewness, i.e. the asymmetry of the distribution around its mean. If the skewness is positive, the confidence interval covers more on the right-hand side of the mean rather than the left one. The kurtosis measures the relative thinness or flatness of the distribution compared to a normal distribution. An empirical test on how the VaR is affected by the standard deviation as well as skewness and kurtosis has been recently stressed by [11], using ten years of daily observations on twelve different foreign exchange spot rates.

Fallacy 4 *Skewness and kurtosis versus are preserved under linear combination: the sum of two positively (negatively) skewed variables is still positively (negatively) skewed.*

Example 5.6. *Linear combinations of equally skewed variables may turn out to be negatively, positively or null skewed.*

Fallacy 4 will be proved even for *identically distributed* variables. Let

$$X = \begin{cases} -1, & p = \frac{1}{4} \\ \frac{1}{3}, & p = \frac{3}{4} \end{cases}, Y = \begin{cases} -1, & p = \frac{1}{4} \\ \frac{1}{3}, & p = \frac{3}{4} \end{cases}, S_a = \begin{cases} -1, & p = 0 \\ \frac{\frac{1}{3}-a}{1+a}, & p = \frac{1}{4} \\ \frac{-1+\frac{1}{3}a}{1+a}, & p = \frac{1}{4} \\ +\frac{1}{3}, & p = \frac{1}{2} \end{cases}$$

where $Sk(X) = Sk(Y) - \frac{2}{\sqrt{3}} < 0$ and $S_a = \frac{X+aY}{1+a}$ with $r_{XY} = -\frac{1}{3}$, so $\beta_{YX} < 0$. Three special cases will be studied:

- $a = \frac{1}{3}$ then $S_{\frac{1}{3}} = \begin{cases} -\frac{2}{9}, & p = \frac{1}{4} \\ 0, & p = \frac{1}{4} \\ +\frac{1}{3}, & p = \frac{1}{2} \end{cases}$ so $Sk(S_{\frac{1}{3}}) < 0$
- $a = 1$ then $S_1 = \begin{cases} -\frac{1}{3}, & p = \frac{1}{2} \\ +\frac{1}{3}, & p = \frac{1}{2} \end{cases}$ so $Sk(S_1) = 0$
- $a = 3$ then $S_3 = \begin{cases} -\frac{2}{3}, & p = \frac{1}{4} \\ 0, & p = \frac{1}{4} \\ +\frac{1}{3}, & p = \frac{1}{2} \end{cases}$ so $Sk(S_3) > 0$

Switching in skewness can be shown up even if the portion a is negligible. For example, let be $a = 3.10^{-6}$ and

$$Y = \begin{cases} -10^{-6}, & p = \frac{1}{4} \\ \frac{1}{3}10^{-6}, & p = \frac{3}{4} \end{cases} \quad \text{where } Sk(Y) = -\frac{2}{\sqrt{3}} < 0.$$

In such case, the re-calibrated portfolio coincides with that previously found S_3. So, skewness switches from negative to positive.

Remark 5.7. *Skewness and kurtosis non-preservation in versus: the relevance of the mixed moments.*

The mathematical explanation of skewness non-preservation in versus is immediate, whenever a more through insight into the formula of the higher order moments is given. By definition, the n-th order moment of the sum $X + Y$ is given by

$$E(X+Y)^n = E(X)^n + E(Y)^n + \sum_{k=1}^{n-1} \binom{n}{k} E\left(X^{n-k}Y^k\right) \tag{6}$$

So, the n-th order moment of the sum depends on both the n-th order pure moments of the addenda, but also on the mixed moments! Consequently, even if the correlation coefficient is $r_{XY} = 0$, i.e. $E(XY) = 0$, the mixed moments may be quite far from zero, so they can play the key-role in the determination of the skewness and kurtosis (and, in general, of all the higher moments) of the sum. This fact is also confirmed by empirical investigations (see [12] for a study over the skewness of a portfolio along a stretch of five years). Moreover, it is worthwhile noting that the condition $r_{xy} = 0$, does not guarantee no-switching in skewness and kurtosis, because the mixed moments $E(X^2Y), E(XY^2), E(X^3Y), E(XY^3), E(X^2Y^2)$ in general differ from zero and they strongly outweighed the pure moments. Since the n-th order moment of the re-balanced portfolio is given by

$$E\left(\frac{X+aY}{1+a}\right)^n = \frac{1}{(1+a)^n}\left\{E(X)^n + a^n E(Y)^n + \sum_{k=1}^{n-1}\binom{n}{k}a^k E\left(X^{n-k}Y^k\right)\right\}$$

the skewness and kurtosis of the augmented portfolio can't be predict without a previous re-calculation of the mixed moments, as well. Unfortunately, easy-to-use formulas for this purpose seem not to exist.

Remark 5.8. *Shortcomings of VaR as a measure of risk has been already highlighted in recent literature. Unfortunately, VaR does not turn out to be a "coherent measure of risk" (for an ample treatment on the VaR shortcomings see* [2], [6], [14]*). And more precisely, the subadditivity property falls short. In light of the previous remark, this last undesirable feature appears not surprising at all. VaR "measures" the fatness of the distribution left-tail, which is strongly influenced by the skewness and kurtosis. Unfortunately, under linear combination, skewness and kurtosis are not preserved (even in sign), so the left tail of the final distribution may change drastically and VaR as well. No property in marginal changes seem possible to be settled out in a non-elliptically contoured world.*

6. Conclusion

VaR is a risk measure focusing on the probability mass spread over the extreme left-tail of the distribution, *i.e.* that part of distribution devoted to capturing information on catastrophic events. The problem tackled in the article is the following: Is this part of distribution very sensitive to even small portfolio re-calibrations? Can IVaR be satisfactorily approximated by formulas involving only the beta coefficients among the assets? These questions seem to be negatively answered outside the "artificial" world of elliptically contoured variables. Unfortunately, portfolios of assets turn out not to possess "intuitive preservation" properties over the tails. For example, a linear combination of assets may not even preserve skewness and kurtosis versus of its addenda. Thus a new-entry asset may change drastically the extreme left-tail of the final distribution and its VaR, as well. In conclusion, outside the elliptical world, approximating formulas involving *only the first two mixed moments* of the final distribution, are to be discouraged.

References

[1] J. V. Andersen and D. Sornette, *Have your cake and eat it too: increasing returns while lowering large risk!*, preprint, (1999), available at hppt://xxx.lanl.gov/abs/cond-mat/9907217.

[2] P. Artzner, F. Delbaen, J.-M. Eber and D. Heath, *Coherent Measures of Risk*, Mathematical Finance, **9** (3), (1999), 203-228.

[3] L. C. Culp, R. Mensink, and A. M. P. Neves, *Value at Risk for Asset Managers*, Derivatives Quarterly, **5** (2), (1999) forthcoming.

[4] K. Dowd, *Beyond Value at Risk: The New Science of Risk Management*, John Wiley & Sons, Chichester and New York, **1998**.

[5] K. Dowd, *A Value at Risk Approach to Risk-Return Analysis*, The Journal of Portfolio Management, (1999), 60-67.

[6] P. Embrechts, A. Mc Neil and D. Straumann, *Correlation: Pitfalls and Alternatives*, Preprint, ETH Zurich, (1999).

[7] M. Garman, *Improving on VaR*, Risk, **9** (5), (1996), 61-63.

[8] M. Garman, *Taking VaR to pieces*, Risk, **10** (10), (1997), 70-71.

[9] W. G. Hallerbach, *Decomposing Portfolio Value-at-Risk: A General Analysis*, Working paper, Tinbergen Institute, (1999).

[10] P. Jorion, *Value at Risk: The New Benchmark for Controlling Market Risk*, The McGraw-Hill Companies, Inc., New York, **1997**.

[11] D. X. Li *Value at Risk Based on the Volatility, Skewness and Kurtosis*, Working Paper, RiskMetrics Group, (1999).

[12] L. Peccati and L. Tibiletti, *On the asymmetry of stock-return distribution*, presented at 14th Meeting of the EURO Working Group of Financial Modelling held in Mantova, Italy, November 25-27, 1993.

[13] L. Tibiletti and E.Volpe, *Higher order moments of a sum of random variables: remarks and applications*, Ratio Mathematica, **11** (1997), 47-57.

[14] J. L. Wirch, *Raising Value at Risk*, North American Actuarial Journal, **3**, (2), (1999), 106-115.

Department of Statistics and Mathematics,
University of Torino,
Piazza Arbarello, 8
I-10122 Torino, Italy

Trends in Mathematics, © 2001 Birkhäuser Verlag Basel/Switzerland

On option expected returns

Rafał Wojakowski and Mark Shackleton

Abstract. We provide a way to derive a continuous-time closed-form formula for the $\mathcal{P}$-expected risk-priced return to holding an option, something of concern to a holder of a naked position. In the proof we rely on Paul Lévy's construction of Brownian motion. Our results have relevance for an arbitrage-free incomplete maket in which a "small" investor can only modify his portfolio at discrete points in time, for whom options are not redundant assets.

1. Introduction

In order to *price* derivatives, it is not necessary to know their *rate of return.* Through the use of a hedge, the options trader eliminates the risk. Consequently, option prices are usually quoted using implied volatilities, indicating that it is the second moment of returns, not the first, that is of concern to professional traders.

Although the trader may well be indifferent to rates of return and likelihoods of exercise, the retail investor who chooses to hold an unhedged option will presumably care about these factors if he has decided to hold the option in the first place. He would only be indifferent to these if he himself could act as a professional. As a result retail investors are often surprised by the indifference shown by professionals toward the very parameters about which they care.

This is the motivation of this paper, the analysis of holding rates of return on option positions, something of concern from the perspective of a non–professional user.

2. Setting

Consider a "small" investor who can access the market only at time $t = 0$ and $0 < \tau < T$, where T is the maturity of an option. The investor is prevented by assumption from rebalancing his portfolio between $t = 0$ and $t = \tau$. Thus, from

---

The authors are grateful to Gordon Blower, Bart Lambrecht, Peter Pope, Gareth Roberts, Stephen Taylor, Adrianos Tsekrekos and Pradeep Yadav for useful discussions, to Math-Finance seminar participants in Cambridge, Glasgow and Lancaster and to Bachelier Finance Society Congress participants in Paris. Usual disclaimers apply.

Contact: Department of Accounting and Finance, Lancaster University, LA1 4YX, UK. `mailto:r.wojakowski@lancaster.ac.uk` or `mailto:m.shackleton@lancaster.ac.uk` URL: `http://www.lancs.ac.uk/staff/wojakows/` or `http://www.lancs.ac.uk/staff/shacklem/` .

his perspective, markets are *incomplete* and for at least one category of investor, options are not redundant assets. Carr and Madan (1999) and others[1] show that in this context options can be used to complete the market and it can be optimal for this investor to hold options in his portfolio. In the general equilibrium they consider, whereas derivatives are held individually, they are not held in aggregate.

Suppose now that the underlying asset follows — under the *actual* probability measure $\mathcal{P}$ — a geometric Brownian motion with drift μ. From the option pricing perspective, from Black and Scholes (1973) and Merton (1973) it is obvious that this risk-priced rate of return μ of is irrelevant. However, this rate of return will affect the rate of return of the option as perceived by our small investor. In this paper we first give a precise derivation of the closed-form, continuous-time formula for the $\mathcal{P}$-expected payoff to holding an option. Second, we relate this quantity to the *risk-priced* i.e. $\mathcal{P}$-expected rate of return on the option.

3. Assumptions

We work with the stock price process $S \triangleq \{S_t\}_{0 \leq t \leq T}$ on some filtered probability space $(\Omega, \mathcal{F}, \{\mathcal{F}_t\}, \mathcal{P})$ with $\{\mathcal{F}_t\}$ satisfying the usual conditions. The risk-priced probability measure $\mathcal{P}$ — also known as standard, objective, subjective, historical or actual — is assumed to exist. The stock price process S solves the stochastic differential equation

$$dS_t = (\mu - \delta)\, S_t dt + \sigma S_t dW_t \ ; \ S_0 > 0 \tag{1}$$

where W is the standard Wiener process under objective probability measure $\mathcal{P}$, μ is the constant drift i.e. risk-priced rate of return, δ is the constant dividend stream per unit of time and $\sigma > 0$ is the volatility assumed to be positive and constant. In the case of foreign exchange options r corresponds to the domestic riskless rate and δ can be interpreted as the foreign riskless rate r_f. Note that we do not require $\mu > 0$ or $\delta > 0$.

The professional market — complete, perfect and arbitrage-free — prices contingent claims at any point in time $t \in [0, T]$ by applying the standard risk-neutral pricing technique. At time t the premium of any European-style contingent claim with terminal payoff c_T is obtained by computing expectation under the equivalent risk-neutral measure $\mathcal{Q}$:

$$\forall_{t \in [0,T]} \qquad c_t \triangleq e^{-r(T-t)} \mathbb{E}^{\mathcal{Q}} \left[c_T \middle| \mathcal{F}_t \right] \tag{2}$$

where r is the constant risk-free rate. To simplify notation in what follows we focus on pricing claims at $t = 0$.

[1]See e.g. Ross (1976), Breeden and Litzenberger (1978), Hakansson (1978), Brennan and Solanki (1981) and Franke, Stapleton and Subrahmanyam (1998). Bertrand, Lesne and Prigent (1998) further generalise results of Carr and Madan (1999) and Brennan and Solanki (1981) in a model with portfolio guarantee constraint. In the related paper by Carr, Jin and Madan (2000) incompleteness, and therefore non-zero option holdings at optimum, are generated by a pure jump Lévy process. See also Bellamy and Jeanblanc (2000).

If a "small" investor can access the market only at time $t = 0$ and $\tau > 0$ the market is *incomplete* from his perspective in the *time dimension*. We will assume that the investor agrees with all market participants that under the historical measure $\mathcal{P}$ the stock has a total return equal to μ. Moreover, it is well known — e.g. Carr and Madan (1999) — that in such setting he will hold options in his optimal portfolio. However, in our context, the market is still *arbitrage-free* for the small investor. Although our investor is excluded from playing the same game as professional traders in the perfect market do, there is no arbitrage opportunity left. In fact he has a discrete-time arbitrage opportunity set. However, he cannot profit from any arbitrage operation because his arbitrage opportunity set is a subset of the trader's continuous-time arbitrage opportunity set. In the latter all arbitrage opportunities are ruled out by assumption, for all $\tau \in [0, T]$. Our small investor would like to know the following: What is, computed at time $t = 0$, the $\mathcal{P}$-expected rate of return on an European-style option which he plans to hold in his portfolio till time $\tau > 0$?

4. Risk-priced expected option payoff

At time $t = 0$ the small investor expects that at time $\tau > 0$ the option will be worth

$$\bar{c}_\tau^{\mathcal{P}} \triangleq \mathbb{E}^{\mathcal{P}}\left[c_\tau \middle| \mathcal{F}_0\right] \tag{3}$$

and, according to assumption (2), he knows that in the future the option will be valued by the market using the equivalent, risk-neutral $\mathcal{Q}$-expectation (2) where the maturity T *payoff* is given by

$$c_T \triangleq f(S_T) \ . \tag{4}$$

For example $f(S_T) = K\mathbf{1}_{S_T>K}$ for binary *cash-or-nothing* call and $f(S_T) = S_T\mathbf{1}_{S_T>K}$ for binary *asset-or-nothing* call. For standard Back-Scholes call we would have $f(S_T) = (S_T - K)^+$.

Proposition 1. *The $\mathcal{P}$-expected value of the option to be held till time $\tau \in (0, T]$ expressed using $\mathcal{Q}$-expectations is*

$$\bar{c}_\tau^{\mathcal{P}} = \mathbb{E}^{\mathcal{Q}}\left[L_\tau^\star c_\tau \middle| \mathcal{F}_0\right] \tag{5}$$

where

$$\forall_{0<t\leq T} \qquad L_t^\star = \exp\left(-\frac{\alpha^2}{2}t + \alpha Z_t\right) , \tag{6}$$

the constant $\alpha \triangleq \frac{\mu - r}{\sigma}$ is the market price of risk and the process $Z_t \triangleq W_t + \alpha t$ is a Brownian motion under $\mathcal{Q}$.

Proof. See Appendix. □

5. Lévy construction of Brownian motion and the fundamental decomposition

In this section we decompose the conditional $\mathcal{Q}$-expectation (5) according to the Lévy construction of Brownian motion. Because $L_\tau^\star$ is $\mathcal{F}_\tau$-measurable, using (4) and (2) in (5) we get

$$\bar{c}_\tau^{\mathcal{P}} = e^{-r(T-\tau)}\chi_f \qquad \text{where} \qquad \chi_f \triangleq \mathbb{E}^{\mathcal{Q}}\left[L_\tau^\star f\left(S_T\right)\middle|\,\mathcal{F}_0\right] . \tag{7}$$

In what follows we will focus on the expectation χ_f for $0 < \tau < T$. Cases $\tau = 0$ and $\tau = T$ are special cases and will be treated separately. For $\tau > T$ one would have to specify the use of exercise proceeds, so in the next section we consider the case $0 < \tau < T$.

5.1. Option held over a finite period of time: $0 < \tau < T$

First note that because the holding period is shorter than the time to maturity, the payoff $f\left(S_T\right)$ is not $\mathcal{F}_\tau$-measurable. Using the definition (6) of $L_t^\star$ we have

$$\chi_f = \mathbb{E}^{\mathcal{Q}}\left[\exp\left(-\frac{\alpha^2}{2}\tau + \alpha Z_\tau\right) f\left(S_T\right)\middle|\,\mathcal{F}_0\right] . \tag{8}$$

The most important fact resulting from the *Lévy construction of Brownian motion*[2] on the open interval $(0, T)$ is that for any interior point $\tau \in (0, T)$, the value of Z_τ can be expressed as

$$Z_\tau = \sqrt{\frac{\tau\left(T-\tau\right)}{T}}X_\tau + \frac{\tau}{T}Z_T , \tag{9}$$

where X_τ is a standard Gaussian random variable *independent* of Z_T and Z_0 and thus from the underlying asset's price at time T given by

$$S_T = S_0 \exp\left(\left(r - \delta - \frac{\sigma^2}{2}\right)T + \sigma Z_T\right) . \tag{10}$$

The second term in (9) can be interpreted as the expected value of Z_τ given the values of the process at $t = 0$ and $t = T$. The independence of X_τ and Z_T yields the following important *decomposition* of χ_f into a *product of two expectations*

$$\chi_f = \mathbb{E}^{\mathcal{Q}}\left[\exp\left(\alpha\sqrt{\frac{\tau\left(T-\tau\right)}{T}}X_\tau\right)\right]\mathbb{E}^{\mathcal{Q}}\left[\exp\left(-\frac{\alpha^2}{2}\tau + \alpha\frac{\tau}{T}Z_T\right) f\left(S_T\right)\middle|\,\mathcal{F}_0\right] . \tag{11}$$

Note that X_τ and Z_T are now *multiplicatively separated.* The first term in (11) turns out to be constant equal to

$$\exp\left(\frac{\alpha^2}{2}\frac{\tau\left(T-\tau\right)}{T}\right) .$$

Therefore, the decomposition (11) can be rewritten in a more compact form

$$\chi_f = \phi \times \phi_f \tag{12}$$

[2]See T. Hida (1980, Proposition 2.2, page 64).

where

$$\phi \triangleq \exp\left(-\frac{(\alpha\tau)^2}{2T}\right) \qquad\qquad \phi_f \triangleq \mathbb{E}^{\mathcal{Q}}\left[\exp\left(\frac{\alpha\tau}{T}Z_T\right)f(S_T)\Big|\mathcal{F}_0\right]. \tag{13}$$

Terms depending on X_τ have been integrated out yielding the constant ϕ. Notice also that the subscript f serves mainly as mnemonic to emphasise the fact that the expectation ϕ_f is to be taken with respect to elements involving some function f.

6. Application: The case of European call

In what follows we will specify the payoff function appearing in (4) to that of European call i.e. $(S_t - K)^+$, where $K > 0$ is the exercise price for maturity $T > 0$. Obviously, the premium c_0 can be expressed using the standard Black-Scholes (1973) formula:

$$c_0 = S_0 e^{-\delta T}\mathcal{N}(d_1) - Ke^{-rT}\mathcal{N}(d_2) \quad \text{where} \quad d_{1,2} \triangleq \frac{\ln\frac{S_0}{K} + \left(r - \delta \pm \frac{\sigma^2}{2}\right)T}{\sigma\sqrt{T}} \tag{14}$$

and $\mathcal{N}(.)$ is the normal cumulative distribution function. The term $\mathcal{N}(d_2) = \mathcal{Q}(\mathbf{1}_{S_T>K})$, where $\mathbf{1}_{S_T>K} \equiv 1$ if $S_T > K$ and $\mathbf{1}_{S_T>K} = 0$ otherwise, is the probability of exercise under the risk-neutral martingale measure $\mathcal{Q}$. Under $\mathcal{Q}$ the *numéraire* is the money market account. Geman, El Karoui and Rochet (1995) show that $\mathcal{N}(d_1)$ can be interpreted as the probability of exercise, but under an equivalent martingale measure $\mathcal{Q}^\star$, which corresponds to choosing the price S of the underlying asset as *numéraire*: $\mathcal{N}(d_1) = \mathcal{Q}^\star(\mathbf{1}_{S_T>K})$.[3]

Distributing $L^\star_\tau$ over $S_T - K$ in the conditional $\mathcal{Q}$-expectation (5) from Proposition 1, we get

$$\bar{c}^{\mathcal{P}}_\tau = e^{-r(T-\tau)}\left\{\underbrace{\mathbb{E}^{\mathcal{Q}}\left[L^\star_\tau S_T \mathbf{1}_{S_T>K}|\mathcal{F}_0\right]}_{\chi_1} - K\underbrace{\mathbb{E}^{\mathcal{Q}}\left[L^\star_\tau \mathbf{1}_{S_T>K}|\mathcal{F}_0\right]}_{\chi_2}\right\}. \tag{15}$$

In what follows we will focus on the two expectations χ_1 and χ_2. Cases $\tau = 0$ and $\tau = T$ are special cases corresponding to Black Scholes (14) and Boness (26) formulae respectively. Using the Lévy construction of Brownian motion and the Girsanov theorem it is then straightforward to obtain the following Proposition:

Proposition 2. *The $\mathcal{Q}$-expectation χ_2 in (15) is equal to*

$$\chi_2 = \mathcal{N}(d_{-,\tau}), \tag{16}$$

---

[3] See also Musiela and Rutkowski (1997).

where

$$d_{-,\tau} \triangleq \frac{\ln\frac{S_0}{K} + \left(r - \delta - \frac{\sigma^2}{2}\right)T + (\mu - r)\,\tau}{\sigma\sqrt{T}} \tag{17}$$

and the term $\mathcal{N}(d_{-,\tau})$ in (16) is equal to the prob ability of exer cising the option at maturity $T : T > \tau$, i.e.

$$\mathcal{N}(d_{-,\tau}) = \mathcal{Q}_\tau\,(S_T > K)\ ,$$

but under a new measure, $\mathcal{Q}_\tau$, equivalent to $\mathcal{Q}$, defined via the R adon-Nikodym density

$$\frac{d\mathcal{Q}_\tau}{d\mathcal{Q}} \triangleq \exp\left(-\frac{\left(\alpha\frac{\tau}{T}\right)^2}{2}T + \left(\alpha\frac{\tau}{T}\right)Z_T\right)\ . \tag{18}$$

Proof. The $\mathcal{Q}$-expectation χ_2 in (15) can be rewritten using our decomposition i.e. $\chi_2 = \phi \times \phi_f$, letting $f(S) = \mathbf{1}_{S_T>K}$ in (12–13). Therefore

$$\chi_2 = \phi \times \mathbb{E}^{\mathcal{Q}}\left[\exp\left(\frac{\alpha\tau}{T}Z_T\right)\mathbf{1}_{S_T>K}\middle|\,\mathcal{F}_0\right]\ .$$

Introducing ϕ in to the $\mathcal{Q}$-expectation yields

$$\chi_2 = \mathbb{E}^{\mathcal{Q}}\left[\exp\left(-\frac{\left(\alpha\frac{\tau}{T}\right)^2}{2}T + \left(\alpha\frac{\tau}{T}\right)Z_T\right)\mathbf{1}_{S_T>K}\middle|\,\mathcal{F}_0\right]\ . \tag{19}$$

Note that now, for any given τ and T, the exponential term defines a new probability measure $\mathcal{Q}_\tau$, equivalent to $\mathcal{Q}$, via the Radon-Nikodym density (18). Girsanov Theorem then states that the process

$$B_t \triangleq Z_t - \left(\alpha\frac{\tau}{T}\right)t$$

is a Brownian motion under $\mathcal{Q}_\tau$. Therefore $Z_T = B_T + \alpha\tau$ and under $\mathcal{Q}_\tau$ the asset price at time T can be written as

$$S_T = S_0\exp\left(\left(r - \delta - \frac{\sigma^2}{2}\right)T + (\mu - r)\,\tau + \sigma B_T\right)\ . \tag{20}$$

The expectation in (19) then becomes equal to the probability of exercising the option at maturity under the new measure $\mathcal{Q}_\tau$

$$\chi_2 = \mathbb{E}^{\mathcal{Q}_\tau}\left[\mathbf{1}_{S_T>K}\middle|\,\mathcal{F}_0\right] = \mathcal{Q}_\tau\,(S_T > K)\ .$$

Using (20) and (17) this can further be expressed as

$$\chi_2 = \mathcal{Q}_\tau\left(\frac{\ln\frac{S_0}{K} + \left(r - \delta - \frac{\sigma^2}{2}\right)T + (\mu - r)\,\tau}{\sigma\sqrt{T}} > -\frac{B_T}{\sqrt{T}}\right) = \mathcal{N}(d_{-,\tau})\ , \tag{21}$$

which completes the proof. □

6.1. The first expectation χ_1

Armed with decomposition (12–13), in which we let $f(S_T) = S_T \mathbf{1}_{S_T > K}$, it is now straightforward to calculate the first expectation χ_1 in (15). The following Proposition holds:

Proposition 3. *The $\mathcal{Q}$-expectation χ_1 in (15) is equal to*

$$\chi_1 = S_0 \exp\left(\left[r - \delta + (\mu - r)\frac{\tau}{T}\right] T\right) \mathcal{N}(d_{+,\tau}) \ , \tag{22}$$

where

$$d_{+,\tau} \triangleq \frac{\ln \dfrac{S_0}{K} + \left(r - \delta + \dfrac{\sigma^2}{2}\right) T + (\mu - r)\,\tau}{\sigma\sqrt{T}} \tag{23}$$

and the term $\mathcal{N}(d_{+,\tau})$ in (22) is equal to the probability of exercising the option at maturity $T : T > \tau$, i.e.

$$\mathcal{N}(d_{+,\tau}) = \mathcal{Q}^\star_\tau (S_T > K) \ ,$$

but under a new measure, $\mathcal{Q}^\star_\tau$, equivalent to $\mathcal{Q}$, defined via the following Radon-Nikodym density

$$\frac{d\mathcal{Q}^\star_\tau}{d\mathcal{Q}} \triangleq \exp\left(-\frac{\left(\frac{\alpha\tau}{T} + \sigma\right)^2}{2} T + \left(\frac{\alpha\tau}{T} + \sigma\right) Z\right) . \tag{24}$$

Proof. The proof is left as an exercise. It is similar to the proof of Proposition 2. However, it appears to be slightly more complex due to careful attention one must pay to manage exponential terms in order to compose the Radon-Nikodym derivative (24). Then applying Girsanov's theorem is straightforward, yielding (22), which completes the proof. □

7. The formula for risk-priced expected option payoff

Propositions 1, 2 and 3 yield the following theorem:

Theorem 4. *The $\mathcal{P}$-expected value of the option to be held till time $\tau : 0 < \tau < T$ is given by the following formula*

$$\bar{c}^{\mathcal{P}}_\tau = S_0 e^{\mu\tau - \delta T} \mathcal{N}(d_{+,\tau}) - K e^{-r(T-\tau)} \mathcal{N}(d_{-,\tau}) \ , \tag{25}$$

where

$$d_{\pm,\tau} \triangleq \frac{\ln \frac{S_0}{K} + \left(r - \delta \pm \frac{\sigma^2}{2}\right) T + (\mu - r)\,\tau}{\sigma\sqrt{T}} .$$

Proof. Using (16) and (22) in (15) gives (25). □

Remark 1. *Our formula (25) is a continuous-time version of Rubinstein's (1984) formula obtained in a discrete-time two-period model. Our results hold for assets paying continuous dividend stream δ per unit of time.*

Remark 2. *A proof for an European put option is similar, starting with payoff* $p_T \triangleq (K - S_T)^+$.

Remark 3. *The following limiting properties of $d_{\pm,\tau}$ hold for $\tau \to 0$*

$$\lim_{\tau \to 0} d_{+,\tau} = d_1 \qquad \lim_{\tau \to 0} d_{-,\tau} = d_2$$

and (25) becomes the standard Black-Scholes (1973) formula (14).

Remark 4. *The following limiting properties of $d_{\pm,\tau}$ hold for $\tau \to T$*

$$d_3 \triangleq \lim_{\tau \to T} d_{+,\tau} = \frac{\ln \frac{S_0}{K} + \left(\mu - \delta + \frac{\sigma^2}{2}\right) T}{\sigma\sqrt{T}},$$

$$d_4 \triangleq \lim_{\tau \to T} d_{-,\tau} = \frac{\ln \frac{S_0}{K} + \left(\mu - \delta - \frac{\sigma^2}{2}\right) T}{\sigma\sqrt{T}}$$

and (25) becomes the Boness (1964) formula

$$\bar{c}_T^{\mathcal{P}} = S_0 e^{(\mu-\delta)T} \mathcal{N}(d_3) - X\mathcal{N}(d_4) \ . \tag{26}$$

In an early attempt to price options Boness (1964) first derived the $\mathcal{P}$-expectation (26) of the terminal payoff to holding an option over its lifetime (see Galai (1978)). As this quantity is expressed in dollars of maturity date, option pricing requires an appropriate discount rate, $r_T^\star$, i.e. the right "risk class." It is important to emphasise that, assuming that $r_T^\star$ is available, discounting (26) will give the right price of the option. In what follows we provide a formula for the $\mathcal{P}$-rate of return on option, $r_\tau^\star$, valid for general time τ.

8. Risk-priced expected option return

Theorem 5. *The expected rate of return on investment in an European call for a general time $\tau : 0 < \tau \leq T$ is given by*

$$r_\tau^\star = \frac{1}{\tau} \ln \frac{\bar{c}_\tau^{\mathcal{P}}}{c_0}, \tag{27}$$

where $\bar{c}_T^{\mathcal{P}}$ is given by our formula (25) and c_0 by the Black-Scholes formula (14). In the case of an European call held till maturity $\tau = T$ the term $\bar{c}_T^{\mathcal{P}}$ is given by the Boness $\mathcal{P}$-expectations formula (26).

Proof. The expected payoff to holding the option $\bar{c}_\tau^{\mathcal{P}}$ from Theorem 4 can be discounted using the rate of return $r_\tau^\star$ corresponding to option's risk class for time horizon τ. It must yield the current price, c_0, of the option. This price is obviously given by the Black-Scholes formula (14). The investor and the market agree on c_0 when the deal is made at time $t = 0$. Equating the two formulations

$$e^{-r_\tau^\star \tau}\mathbb{E}^{\mathcal{P}}\left[c_\tau | \mathcal{F}_0\right] = c_0 = e^{-rT}\mathbb{E}^{\mathcal{Q}}\left[c_T | \mathcal{F}_0\right] \tag{28}$$

gives

$$r_\tau^\star = r\frac{T}{\tau} + \frac{1}{\tau}\ln\frac{\mathbb{E}^{\mathcal{P}}\left[c_\tau | \mathcal{F}_0\right]}{\mathbb{E}^{\mathcal{Q}}\left[c_T | \mathcal{F}_0\right]},$$

where using the right equality from (28) and the definition (3) of $\bar{c}_\tau^{\mathcal{P}}$ yields (27).

□

9. Concluding remarks

First, using a change of measure technique, we provide a continuous-time proof of a formula yielding the risk-priced $\mathcal{P}$-expected option payoff. In our formulation we allow dividends and interpret special cases of Black-Scholes (1973) and Boness (1964) formulae. Second, we derive the formula yielding the risk-priced return on an option, valid for general time horizons less then or equal to the time to maturity. Our setup does not preclude or eliminate the necessity to rely on the Black-Scholes risk-neutral pricing. We extend their setting to include the actual probability measure $\mathcal{P}$ which is necessary when evaluating actual rates of return and, as in our case, markets are incomplete in the time dimension.

Appendix A. Proof of Proposition 1

To shift from $\mathcal{P}$ to $\mathcal{Q}$ we use the formula for change of measure in conditional expectations

$$\mathbb{E}^{\mathcal{P}}\left[c_\tau | \mathcal{F}_0\right] = \mathbb{E}^{\mathcal{Q}}\left[\frac{d\mathcal{P}}{d\mathcal{Q}}c_\tau \middle| \mathcal{F}_0\right] \Big/ \mathbb{E}^{\mathcal{Q}}\left[\frac{d\mathcal{P}}{d\mathcal{Q}} \middle| \mathcal{F}_0\right], \tag{29}$$

where:

$$\frac{d\mathcal{P}}{d\mathcal{Q}} \triangleq \exp\left(\frac{\alpha^2}{2}T + \alpha W_T\right) = \exp\left(-\frac{\alpha^2}{2}T + \alpha Z_T\right) \tag{30}$$

is the Radon-Nikodym derivative. Under $\mathcal{P}$ the process

$$\forall_{0<t\leq T} \qquad L_t \triangleq \exp\left(-\frac{\alpha^2}{2}t - \alpha W_t\right)$$

is a martingale. Similarly, under $\mathcal{Q}$ the process $L_t^\star$ defined by (6) is a martingale $\forall_{0<s\leq t\leq T}\ \mathbb{E}^{\mathcal{Q}}\left[L_t^\star | \mathcal{F}_s\right] = L_s^\star$. The martingale property implies

$$\mathbb{E}^{\mathcal{Q}}\left[\exp\left(\frac{\alpha^2}{2}T + \alpha W_T\right)\middle| \mathcal{F}_0\right] = 1\,.$$

Therefore, the denominator term in (29) vanishes and since c_τ is $\mathcal{F}_\tau$-measurable, we have

$$\bar{c}_\tau^{\mathcal{P}} = \mathbb{E}^{\mathcal{Q}}\left[\mathbb{E}^{\mathcal{Q}}\left[L_T^\star\, c_\tau | \mathcal{F}_\tau\right] | \mathcal{F}_0\right] = \mathbb{E}^{\mathcal{Q}}\left[c_\tau \mathbb{E}^{\mathcal{Q}}\left[L_T^\star | \mathcal{F}_\tau\right] | \mathcal{F}_0\right]\,.$$

Finally, using the martingale property of $L_t^\star$ again, we get (5).

References

BELLAMY, N., AND M. JEANBLANC (2000): "Incompleteness of Markets Driven by a Mixed Diffusion," *Finance and Stochastics*, **4**(2), 209–222.

BERTRAND, P., J. P. LESNE, AND J.-L. PRIGENT (1998): "Gestion de Portefeuille avec Garantie: Allocation Optimale en Actifs Dérivés," in *Journées Internationales de Finance 1998 (AFFI Conference Proceedings, Ecole Supérieure des Affaires - Lille 2)*, Paris. Association Française de Finance.

BLACK, F., AND M. SCHOLES (1973): "The Pricing of Options and Corporate Liabilities," *Journal of Political Economy*, **81**(3), 637–654.

BONESS, A. J. (1964): "Elements of a Theory of Stock Option Value," *Journal of Political Economy*, **72**(2), 163–175.

BREEDEN, D. T., AND R. H. LITZENBERGER (1978): "Prices Of State-Contingent Claims Implicit In Option Prices," *Journal of Business*, **51**(4), 621–652.

BRENNAN, M. J., AND R. SOLANKI (1981): "Optimal Portfolio Insurance," *Journal of Financial and Quantitative Analysis*, **16**(3), 279–300.

CARR, P., X. JIN, AND D. MADAN (2000): "Optimal Investment in Derivatives Securities," *Forthcoming in Finance and Stochastics.*

CARR, P., AND D. MADAN (1999): "Optimal Positioning in Derivative Securities," University of Maryland working paper.

FRANKE, G., R. C. STAPLETON, AND M. G. SUBRAHMANYAM (1998): "Who Buys and Who Sells Options: The Role of Options in an Economy with Background Risk," *Journal of Economic Theory*, **82**(1), 89–109.

GALAI, D. (1978): "On the Boness and Black Scholes models for valuation of call options," *Journal of Financial and Quantitative Analysis*, **13**(1), 15–27.

GEMAN, H., N. EL KAROUI, AND J.-C. ROCHET (1995): "Changes of Numéraire, Changes of Probability Measure and Option Pricing," *Journal of Applied Probability*, **32**, 443–458.

HAKANSSON, N. H. (1978): "Welfare Aspects Of Options And Supershares," *Journal of Finance*, **33**(3), 759–776.

HIDA, T. (1980): *Brownian Motion*, vol. 11 of *Applications of Mathematics*. Springer-Verlag, Berlin Heidelberg New York.

MERTON, R. C. (1973): "The Theory of Rational Option Pricing," *Bell Journal of Economics*, **4**(1), 141–183.

MUSIELA, M., AND M. RUTKOWSKI (1997): *Martingale Methods in Financial Modelling*, vol. 36 of *Applications of Mathematics: Stochastic Modelling and Applied Probability*. Springer-Verlag, Berlin Heidelberg New York, 1998 edn.

ROSS, S. A. (1976): "Options And Efficiency," *Quarterly Journal of Economics*, **90**(1), 75–89.

RUBINSTEIN, M. (1984): "A Simple Formula for the Expected Rate of Return of an Option over a Finite Holding Period," *Journal of Finance*, **39**(5), 1503–1509.

Department of Accounting and Finance, Lancaster University, LA1 4YX, UK

MIX
Papier aus verantwortungsvollen Quellen
Paper from responsible sources
FSC® C105338

FSC
www.fsc.org

If you have any concerns about our products,
you can contact us on
ProductSafety@springernature.com

In case Publisher is established outside the EU,
the EU authorized representative is:
Springer Nature Customer Service Center GmbH
Europaplatz 3, 69115 Heidelberg, Germany

Printed by Libri Plureos GmbH
in Hamburg, Germany